D1174190

# Computer
# Image Processing
# and Recognition

*This is a volume in*
COMPUTER SCIENCE AND APPLIED MATHEMATICS

*A Series of Monographs and Textbooks*

*Editor:* WERNER RHEINBOLDT

A complete list of titles in this series appears at the end of this volume.

# Computer Image Processing and Recognition

ERNEST L. HALL

*Department of Electrical Engineering*
*The University of Tennessee*
*Knoxville, Tennessee*

 1979

ACADEMIC PRESS

A Subsidiary of Harcourt Brace Jovanovich, Publishers

New York   London   Toronto   Sydney   San Francisco

ACADEMIC PRESS, INC.
111 Fifth Avenue, New York, New York 10003

*United Kingdom Edition published by*
ACADEMIC PRESS, INC. (LONDON) LTD.
24/28 Oval Road, London NW1 7DX

Library of Congress Cataloging in Publication Data

Hall, Ernest L
    Computer image processing and recognition.

    (Computer science and applied mathematics)
    Includes bibliographies and index.
    1. Image processing.  2. Optical pattern
recognition.  I.  Title.
TA1632.H34    621.3819'598    79–6937
ISBN 0–12–318850–4

PRINTED IN THE UNITED STATES OF AMERICA

79 80 81 82    9 8 7 6 5 4 3 2 1

*To Donald, Charles, Jeannine, Michael, and Bettie*

# Contents

## 4  Image Enhancement and Restoration

## 5  Reconstruction from Projections

## 6  Digital Television, Encoding, and Data Compression

## 7  Scene Understanding

# Preface

The purpose of this book is to provide a unified, systematic introduction to concepts and techniques that have proven useful in the computer analysis of pictures.

This book is an outgrowth of courses taught by the author in the Electrical Engineering and Computer Science Departments at the University of Tennessee, the University of Southern California, Yale University, and the University of Missouri over the past several years. The book is well suited to a three-quarter or two-semester sequence in picture processing and pattern recognition at the senior or graduate level with the extent and depth of coverage selected by the instructor.

The book is especially intended as an aid to the engineer and computer or physical scientist who needs to design or use a machine that performs an intelligent task based upon pictorial information. In some cases task evaluation will be made by human observation and subjective judgment. For others the evaluation may be made objectively. The number and variety of intelligent tasks is limitless; however, a certain set quite often occurs. This includes enhancements of degraded images, compression of pictorial information, recognition of patterns in scenes, the reconstruction of a picture from projections, and descriptions of objects in a scene.

Although the book is intended mainly for seniors and graduate students in electrical engineering or computer science and researchers in image processing and recognition, this material has also been taught successfully to physicists, to mechanical, civil, nuclear, and industrial engineers as well as to medical students, physicians, and psychologists interested in computer image processing. Some familiarity with the rudiments of linear algebra, communication theory, optics, Fourier analysis, and probability theory is helpful but not required since supplemental study of the references given at the end of each chapter is sufficient.

Information regarding the availability of a solution manual and a set of slides of the pictorial illustrations may be obtained from the author.

# Acknowledgments

I am deeply indebted to my students and colleagues whose encouragement, discussions, inspiration, and support have contributed to the preparation of this book. These contributions have taken many forms—building imaging hardware, many hours at the computer, photographic and illustration work, discussions and analyses of methods, problem solving, facilities and research support, and technical interchange meetings throughout the world. I particularly appreciate the inspiration and guidance provided by Rafael C. Gonzalez. Joseph M. Googe and Fred N. Peebles have provided an academic atmosphere at the University of Tennessee which encourages excellence in teaching and research. To my faculty associates in the Image and Pattern Analysis Laboratory, especially Donald W. Bouldin, Michael G. Thomason, Robert T. Gregory, James R. Carter, Robert R. Rochelle, and Alejandro Barrero, and to our students, in particular Firooz Sadjadi, Juin-Jet Hwang, Ted G. Saba, David Bryant, David Davies, Michael E. Casey, Chien-Chyun Lee, Eddie P. Tinnel, Kenneth Freshwater, James Heller, Luis Bryan, Mukesh Sundaram, David Matherly, Pin Chou, and Shu Chiu, who helped substantially in the completion of the manuscript, I am sincerely grateful.

I especially wish to thank my colleagues who reviewed chapters of the manuscript: Robert Y. Wong of the California State University at Pomona; Charles F. Hall of the Space and Missile Systems Organization, USAF; Gunar Robinson and Yik S. Kwoh of the University of Southern California; Richard P. Kruger of the Los Alamos Scientific Laboratory; and Rafael C. Gonzalez, Michael G. Thomason, and Donald W. Bouldin of the University of Tennessee.

I also wish to thank Gale D. Slutzky and Richard G. Gruber for their excellent work on the illustrations and for editorial assistance, and Mary Bearden and her staff for their excellent assistance in the typing of the manuscript.

Several individuals deserve a special acknowledgment for indirect contributions to this book, especially Azriel Rosenfeld, King Sun Fu, and the many others who have provided the fundamental concepts and emphasis of the field of computer image processing and pattern recognition; Joseph D. Rouge for my introduction to scene matching; James T. Karem for emphasizing the three dimensional; Gerald Huth for my introduction to three-dimensional reconstruction; Ali Habibi, Gunar Robinson, and William K. Pratt, who helped me appreciate image coding; Harry C. Andrews for his emphasis of image restoration; and Richard P. Kruger for his emphasis of image enhancement and automated measurements. I shall always be indebted to my professors Dennis Tebbe, Robert W. McLaren, and Samuel J. Dwyer at the University of Missouri–Columbia.

Knowledge cannot be discovered, understood, or disseminated without research, and research may not be conducted without knowledgeable and interested sponsorship. I gratefully acknowledge the following sponsors: the USAF Space and Missile Systems Command, the Advanced Research Projects Agency, the National Institute for Occupational Safety and Health, the National Science Foundation, the National Heart Lung Institute, the Oak Ridge National Laboratory, and the Jet Propulsion Laboratory.

Finally, my special thanks go to the staff of Academic Press for their continued support and encouragement. The excellent editing, which transformed a rough manuscript into a beautiful text, will long be appreciated.

# List of Symbols

| Symbol | Definition |
|--------|------------|
| | **Chapter 2** |
| $Q$ | Radiant energy |
| $w$ | Radiant density |
| $P$ | Radiant power |
| $M$ | Radiant emittance |
| $E$ | Irradiance |
| $I$ | Radiant intensity |
| $L$ | Radiance |
| $\rho(\lambda)$ | Spectral reflectance |
| $\tau(\lambda)$ | Spectral transmittance |
| $\alpha(\lambda)$ | Spectral absorption |
| $K$ | Luminosity or luminous efficiency |
| $E$ | Radiometric efficiency |
| $\bar{x}(\lambda), \bar{y}(\lambda), \bar{z}(\lambda)$ | Tristimulus vector values |
| $d$ | Euclidean distance |
| $L$ | Lightness in *Lab* system |
| $a$ | Redness–greenness in *Lab* system |
| $b$ | Yellowness–blueness in *Lab* system |
| $C$ | Contrast |
| $C_R$ | Contrast ratio |
| $B$ | Luminance level |
| $C(u)$ | Contrast transfer function |
| $L(x)$ | Fourier series of rectangular luminance profile |
| $G(u,v)$ | Fourier transform of $g(x,y)$ |
| $F(u,v)$ | Fourier transform of $f(x,y)$ |
| $H(u,v)$ | Optical transform function, Fourier transform of $h(x,y)$ |
| $l(x)$ | Line spread function |
| $h(r)$ | Point spread function |

| Symbol | Definition |
|--------|------------|
| $r$ | Radial distance |
| $\omega$ | Radial frequency |
| $J_0(x)$ | Bessel function of order zero |
| $A$ | Acutance |
| $L(u)$ | Fourier transform of line spread function |
| $\mathbf{e}$ | Retinal receptor output vector |
| $\mathbf{g}$ | Retinal inhibition output vector |
| $\mathbf{B}$ | Retinal connection matrix |
| $\mathbf{I}$ | Identity matrix |
| $C(\lambda)$ | Spectral distribution of light |
| $\phi_i(\lambda)$ | Spectral response of cones |
| $t_j$ | Tristimulus value |
| $\mathbf{S}$ | Perceived tristimulus vector |
| $p_j(\lambda)$ | Primary spectral functions |
| $q_j(\lambda)$ | Primary spectral functions |
| $\alpha$ | Stereo angle |
| $D$ | Optical density |
| $\eta$ | Index of refraction of medium |
| $d$ | Pupil diameter |
| $M(u,v)$ | Modulation transfer function |
| $\mathbf{M}$ | Visual match |
| | **Chapter 3** |
| $\mathbf{v}_i$ | Imaging point |
| $\mathbf{v}_o$ | Object point |
| $\hat{\mathbf{v}}$ | Homogeneous coordinate |
| $\mathbf{P}$ | Linear transformation matrix |

| Symbol | Definition | Symbol | Definition |
|--------|-----------|--------|-----------|
| $\hat{a}$ | Normal vector to plane | $\overline{\varepsilon^2}$ | Mean square error |
| H | Perspective transformation matrix | $A_i$ | Eigenvector of a covariance matrix |
| T | Translation matrix | $\lambda_i$ | Eigenvalue of covariance matrix |
| S | Scaling matrix | | |
| R | Rotation matrix | $\mu$ | Sample mean |
| L | Translation matrix | $V$ | Relative variance |
| $f(x,y)$ | Two-dimensional function | H | Walsh–Hadamard matrix |
| $F(u,v)$ | Fourier transform of two-dimensional function | $\equiv$ | Congruence |
| $P_B(u,v)$ | Pulse or aperture function | | **Chapter 4** |
| $F_p(u,v)$ | Fourier transform of pulse function | $C$ | Contrast |
| $C_{mn}$ | Fourier transform coefficients | $l(x,y)$ | Transmitted luminance |
| | | $d(x,y)$ | Displayed luminance |
| $S_f(u,v)$ | Power spectrum of stationary stochastic process | $t(x,y)$ | Transmittance function |
| | | $x$ | Continuous random variable |
| $p(f)$ | Probability function | $S_K$ | Histogram-equalized levels |
| $R_j$ | Representation value in quantization | $\Lambda$ | Diagonal matrix with the eigenvalues of the covariance matrix |
| $D_j$ | Decision value in quantization | | |
| $[f]$ | Sampled, quantized image matrix | n | Noise vector |
| | | $E$ | Squared error |
| $f_{ij}$ | Sampled, quantized image matrix element | $\lambda$ | Lagrangian multiplier |
| | | $H$ | Entropy |
| f | Sampled image vector | $\Lambda^\#$ | Diagonal covariance matrix |
| $E$ | Energy of a vector | $S_f(u,v)$ | Power spectra |
| $[C]$ | Bidiagonal matrix | $S_n(x,y)$ | Circulant noise correlation |
| g | Smoothing relationship | $\Sigma_f$ | Covariance matrix |
| p(f) | Joint density function | f | Mean vector |
| $\bar{f}$ | Mean or expected value of sampled image vector | $\bar{\sigma}^2$ | Variance |
| $\Sigma$ | Covariance matrix | | **Chapter 5** |
| $R(f)$ | Correlation matrix | | |
| $\mu(K)$ | Ensemble mean vector | f | Column vector with $i, j$ elements |
| $R(m,n)$ | Autocorrelation function | | |
| $E(f)$ | Mean of random variable | g | Column vector with $m, n$ elements |
| $f$ | Random variable | | |
| $\text{var}(f)$ | Variance of random variable | $J(f)$ | Criteria function |
| | | C | Semidefinite matrix |
| $[L]$ | Orthonormal matrix | $\xi$ | Damping factor |
| F | Orthonormal transform | $T$ | Total number of bits |
| X | Random vector | $L$ | Number of possible images |
| $\mu$ | Mean vector | $I$ | Light emitted |
| S | Scatter matrix or autocorrelation matrix | $g(s,\theta)$ | Projection |
| R | Correlation matrix | | **Chapter 6** |
| D | Standard deviation matrix | $M$ | Set of messages |
| Y | Random vector | $X$ | Source alphabet |
| $\varepsilon$ | Error vector | $Y$ | Channel encoded symbols |

| Symbol | Definition |
|--------|-----------|
| $\hat{Y}$ | Received channel symbols |
| $\hat{X}$ | Received source symbols |
| $\hat{M}$ | Decoded message |
| SER | Signal to error ratio |
| $\text{Prob}(0) = p$ | Probability of binary 0 |
| $h(p)$ | Information measure |
| $H(J)$ | Average information or entropy |
| $H(X, Y)$ | Joint entropy |
| $H(X/Y)$ | Conditional entropy |
| $I(X, Y)$ | Mutual information |
| $L(n)$ | Length of $n$th sequence |
| $C$ | Channel capacity |
| $\pi_2$ | Transition matrix of binary channel |
| $F$ | Word distortion measure |
| $\rho_n(\mathbf{x}, \mathbf{y})$ | Single letter fidelity criterion |
| $d(Q)$ | Average distortion |
| $Q_D$ | Set of all $D$-admissible channels |
| $R(D)$ | Rate distortion function |
| $\Sigma_\mathbf{X}$ | Covariance matrix of $\mathbf{X}$ |
| $\Sigma_\mathbf{Y}$ | Covariance matrix of $\mathbf{Y}$ |
| $S(\omega_r)$ | Power spectrum |

**Chapter 7**

| Symbol | Definition |
|--------|-----------|
| $S(\mathbf{X}, \mathbf{X}_1)$ | Similarity function |
| $\chi_i$ | Disjoint subsets |
| $c$ | Number of clusters |
| $e(T)$ | Probability of error of threshold value |
| $\Phi$ | Transformation operator |
| $\nabla^2$ | Laplacian operator |
| $T$ | Threshold value |
| $\nabla f$ | Gradient vector |
| $G$ | Discrete gradient vector |
| $\mathbf{p}$ | Parameter vector |
| $\mathbf{F}'(\mathbf{P}^{(k)})$ | Jacobian matrix |
| $\Delta^{(k)}$ | Increment vector |
| $\mu_{pq}$ | Central moments |
| $\eta_{pq}$ | Normalized central moments |

| Symbol | Definition |
|--------|-----------|
| $E$ | Euler number |
| $\lambda$ | Empty sentence |
| $V^+, V^*$ | Sets of sentences |
| $P$ | Set of productions |
| $\rightarrow$ | Indicate replacement of a string by another string |
| $\underset{G}{\Rightarrow}$ | Indicates operations in the grammar G |
| $G$ | Formal string grammar |
| $T$ | Tree or a finite set of one or more nodes |
| $R$ | Ranking function |
| $\alpha, \beta, \phi$ | Sublevels |
| $\omega$ | Web |

**Chapter 8**

| Symbol | Definition |
|--------|-----------|
| $\lambda_n$ | Likelihood ratio |
| $R_s(u, v)$ | Statistical correlation measure |
| $D_s(i, j), D_w(i, j)$ | Spatial filter functions |
| $\Sigma_\mathbf{s}, \Sigma_\mathbf{w}$ | Image covariance matrices |
| $\rho$ | Correlation between adjacent image elements |
| $\delta(u, v)$ | Error function |
| $E_\mathbf{R}$ | Spectral energy |
| $E_\mathbf{A}$ | Aliasing error energy |
| $s(x, y)$ | Sampling signal |
| $F_K(u, v)$ | Fourier spectrum of the sampled image field |
| $\varepsilon_\mathbf{A}$ | Aliasing error |
| $\bar{s}$ | Mean intensity of the image elements |
| $\bar{w}$ | Mean intensity of the window |
| $\sigma$ | Standard deviation |
| $\bar{e}$ | Mean |
| $E_n$ | Cumulative error |
| $\bar{E}_n$ | Expected error |
| $f(x, y)$ | Gray-scale image |
| $R_b$ | Correlation with a background level |
| $P_f$ | Probability of a false fix |
| $P_d$ | Probability of a match |

# 1 | Introduction

## 1.1 Computer Image Processing and Recognition

It is estimated that 75% of the information received by a human is visual. When you receive and use visual information, we refer to this process as sight, perception, or understanding. When a computer receives and uses visual information, we call this *computer image processing and recognition. Image* rather than picture is used in the title because computers store numerical images of a picture or scene. *Processing* and *recognition* refer to two broad classes of techniques that have evolved in this field. The first part of this book is devoted to representation, enhancement and restoration, and image compression and reconstruction, which are image processing topics. The final chapters on scene understanding and matching are image recognition topics.

The history of computer image processing and recognition is relatively brief. One could argue that it is no older than the first electronic computer, ENIAC, built by Eckert and Mauchy in 1946. However, staunch proponents might argue that the 1801 weaving loom of Jacquard was actually processing image information or that the telegraph transmission of images is a good example of digital transmission of image information.

Many computer picture processing techniques have been developed, exploited, or applied only in the last decade. The modern advancement in this area is mainly due to the recent availability of image scanning and display hardware at a reasonable cost, and the relatively free use of computers. However, certain economical computer processing techniques have also contributed to this development. Without the popularization of the fast Fourier transform algorithm by Cooley and Tukey and others in 1965, transform processing of images would probably have remained in the domain of optics. Also, the race to the moon provided a major impetus to computer picture processing. The Jet Propulsion Laboratory was assigned the task of providing television coverage of man's landing on the moon. The images had to be of standard broadcast quality. To provide this quality, the television video had to be scanned at a slow rate, digitized, coded, and transmitted. Error correction and several image restoration techniques had to be accomplished at the earth stations to provide suitable quality images. Transmission of image information on earth was also being studied. The possibility of a Picture-phone in every home was being carefully studied by Bell Telephone Laboratories.

Simultaneously, it was noted that certain measurements which humans were making from pictures seemed simple and could possibly be automated. An example of this is the *karyotyping* or classification of chromosomes from microscope images. In other applications, workers noticed that raw pictorial information could be enhanced by computer processing to produce a greatly improved image. For example, an expensive seismic record that should provide a map of underground structure could be so degraded by surface scatter as to make the trace worthless. A simple compensation technique could be used to cancel out this irrelevant information and provide an accurate representation of the underground structure, which might show a likely oil deposit. These applications and many more provided the motivation for the recently evolved discipline of computer image processing.

## 1.2 Applications

As with any emerging discipline, one runs a risk in classifying the major problems or subdivisions of the area. It is probable that certain problems have not been discovered or formulated yet. This book will consider five of the major problems that have evolved in computer picture processing: *enhancement, communications, reconstruction, segmentation,* and *recognition*. Even though it is certain that new problems will arise, it is just as certain that these problems will remain important in most image processing applications. Furthermore, an understanding of these problems and their solutions may be necessary for future developments. It should also be pointed out that the optimum methods for image enhancement, coding, feature extraction, or recognition have yet to be developed. Therefore, the solutions presented will provide intuition, guidelines, and facts, but cannot provide the "answer." Let us now explore each of the major problems in detail.

### 1.2.1 Image Enhancement and Restoration

Image enhancement techniques are designed to improve image quality for human viewing. This formulation tacitly implies that an intelligent human viewer is available to recognize and extract useful information from an image. This viewpoint also defines the human as a link in the image processing system. Since human subjective judgment may be either wise or fickle, certain difficulties may arise. Usually, a careful study of the particular problem, or subjective testing of a group of human viewers can circumvent these difficulties. The psychopictorial phenomena involved should always be considered. Image enhancement is only necessary when a *human* wants to improve the quality of an image.

A particular class of enhancement problems usually referred to as image restoration is much more tractable. An ideal image is considered which has been degraded in some manner. The problem is to restore the original image or compensate for the degradation. Usually, an optimum or adequate solution may be determined. For example, systematic geometric distortions may be eliminated, or if the statistics of the signal and noise are given, then an optimum Wiener filter may be determined.

## 1.2.2 Three-Dimensional Reconstruction

Since most physical objects are spatially three dimensional, the reconstruction of these objects is particularly fascinating. The basic techniques for reconstructing a three-dimensional object from its two-dimensional projections are of wide application. The implementation of these techniques for x-ray imaging has started a new medical discipline called computer tomography, and several devices are now available. The methods were developed in radio astronomy and apply to problems in a wide variety of fields, including art and industrial testing.

## 1.2.3 Digital Television and Image Compression

Digital television is an exciting application of computer image processing and is used to introduce the general image communication problem.

Image communication theory is concerned with two major questions: What information must be transmitted? What is the most effective method for transmitting the information? A great deal of attention has been given to the second question, since it is a general communication problem; therefore we shall not consider the design of communications systems or channel coding in detail. The first question is especially relevant, interesting, and unique for image transmission. Again, the human must be considered as an integral part of the image communication system. The psychophysical properties of the human visual system may often be exploited to great advantage. A striking example is *Roberts's coding scheme* in which random noise is added to a received image to apparently increase the quality while maintaining a low transmission rate. The psychophysical property exploited by this coding method is that humans generally prefer random noise to systematic noise. The added random noise served to break up the "false contours" produced by quantization.

An excellent theoretical formulation for the image communication problem is available in Shannon's rate distortion theory. In this theory, one considers a stationary, stochastic source to be the input to a noisy communications channel. A distortion measure such as the mean square error between the input and output of the channel is defined. A rate distortion function may be defined as the smallest mutual information rate between the input and output, such that the average distortion is less than a specified value. A coding theorem may then be given, stating that suitably encoded information may be transmitted over any channel with distortion less than the specified amount, *provided* that the *rate distortion function* is less than the *channel capacity*. The difficulty in practice is that a suitable objective distortion measure that agrees with subjective human evaluation is difficult to determine.

### 1.2.4   Segmentation and Description

The problem of describing verbally or by objective measurements the information contained in an image or scene requires a high level of intelligence for man or machine. Most of our knowledge about the world has been and continues to be obtained from observations. Therefore, the potential applications of *computer segmentation* and *description* is staggering. Some of the applications have been realized. Computer processing of chest x rays, chromosome karyotyping, and blood cell analyses are a few medical examples. Many other areas have not even been considered due to the apparent difficulty of the solutions.

In the last few years, it has been recognized that the problem may generally be divided into two stages—segmentation followed by description. Segmentation consists of isolating a meaningful object or region. Description is the measurement of the properties of a single region or the relationships between regions.

The segmentation problem may be approached as a rather special clustering problem in which points in $n$-dimensional space with similar properties are grouped together; however, the points must also cluster in a contiguous region in two- or three-dimensional space. The description problem involves both *measurements* (unary relations) and *relationships* (binary or higher-order relations). Structural descriptions are almost always of interest and in certain cases syntactic descriptions may be used to describe a scene concisely.

### 1.2.5   Scene Matching and Recognition

Visual recognition is such a natural procedure for humans that it is often proposed as an inherent function of the eye–brain system. Given a picture of an object such as a ball, recognize any new picture of the object. The applications of recognition vary from "simple" character recognition to complex scene matching for medical diagnosis, industrial inspection, or earth resource study from satellite images.

Starting from the simple "template matching" approach, a variety of methods for efficient scene matching are presented. A novel approach involving a coarse to fine search using invariant measurements attempts to illustrate that high speed, accuracy, and invariance to changes can be accomplished with computer techniques.

## Introductory Bibliography

As with any emerging discipline, several thousand papers have been published on techniques and applications of computer image processing and recognition. A selected bibliography is presented to indicate some excellent available literature. The references have been divided into six categories: *computer image processing, computer pattern recognition, visual perception, computer graphics, artificial intelligence,* and *optics and electro-optics.* A collection of noteworthy special issues of journals devoted to the subject is also included. Specific references are included with each chapter.

### Computer Image Processing

Aggarwal, J. K., Duda, R. O., and Rosenfeld, A., eds. (1977). "Computer Methods in Image Analysis." IEEE Press, New York.

Andrews, H. C. (1970). "Computer Techniques in Image Processing." Academic Press, New York.

Andrews, H. C., and Hunt, B. R. (1977). "Digital Image Restoration." Prentice-Hall, Englewood Cliffs, New Jersey.

Andrews, H. C., ed. (1978). "Tutorial and Selected Papers in Digital Image Processing." IEEE Computer Society, Long Beach, California.

Gonzalez, R. C., and Wintz, P. A. (1977). "Digital Image Processing." Addison-Wesley, Reading, Massachusetts.

Grasseli, A., ed. (1969). "Automatic Interpretation and Classification of Images." Academic Press, New York.

Huang, T. S., ed. (1975). "Picture Processing and Digital Filtering." Springer-Verlag, Berlin and New York.

Huang, T. S., and Tretiak, O. J., ed. (1972). "Proceedings of the 1969 Symposium on Picture Bandwidth Compressing." Gordon & Breach, New York.

Lipkin, B. S., and Rosenfeld, A., ed. (1970). "Picture Processing and Psychopictorics." Academic Press, New York.

Pearson, D. E. (1975). "Transmission and Display of Pictorial Information." Wiley, New York.

Pratt, W. K. (1978). "Digital Image Processing." Wiley, New York.

Rosenfeld, A. (1969). "Picture Processing by Computer." Academic Press, New York.

Rosenfeld, A., and Kak, A. C. (1976). "Digital Picture Processing." Academic Press, New York.

Tippet, J. T., ed. (1965). "Optical and Electro-Optical Information Processing." MIT Press, Cambridge, Massachusetts.

## Computer Pattern Recognition

Agrawala, A. K., ed. (1977). "Machine Recognition of Patterns." IEEE Press, New York.
Andrews, H. C. (1972). "Introduction to Mathematical Techniques in Pattern Recognition." Wiley, New York.
Bongard, N., ed. (1970). "Pattern Recognition." Spartan Books, Washington, D.C.
Cacoullow, T. (1973). "Discriminate Analysis and Applications." Academic Press, New York.
Chen, C. H. (1973). "Statistical Pattern Recognition." Spartan Books, Washington, D.C.
Duda, R. O., and Hart, P. E. (1973). "Pattern Classification and Scene Analysis." Wiley, New York.
Everitt, B. (1974). "Cluster Analysis." Wiley (Halsted Press), New York.
Fu, K. S. (1968). "Sequential Methods in Pattern Recognition and Machine Learning." Academic Press, New York.
Fu, K. S. (1974). "Syntactic Methods in Pattern Recognition." Academic Press, New York.
Fu, K. S., ed. (1976). "Digital Pattern Recognition." Springer-Verlag, Berlin and New York.
Fu, K. S., ed. (1977). "Syntactic Pattern Recognition Applications." Springer-Verlag, Berlin and New York.
Fukunaga, K. (1972). "Introduction to Statistical Pattern Recognition." Academic Press, New York.
Gonzalez, R. C., and Thomason, M. G. (1978). "Syntactic Pattern Recognition—An Introduction." Addison-Wesley, Reading, Massachusetts.
Inbar, G. F., ed. (1975). "Signal Analysis and Pattern Recognition in Biomedical Engineering." Wiley, New York.
Kanal, L., ed. (1968). "Pattern Recognition." Thompson, Washington, D.C.
Klinger, A., Fu, K. S., and Kunii, T. L., eds. (1977). "Data Structures, Computer Graphics and Pattern Recognition." Academic Press, New York.
Meisel, W. S. (1972). "Computer-Oriented Approaches to Pattern Recognition." Academic Press, New York.
Mendel, J. M., and Fu, K. S., eds. (1970). "Adaptive Learning and Pattern Recognition Systems." Academic Press, New York.
Morrison, D. F. (1976). "Multivariate Statistical Methods." McGraw-Hill, New York.
Nilsson, N. J. (1965). "Learning Machines." McGraw-Hill, New York.
Sebestyen, G. S. (1962). "Decision-Making Processes in Pattern Recognition." ACM Monograph Series. Macmillan, New York.
Tatsuoka, M. M. (1971). "Multivariate Analysis." Wiley, New York.
Tou, J. T., and Gonzalez, R. C. (1974). "Pattern Recognition Principles." Addison-Wesley, Reading, Massachusetts.
Tsypkin, Y. Z. (1971). "Adaptation and Learning in Automatic Systems." Academic Press, New York.
Uhr, L., ed. (1966). "Pattern Recognition." Wiley, New York.
Ullman, J. R. (1973). "Pattern Recognition Techniques." Crane, Russak, New York.
Watanabe, S. "Knowing and Guessing." Wiley, New York.
Watanabe, S., ed. (1969). "Methodologies of Pattern Recognition." Academic Press, New York.
Young, T. Y., and Calvert, T. W. (1974). "Classification, Estimation and Pattern Recognition." Am. Elsevier, New York.

## Visual Perception

Committee on Colorimetry of the Optical Society of America (1963). "The Science of Color." Opt. Soc. Am., Washington, D.C.
Cornsweet, T. N. (1970). "Visual Perception." Academic Press, New York.
Dodwell, P. C. (1970). "Visual Pattern Recognition." Holt, New York.

Julesz, B. (1971). "Foundations of Cyclopian Perception." Univ. of Chicago Press, Chicago, Illinois.
Wyszecki, G. W., and Stiles, W. S. (1967). "Color Science." Wiley, New York.

## Computer Graphics

Chasen, S. H. (1978). "Geometric Principles and Procedures for Computer Graphic Applications." Prentice-Hall, Englewood Cliffs, New Jersey.
Gilio, W. K. (1978). "Interactive Computer Graphics." Prentice-Hall, Englewood Cliffs, New Jersey.
Newman, W. M., and Sproull, R. F. (1973). "Principles of Interactive Graphics." McGraw-Hill, New York.
Rogers, D. F., and Adams, B. (1977). "Principles of Interactive Computer Graphics." McGraw-Hill, New York.

## Artificial Intelligence

Arbib, M. A. (1972). "The Metaphorical Brain, An Introduction to Cybernetics as Artificial Intelligence and Brain Theory." Wiley, New York.
Jackson, P. C. (1974). "Introduction to Artificial Intelligence." Mason-Charter, New York.
Weizenbaum, J. (1976). "Computer Power and Human Reason." Freeman, San Francisco, California.

## Optics and Electro-Optics

Carlson, P. F. (1977). "Introduction to Applied Optics for Engineers." Academic Press, New York.
Goodman, J. W. (1968). "Introduction to Fourier Optics." McGraw-Hill, New York.
Papoulis, A. (1968). "Systems and Transforms with Application in Optics." McGraw-Hill, New York.
Preston, K., Jr. (1972). "Coherent Optical Computers." McGraw-Hill, New York.
"RCA Electro-Optics Handbook," Tech. Ser. EOH-11. RCA Corp., Princeton, New Jersey.
Takanori, O. (1976). "Three Dimensional Imaging Techniques." Academic Press, New York.

## Special Journal Issues

*Computer Graphics and Image Processing* (1972–1978). Vols. 1, 2, 3, 4, and 5.
Special Issue Digital Communications. (1971). *IEEE Trans. Commun. Technol.* **COM-19**, No. 6, Part 1.
Special Issue on Digital Filtering and Image Processing. (1975). *IEEE Trans. Circuits Syst.* **CAS-2**.
Special Issue on Digital Image Processing. (1974). *IEEE Comput.* **7**, No. 5.
Special Issue on Digital Picture Processing. (1972). *Proc. IEEE* **60**, No. 7.
Special Issue on Redundancy Reduction. (1967). *Proc. IEEE* **65**, No. 3.
Special Issue on Two Dimensional Digital Signal Processing. (1972). *IEEE Trans. Comput.* **C-21**, No. 7.

# 2 | Image Formation and Perception

## 2.1 Introduction

Computer picture processing is concerned with *physical images*, and therefore the elements of image formation and perception often play key roles in the formulation and solution of picture processing problems. In *x-ray image analysis*, for example, the atomic interactions of x rays with matter set the scene in terms of *information content*, *contrast*, and *noise* before an x-ray photograph is produced. Similarly, since the human perception of an edge enhances the boundary, the designer of an edge detection program must carefully consider that the computer must process the information on the film rather than the information seen on the film. In this chapter, the physical fundamentals of image formation will be introduced as well as the psychological fundamentals of image perception by humans.

## 2.2 Image Formation

*Light* is radiant energy that produces a visual sensation upon stimulation of the retina of the normal human eye. The psychophysical concept of light is the bridge between the physics of light and the reaction of man to visual stimuli. A physical image, an energy distribution in three-dimensional space and time, is observed from the two retinal images as a three-dimensional scene. Psychopictorics is the discipline that connects the psychology of light physics with computer techniques. It is therefore appropriate to begin the discussion of computer image processing by reviewing the physical image-formation process and the psychology and physiology of vision. Obviously, the principles of image processing will apply to images other than the visible images, such as those formed by synthetic aperture radar, seismographs, x rays, infrared radiation, or ultrasound, if they are converted to light images.

8

### 2.2.1   Radiometry

Light occupies a narrow region of the electromagnetic spectrum, from about 350 nm (violet) to 780 nm (red). As with all forms of energy, radiant energy $Q$ may be quantitatively measured and expressed in joules or other convenient units. For example, one may determine the total energy radiated by a source. Several definitions are useful in describing the spatial and temporal variations of radiant energy (Klein, 1970). One may measure the *radiant density w* or total energy contained in a volume $V$ of space:

$$w = dQ/dV.$$

Another useful quantity is the *radiant power* or flux $P$, specified in watts:

$$P = dQ/dt.$$

The radiant power emitted per unit area $A_s$ of a source is called the *radiant emittance M*:

$$M = dP/dA_s.$$

The radiant power incident per unit area $A_r$ of a receiver is called the *irradiance E* of the receiver surface:

$$E = dP/dA_r.$$

The *radiant intensity I* is the power radiated per unit solid angle of the source and may be expressed in watts per unit solid angle $\omega$:

$$I = dP/d\omega.$$

The unit solid angle $\omega$, expressed in steradians, is the solid angle subtended by $1\,m^2$ of surface of a sphere having a radius of $1\,m$. Finally, the power radiated into a unit solid angle by a unit projected area $A_p$ of a source is called *radiance L*:

$$L = dI/dA_p.$$

To measure radiance, one must collect the radiant power from the source and divide by the solid angle subtended by the receiver and by the projected area of the element of the source on which the solid angle is centered.

The spectral composition of radiant energy $Q(\lambda)$ or power $P(\lambda)$ may be specified as the energy or power at each wavelength $\lambda$. Relative spectral distributions that specify the relative spectral composition of radiant energy or power are usually sufficient for most problems and are more frequently encountered. Given the spectral composition $P(\lambda)$, the total *radiant power P* may be determined from

$$P = \int_0^\infty P(\lambda)\,d\lambda.$$

Whenever radiant flux $P_I$ is incident upon an object, some of it $(P_R)$ may be

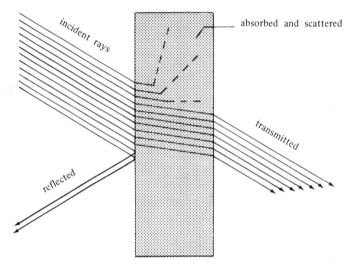

*Figure 2.1*   Incident, reflected, absorbed, scattered, and transmitted light.

reflected by the surface, some ($P_A$) absorbed, and some ($P_T$) transmitted, as illustrated in Fig. 2.1:

$$P_I = P_R + P_A + P_T.$$

The *spectral reflectance* $\rho(\lambda)$, *transmittance* $\tau(\lambda)$, and *absorption* $\alpha(\lambda)$ may be determined at each wavelength $\lambda$ by the following relationships:

$$\rho(\lambda) = P_R(\lambda)/P_I(\lambda), \qquad \tau(\lambda) = P_T(\lambda)/P_I(\lambda), \qquad \alpha(\lambda) = P_A(\lambda)/P_T(\lambda).$$

For accurate use of these quantities, the spatial conditions must also be specified, especially the direction of incidence.

The previous concepts of radiometry are useful; however, light cannot be adequately described as radiant energy. Daily experience readily shows that radiance is not simply related to perceived brightness. A small night-light that is barely visible during the day may be bright enough to read by at night. Since light produces a visual sensation upon stimulation of the retina of the normal human observer, perhaps this difficulty can be avoided by characterizing the normal human retina. An immediate difficulty arises since one cannot directly measure the retinal response as one would the response of a photoelectric cell. Some indirect method must be employed. Furthermore, the simplest response to elicit is that to brightness, rather than that to color. Such considerations led to the development of photometry.

### 2.2.2  Photometry

*Photometry* is the science which relates perceived brightness in the normal observer to radiant energy. The first standard to be developed was the standard luminous intensity or, simply, intensity. The procedure consisted of defining a known source, a candle. Unknown sources could then be compared by an observer and distance or wedge filters varied until the observer judged the sources to be of equal intensity. The unit of luminous intensity, the candela, is defined as the luminous intensity of $\frac{1}{60}$ cm$^2$ of the projected area of a blackbody radiator operating at the temperature of solidification of platinum (2045°K). Next, methods for comparing the luminance of different colors were developed. To minimize the difficulties in comparing two chromaticities, a step-by-step comparison of two slightly different wavelengths was made throughout the spectrum under good lighting conditions. The observed light field was divided into two parts, one side containing a source of known radiance and wavelength, the other containing a source of slightly different wavelength. The observer was asked to adjust the second source until the visual difference was minimized. The radiance differences were then plotted to yield a *luminosity function* that may be considered as the spectral response curve for the normal human observer. With a luminosity curve, one can easily relate luminous intensity to radiant intensity. When the same experiment was conducted under poor lighting conditions, i.e., for dark-adapted eyes, two main differences were noted. First, the peak of the luminosity function shifted from 555 nm (green region) to 507 nm (blue region). Also, a tremendous increase in peak sensitivity from 680 lm/W for the good lighting conditions (photopic vision) to 1746 lm/W for the dark-adapted eye (scotopic vision) occurred. The CIE standard luminosity functions (Committee on Colorimetry of the Optical Society of America, 1963) are shown in Fig. 2.2. The abbreviation CIE consists of the initials of and is the recommended method of referring to the *Commission Internationale de*

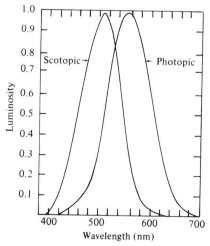

*Figure 2.2*   Normalized luminosity functions for good lighting (photopic) and dark adapted (scoptic) vision. [From the Committee on Colorimetry of the Optical Society of America (1963, p. 225).]

*l'Éclairage*, the International Commission on Illumination. Thus, the standard observer defined by the luminosity functions is an international standard.

The adoption of these standard data was somewhat arbitrary, since different data would be produced for different observers, and a change in observation conditions would yield different data even for the same observers. The usefulness of the standard luminosity curve is that it provides a convenient basis for general agreement concerning photometric concepts, such as the visual effectiveness of radiant energy. *Luminosity* (lumens/watt) measures the efficiency of the time rate of radiant energy for the production of brightness, but depends upon viewing distance since it does not account for the area of the retina. If brightness depends upon the time rate of energy incident upon a unit area of the retina, then a photometric unit which would correspond to irradiance (watts/square meter) would be useful. This unit, which is called *illuminance*, should correlate with irradiance and should have the advantage of being more independent of observation conditions such as viewing distance. The ratio of illuminance to irradiance can still be expressed as luminosity (lumens/watt) since the common area cancels when the ratio is taken. A complete set of photometric units corresponding to radiometric units, as shown in Table 2.1, can thus be established and related by the standard luminosity function. The luminosity $K$ may be defined as the ratio of any photometric quantity to the corresponding radiometric quantity. For example, the luminosity $K$ or luminous efficiency of radiant spectral power $P(\lambda)$ is given by

$$K = K_m \int_0^\infty \bar{y}(\lambda) P(\lambda) \, d\lambda \Big/ \int_0^\infty P(\lambda) \, d\lambda,$$

where $K_m$ is the maximum possible luminosity of radiant energy and $\bar{y}$ is the photopic luminosity function.

The proper radiometric or photometric unit for a given situation may be determined from consideration of the type of photodetector used. If the receptor is the human eye, then the response depends upon either the spectral composition of luminance or illuminance as appropriate. If the detector is a selective photocell, then its response depends upon the spectral composition of radiant flux, in watts. If it is nonselective, the response depends only upon the radiant flux. A photographic plate responds to the spectral composition of energy per unit area.

**Example:** *Radiometric and photometric luminous efficiency of a blackbody radiator.* To illustrate the radiometric and photometric concepts, we may compute the luminous efficiency of a perfect radiator.

The power emitted by a blackbody radiator, according to the *Stefan–Boltzmann law*, is proportional to the fourth power of its absolute temperature,

$$W = 5.67 \times 10^{-12} T^4,$$

Table 2.1

Radiometric and photometric units[a]

| Radiometry (Physical units) | | | Photometry (Psychophysical units) | | |
|---|---|---|---|---|---|
| Symbol | Definition | SI unit | Symbol | Definition | SI unit |
| $Q$ | Radiant energy | joule | $Q$ | Luminous energy | lumen second (talbot) |
| $w$ | Radiant density | joules per cubic meter | $w$ | Luminous density | lumen seconds per cubic meter (talbot/m$^3$) |
| $P$ | Radiant flux or power | watt | $P$ | Luminous flux or power | lumen |
| $M$ | Radiant emittance | watts per square meter | $M$ | Luminous emittance | lumens per square meter (lux) |
| $E$ | Irradiance | watts per square meter | $E$ | Illuminance | lumens per square meter (lux) |
| $I$ | Radiant intensity | watts per steradian | $I$ | Luminous intensity | candela (lumens per steradian) |
| $N$ | Radiance | watts per steradian and square meter | $B$ | Luminance | lumens per steradian and square meter (nit) |
| $\rho(\lambda)$ | Spectral reflectance | | $r(\lambda)$ | Luminous reflectance | |
| $\tau(\lambda)$ | Spectral transmittance | | $t(\lambda)$ | Luminous transmittance | |
| $\alpha(\lambda)$ | Spectral absorptance | | $a(\lambda)$ | Luminous absorptance | |

[a]Ratio of photometric quantity to corresponding radiometric quantity (standard units) is luminosity or luminous efficiency $K(\lambda)$ (l/W) or special luminous efficacy $K(\lambda)$ (l/W). Luminous efficiency is the ratio of the spectral luminous efficacy to its maximum value and thus is a numerical quantity.

where $W$ is the radiant emittance in watts/square centimeter and $T$ is the temperature in degrees Kelvin.

The fraction of this power at a particular wavelength $\lambda$ is governed by *Planck's law*:

$$W(\lambda) = (2.39 \times 10^{-11})/\lambda^5 [\exp(1.432/T\lambda) - 1],$$

where $\lambda$ is the wavelength in centimeters.

Suppose we are interested in the fraction of the power contained in the visible spectrum, which is approximately 400 to 700 nm. We may calculate the visible portion by integrating Planck's equation over these limits:

$$W_{\text{vis}} = \int_{4 \times 10^{-5}}^{7 \times 10^{-5}} W(\lambda) \, d\lambda.$$

We may define a *radiometric efficiency* as the ratio of the power in the visible spectrum to the total power:

$$E = \frac{64.77}{T^4} \int_{4 \times 10^{-5}}^{7 \times 10^{-5}} \frac{d\lambda}{\lambda^5 [\exp(1.432/T\lambda) - 1]}.$$

However, this computation does not take into account the spectral sensitivity of the eye. By introducing the standard luminosity function, we may compute the *luminous efficiency*

$$K = \frac{64.77}{T^4} \int_{4 \times 10^{-5}}^{7 \times 10^{-5}} \frac{\bar{y}(\lambda) \, d\lambda}{\lambda^5 [\exp(1.432/T\lambda) - 1]}.$$

The radiometric efficiency and luminous efficiency were computed using

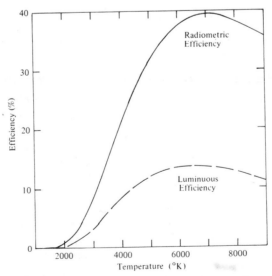

*Figure 2.3* Efficiency of blackbody radiator computed radiometrically and photometrically.

Simpson's rule (Dorn and McCracken, 1974) and are shown in Fig. 2.3. Note that the luminous efficiency is considerably lower than the visible radiometric efficiency, which emphasizes the fact that only a fraction of the light available is visually effective. Also note that the fraction of energy emitted below 1500°K is negligible. Furthermore, at the melting point of tungsten ($\simeq$3600°K), only about 15% of the radiant energy is visible.

### 2.2.3  Colorimetry

The sensation of color is determined by the light frequencies contained in the visual image. As defined by the Committee on Colorimetry of the Optical Society of America (1963): "*Color* consists of the characteristics of light other than spatial and temporal inhomogeneities."

Color is a psychophysical concept depending both upon the spectral distribution of the radiant energy of the illumination source and upon the visual sensations perceived by the observer. The dependence upon source illumination is readily observed. The sky appears black at night, various shades of blue during the day, and orange or red at sunset. Color perception depends mainly upon the physics of light and the physiology of the visual system, which results in psychological color sensations of *hue, saturation,* and *brightness,* as illustrated in Plate I.* Hue is the color sensation associated with different parts of the spectrum such as red, blue, green, or yellow. A psychophysical variable related to hue is the dominant wavelength of the light. For monochromatic light the perceived hue and dominant wavelength are directly related. Saturation is the color sensation corresponding to the degree of hue in a color. Purity is the psychophysical quality most closely related to saturation. Most colors can be matched by an additive mixture of monochromatic and white light. Purity is related to the amount of white which must be added to monochromatic light to obtain a match. Monochromatic light has 100% purity and white light has zero. Like hue and dominant wavelength, saturation and purity are related but not identical. Pure yellow is less saturated than pure violet. Brightness is the primary visual sensation, and luminance is the corresponding psychophysical variable. The three color sensations may be separately identified but are not independent. A change in brightness can cause a change in saturation or even hue.

Since luminance is one characteristic of color, photometry is included in colorimetry, which is the science of color measurement. All of the concepts of photometry therefore apply to colorimetry, which adds the concept of chromaticity.

The *tristimulus theory* is the best approximation to a hypothesis which explains color vision to a reasonable extent. It has been determined that combinations of the three primary colors can match any unknown color $C_1$

*The plates appear following p. 334.

for observers with normal color vision:

$$C_1(\lambda) \overset{\mathrm{M}}{=} rR(\lambda) + gG(\lambda) + bB(\lambda),$$

where $\overset{\mathrm{M}}{=}$ indicates a human match rather than mathematical equality, and $R$ is the red, $G$ the green, and $B$ the blue spectral primary function.

In some cases, it is found that either one of the primaries or white has to be added to the unknown to effect the match. Since white may be represented as $W = R + G + B$, the addition of an amount $w$ of white results in

$$C_1(\lambda) + wW(\lambda) \overset{\mathrm{M}}{=} rR(\lambda) + gG(\lambda) + bB(\lambda)$$

or

$$C_1(\lambda) \overset{\mathrm{M}}{=} (r - w)R(\lambda) + (g - w)G(\lambda) + (b - w)B(\lambda)$$

which may be written

$$C_1(\lambda) \overset{\mathrm{M}}{=} r_c R(\lambda) + g_c G(\lambda) + b_c B(\lambda).$$

The *coefficients* $(r_c, g_c, b_c)$ represent the color mixture of $C_1$ with respect to $R$, $G$, $B$, and $W$. Note that only one coefficient need be positive and that negative coefficients should not lead to the erroneous conclusion that some colors cannot be matched by the primaries.

Given the *tristimulus values* for a color obtained by matching the color with one set of primaries, it is possible to convert the tristimulus values to those which would have been obtained with another set of primaries provided that no single primary spectral function can be matched with a linear combination of the other two.

To develop this relationship, suppose that a test color $C(\lambda)$ is matched by a combination of primaries $R, G, B$,

$$C(\lambda) \overset{\mathrm{M}}{=} rR(\lambda) + gG(\lambda) + bB(\lambda).$$

Now suppose the same color is matched by a combination of different primaries,

$$C(\lambda) \overset{\mathrm{M}}{=} xX(\lambda) + yY(\lambda) + zZ(\lambda).$$

To relate the two sets of tristimulus coefficients, we first match the $R(\lambda), G(\lambda), B(\lambda)$ primaries with combinations of the $X(\lambda), Y(\lambda), Z(\lambda)$ primaries

$$R(\lambda) \overset{\mathrm{M}}{=} a_{11}X(\lambda) + a_{12}Y(\lambda) + a_{13}Z(\lambda),$$

$$G(\lambda) \overset{\mathrm{M}}{=} a_{21}X(\lambda) + a_{22}Y(\lambda) + a_{23}Z(\lambda),$$

$$B(\lambda) \overset{\mathrm{M}}{=} a_{31}X(\lambda) + a_{32}Y(\lambda) + a_{33}Z(\lambda).$$

Using the law of color additivity, the color $C(\lambda)$ may be expressed as

$$C(\lambda) \stackrel{M}{=} ra_{11}X(\lambda) + ra_{12}Y(\lambda) + ra_{13}Z(\lambda)$$
$$+ ga_{21}X(\lambda) + ga_{22}Y(\lambda) + ga_{23}Z(\lambda)$$
$$+ ba_{31}X(\lambda) + ba_{32}Y(\lambda) + ba_{33}Z(\lambda).$$

Comparing the two matches of $C(\lambda)$ with a combination of the $X, Y, Z$ primaries leads to the relation

$$\begin{bmatrix} x \\ y \\ z \end{bmatrix} = \begin{bmatrix} a_{11} & a_{21} & a_{31} \\ a_{12} & a_{22} & a_{32} \\ a_{13} & a_{23} & a_{33} \end{bmatrix} \begin{bmatrix} r \\ g \\ b \end{bmatrix}.$$

The above matrix may be used to convert the tristimulus values obtained with the $R, G, B$ primaries to equivalent tristimulus values obtainable with the $X, Y, Z$ primaries.

The accurate specification of color is very difficult since the perceived color characteristics cannot be directly measured, and color is influenced by spatial factors such as background texture. However, the importance of color specification is clearly demonstrated by the many color systems available. Several of these (the Munsell, the Ostwald, the 1931 and 1960 CIE, the UCS, and *Lab*) will be described.

The color system developed by Albert H. Munsell, a painter and art teacher, is based on three variables, hue, chroma, and value, as shown in Plate II. Hues are arranged in spectral order around a circle. Chroma corresponds approximately to saturation, and value is related to brightness. The origin of the coordinate system corresponds to black, which is the darkest and least saturated color.

In practice, the *Munsell system* is an atlas of separate pages of color samples, arranged in a conical tree about the value axis.

Approximately ten shades each of chroma, hue, and value are used with the differences between neighboring samples chosen to represent psychologically equal intervals. The Munsell system is an atlas of surface colors. Its reliability depends upon the use of a standard white illumination source and to some extent upon the surface texture of the sample.

The *Ostwald system* uses the psychophysical variables of dominant wavelength, purity, and luminance rather than the psychological variables of hue, saturation, and brightness, which are approximated by the Munsell system. The Ostwald coordinate system is arranged with hues of maximum purity forming an equatorial circle and with complementary colors opposite. The axis of the circle starts with black at the bottom and goes to white at the top. Purity increases outward from the origin as shown in Fig. 2.4. Each sample in the Ostwald color tree is identified by three numbers corresponding to the proportion of black, white, and pure color.

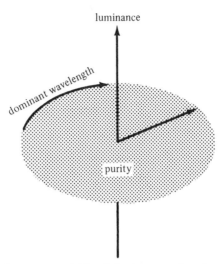

*Figure 2.4*  Ostwald color system variables of dominant wavelength, purity, and luminance.

The *CIE system* provides an internationally standard method for specifying a color. It deals with dominant wavelength, purity, and luminance like the Ostwald system, but can also be related to the Munsell system. The CIE system was developed to provide color matching characteristics for the average visual system. Using a colorimeter, a number of observers made a series of color matches against a spectrum of monochromatic colors. The mixtures of three primary colors, red, green, and blue, were recorded and used to produce color mixture curves.

These curves contained negative values. For computational convenience, it was desirable to obtain linear transformations of the color-mixture curves which had no negative values. The transformed curves contained the same information and were called *hypothetical components*, since they could not be realized experimentally. It was also possible in this transformation to require the green curve to exactly correspond to the *luminosity function* $\bar{y}(\lambda)$ as previously defined. Finally, these curves were averaged for many observers to obtain standard observer curves. The standard CIE color mixtures are shown in Fig. 2.5 and are the tristimulus values $\bar{x}(\lambda), \bar{y}(\lambda), \bar{z}(\lambda)$ of the spectrum for the standard observer. The tristimulus values $(X, Y, Z)$ of any color $C_1$ are defined as the magnitudes of the three standard stimuli necessary in a mixture to match the color $C_1$. For example, if the color $C_1$ has spectral composition $P(\lambda)$, then

$$X = K_{max} \int_0^\infty \bar{x}(\lambda) P(\lambda) d\lambda, \quad Y = K_{max} \int_0^\infty \bar{y}(\lambda) P(\lambda) d\lambda,$$

$$Z = K_{max} \int_0^\infty \bar{z}(\lambda) P(\lambda) d\lambda,$$

where $K_{max}$ is the maximum luminosity possible in the matching situation.

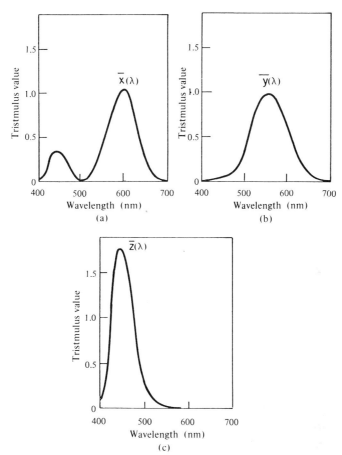

*Figure 2.5*  CIE standard tristimulus functions. [From the Committee on Colorimetry of the Optical Society of America (1963, p. 242).]

Since any color may be specified by a triple $(X, Y, Z)$, one may represent any color by a point within a *color solid* in a three-dimensional space. The shape of the solid depends upon the choice of the tristimulus vectors $\bar{x}(\lambda), \bar{y}(\lambda), \bar{z}(\lambda)$ and the orientation of these vectors in space. It is also possible to represent the main components of chromaticity, dominant wavelength, and purity as a closed area in a plane. The standard method of representing chromaticity information of a color $(X, Y, Z)$ is by defining

$$x = X/(X + Y + Z), \qquad y = Y/(X + Y + Z), \qquad z = Z/(X + Y + Z).$$

The unit plane for which $x + y + z = 1$ is referred to as the *CIE (1931) chromaticity diagram* and normalized amounts of all real colors are repre- sented by points $(x, y)$ within this horseshoe-shaped closed region, as shown in Plate III. Note that the boundary of the region is determined by the tristimulus values of the standard observer to equal amounts of radiant power from narrow spectral bands. The chromaticity diagram may be used to determine *dominant wavelength, purity, complementary wavelength, color temperature,* and other important facts for an unknown color represented by a point $(x,y)$.

The CIE chromaticity diagram shown in Plate III represents all colors of the spectrum. The perimeter points correspond to dominant monochromatic wavelengths, and the achromatic or white point is near the center. Colors increase in saturation with their distance from this central point. The large area associated with greens reflects the fact that the eye is most sensitive to them. The nonspectral purples lie on the line connecting the blue and red ends of the curve. Each pie-like sector contains a small range of hues, graded by saturations. Hues at opposite ends of any straight line through the center are complementary. An additive mixture of any two colors is represented by an intermediate point somewhere on the straight line between the points representing the two colors.

The 1931 CIE system, although still widely used, has a major deficiency. Colors on the CIE diagram which are separated by an equal numerical difference are not equally different in appearance. This is illustrated in Fig. 2.6 in which ellipses of just noticeable differences (enlarged 10 times) are shown on the $x$–$y$ CIE diagram. The *1960 CIE uniform chromaticity scale* (UCS) system is derived from the 1931 CIE diagram by stretching the axis

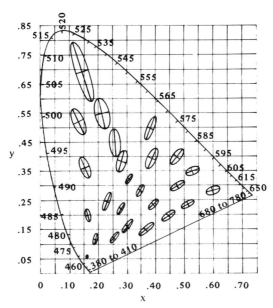

*Figure 2.6* Standard deviations of color matching represented as just noticeable difference regions enlarged 10 times on the chromaticity diagram. [Adapted from the Committee on Colorimetry of the Optical Society of America (1963, p. 251).]

unevenly in an attempt to make linear distances correspond more closely to perceived differences. This is accomplished by converting the tristimulus values $X, Y, Z$ to new tristimulus values $U, V, W$ by

$$U = \tfrac{2}{3} X, \qquad V = Y, \qquad W = -\tfrac{1}{2} X + \tfrac{3}{2} Y + \tfrac{1}{2} Z.$$

The inverse tristimulus transformation is

$$X = \tfrac{3}{2} U, \qquad Y = U, \qquad Z = \tfrac{3}{2} U - 3V + 2W.$$

Chromaticity coordinates may then be calculated as

$$u = U/(U + V + W), \qquad v = V/(U + V + W),$$
$$w = W/(U + V + W) = 1 - u - v.$$

This gives the $u, v$ chromaticity coordinates as

$$u = 4x/(-2x + 12y + 3), \qquad v = 6y/(-2x + 12y + 3).$$

The inverse transformations are

$$x = \tfrac{3}{2} u/(u - 4v + 2), \qquad y = u/(u - 4v + 2).$$

The UCS, or $U^* V^* W^*$ system, is a further attempt to obtain a meaningful color distance. This system is a CIE recommended uniform perceptual color space. The transformation is defined by

$$U^* = 13 W^* (u - v_0), \qquad V^* = 13 W^* (v - v_0), \qquad W^* = 25(100y/y_0)^{1/3} - 17,$$

where $y$ is the luminance of the color, $u_0$ and $v_0$ are its measured 1960 CIE–UCS chromaticity of the reference white or perceived achromatic point, and $y_0$ is the luminance of this standard white in the 1931 CIE coordinates.

The calculation of a measure $D$ of the perceived color difference between two samples viewed in close proximity is simply the Euclidean distance

$$d = \left[ (U_1^* - U_2^*)^2 + (V_1^* - V_2^*)^2 + (W_1^* - W_2^*)^2 \right]^{1/2},$$

where $(U_1^*, V_1^*, W_1^*)$ and $(U_2^*, V_2^*, W_2^*)$ are the coordinates of the two color samples. For reflecting samples, $u_0$ and $v_0$ are computed for the illuminant.

This system may be considered as an extension of the CIE 1960 system to include the brightness dimension. The scaling of the chromaticities by $13 W^*$ increases the distance for brighter colors. A close relation exists between the $U^* V^* W^*$ system and the Munsell system.

A final system that will be considered is a simplified version of the Adams–Nickerson space, or *Lab*, and is presently being considered for adoption by the CIE. The system is defined by

$$L = 25(100 Y / Y_0)^{1/3} - 16,$$
$$a = 500 \left[ (X/X_0)^{1/3} - (Y/Y_0)^{1/3} \right],$$
$$b = 200 \left[ (Y/Y_0)^{1/3} - (Z/Z_0)^{1/3} \right],$$

where $(X, Y, Z)$ are the CIE 1931 tristimulus values and $(X_0, Y_0, Z_0)$ are the tristimulus values of the reference white. The three coordinates $L, a, b$ correspond to lightness, redness–greenness, and yellowness–blueness, respectively. An example of four constant luminance slices is shown in Plate IV. Note that the perimeters of each plane are determined by the limited gamut of colors of the television display.

This brief introduction to radiometry, photometry, and colorimetry has presented only the minimum fundamental concepts and has hopefully whetted the reader's appetite for further investigation of the references. Also, at this point, one may have developed several questions. Why does radiant energy not correspond to brightness? What determines the shape of the luminosity and tristimulus functions? What determines the difference in the photopic and scotopic luminosity curves? To answer such questions, one must look more closely at the structure of the human visual system and the visual perception process.

## 2.3   Visual Perceptual Processing

An understanding of the perceptual processing capabilities of the human provides an ever-present motivation and certain guidelines for developing similar *paradigms* for the computer. If a task can be performed by a human it is difficult to say that the task is impossible to accomplish with a machine. On the other hand, studies of computer image processing and recognition have been recognized as being important in the psychology of perception and promise increased contributions in the future.

In this section, a brief introduction is given to the characteristics of the visual system and visually related parameters which are of established importance in the design of picture processing techniques. Since many picture processing results are evaluated by visual observation and subjective judgment, it is important to realize the modifications the human visual system makes on the perceived image.

The study of perceptual processing and visual pattern recognition is a developed discipline and several excellent texts are available. An excellent collection of papers with a historical overview is by Dodwell (1971). A systematic and interesting text, which stresses physiological factors, is that of Cornsweet (1970). Other excellent texts include Davson (1962, 1963) and Graham (1965).

When one first seriously considers visual phenomena, it often appears that factors such as motivation predominate. A strong contrary premise is presented by Cornsweet. He indicates that in the study of most perceptual phenomena, 99.9% of the magnitude of the phenomena can be accounted for

by physical and physiological factors, with only 0.1% attributable to psychological factors. In the following sections, some justification for this premise
will be presented.

### 2.3.1   Structure of the Visual System

The complex aspects of visual perception are best introduced by a brief
description of the anatomy of the human visual system (HVS). Anatomically,
the HVS may be divided into four elements: the eyes, the neural communications pathway from the eyes, the lateral geniculate body, and the visual
cortex, as shown in Fig. 2.7. A scene is imaged onto the left and right retinal
receptors. The retinal receptors through a photochemical reaction convert the
light energy into electrical pulses. These pulses are transmitted via the optic
nerve to the optic chiasma, through the lateral geniculate bodies, and finally
to the visual cortex (Area 17) in the occipital lobe of the brain.

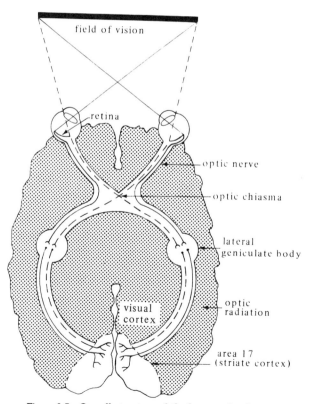

*Figure 2.7*   Overall structure of the human visual system.

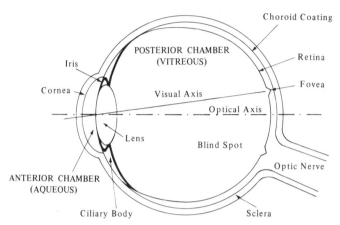

*Figure 2.8*  Simplified cross section of the human eye.

A simplified cross section of the human eye is shown in Fig. 2.8. The opaque outermost layer, the *sclera*, is a nearly spherical shell with a radius of about 11 mm and is 1 mm thick. At the front of the eye, the sclera merges into the transparent *cornea*, which bulges forward with a radius of curvature of about 8 mm and thickness of about 1 mm. At the rear, the *optic nerve* penetrates the sclera on the nasal side of the principal axis. The *ciliary body* is located behind the point at which the cornea and sclera join. Just in front of the ciliary body is the iris. The *iris* is the nearly circular aperture which constitutes the *pupil* of the eye. The size of the pupillary opening can be changed from about 2 to 8 mm by dilation and contraction of the muscles controlling the iris, which occurs as a function of illuminance and psychological responses.

Directly behind the pupil, a double convex lens is suspended in such a way that its shape can be changed. These shape changes vary the effective focal length of the lens and permit accommodation for varying viewing distances. Two cavities separated by the lens are also present. The anterior chamber is filled with waterlike aqueous humor, while the posterior chamber is filled with a viscous, vitreous humor. The innermost layer of the eye is covered by the light-sensitive retina, which contains the photoreceptors and a highly organized neural structure shown in Plate V.

The retinal surface contains a mosaic of unequally distributed photoreceptor cells called rods and cones. The concentration of rods and cones at various distances from the center of the retina is shown in Fig. 2.9. Approximately in the center is a small depression about 0.4 mm wide called the fovea, which contains only cones. Because of the nature and high density of cones, color and detail recognition is best in the *foveal* area. Rods are about 1–2 μm thick and some 60 μm long. The cones vary in size, but those occupying the center of the fovea are also about 1–2 μm thick. It has been

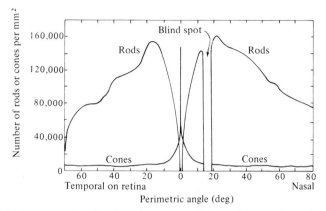

*Figure 2.9*  Concentration of rods and cones at different distances from the center of the fovea. (Note that the highest density is roughly equivalent to a 400 by 400 array of receptors in a millimeter). [From Pirenne (1967).]

estimated that there are about 5–7 million cones and between 75 and 150 million rods. In the fovea, the connection between cones and optical fibers is approximately one to one. As the distance from the fovea increases, the number of connections increases, and the degree of summing increases with a ratio as high as 140 to 1 in the periphery. The rods are sensitive to very low illumination and are responsible for *scotopic* vision. The order of minimum detectable luminance is about 1 nL. The cones, which are very tightly packed in the fovea, lie in line with the visual axis and are responsible for the most acute vision, *photopic* vision. Cone minimum sensitivity is on the order of a microlambert. The visual pain threshold is about 16 L.

The pupil of the eye contracts when exposed to a brighter light. The effect is to reduce the amount of light that falls on the retina. However, as time passes, the retina adapts to the new level and the pupil returns to its original size. The pupil can control the amount of light entering by about a factor of 30. The retina can also adapt to changes in light intensity, but by a factor of about $10^{10}$.

The network of nerve cells and ganglia is illustrated in Plate V. The somewhat reversed pathway by which the light must pass through the network before reaching the photosensitive receptors is believed due to the evolutionary folding of a plane of receptors into the present spherical shape.

The neural responses of the retinal receptors are transmitted through a layer of bipolar cells to large ganglion cells, also in the eye, where they are combined and encoded for transmission through the optic tract. At the blind spot, the one hundred million receptors connect to about one million optic nerve fibers. The optic nerve fibers via the *optical chiasma* divide the image field and project retinal regions to the *lateral geniculate* cells. The geniculate cells then connect with cells in the visual cortex.

The importance of the retinal receptors is emphasized by considering color blindness. Color blindness is the common name for any pronounced deviation from normal color vision. According to Konig's hypothesis (Wyszecky and Stiles, 1967), color blindness occurs whenever one or more of the three types of photoreceptors is inoperative. The normal observer, called a *trichromat*, can match any color with a combination of red, green, and blue colored lights. Under controlled conditions most trichromats will use about the same combinations of selected lights to match any particular color. An individual who arrives at a match with a significantly different combination may be an *anomalous trichromat* (less severe departure from normal vision), a *dichromat* (partial color blindness), or a *monochromat* (complete color blindness).

Theory predicts three types of dichromats corresponding to one of the three receptors being inoperative. The dichromat can match all colors with mixtures of two primary lights rather than three. One common form of dichromatism is red–green blindness in which the green cones may be inoperative. This individual can perceive only two colors—yellow and blue. A neutral gray is seen instead of blue–green and purple. Dark reds, greens, and grays are confused and sensitivity to brightness is decreased by about one-half. Dichromatism occurs in about 2% of white males but in only about 0.03% of white females.

Monochromatism, in theory, can also occur in three forms corresponding to the combinations of two inoperative receptors. However, monochromatism is very rare with only about one person in 30,000 afflicted. Another form of monochromatism occurs if all the foveal cones are inoperative. In this form, the monochromat is unable to distinguish any colors and visual acuity is also severely impaired. A monochromat can see only differences in brightness and can match any color with a single light. Relative luminosity is the only criterion that can be distinguished.

### 2.3.2  Visually Related Parameters

Certain visually related phenomena and parameters are repeatedly used in picture processing to describe properties of the visual system, important information in images, and characteristics of imaging devices. Several basic concepts including brightness and contrast, resolution and sharpness, and texture will be introduced in this section. Each of these concepts will be continuously built upon throughout the text.

#### Brightness and Contrast

*Brightness* is the psychological concept or sensation associated with the amount of light stimulus. Due to the great adaptive ability of the eye, absolute brightness cannot be accurately judged by a human. Light source *intensity* depends upon the total light emitted and the size of the solid angle

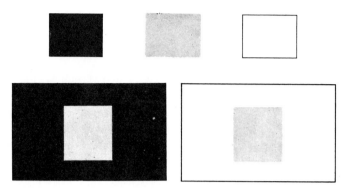

*Figure 2.10*   The perceived grayness of a surface is influenced by its background.

from which it is emitted. Two sources of equal intensity do not appear equally bright. Luminance, the intensity per unit area, is a psychophysical property that can be measured.

*Lightness* is related to the observer's recognition of a difference in whiteness, blackness, or grayness between objects and is therefore different from brightness. The term *contrast* is used to emphasize the difference in luminance of objects. The perceived grayness of a surface depends upon the local background as shown in Fig. 2.10. This phenomenon is called *simultaneous contrast*. Generally, if the ratio of contrast between an object and its local background or surround remains constant, the lightness perceived will remain constant.

Several measurements of contrast are in general use. In psychology, contrast $C$ refers to the ratio of the difference in luminance of an object $B_0$ and the immediate surround $B$, as shown in Fig. 2.11:

$$C = (B_0 - B)/B = \Delta B/B.$$

With this definition, contrast may be either positive or negative, with negative contrast meaning that the object is less luminant than the background. The size and shape of the object and surround should be specified to avoid ambiguity.

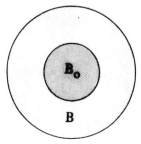

*Figure 2.11*   Contrast between an object and its surround.

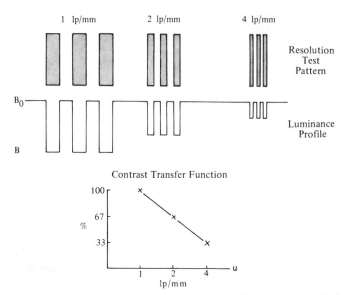

*Figure 2.12*   Line pair grating showing the determination of a contrast transfer function.

A contrast measurement commonly used in optics describes the contrast $C$ of a spatial frequency grating as the ratio of the deviation in a periodic luminance to the average luminance. For a grating as shown in Fig. 2.12, the contrast is defined as

$$C = (B_0 - B)/(B_0 + B),$$

where $B_0$ is the maximum and $B$ the minimum luminance levels of light passing through the grating. Note that if the type grating, such as square wave or sinusoidal, is specified then this use of contrast takes on an unambiguous meaning.

Finally, in photographic work the subject contrast $C$ is defined as

$$C = D_0 - D,$$

where $D_0$ represents the reflectance or transmittance photographic density of the subject and $D$ the density of the surround. Since photographic density is proportional to the logarithm of the luminance of the light used to expose the film, as shown in Fig. 2.13, this contrast measure is easily related to the previous measurements.

Contrast ratio is also used to describe differences in luminance, usually in an overall scene. The *contrast ratio* $C_R$ is defined as the ratio of the maximum to the minimum luminance in the scene. The contrast ratio is commonly used to describe the ratio of the largest and smallest luminances produced by an image display, and ratios as large as 20 may be encountered for certain displays.

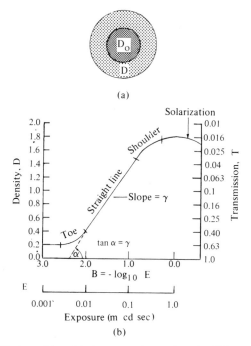

*Figure 2.13* (a) Photographic subject contrast is equal to the difference in density and (b) characteristic Hurter–Driffield (HD) curve for film.

The *just noticeable difference* in luminance is an important concept since if an observer cannot perceive a change, its importance is questionable. Classical experiments that determined the just noticeable difference in luminance between a small region and its surround for varying background luminances are described by Graham (1965). The resulting luminance discrimination curve is shown in Fig. 2.14a. Note the decrease in the *Weber ratio* $\Delta L/L$ over a wide range of increasing luminance. Also, over a limited luminance range (0.1–1000) the ratio is nearly constant. This permits a quick estimate of the number of distinguishable levels. Assume that the ratio is constant at 2% over the luminance range of interest. One may then argue that only 50 different luminance levels may be observed over the luminance range. For the total observation range the intensity discrimination curve in Fig. 2.14a, shows that the ratio is not constant, especially at low luminance levels. Thus, a more accurate estimate must be obtained. A solution equation was derived from the differential equation of the photochemical system by Hecht (Graham, 1965). The solution is of the form

$$\Delta L/L = K_1\left(1 + 1/\sqrt{K_2 L}\,\right)^2,$$

where $K_1$ and $K_2$ are related to the parameters describing the photochemical

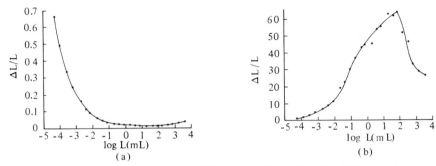

*Figure 2.14* (a) Discrimination curve and (b) number of just noticeable differences at various luminance levels. [After Hecht (1934).]

response. Note that for large luminance values the ratio is constant and that for small values a large Weber ratio is predicted.

The reciprocal of the Weber ratio may be interpreted as an incremental number of just noticeable differences or distinguishable levels at each luminance level. A curve of this relationship is shown in Fig. 2.14b, which emphasizes the loss of discrimination at low levels, and that the incremental number is about 500 levels over a large image. For a given luminance range, the average of the incremental number of levels may be used as an estimate for the number of distinguishable levels.

The number of distinguishable color levels may be estimated for individual colors in a similar manner. However, if one assumes all combinations are distinguishable, then the total number of distinguishable colors is the product of the number of individual colors. Thus, several million possible colors may be distinguishable. However, the number of distinguishable chrominance levels is less than the number of luminance levels due mainly to the decreased dynamic range of cones. It is safe to say that at least several thousand different colors are distinguishable and, as a lower limit, Munsell (1946) specifically names over 400.

### Resolution and Sharpness

*Resolution* is one of the most complex terms encountered in visual system parameters. It may be defined in terms of modulation transfer functions, optical line pairs, television lines, or spot size. Furthermore, each definition is internally consistent, but correlations among the various definitions need to be clearly stated.

The psychological measure of resolution is the visual system modulation transfer function, which exhibits a bandpass filter characteristic. This response can be observed on the patterns shown in Fig. 2.15. These Campbell and Robson type patterns are computer generated sinusoidally modulated patterns. Note that visibility of the pattern falls off at both low and high spatial frequencies. The frequency of the peak response varies with pupil

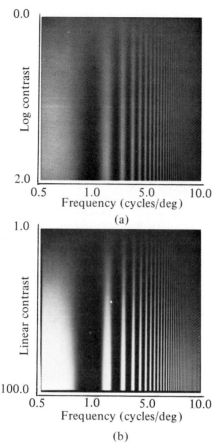

*Figure 2.15*   Sinusoidally modulated pattern illustrating frequency sensitivity of HVS. The images should be viewed at a distance equal to 15 times their width. (Courtesy of C. F. Hall.)

diameter between 3 and 7 cycles/deg. The upper cutoff frequency is about 20 cycles/deg.

A visual indication of the low-frequency rolloff or high-pass filter effect is given by the Mach band phenomenon shown in Fig. 2.16. The system overshoot is visually present as the light bands in dark regions and dark bands in light regions. This response is similar to an electronic differentiation, an optical high-pass filter, or photographic *unsharp masking*. Unsharp masking consists of exposing a film through a superimposed negative and slightly defocused positive transparency of the same scene. The resulting image has enhanced edges.

Spatial frequencies in optical systems are described in line pairs. A *line pair* is defined as one dark line and one white line of equal width. Spatial frequency may be described by the number of line pairs per millimeter. The simplest method for characterizing the resolution of an optical system is to

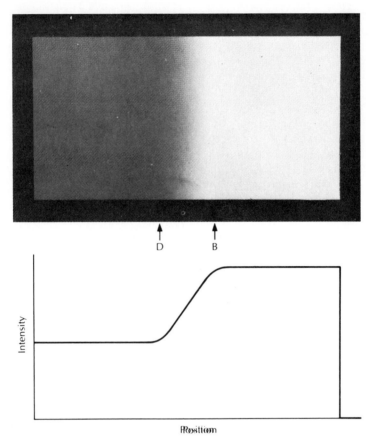

*Figure 2.16* Illustration of Mach bands. (Note the appearance of dark and light bands at boundaries although intensity varies uniformly.) [From Cornsweet (1970, p. 276).]

determine a *contrast transfer function* (CTF). To determine the CTF, one uses a line-pair test grating such as that shown in Fig. 2.12 as the input image. Assume that the luminance profile of the output image is still rectangular but that the amplitudes of certain frequencies has been decreased.

A contrast transfer function $C(u)$ may then be developed from the luminance profile by computing the ratio

$$C(u) = \Delta B(u)/\Delta B(0),$$

where $\Delta B(u)$ represents the difference in luminance of the line pattern of spatial frequency $u$ for each frequency in the test pattern, as illustrated in Fig. 2.12.

The contrast transfer function specifies the percent contrast of the line pattern at various spatial frequencies. A limiting resolution may also be determined from $C(u)$ as the spacing at which $C(u)$ is less than a certain percentage, such as 10%, or as the spacing at which the lines are just visually discernible.

A rectangular luminance profile of amplitude range $B_0 - B$ and line pair spacing $x_0$ with infinite duration also may be expressed by its Fourier series $L(x)$

$$L(x) = \frac{B_0 + B}{2} + \frac{B_0 - B}{2} \sum_{n=1}^{\infty} \frac{\sin \pi n}{\pi n} \cos \frac{2\pi n x}{x_0}.$$

However, several harmonic terms are needed to represent the contrast function.

Although the contrast transfer function is useful, a problem is encountered if the output luminance profile is not rectangular. Since most optical systems are diffraction limited and cathode-ray tube (CRT) spot brightness is Gaussian in distribution, it is difficult to achieve a rectangular profile. A more serious defect is that the contrast transfer function cannot be directly used as a transfer function of a linear, position invariant system.

One solution to this problem is to use a test pattern in which the transmittance varies sinusoidally. Since the transfer function of any linear, position invariant system may be determined from its response to sinusoidal inputs, this approach offers not only a measure of resolution, but also a method to characterize the system response to any input pattern. The Fourier series of an infinite duration sinusoidal grating of spacing $x_0$ and amplitude range $B_0 - B$ is

$$L_1(x) = \frac{B_0 + B}{2} + \frac{B_0 - B}{2} \sin \frac{2\pi x}{x_0}.$$

Clearly, the Fourier series of the rectangular and sinusoidal patterns are related and one may be determined from the other; however, the response to sinusoidal inputs provides a more useful function called the *optical transfer function* (OTF).

The OTF may be developed using a sinusoidal test pattern in a manner similar to that used for the contrast transfer function; however, other available methods may be easier for certain systems.

The optical analog of an impulse response is a point-spread function. A linear position invariant optical system is completely characterized by its *point-spread function* $h(x, y)$. Furthermore, the output $g(x, y)$ from the system to an input $f(x, y)$ may be determined from the convolution (Goodman, 1968)

$$g(x, y) = \int_{-\infty}^{\infty} \int f(\alpha, \beta) h(x - \alpha, y - \beta) \, d\alpha \, d\beta.$$

The point-spread function may be determined as the system response to an impulse point of light. It may also be desirable to consider the two-dimensional Fourier transform of the point-spread function to determine the frequency response. Also, the Fourier transform $G(u, v)$ from the system to an input with transform $F(u, v)$ is simply

$$G(u, v) = F(u, v) H(u, v),$$

where the optical transfer function $H(u,v)$ is the two-dimensional Fourier transform of the point-spread function $h(x,y)$. The magnitude or modulus of $H$ is called the *modulation transfer function* (MTF), $M(u,v)=\|H(u,v)\|$.

If the system is isotropic (rotationally invariant), then a two-dimensional analysis is not necessary, since the point-spread function may be determined from the *line-spread function* $l(x)$ (Papoulis, 1968), where

$$l(x)=\int_{-\infty}^{\infty}h(x,y)\,dy$$

is the response to a line mass $\delta(x)$ along the $x=0$ axis. In fact, it may be shown that the point-spread function, where

$$r=\left(x^{2}+y^{2}\right)^{1/2}$$

is

$$h(r)=\frac{1}{2\pi r}\left(-\frac{d}{dr}\int_{-\infty}^{\infty}l(r)\,dy\right).$$

The corresponding system function is the Hankel transform of $2\pi h(r)$:

$$H(u,v)=2\pi\int_{-\infty}^{\infty}rh(r)J_{0}(\omega r)\,dr,$$

where $\omega=(u^{2}+v^{2})^{1/2}$ and $J_{0}(x)$ is the Bessel function of order zero.

The resolution limit of a diffraction-limited optical system is normally specified by the *Rayleigh criterion*. The Rayleigh limit states that two equally bright point or line sources can just be resolved if the central maximum of the diffraction pattern of one source coincides with the first minimum of the other, as shown in Fig. 2.17. This criterion holds equally well whether these two diffraction patterns are produced by a narrow slit or a small circular aperture, although the mathematical descriptions of these patterns is slightly different.

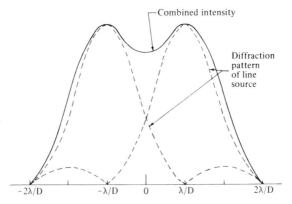

*Figure 2.17*  Rayleigh criterion for resolution.

*Table 2.2*

*Relationships among resolution criteria*[a]

|  | TV 50% | 10% MTF | 50% MTF | Optical |
|---|---|---|---|---|
| TV 50% (3 dB) | 1 | 1.14 | 0.63 | 0.59 |
| 10% MTF | 0.88 | 1 | 0.55 | 0.52 |
| 50% MTF | 1.6 | 1.82 | 1 | 0.94 |
| Optical (1/e) | 1.7 | 1.94 | 1.06 | 1 |

[a]To convert from the criterion at left to the criterion above, multiply by the given constant.

The modulation transfer function, or MTF, may be computed as the modulus or magnitude of the OTF and may be used not only to give a resolution limit at a single point, but also to characterize the response to an arbitrary input. The MTFs published for the HVS and various types of equipment may have been derived from sinusoidal test patterns, Fourier transforms of line-spread functions, or from point-spread functions. Therefore, some variance may be expected. However, the MTF is generally the best method available to describe the resolution of an imaging system.

*Resolution* for TV systems is usually given in TV lines, which are not equal to optical line pairs. For TV lines, both black and white lines are counted. Also, since CRT spot brightness tends to follow a Gaussian spatial distribution, some particular luminance level, such as half the peak value, must be specified to determine line width. For example, 1000 TV lines equal approximately 416 optical line pairs. The relationships between several resolution criteria and conversion factors are shown in Table 2.2 (Sherr, 1970). As may be seen from the conversion factors, limiting resolution may be easily confused if the criterion is not specified. Therefore, one should use the MTF whenever possible to specify resolution.

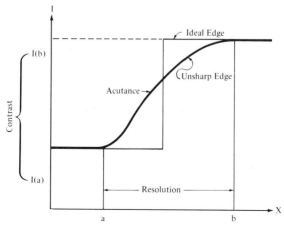

*Figure 2.18* Idealized relationships between contrast, acutance, and resolution.

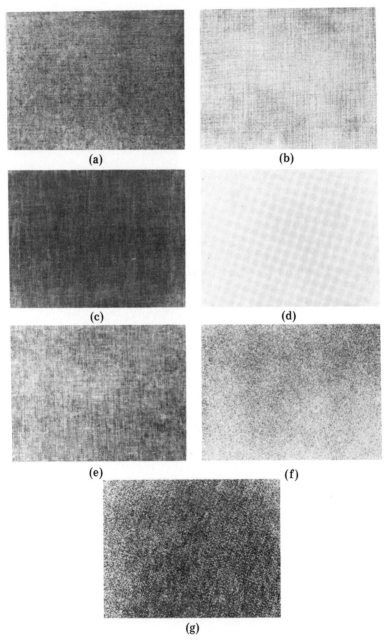

*Figure 2.19* Texture patterns: (a) tapestry; (b) tweed; (c) drawn cotton; (d) dot screen; (e) rough linen; (f) gravel; and (g) reticulated grain. (Paterson texture screens, reprinted by permission of Paterson Products Ltd., London, England.)

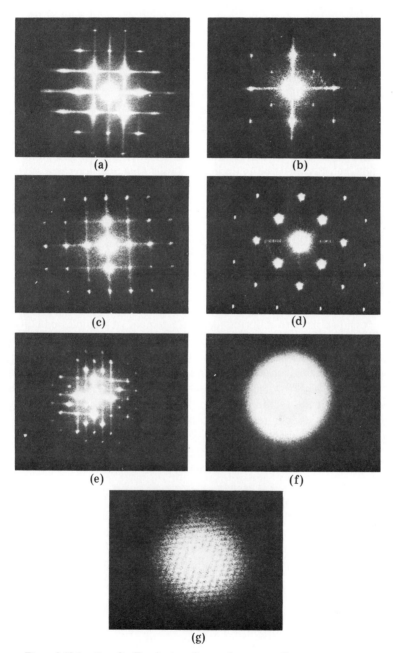

*Figure 2.19 (continued)* Fourier transforms of corresponding texture patterns.

Resolution is related to the ability to discriminate fine detail in the field of view. Edge *sharpness* or acutance is related to resolution, but should be differentiated from it. An idealized relationship is shown in Fig. 2.18, in which resolution is related to the width of an unsharp edge while acutance is a function of the shape of the edge, and contrast a function of the height.

One might expect that increasing values of the derivative of the edge would correspond to better reproduction of higher spatial frequencies, and therefore greater sharpness. However, neither the maximum gradient nor the average gradient has been found to correlate with the sensation of sharpness. Since it is possible to have the same maximum or average gradient for many functions, the lack of correlation is understandable. One characteristic of the edge function that does have a reasonable relationship to edge appearance and that will be used as the definition of edge sharpness or acutance is the mean square gradient

$$A = \int_a^b \left( \frac{df}{dx} \right)^2 dx \Big/ \left[ f(b) - f(a) \right].$$

### Texture

*Texture*, in common usage, refers to the character of a woven fabric or textile. It is also used to describe any arrangement of particles or constituent parts of any material such as wood, concrete, or metal. Texture is often correlated with tactile sensations and described as smooth, coarse, or grainy. For image processing, *visual texture* may be defined as a repetitive arrangement of a basic pattern. It is a regional property and clearly depends on the size and shape of the region. An edge between two texture patterns might be determined by moving a region across the patterns and noting a significant change in a texture measure over the region.

A texture pattern normally has some degree of randomness although it is often useful to consider nonrandom repetitions, for example, character arrays, as texture patterns.

One method of texture analysis is based upon Fourier transforms as illustrated in Fig. 2.19. A set of texture patterns and the corresponding optically computed Fourier transforms are shown. The texture patterns are commercially available from Paterson Products, London, England. The fourth pattern, a dot screen, is greatly simplified in the Fourier domain. As will be demonstrated in later discussions of sampling, the pattern consists of a single spectral response called the main lobe which is the transform of a single dot, replicated at spacings inversely proportional to the spacings of the dots. The other texture patterns, which includes tapestry, drawn cotton, rough linen, and tweed, exhibits a similar transform pattern except that the spacings of the replica spectra are somewhat random because the basic pattern repetitiveness is somewhat random. The last two patterns of gravel and sand also exhibit this property but in a somewhat circularly symmetric manner.

Texture patterns are often encountered in natural images, and some studies of texture have provided insight into human perception. In a set of remarkable experiments (Julesz, 1960, 1962), some of the primitive mechanisms for separating textural fields were examined. In these experiments, sets of random dots were used in order to do away with the normal visual cues a person uses to discriminate patterns. Additionally, points could be connected using different probability functions. It was recognized that differences between two fields could always be detected by careful analysis. The area under study was the primitive and spontaneous process that enables a person to detect texture differences without going through a cognitive process about how they differ.

It was determined that textures that differed in their first- and second-order probability distributions were easily separable, but those which differed only in their higher-order statistics were difficult to separate or discern. While this is true to some extent, and from a picture processing point of view worth keeping in mind, there was a more primitive mechanism at work that allowed separation of textures at the higher-order statistical level.

It was found that in visual discrimination, clusters or lines formed by proximate points of uniform brightness played a decisive role. Julesz called this process *connectivity detection*. For this to function, the distance between equal brightness elements must be minimized for the eye–brain combination to organize them into uniform detectable segments. Experimentation indicates the visual system acts as a slicer separating brightness into two categories, dark and light, but that it cannot quantize into several levels and form categories or groups that are not adjacent in value. The basic discrimination process assigns great importance to line structure, but, when other properties are present, such as width, brightness, and orientation, it is difficult to

*Figure 2.20*   Random dot stereograms of the word "yes." (Reprinted from "Foundations of Cylcopean Perception" by Bela Julesz, by permission of The University of Chicago Press, Chicago, © copyright 1971 by Bell Telephone Laboratories, Inc.)

separate two similar, closely related textural fields. Clusters formed by proximate points of uniform brightness are immediately noticed. Clusters having
unequal brightness are ignored.

One of the most significant outcomes of Julesz's experiments is the demon-
stration that human depth perception does not depend upon object recognition. This process is illustrated in Fig. 2.20 in which two random dot
stereograms are shown. If these images are viewed with a stereoscope a
three-dimensional object may be visualized. This work demonstrates the
significant contribution that can be made by applying engineering and
computer science skills to the task of understanding perception.

## 2.4   Structural Models for Perceptual Processing

A review of structural visual system models is now presented to provide the
designer with a method of analyzing visual phenomena. Note that the
questions asked in the study of perceptual processing are also important
questions in machine design. How does the human *abstract*, from continually
changing visual impressions, certain *attributes* or *features* to which to respond? How does the human *recognize* patterns? When are images effectively
equivalent in eliciting a response?

The entire repertoire of visual functions in humans is dependent upon all
the elements, but particularly upon the activities of the occipital cortex.
However, explanations of many of the intellectual functions may be made by
study of composites of simpler processes.

Visual perception can only be understood if we are prepared to analyze its
functions in several fairly distinct stages. Even then visual models cannot give
a complete explanation of perception. However, the formulation and analysis
of simple visual models that agree with some corresponding empirical evidence are valuable because the models aid in the solution of certain problems
associated with perception. With any imaging model, the statement that the
output image is $f(x, y)$ is not equivalent to the statement that the observer
sees $f(x, y)$.

A complete model of the visual system must include models of the eye, the
optic tract, and the visual cortex. Although the state of development of a
complete visual system model is still in its embryonic states, the author agrees
with Cornsweet's (1970) comment that although the human visual system is
obviously complex, it is possible that it is really composed of a very large
number of repetitions and slight variations of a few simple mechanisms.

A simplified functional description of the perceptual process is shown in
Fig. 2.21. The stimulus image reflected from a scene is transmitted to the
image-forming elements of the eyes. These elements focus the image on the
retinal receptors. The retinal neural connections may then perform certain

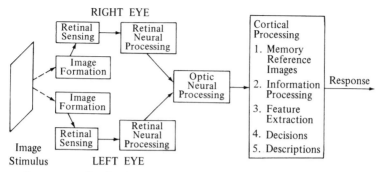

*Figure 2.21*   Simplified functional block diagram of the perceptual process.

processing operations on the neural signals. Further neural processing may be performed in the optic neural connections from the eyes. Cortical processing tasks include providing memory reference images, various types of information processing, feature extraction, decisions, and descriptions of various elements of the scene. The results of this process may be indicated in several ways by the response.

A model of the visual system is shown in block diagram form in Fig. 2.22. The stimulus image is represented by an image function. The image forming elements of the eye are represented by an optical filter. The retinal receptor response is divided into two major pathways, one for rods and one for cones. Each pathway contains a selective spectral filter, a low-pass temporal filter, and a low-pass spatial filter. These are followed by a threshold function, a log conversion, and a high-pass spatiotemporal filter. The neural connections in each channel are represented by a spatial optical filter and summations. The neural processing in the optic nerve is again represented by a summation. The transmission link to the cortex may be represented by a set of transmission lines. Finally, the cortical computations may be modeled as summations. The stage of development of the model of the visual system is far from complete, although a large amount of work has been accomplished. At present, certain elements such as the image forming mechanisms can be modeled very accurately; other elements, for example the cortical computations, are only modeled in a conjectural manner. Therefore, one may expect to find insight, but not a complete solution from the following description of the visual model.

### 2.4.1   Models of the Ocular Media

Only since the 17th century have we appreciated the eye as an image-catching device. Vision begins with light entering the eye. These light patterns, carrying information in spatial, spectral, and temporal variations, must travel through the various parts of the eye to the light-sensitive retina.

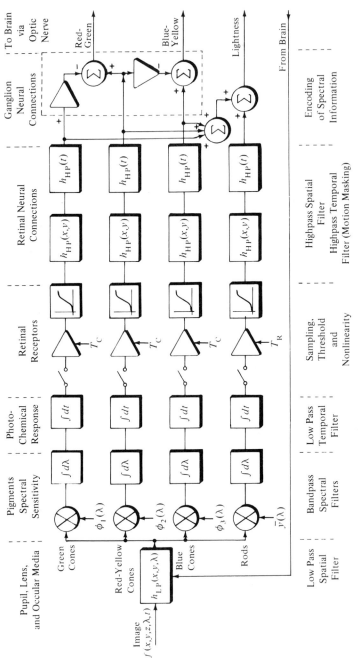

*Figure 2.22* Visual system model—monocular.

This process is analogous to the optical system of a camera containing an aperture, lens, and film (Campbell, 1968). The success of the optical model of the eye is readily apparent to anyone wearing eyeglasses. The delicate task of controlling and focusing light is accomplished by a sophisticated lens system in the eye. The cornea and aqueous humor refract light and also affect image formation. The neural networks control the amount of light entering the eye by regulating the size of the pupil opening through a feedback accommodation signal. The diffraction limit produced by the pupil size must also be accounted for by the eye's focusing signals. The partially focused light emerging from the pupil next encounters the remarkably adaptable lens. Unlike a camera lens, which must be moved forward and backward to change focus, the eye's lens changes shape by means of a feedback signal. Also, a closely related action is involved in depth perception, in which the image of an object must be focused on the fovea in each eye. The eye's lens consists of over 2000 fine layers, or lamellae, like the skins of an onion, encased in a clear membrane. Because new layers are continuously being created, the lens grows large and less pliable with age. Following the lens, the vitreous humor acts as a final refractive medium. The optical system of the eye is also subject to various aberrations. Spherical aberrations that produce a shorter focus at the periphery of a lens than in the central region are largely corrected by the cornea, which has a lower curvature at the periphery than at the center. Chromatic aberrations produce different focal lengths for different colors and are virtually uncorrected. The focus is correct at the red end, slightly short in the blue-green region, and shortest at the shorter wavelengths in the blue, violet, and ultraviolet portions of the spectrum. The human lens, which has a sharp cutoff in the near ultraviolet, eliminates part of the error.

A simplified optical model, which neglects the feedback mechanisms, of the image forming elements of the eye is shown in Fig. 2.23. The cornea, aqueous humor, lens, and vitreous humor have all been considered as a single lens located at a distance $Z_3$ in front of the retinal plane. The pupil aperture is located a distance $Z_2$ in front of the lens. The distance $Z_1$ corresponds to the viewing distance. Using the operational notation of Vander Lugt (1966), an equivalent mathematical model may be developed as shown in Fig. 2.24. This model may be used to describe changes in viewing distance, pupil size, lens focal length, and other factors. It may also be used for predicting their effects.

The use of the model will now be illustrated. In the simplest geometric case, we may consider the model shown in Fig. 2.23. Even with this simple model, we may easily compute several important factors. For example, a lowercase pica type letter is about 1.5-mm high. When viewed at a normal reading distance of 40 cm, it produces a retinal image of size 64 $\mu$m. Since the fovea is about 400-$\mu$m wide, the retinal image of the letter could easily be viewed with highest precision. On the other hand, the retinal image of a page 21.5-cm wide at the same viewing distance is greater than 9 mm and therefore could not be viewed just with the foveal area. In fact, the largest image whose retinal image would fall within the fovea at 40 cm is about 9 cm.

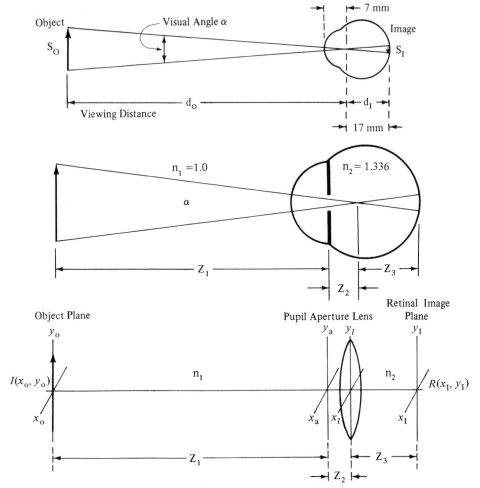

Figure 2.23 Simplified optical models of the image forming elements of the eye. Visual angle: $\tan \frac{1}{2}\alpha = \frac{1}{2}S_O/d_O = \frac{1}{2}S_I/d_I$. Image size: $S_I = S_O d_I/d_O = 2d_I \tan \frac{1}{2}\alpha$.

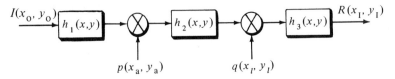

Figure 2.24 Equivalent mathematical model:

$$R(x_I,y_I) = h_3(x,y)*(q(x_I,y_I)\{h_2(x,y)*p(x_a,y_a)[h_1(x,y)*I(x_0,y_0)]\}),$$

where * represents two-dimensional convolution.

Another useful example is the use of this model to determine spatial frequency or to convert line pairs/millimeter to cycles/degree. Suppose a test pattern consists of equal width dark and light lines with $u_0$ line pairs/millimeter. One line pair is $(1/u_0)$-mm high, at a viewing distance $d_0$, the number of line pairs/millimeter on the retina $u_R$ is simply $u_R = u_0 d_0 / d_1$. For example, 2 line pairs/mm at 40 cm produces about 47 line pairs/mm on the retina. We may also wish to convert this to cycles/degree of visual angle. The visual angle $\alpha$ for one cycle is given by: $\tan \alpha = 1/d_0$. Therefore, $1/\alpha$ is the number of cycles/degree of viewing angle, and is about 6 cycles/deg. Note that the number of cycles/degree is the same either at the object or the retina.

The next level of complexity introduces the diffraction limit produced by the finite pupil size. Experimental measurements of the optical line-spread function have been made for various pupil diameters and degrees of focusing (Davidson, 1968). The experimentally determined line-spread functions $l(x)$ were found to be a monotonically decreasing function closely described by

$$l(x) = \exp(-\alpha |x|),$$

where $\alpha$ depends on the pupil size. For a pupil diameter of 3 mm, $\alpha = 0.7$. The Fourier transform $L(u)$ of $l(x)$ was also computed as

$$L(u) = 2\alpha / (\alpha^2 + \omega^2)$$

to describe the modulation transfer function of the occular media of the eye for various pupil diameters. By evaluating this function with $\alpha = 0.7$, one may see that the half-power frequency of this low-pass filter is at approximately 6.6 cycles/deg.

### 2.4.2 Individual Characteristics of Visual Photoreceptors

Now we shall concentrate on the individual characteristics of the visual photoreceptors, i.e., *rods* and *cones*. These two classes of receptors may be distinguished on a number of different criteria. One basic difference is that rods are the receptors that contain a particular kind of visual pigment called *rhodopsin*, or visual purple. Cones contain different pigments, *chlorabe*, *erythrolable*, and *iodopsin*, and account for color sensitivity.

The method of absorption of light quanta by the visual system is through a photochemical process. Human rods contain pigment molecules imbedded in layers of membranes. In a dark-adapted eye, nearly all the pigment molecules are in a relaxed state. When a quantum strikes the molecule and is absorbed by it, the absorbed energy changes the shape of the molecule by altering the chemical bonds through a process called *isomerization*. The isomerized molecule undergoes a spontaneous series of changes and finally returns to the initial stable state. However, if the molecule should absorb another quanta during the change, it may be driven back to the initial state without undergoing all transitions. This photoreversal is only important when the probability of a molecule capturing more than a single quantum is high, which requires levels of intensity considerably greater than those normally experienced. Thus, only two transitions are of primary importance, the isomerization transition due to absorption of light and the return of the isomerized molecule to the resting state.

Visual pigment in the resting state is capable of absorbing light. The different number of quanta absorbed for different wavelength light quanta is called the absorption spectrum. This curve may be obtained by extracting visual pigment from a dark-adapted eye and determining the spectrum using a spectrophotometer.

A simple model for the steady state receptor response of rods is given by the scotopic or photopic luminosity function depending upon the luminance level. The model consists of a spectrally sensitive absorptive filter placed in front of the photoreceptors. Luminance response depends only on the amount of energy absorbed, not its origin.

A model for the spectral sensitivity of cones may also be based upon the absorption spectra of pigments. In a normal human eye, three distinct pigments have been extracted, and psychologically determined estimates of the absorption spectra of the three visual pigments are shown in Fig. 2.25. The peak responses are in the blue–violet, green, and yellow regions. Although the red receptor peaks in the yellow, it extends far enough into the longer wavelength to sense red easily. Also, note the difference in magnitudes in the responses. The spectral absorption of the visual pigments describes a

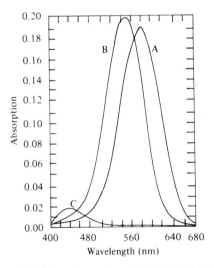

*Figure 2.25* Estimates of the absorption spectra of the three-color cone pigments in a human subject with normal color vision. The *C* curve is three times higher than the average. (From The Receptors for Human Color Vision, by G. Wald in *Science* **145**, 1007–1017, Figure 4, September, 1964. Copyright 1964 by the American Association for the Advancement of Science.)

portion of the receptor response; however, receptor response and connections significantly affect this response.

The next element in the model accounts for the conversion of light quanta to neural response. The neuron response must be represented by a nonlinear element. The basic nonlinearity is the threshold cutoff below which the response is zero. The *suprathreshold response* is also nonlinear with a predominant saturation effect. Many different functions, including logarithm and root functions, have been proposed to describe the suprathreshold response. Logarithms will be used in the model. The response of individual neurons in animals is now studied by placing a microelectrode at the neuron and observing the electrical response.

The neural response may also be derived from a psychovisual experiment. The *Weber–Fechner law* is based on the observation that the just noticeable difference in luminance is a fixed fraction of the stimulus luminance. In this experiment, an observer is asked to adjust a controllable illuminated region until it is just noticeably brighter than a reference region. This experiment is repeated for reference image illuminations from very dark to very bright. The ratio of differential illumination is then plotted as a function of $L$, as shown in Fig. 2.14b. Since the curve is relatively constant over a wide range of $L$, it is reasonable to assume the Weber–Fechner law:

$$\Delta L / L = k \Delta B.$$

Integrating each side of this expression gives

$$B = c \log L,$$

where $B$ is the lightness, $L$ is the stimulus luminance, and $c$ is a constant. The above derivation is not rigorous, and there are other reasonable approaches; however, classically, the brightness function has been taken to be logarithmic. The negative portion of the log function does not come into consideration because of the threshold or minimum energy required to obtain a neural response.

The minimum energy required to obtain a neural response from a subject was determined by the significant experiment of Hecht, Schlaer, and Pirenne in 1942 (Cornsweet, 1970). In an experiment of this type, the observer is given a set of varying illuminations and is required to respond "seen" or "not seen." Since there is some probability that a person will respond positively when no source is present, some threshold value must be selected. This was chosen to be 60%. The amount of energy striking the cornea was determined to lie between 54 and 148 quanta. After allowing for absorption and scattering, the useful energy was about 5–14 quanta. The image of the flash of light fell on about 500 rods. It was therefore considered unlikely that a rod received more than 1 quantum. This suggests that a group of rods must be stimulated in order for a light to be seen with a total impinging energy of 5–14 quanta. Further importance of neural interconnections is emphasized by the next element of the model.

### 2.4.3 Models of Receptive Fields in the Visual Cortex

Incoming light passes through approximately 10 layers of neural cells before reaching the receptor cells. This reversed information pathway is made up of *bipolar* and *ganglion* cells. A receptor cell may send nerve endings to several bipolar cells. Conversely, several receptor cells may be connected to a single ganglion cell. Any one of the synapses connected to a particular cell may perform either an excitatory function, increasing the cell's firing rate, or an inhibitory function, decreasing the rate. The cluster of receptors feeding into a single cell anywhere along the visual pathway is called the receptor field of that cell.

Hubel and Weisel determined that receptive fields of different types occurred in cortical units of the visual system. These receptive fields may be divided into *simple, complex,* and *hypercomplex* types. Simple units respond to retinal stimulation patterns of definite location, width, and orientation, such as elliptic or rectangular fields with an "on" or excitatory region and an "off" or inhibitory region.

In studies of receptive fields of retinal ganglion cells in mammalian eyes, two distinct types were found. One cell type, *on–center,* was associated with a receptive field consisting of a small circular on–off excitatory area and a surrounding circular off–on inhibitory area. The other type, *off–center,* was associated with a reversed concentric receptive field. Note that for both types, the response depends upon the size of the illuminated region. These ganglion cells respond to contrast in illumination between on and off areas in the receptive fields. A structural model of the receptive fields at different levels is shown in Fig. 2.26.

The spatially structured on–off receptive fields may be modeled as either a spatial or spatial frequency dependent element. The effect of the computation is to enhance the contrast of edges, i.e., areas of transition from low to high or high to low illumination in the image. A simple, linear edge enhancement model is a high-pass spatial filter. Note that the *Mach band phenomena* is predicted by this simple model. Excellent presentations of Mach band cancellation by using the inverse of the high-pass response are given by Cornsweet (1970), Davidson (1968), and Stockham (1972). One description of this element of the overall model is a high-pass spatial filter.

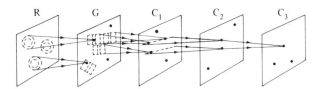

*Figure 2.26* Models for simple, complex, and hypercomplex units in the visual system.

### 2.4.4 Retinal Neural Networks

Furman (1965) proposed four types of neural networks: a forward inhibition model, a backward inhibition model, a forward shunting model, and a backward shunting model. These four have essentially the same spatial high-pass effect. Only the backward inhibition model will be considered in detail.

The backward inhibition model for two receptors is shown in Fig. 2.27. In the figure, the receptors have a logarithmic response to the incoming light intensity; thus, they correspond to the logarithmic operation in the proposed model. In Fig. 2.27, $e_i$ is the frequency at which receptor $i$ would produce pulses if the receptor $i$ alone was illuminated (the level of excitation of the receptor $i$); $g_i$ is the frequency of the pulses of the receptor $i$ after the retinal neural network (pulses/sec); $b_{ij}$ is the inhibitory coefficient representing the strength of the inhibition that $g_j$ exerts on $g_i$. If one neglects the inhibitory threshold and the case in which frequencies of pulses are below inhibitory threshold, the following equations hold:

$$g_1 = e_1 - b_{12}g_2, \qquad g_2 = e_2 - b_{21}g_1.$$

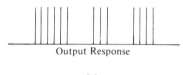

Output Response

(a)

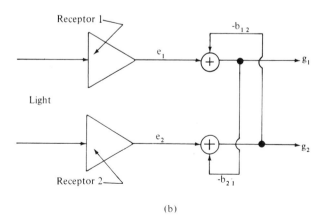

(b)

*Figure 2.27* Backward inhibition model for two receptors in the retinal neural network. [From C. F. Hall and E. L. Hall (1977).]

The general equation considering other receptors is

$$g_i = e_i - \sum_{j=1}^{n} b_{ij} g_j, \qquad i=1,2,\ldots,n.$$

A more detailed explanation of this network may be found in Cornsweet (1970).

The equations may also be written in concise matrix form with the definitions

$$\mathbf{e} = \begin{bmatrix} e_1 \\ e_2 \\ \vdots \\ e_n \end{bmatrix}, \qquad \mathbf{g} = \begin{bmatrix} g_1 \\ g_2 \\ \vdots \\ g_n \end{bmatrix}, \qquad \text{and} \qquad \mathbf{B} = \begin{bmatrix} b_{11} & b_{12} & \cdots & b_{1n} \\ b_{21} & b_{22} & \cdots & b_{2n} \\ \vdots & & & \\ b_{n1} & b_{n2} & \cdots & b_{nn} \end{bmatrix}.$$

With this matrix notation, the above equation will be

$$\mathbf{g} = \mathbf{e} - \mathbf{Bg}.$$

Then

$$(\mathbf{I} + \mathbf{B})\mathbf{g} = \mathbf{e} \qquad \text{or} \qquad \mathbf{g} = (\mathbf{I} + \mathbf{B})^{-1}\mathbf{e}.$$

Now assume that there is no self-inhibitory interaction and that the inhibitory interaction is an exponentially decreasing function of the distance of the receptors:

$$b_{ij} = \begin{cases} 0, & i=j, \\ a_0 \exp(-a|i-j|), & i \neq j. \end{cases}$$

Under this assumption, the matrix $\mathbf{I} + \mathbf{B}$ is

$$\mathbf{I} + \mathbf{B} = \begin{bmatrix} 1 & a_0 e^{-a} & a_0 e^{-2a} & \cdots & a_0 e^{-na} \\ a_0 e^{-a} & 1 & a_0 e^{-a} & \cdots & a_0 e^{-(n-1)a} \\ \vdots & & & & \\ a_0 e^{-na} & a_0 e^{-(n-1)a} & a_0 e^{-(n-2)a} & \cdots & 1 \end{bmatrix}.$$

This matrix is *Toeplitz* since all diagonal terms are equal, and, furthermore, may be interpreted as corresponding to a linear shift-invariant system. This convolutional matrix has an impulse response representation

$$g(x) = (1 - a_0)\delta(x) + a_0 \exp(-a|x|).$$

The Fourier transform of the impulse response is

$$G(\omega) = (1 - a_0) + a_0 [2a/(a^2 + \omega^2)],$$

where $\omega$ indicates radial frequency. Thus, $(\mathbf{I} + \mathbf{B})^{-1}$ has the frequency characteristic

$$H_2(\omega) = \frac{1}{G(\omega)} = \frac{a^2 + \omega^2}{2a_0 a + (1 - a_0)(a^2 + \omega^2)}.$$

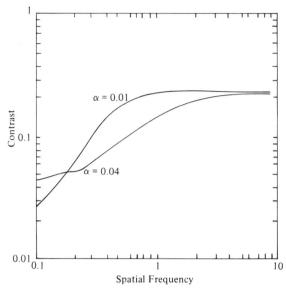

*Figure 2.28* High-pass spatial filter response $H_2(\omega)$ produced by neutral inhibition strength values $\alpha$ and a distance factor $\alpha_0 = 0.8$. [From C. F. Hall and E. L. Hall (1977).]

This function represents a high-pass filter with

$$H_2(0) = a/[2a_0 + a(1 - a_0)] \qquad \text{and} \qquad \lim_{\omega \to \infty} H_2(\omega) = 1/(1 - a_0).$$

The shapes of this function for several values of $a_0$ and $a$ are shown in Figs. 2.28 and 2.29. From comparing this function to the experimental results of Davidson (1968), it appears that suitable parameter values are $a = 0.01$ and $a_0 = 0.20$.

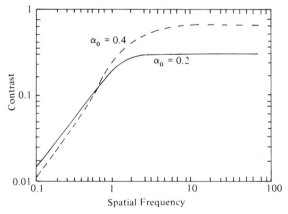

*Figure 2.29* High-pass spatial filter response $H_2(\omega)$ produced by neural inhibition for two distance factors $\alpha_0$ and an inhibition strength factor $\alpha = 0.01$. [From C. F. Hall and E. L. Hall (1977).]

### 2.4.5  Spatial Frequency Response Models

*First Approximation Model*

A first approximation model for the HVS spatial frequency response can be made by considering the system to be linear, isotropic, time- and space-invariant, monocular, monochromatic, and photopic.

The use of a MTF to estimate the system response to an input image assumes that each of the Fourier components transferred through the HVS are added in a simple algebraic fashion at some point in the system. In general, if the intensity of the light radiated from an object is increased, the magnitude of the response of the HVS should increase in direct proportion. In other words, the system should be linear and, thus, obey the principle of superposition.

An isotropic system has characteristics that are invariant to direction. The visual system is not isotropic (Campbell *et al.*, 1966). The response of the system to a rotated contrast grating is a function of the frequency of the grating as well as the angle of orientation. The sensitivity of the system decreases to a minimum at 45° and then rises again reaching the original level at 90° rotation. At the point of maximum deviation, 45°, a frequency of 50 cycles/deg would exhibit a 3-dB loss. A frequency of 10 cycles/deg is reduced by only 15% when rotated from the vertical or horizontal to a 45° position. This anisotropic behavior will not be included in our first models. It may be added easily when the application warrants the increased complexity.

The HVS is not spatially homogeneous, either in optics or receptors. However, optical spatial invariance is a good assumption near the optic axis and, even though the densities of rods and cones vary greatly with retinal position, the foveal region is relatively homogeneous (Brown, 1965). The situation in the case of temporal homogeneity is even more complicated. The apparent existence of sustained and transient channels in the visual system (Tolhurst, 1975a, b) produces complex and interrelated responses to psychophysical stimuli which are difficult to interpret (Tynan and Sekuler, 1974).

Monocular vision is usually assumed when developing models for the HVS.

There are several ways to define monochromatic vision: vision by rods only, inability to distinguish differences in hue, or sensitivity to a single narrow band of spectral energies are a few examples. We will adopt the most pragmatic one and define monochromatic vision as the inability to discriminate differences in hue.

The assumptions detailed previously are valid for the system which produces the MTF derived by means of contrast gratings. The question then becomes, "How can the system which produced this response be modeled?" Of course, one can just represent the system as a band-pass filter (Fig. 2.30a). This is difficult to defend from a physiological standpoint since the response is a compound one due to several mechanisms within the HVS.

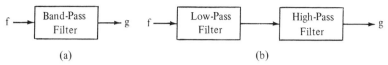

*Figure 2.30*   (a) Simple band-pass filter model of the HVS and (b) a low-pass, high-pass filter model of the HVS. [From C. F. Hall and E. L. Hall (1977).]

A slightly more detailed model would be a combination of a low-pass and a high-pass filter (Fig. 2.30b). The low-frequency portion, which produces the well-known Mach band phenomenon, is a result of lateral inhibition. The high-frequency portion could be determined by several mechanisms (or a combination thereof). The optics of the eye including retinal size obviously limit the high-frequency response of the system. The size and density of the photoreceptors will also limit visual acuity or high spatial-frequency response. Neural summation due to many receptors connecting to a single ganglion cell in the retina could also account for this end of the response curve. In addition, light scattering within the aqueous humor is probably an increasing function of frequency and would cause a lower sensitivity at these frequencies due to the increased losses.

Although the model depicted in Figure 2.30b can be used to describe a system which produces an MTF similar to the HVS, it will not account for the phenomenon of "brightness constancy." In general, the brightness of an object tends to remain constant, despite variations in the illumination falling on it. This phenomenon can only be accounted for by introducing a nonlinearity in the system.

### Simple Nonlinear Models

The linear model presented in the previous section was found to be inadequate. In particular, certain characteristics of the HVS imply the existence of a nonlinearity in the system. This nonlinearity is in addition to the threshold nonlinearity that obviously exists. Both of these nonlinearities may be a result of the same mechanism. If we make this assumption, then the nonlinear system can be modeled in two ways, as shown in Fig. 2.31.

The question immediately arises how an MTF, which requires a linear system, can be obtained for such a system? The answer to this question

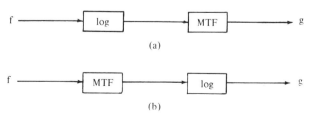

*Figure 2.31*   Two simple nonlinear models of the HVS. (From C. F. Hall and E. L. Hall (1977).]

follows from the manner in which an MTF for the HVS is derived. A sine-wave grating is presented to the human observer via an appropriate display device. The contrast of this fixed frequency grating is reduced until the grating is barely visible. For small variations about this visible threshold, it can be argued that linearity is a valid assumption. Thus, the results of these measurements from observations with several frequencies of gratings can be combined (due to superposition) and an overall MTF obtained. It is only when we increase the contrast significantly above threshold that the second nonlinearity becomes a factor.

Campbell, Carpenter, and Levinson have hypothesized that the model shown in Fig. 2.30a is appropriate since the visibility of aperiodic patterns appears to require that the threshold mechanism occurs after a "pooling" of the outputs from the spatial frequency channels (Campbell *et al.*, 1969). Such a mechanism could be located in the lateral geniculate nucleus (LGN) or the visual cortex. Evidence indicates that little processing of visual data occurs in the LGN (Kulikowski and Tolhurst, 1973); therefore the most likely location would be in the visual cortex. Bodis-Wollner has found that contrast sensitivity is significantly decreased (particularly at medium and high frequencies) in patients with neurological disorders of the visual cortex (Bodis-Wollner, 1972). Whether this indicates a feedback path to a nonlinear mechanism located earlier in the system or the existence of a nonlinearity in the cortical area itself can only be conjectured. The model illustrated in Fig. 2.31a has stronger anatomical support.

The photoreceptors within the retina are threshold devices. In the case of the receptors in the eye of *Limulus*, a logarithmic characteristic is exhibited as well (Cornsweet, 1970). The receptors can also be interconnected in such a way as to produce lateral inhibition. Lateral inhibition produces a high-pass filter effect as noted earlier. Consider now the phenomenon of brightness constancy in the context of this model.

Figure 2.32a contains the intensity profiles of two hypothetical images. The difference between the two profiles is that the intensity of the second is five times greater everywhere. A human observer would not detect this difference, and in fact the two images would be perceived as having the same "brightness." A linear model for the HVS would filter out the average intensity level, passing the higher frequency transition of the step change in intensity. Since the step itself has been increased by a factor of 5, the change would be preserved by the linear system. If, however, the model is nonlinear and, in particular, logarithmic, then the situation depicted in Fig. 2.32b occurs. As can be seen the step in intensity is now equal in amplitude for both intensity levels. After high-pass filtering, this constant amplitude step is the only thing that passes through the system; thus, brightness constancy results. We see then that the model of Fig. 2.32a not only is physiologically sound, but it predicts brightness constancy as well. Stockham (1972) has successfully

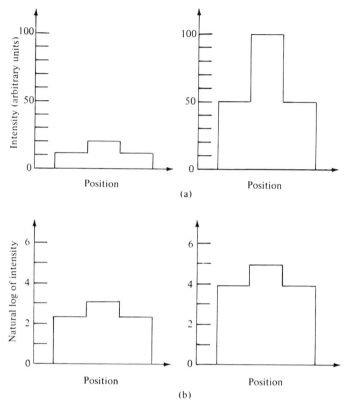

*Figure 2.32*   Intensity profiles illustrating "brightness constancy." [After Cornsweet (1970, p. 355).]

applied a version of this model to image processing. In addition, Mannos and Sakrison (1974) have found the same basic model to be appropriate for image coding.

Davidson (1968) performed an interesting contrast grating experiment with exponential sine-wave gratings. He reasoned that this stimulus should linearize the HVS. The results are shown in Fig. 2.33. (Note that the data have not been corrected for the transfer function of the experimental apparatus.) The variation in the high-frequency portion of the curves at different brightness levels is of primary interest.

Recently, Henning *et al.* (1975) have reported experimental results that also indicate a nonlinear distortion of signals at high, but not low, spatial frequencies. Their results do not support a model consisting of a logarithmic nonlinearity followed by linear independent frequency channels. This result and the high-frequency variations in Davidson's data can be explained by the next model.

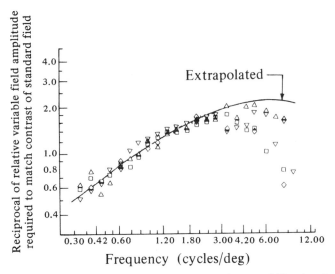

*Figure 2.33*  HVS response to exponential sine-wave gratings. △, 1.86 units of contrast; ▽, 5.86 units of contrast; ◊, 7.73 units of contrast; and □, 9.61 units of contrast. [From Davidson (1968, p. 1305).]

### A Nonlinear Model

Another nonlinear model is shown in Fig. 2.34. The model has simply placed the low-pass filter in front of the nonlinearity. This modification can be justified by the presence of the optical aberrations in the eye.

This model was analyzed by Hall and Hall (1977). The results show that the frequency response of this model $H(\omega)$ varies with input contrast in a manner similar to the experimental variation found by Davidson.

A plot of this characteristic function for several values of contrast is shown in Fig. 2.35a. These curves compare favorably with the experimental results of Davidson. Thus, the variation of the describing functions at high spatial frequencies appears to be due to a predictable response of the HVS.

A pictorial example of the effects of the nonlinear model is also shown in Fig. 2.35. An original scene is shown in Fig. 2.35b and the HVS-processed result is shown in Fig. 2.35c. The low-pass filter $H_{lp}(\omega)$ was selected to emphasize the low-pass filtering of the visual system. The specific function used was

$$H_{lp}(\omega) = 0.14/(0.49 + \omega^2),$$

*Figure 2.34*  Nonlinear HVS spatial frequency model.

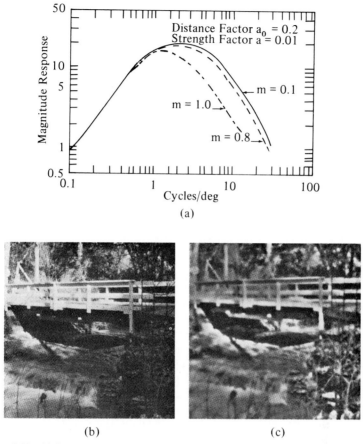

(a)

(b)                                            (c)

*Figure 2.35*   (a) Response of the nonlinear model. (b) Original image. (c) Image processed by the nonlinear visual system model. [From Hall (1978).]

which corresponds to a 3-mm pupil and has a half-power frequency of 6.6 cycles/deg. The high-pass filter $H_{hp}(\omega)$ was

$$H_{hp}(\omega) = (10^{-4} + \omega^2)/4 \times 10^{-3} + 0.8\omega^2,$$

which approximates the previous model with parameters $a_0 = 0.2$ and $a = 0.01$. A comparison of Figs. 2.35b and 2.35c reveals the primary characteristics of the spatial frequency response of the visual system. Edges are enhanced by the high-pass filter characteristic; however, the overall scene is also blurred by the low-pass filter characteristic. The low-pass characteristic limits the information and can be compensated. Any reader wearing glasses can illustrate this hypothesis by observing a scene with and without his compensating spatial filters. The digital compensation techniques will be described in Chapter 4.

### 2.4.6  Temporal Frequency Response Model

The photochemical process is especially important in predicting the temporal response of the visual system. The visual system has a limited response to temporal changes in luminance. As the temporal frequency of a light source is increased for any given size and luminance of a stimulus, a point is reached at which the perceived response indicates no further changes. This point, which is called the *critical fusion frequency* (CFF), increases with luminance for a given size stimulus and increases with stimulus size for a given luminance. For normal viewing luminance. The CFF is between 20 and 40 Hz. To avoid this problem, motion pictures are shown at 24 frames/sec, but the shutter flashes each frame three times, providing a viewing rate of 72 frames/sec. Broadcast television produces 60 frames/sec by interlacing the lines on two 30-frame/sec presentations.

The temporal effect of the photochemical reaction may be modeled by a temporal low-pass filter. The time required for light absorbed by the visual receptors to initiate a chemical reaction which generates a nerve pulse in the associated neuron is about 30 msec. The high-frequency attenuation of a low-pass filter easily describes such effects as the CFF. Also, the temporal summation effect, called *Bloch's law*, which states that the total number of quanta required to detect a flash is constant for all flashes shorter than some critical duration of about 30 msec, is predicted by an integrating-capacitor-type model.

The same question arises about the location of elements responsible for the temporal frequency response in the visual system. In the usual model the elements consist of a nonlinearity followed by the temporal band-pass filter. The proposed model consisting of a low-pass temporal filter; the nonlinearity and the high-pass filter may again be appropriate.

There is again a physiological reason for placing the low-pass filter before the nonlinearity. The temporal low-pass filter arises from the 30-msec time required for light absorbed by the visual receptors to initiate the photochemical reaction which generates the neural pulse. Also, there is again an inconsistency encountered if one assumes the log–band-pass characteristic.

If one postulates the existence of the excitational interaction among nerve cells as Yasuda and Hiwatashi (1968), then the neural network has a band-pass characteristic. However, the difficulty with this assumption is that, in order to explain the temporal characteristics of the HVS, one has to assume that the response-time delay of the excitational interaction is ignored whereas that of the inhibition is considered. Noticing that the excitational interaction and the inhibitory interaction produce the spatial low-pass effect and the high-pass effect, respectively, the above statement means that the low-pass effect does not have a time delay whereas the high-pass effect does

### Combined Spatiotemporal Response

Robson (1966) experimentally measured spatiotemporal frequency responses of the visual system. He used a 2.5° by 2.5° grating target in which the luminance at right angles to the bars was

$$L(v,f) = L_0\big[\,1 + m(\cos 2\pi v y)(\cos 2\pi f t)\,\big].$$

$L_0$ was kept constant at 20 cd/m$^2$ and the value of $m$ at subjective disappearance of bars was measured for different frequencies $v$ and $f$. The inverse of that value of $m$ was defined as the contrast sensitivity. The results are shown in Figs. 2.36 and 2.37.

Since Robson did not vary the input contrast, the predictable variation of the resultant curves at high frequencies was of course not observed. However, another interesting fact of the proposed model may be correlated with Robson's results. This is the separability of the spatiotemporal response. As

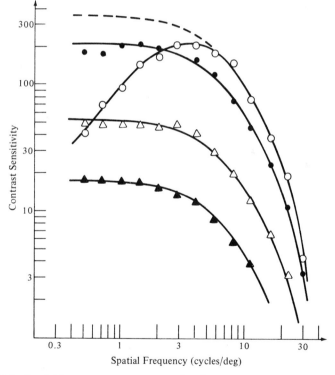

*Figure 2.36* Spatial frequency response curves for different temporal frequencies: ○ 1 Hz, ● 6 Hz, △ 16 Hz, and ▲ 22 Hz. The broken line would apply if all curves were to be generated from a single low-pass curve by shifts on the vertical direction only. [From Budrikis (1972).]

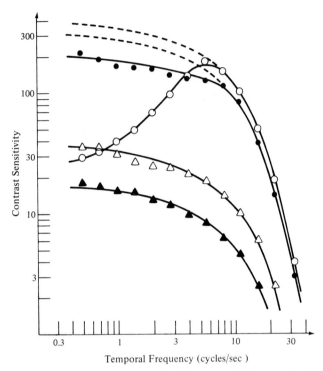

*Figure 2.37* Temporal frequency response for different spatial frequencies: ○ 0.5 cycle/deg, ● 4 cycles/deg, △ 16 cycles/deg, ▲ 22 cycles/deg. The broken lines would apply if all curves were to be generated from a single low-pass curve by shifts on the vertical direction only. [From Budrikis (1972).]

pointed out by Budrikis (1972), if the human visual system response were product separable then the family of response curves of spatial or temporal frequency with one of the frequencies fixed would differ from each other by constant factors, or in the logarithmic plots, by constant displacements in the vertical direction. The experimental curves do exhibit this property except at frequencies below about 5 cycles/deg and 6 Hz. Note that this is not predicted by the log–band-pass-type model.

A possible explanation of this effect is predictable from the nonlinear model. The basic mechanism is that if the output of the nonlinearity is at sufficiently high frequencies so as to be within the passband of the high-pass spatiotemporal filter then the effect of this high-pass filter is negligible. In this situation, the spatiotemporal response is product separable since the low-pass spatial and temporal filters are independent. However, if the output of the nonlinearity contains low-frequency signals that are affected by the stop band of the high-pass filters then the overall response is not product separable since the high-pass spatiotemporal filters are not independent.

### 2.4.7   Models of Color Vision

A model of color vision could enable one to predict perceived color sensations. At the present time not enough is known about the eye–brain combination to permit a complete mathematical model; however, the study of simplified models which predict some responses is worthwhile since they lead to a greater understanding of the color vision process and could lead to the discovery of the complete model.

Certain viewing situation effects can be accurately measured with the techniques of colorimetry and therefore isolated from the study of the visual system. The spectral distribution of the illuminant, the viewing situation, and the geometry affect the perceived scene but can be considered separately by considering the light distribution impinging on the eye. This removes from consideration perceptual variations due to different illuminations of the same scene, whether light is transmitted or reflected from a scene, and whether or not the scene is viewed through an external aperture.

The usefulness of a color vision model is dependent on its success in predicting the perceived color sensations. An exact model should also demonstrate color phenomena; i.e., it should "see" what the visual system sees and "not see" what the visual system does not see. Although the ideal model is as yet undetermined, even limited models could be very useful. A model that could accurately predict the perceived sensations of brightness, hue, and saturation would be of universal use in all aspects of communicating color information. The systems such as the Munsell and CIE previously described are giant steps toward this goal. Color matching is even a more limited goal and can be predicted, although not with complete subjective agreement. A complete model might also predict complex visual phenomena. One such phenomenon is *color constancy*, which is the ability of the visual system to compensate for ordinary changes in illumination and viewing conditions. This compensation ability implies a certain invariance property of the perceived response. An example is the invariance of the color of grass throughout most of a clear day even though the illumination is changed by atmospheric scattering. *Simultaneous color contrast* is another phenomenon that should be predicted by a good model. This visual phenomenon is the effect of two identical colors appearing quite differently if viewed on different background colors. The spreading effect of white and black background into adjacent regions seems to be a contradiction to simultaneous color contrast. *Color Mach bands* are shown in Plate VI and should also be predicted by a good model. The eight colors shown shift from cyan to yellow in uniform bands and yet the perceived cyan appears to have a darker hue just before the transition to yellow. The phenomenon can be appreciated best by masking off adjacent fields with white cards which will suppress the banding effect. Although the Mach phenomenon has been extensively studied for brightness, color Mach bands have only recently been investigated. They appear strongest for yellow–cyan transitions and may be absent for other color combinations.

If combined temporal–color effects are also considered, the phenomenon of afterimages may be important. A negative afterimage is a complementary color image produced by fatigue after viewing a positive color. This phenomenon is one instance of the complementary color ability of the visual system. Although a composite color cannot be divided into its components, the complement can be formed.

A simplified model of color vision will now be presented, which provides first-order prediction of some major phenomena of color perception including color constancy and simultaneous contrast. The color model presented is a physiologically based extension of the previously considered spatial frequency model. For the previous model, the lens, retinal receptors and retinal neural connections were included. For the color model, the three types of color-sensitive cones must be included, as well as the encoding of these signals by the ganglion cells before transmission to the higher visual centers.

### Color Matching, Simple Model

A simplified model of the color process is shown in Fig. 2.38. The spectral distribution of light $C(\lambda)$ impinging on the eye is detected by three color-sensitive receptors, corresponding to the three types of cones. The spectral responses of the cones $\phi_i(\lambda)$ are multiplied by the incoming light and integrated over all visible wavelengths to produce the *tristimulus values* $t_j$. This model accounts mainly for the spectral filtering of the receptors. Even this simple model is useful since it may be used to derive the conditions for color matching.

*Color matching* is the process of adjusting a color mixture until it is visually indistinguishable from a sample color. Consider the matching of a unit intensity monochromatic light of wavelength $\lambda_0$. The mathematical representation of this light is simply $\delta(\lambda - \lambda_0)$. To match this light to a mixture of

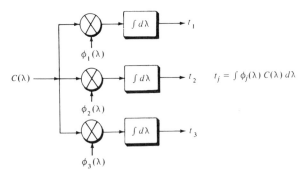

*Figure 2.38*   First-order model for color vision. [From Wallis (1975).]

three primary colors $p_j(\lambda)$ the same perceived response must be obtained from both the mixture and monochromatic light. Using the simplified model gives the matching relationship for each receptor,

$$\int_\lambda \phi_i(\lambda)\delta(\lambda-\lambda_0)\,d\lambda = \int_\lambda \phi_i(\lambda)\sum_{j=1}^3 t_j p_j(\lambda)\,d\lambda,$$

for $i=1, 2, 3$. Evaluating this expression gives

$$\phi_i(\lambda_0) = \sum_{j=1}^3 t_j \int_\lambda \phi_i(\lambda)p_j(\lambda)\,d\lambda,$$

for $i=1, 2, 3$. These equations may be written in matrix form as

$$\begin{bmatrix} \phi_1(\lambda_0) \\ \phi_2(\lambda_0) \\ \phi_3(\lambda_0) \end{bmatrix} = \begin{bmatrix} a_{11} & a_{12} & a_{13} \\ a_{21} & a_{22} & a_{23} \\ a_{31} & a_{32} & a_{33} \end{bmatrix} \begin{bmatrix} t_1 \\ t_2 \\ t_3 \end{bmatrix} = \mathbf{AT},$$

where

$$\mathbf{A} = \{a_{ij}\} \quad \text{and} \quad \mathbf{T} = (t_1, t_2, t_3)'; \quad a_{ij} = \int_\lambda \phi_i(\lambda)p_j(\lambda)\,d\lambda.$$

Therefore, the tristimulus values or weights required to match a spectral line are always linear combinations of the receptor sensitivity functions at the wavelength of the monochromatic line.

If the same spectral line $\delta(\lambda-\lambda_0)$ is matched with a different set of primaries $q_j(\lambda)$, then the perceived tristimulus values $S_i$ are linearly related to the previous tristimulus values $t_i$. Since

$$\begin{bmatrix} \phi_1(\lambda_0) \\ \phi_2(\lambda_0) \\ \phi_3(\lambda_0) \end{bmatrix} = \begin{bmatrix} b_{11} & b_{12} & b_{13} \\ b_{21} & b_{22} & b_{23} \\ b_{31} & b_{32} & b_{33} \end{bmatrix} \begin{bmatrix} s_1 \\ s_2 \\ s_3 \end{bmatrix} = \mathbf{BS},$$

where

$$\mathbf{B} = \{b_{ij}\}, \quad \mathbf{S} = (s_1, s_2, s_3)', \quad \text{and} \quad b_{ij} = \int \phi_i(\lambda)q_j(\lambda)\,d\lambda,$$

equating the two expressions and computing the inverse of the **B** matrix (assuming that it exists) gives

$$\mathbf{S} = \mathbf{B}^{-1}\mathbf{AT}.$$

Therefore, the tristimulus values obtained with any two sets of primaries are linearly related provided that the **A** and **B** matrices are nonsingular, which occurs if the sets of primaries $\{p_j(\lambda)\}$ and $\{q_j(\lambda)\}$ are each linearly independent.

### Structural Color Vision Model

A more detailed model of color vision is shown in block diagram form in Fig. 2.22. The components of the model correspond to the known functional elements of the visual system and the placement of components is inferred from psychovisual experiments.

The stimulus image is shown as a function of spatial, spectral, and temporal variables. This image function is filtered by the low-pass spatial filter corresponding to the pupil, lens, and ocular media before reaching the retinal receptors. A vector channel with components corresponding to the four types of retinal receptors is encountered next. The variable spacing of these receptors is important and is considered later in the spatial filtering portion of the model. The spectral sensitivities of the receptors is modeled by the spectral band-pass filter as in the simple model. The temporal effect of the photochemical response is represented by a temporal low-pass filter. The sampling, threshold, and nonlinearity of the receptors provide the next elements of this model. The next elements in the rod channel are the high-pass spatial filter, which corresponds to the retinal–neural connections and predicts the Mach band phenomenon, and the high-pass temporal filter which models motion masking. However, the corresponding connections for cones do not occur at the retina'–neural level but at the ganglion level. Furthermore, the *opponent theory* indicates that one primary effect of these ganglion connections is to transform the spectral responses into difference signals corresponding to red–green, blue–yellow, and brightness signals. Therefore, in the chromaticity channels the spatial and temporal filters are placed after this transformation.

This basic structure is replicated many times in each eye. Estimates for the number of replications can be derived from the number of receptors and from the number of neurons in the optic nerve. A quick estimate may be made as follows: Suppose that the one million neurons in the optic nerve are divided evenly for the right and left eyes, and that the half million channels for each eye are divided evenly between rods and cones. Then we may consider two square arrays of size 500 by 500 for each eye; one corresponding to rods and the other to cones. The cone array is packed in the *foveal area* in a square of about 0.7 mm on each side. The rod array covers the entire retinal surface which, for simplicity, may be considered a square with sides of length 7.0 mm. This order of magnitude difference in resolution provides instantaneous global and local recognition.

### 2.4.8  Binocular Vision Models

To explore binocular vision, it is interesting to consider first the geometry for *stereopsis*. The general geometry for producing two images, $I_L$ and $I_R$, of the same scene is shown in Fig. 2.39. The first problem is to determine the

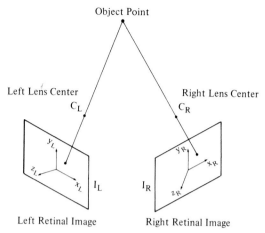

*Figure 2.39* General geometry for stereopsis.

transformation **T** that will permit image $I_R$ to be brought into correspondence with $I_L$. In general, there are translation, rotation, scale, and perspective differences between the two images. The solution to this problem in three-dimensional coordinates is nonlinear because of the perspective transformation. Considered in four-dimensional, *homogeneous coordinates*, however, the solution transformation **T** is linear (Rogers and Adams, 1976). Recall that the homogeneous coordinates of a physical point $(x,y,z)$ are $(x,y,z,w)$, where $w$ is an arbitrary scale factor. The desired transformation may be written

$$
\begin{bmatrix} x_L \\ y_L \\ z_L \\ 1 \end{bmatrix} = \begin{bmatrix} T_{11} & T_{12} & T_{13} & T_{14} \\ T_{21} & T_{22} & T_{23} & T_{24} \\ T_{31} & T_{32} & T_{33} & T_{34} \\ T_{41} & T_{42} & T_{43} & T_{44} \end{bmatrix} \begin{bmatrix} x_R \\ y_R \\ z_R \\ w \end{bmatrix} .
$$

The desired transformation matrix may be partitioned into four submatrices: The elements $T_{14}, T_{24}, T_{34}$ describe the translation differences between the points while elements $T_{41}, T_{42}, T_{43}$ describe the general three-point perspective transformation. The $3 \times 3$ matrix describes the rotation and scale while element $T_{44}$ is an overall scale factor. The general transformation involves two coordinate systems, one for the left and the other for the right image.

The stereoscope is a binocular instrument constructed to view stereoscopic pictures produced by photography. Each eye sees a separate image or slightly different picture of the same object, which when viewed through the stereoscope are blended into one and, apart altogether from perspective or light or shade, stand out in bold relief, and the appearance of distance or other background is thrown back as in reality. In a stereographic viewer, a separate perspective view of a scene must be created for each eye. Then the views must be laterally displaced in order to place each view directly in front of the appropriate eye.

For a human with normal eyesight, the strongest *stereo fusion* occurs at a distance of about 50 cm in front of the eyes. For an eye separation of 5 cm, the stereo angle is

$$\alpha = 2\tan^{-1}(2.5/50) = 5.72°.$$

For a stereo viewer, let $d$ be the scaled separation distance between the eyes. If the viewer has a focal length $f$ then the value of $d$ is fixed by the requirement that

$$2\tan^{-1}(d/2f) = 5.72°.$$

Thus $d/f = \frac{1}{10}$ to maintain the correct stereo angle.

To obtain this scaled eye separation distance from a single view of the object, a $d/2 = f/20$ horizontal translation is performed before creating the left eye perspective and a $-d/2 = -f/20$ translation is performed before creating the right eye perspective.

To create a stereo pair from the three-dimensional coordinates, two images are produced using the transformations

$$\mathbf{T}_{LE} = \begin{bmatrix} 1 & 0 & 0 & 0 \\ 0 & 1 & 0 & 0 \\ 0 & 0 & 1 & -1/f \\ f/20 & 0 & 0 & 1 \end{bmatrix}, \quad \mathbf{T}_{RE} = \begin{bmatrix} 1 & 0 & 0 & 0 \\ 0 & 1 & 0 & 0 \\ 0 & 0 & 1 & -1/f \\ -f/20 & 0 & 0 & 1 \end{bmatrix}.$$

This creates two separate perspective views, one for the left eye and one for the right. The final step is a translation in the $x$ direction of $L$ for the right eye view and $-L$ for the left eye view where $L = \frac{1}{2}d$. For more information on stereo vision, including models, consult Julesz (1971).

## 2.5   Computations Associated with Perception of Shapes and Patterns

### 2.5.1   Invariance Properties

In a discussion of shape recognition, Deutsch (1955) described four *invariance properties* of human perception.

1. Shapes may be recognized independent of their location in the visual field. It is not necessary to fixate on the center of a figure in order to recognize it nor need the eyes be moved around the contours of a figure. Thus shape recognition is invariant to translation provided the object remains in the visual field.

2. Recognition can be effected independently of the angle of inclination of a figure in the visual field. The angle refers to the angular orientation of

the figure in two dimensions although some invariance to the depth angle is also observed. Therefore the recognition of a figure is invariant to rotation in two dimensions, although the angle of rotation may also be recognized.

3.  The size of a figure does not interfere with the recognition of its shape provided the entire figure is within the visual field. Thus visual recognition exhibits an invariance to scale.

4.  Mirror images appear alike. Both humans and animals tend to confuse these.

Deutsch also points out that certain primitive organisms find it more difficult to distinguish between squares and circles than between rectangles and squares, which casts some doubt on recognition theories based on angular properties of figures. He also argues that the invariance properties are common to all parts of the visual cortex.

The limitations of the proposed model are many; however, it does illustrate a logical approach to recognition. First, the desired recognition properties are determined; then a system that satisfies these properties is designed. The particular system described was devoted to shape recognition. Other systems need to be developed that are invariant to intensity changes and more complex transformations. The power and simplicity of Deutsch's approach was a major contribution from psychology to the understanding of recognition systems.

### 2.5.2  Neural Computations

It is interesting to speculate on the computation that can be performed by the human visual system. It may be assumed that the functional columns in the visual input cortex are connected with the visual association cortex and other areas. Furthermore, it is presumed that it is not the activity of a single cell, but rather the action of the entire array of columns that performs the required computation. Thus the question arises: What computations can suitably connected arrays of columns of neuromines perform?

Kabrisky (1966) has demonstrated that a two-dimensional cross-correlation computation could be performed by such a model in a manner analogous to the performance of an optical cross correlator. Tallman (1970) suggested that a two-dimensional Fourier transform could also serve as a computational model. Carl (1969) proposed a similar model, using a two-dimensional Hadamard transform. The mathematical consequences of the previous assumptions will now be illustrated.

Let $A$ denote the state of a neuron. If the linear sum of the inputs exceeds the threshold value, $A$ may be measured directly as the pulse repetition rate of the electrical discharges on the axon. If the threshold has not been reached, $A$ is the negative value of the excitatory input required to reach the threshold. Thus $A$ can be positive or negative, depending on whether the neuron is firing or is inhibited. The input to this neuron may be represented by $f(i,j)$, where $i$

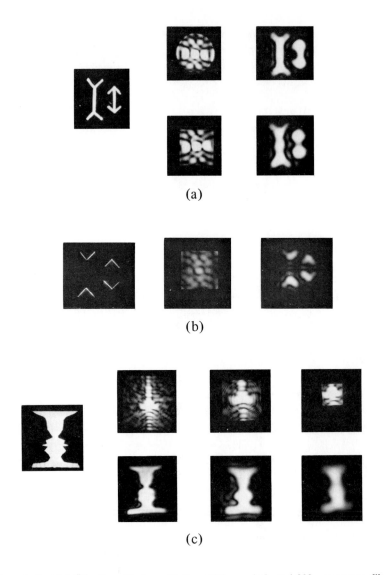

(a)

(b)

(c)

*Figure 2.40* (a) Müller-Lyer illusion with lines (200-μm circle and 200-μm square filter. (b) Müller-Lyer illusion without lines (200-μm square filter). (c) Candlestick–faces illusion (300-, 200-, and 100-μm square filters). (d) Horizontal–vertical illusion (300- by 180-μm rectangular filter). (e) Principle of closure illustrated by dotted *G* (200-μm circle and square filters). (f) Principle of closure illustrated by dotted shape (200-μm circle and square filters). [From Ginsburg (1971).]

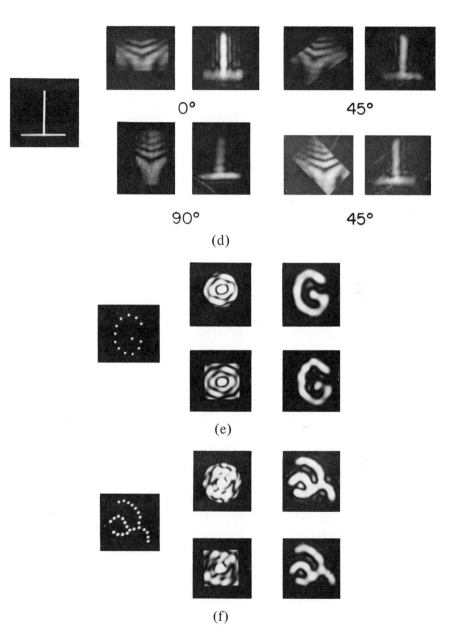

0°  45°

90°  45°

(d)

(e)

(f)

*Figure 2.40  (continued)*

and $j$ denote spatial position on a cortex column. These assumptions permit the mathematical definition of the formal neuron as

$$A = c \sum_i \sum_j A_{ij} f(i,j),$$

where $c$ is a proportionality constant and $A_{ij}$ is the weight of the $(i,j)$ input.

If the constants are chosen appropriately, this computational form could compute a convolution, correlation, Fourier transform, or Hadamard transform. For example, if

$$A_{ik} = \exp(-j2\pi lk/N),$$

then the computation is a discrete Fourier transform.

If computations of this type are made by the visual system then simulations should demonstrate properties of the visual system. This thesis was pursued by Ginsburg (1971) and some of his results in classifying ambiguous illusions are shown in Fig. 2.40. The Müller–Lyer illusion in which two lines of identical length appear to be of different length is clarified by the low-pass spatial filter. The line that appears longest is indeed longest after filtering. The candlestick-face illusion is also clarified after sufficient spatial filtering. The principle of closure is also demonstrated for the dot letter G.

### 2.5.3 Layered Machines

The concept of a layered machine capable of performing the complex tasks of recognition follows directly from the receptive field studies shown in Fig. 2.26. Demonstrations of the power of this approach are given by Uhr (1971) and Marco and Giebel (1970). Two significant questions about the layered machine are now under investigation and could significantly affect our understanding of perceptual processing.

The first question is related to the design of the processing structures between layers. How are the layers connected to perform a given task? What architecture would perform a significant number of perceptual tasks?

The second question is related to the hardware implementation. Significant developments in optical processing, Boolean optical processing, and parallel computer processing have been made. The construction of intelligent systems may not be too far away.

### Bibliographical Guide

The complexities of imaging are clarified to a large extent in the following references: An inexpensive, beautifully illustrated overview is presented by

Rainwater (1971). Definitive technical information on radiometry, photometry, and colorimetry may be found in the books by the Committee on Colorimetry of the Optical Society of America (1963) and Wyszecki and Stiles (1967). Important factors for display systems are described by Klein (1970) and Sherr (1970).

The scientific study of perceptual processing often results in the dual benefits of increased understanding of humans and guidelines for computer processing. The text *Visual Perception* by Cornsweet (1970) provides an excellent introduction to the field. A comprehensive reference is *Vision and Visual Perception*, edited by Graham (1965). Other excellent references are Davson (1962, 1963) and Dodwell (1970).

The spatial sensitivity of the human visual system is of fundamental importance and there is now a high correlation between physiologically and psychologically derived experimental results. As an introduction to the psychophysical experimentation, the reader may consult the works of Lowry and DePalma (1961), Schade (1956), Spitzberg and Richards (1975), Tynan and Sekuler (1974), Robson (1966), Campbell *et al.* (1966, 1969), Campbell (1968), Davidson (1968), Westhemier and Campbell (1962), and Yasuda and Hiwatashi (1968).

An introduction to the neurophysiological experimentation in animals may be found in Hubel and Wiesel (1959).

The mathematical techniques for two-dimensional spatial analysis are described in Goodman (1968), Papoulis (1968), Vander Lugt (1966), and Klein (1970). Resolution criteria are introduced in Sherr (1970) and Heyning (1967).

Texture generation and stereoscopic vision are treated by Julesz (1960, 1962). His book *Foundations of Cyclopean Perception* (Julesz, 1971) contains some remarkable images, which must be viewed to be fully appreciated.

Structural models of the visual system may some day be the key to understanding human visual processing. An excellent introduction is given in Cornsweet (1970). Further studies on spatial response are described by Furman (1965), Stockham (1972), Mannos and Sakrison (1974), Hall and Hall (1977), and Hall (1978). Investigations of temporal response are described by Robson (1966) and Budrikis (1972). Color perception models are described by Wallis (1975) and Land (1977). Binocular vision models are described in Julesz (1971). The mathematical techniques are described in Rogers and Adams (1976).

Very little is known about the computations associated with perception. The importance of invariance properties is described by Deutsch (1955). Neural computations that could perform some of the perceptual tasks are described in Kabrisky (1966), Tallman (1970), Ginsburg (1971), and Carl (1969). Layered machine computations are described by Uhr (1971) and Marco (1973).

## PROBLEMS

**2.1** Most artificial sources of radiant energy consist of incandescent solids whose emittance depends upon the temperature. Planck's formula, which describes the spectral emittance of complete radiators, is

$$W_\lambda = C_1 \lambda^{-5} / [\exp(C_2/\lambda T) - 1].$$

For small values of $\lambda T$, Planck's formula becomes equivalent to Wein's formula

$$W_\lambda = C_1 \lambda^{-5} \exp(-C_2/\lambda T).$$

Assume that Wein's law holds and compute the total emittance of a blackbody source at $T = 2042°\text{K}$ (the freezing point of platinum). Use

$$C_1 = 3.735 \times 10^{-5} \quad \text{erg cm}^2/\text{sec},$$

$$C_2 = 1.438 \quad \text{cm }°\text{K}.$$

**2.2** The spectral transmittance $\tau(\lambda)$ of an object is defined as the ratio of transmitted power divided by the incident power, i.e., $\tau(\lambda) = P_t(\lambda)/P_i(\lambda)$. It is often convenient to describe the transmittance in terms of density $D$, where $D = -\log_{10}\tau$. Compute $D$ for transmittance values of 1, 0.1, 0.01, and 0.001.

**2.3** Use the simple optical model of the eye and determine

(a)  the visual angle of the 0.4-mm-wide foveal area;

(b)  the visual angle and retinal image size of a 6-ft-tall person at a distance of 20 ft;

(c)  the visual angle between two cones spaced 1 $\mu$m apart.

**2.4** (After Heyning)  Assume that the pupil may be represented by a slit aperture of diameter $D$. The luminance pattern at the retina may then be described by

$$L = (L_0 D/z)^2 (\sin^2 \theta)/\theta^2,$$

where $\theta = (\pi n D/\lambda)\sin\alpha$, $n = 1.336$ is the index of refraction of the medium, $\lambda = 577 \times 10^{-6}$ m the wavelength, $\alpha$ the angle subtended between the aperture center and any other point located radially outward from the center, $D = 3$ mm the pupil diameter, $z$ the length of the optical path from aperture to image plane, and $L_0$ the luminance at unit distance from aperture. Determine the Rayleigh criteria for the eye. (*Hint:* If $\theta = \pi$, $\sin\alpha \simeq \alpha \simeq \lambda/nD$.)

**2.5** Consider the simplified visual system model that neglects spectral and temporal effects and combines the effect of the ocular media and inhibition into a combined MTF which follows the log function for suprathreshold vision as shown in Fig. 2.41.

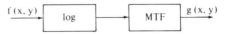

*Figure 2.41*  A simple nonlinear model of the human visual system.

Define two processing operations on the input image that would cancel the nonlinear effects of the log function and the decreased spatial frequency response due to the MTF.

**2.6** The spatial frequency response may be specified in terms of the point-spread function, line-spread function, or modulation transfer function. The terms are defined for a linear, position-invariant system $L$ as follows:

point-spread function,

$$h(x,y) = L[\delta(x,y)];$$

line-spread function,

$$l(x) = L[\delta(x)] = \int_{-\infty}^{\infty} h(x,y)\,dy;$$

modulation transfer function, $H(u,v)$,

$$g(x,y) = L\{\exp[j(ux+vy)]\} = H(u,v)\exp[j(ux+vy)]$$

or

$$H(u,v) = \int\int_{-\infty}^{\infty} h(x,y)\exp[-j(ux+vy)]\,dx\,dy.$$

(a) Assume that the eye's response is isotropic and that the line-spread function has been determined empirically to be

$$l(x) = \exp(-0.7|x|),$$

where $l(x)$ is the illuminance at a distance $x$ in mimutes of arc from the center of the visual axis. Determine the point-spread function and modulation transfer function.

(b) Assume that the modulation transfer function of the eye is of the form

$$H(u,v) = H(r) = \begin{cases} 2/\pi \cos^{-1} r, & 0 \le r \le 1, \\ 0, & r > 1, \end{cases}$$

where $r$ is the normalized distance from the center of the visual axis. Determine the corresponding point-spread function.

(c) Assume that the modulation transfer function of the eye is of the form

$$H(u,v) = H(r) = \varepsilon^{-\alpha r^2}, \quad \alpha > 0.$$

Determine the corresponding point-spread function. (*Hint:* Consider the Hankel transform.)

# References

Bodis-Wollner, I. (1972). Visual Acuity and Contrast Sensitivity in Patients with Cerebral Lesions. *Science* **178**, 769–771.

Brown, J. L. (1965). The Structure of the Visual System. In "Vision and Visual Perception" (C. H. Graham, ed.), pp. 39–59. Wiley, New York.

Budrikis, Z. L. (1972). Visual Fidelity Criterion and Model. *Proc. IEEE* **60**, No. 7, 771–779.

Campbell, F. W. (1968). The Human Eye as an Optical Filter. *Proc. IEEE* **56**, No. 6, 1009–1015.

Campbell, F. W., Kulikowski, J. J., and Levinson, J. Z. (1966). The Effect of Orientation on the Visual Resolution of Gratings. *J. Physiol. (London)* **187**, 427–436.

Campbell, F. W., Carpenter, R. H. S., and Levinson, J. Z., (1969). Visibility of Aperiodic Patterns Compared with that of Sinusoidal Gratings. *J. Physiol. (London)* **204**, 283–298.

Carl, J. W. (1969). Generalized Harmonic Analysis for Pattern Recognition: A Biologically Derived Model, Rep. No. GE/BE 70S-1, Air Force Inst. Technol. Wright-Patterson AFB, Ohio.

Committee on Colorimetry of the Optical Society of America (1963). "The Science of Color." Opt. Soc. Am., Washington, D.C.

Cornsweet, T. N. (1970). "Visual Perception." Academic, New York.

Davidson, M. S. (1968). Perturbation Approach to Spatial Brightness Interaction in Human Vision. *J. Opt. Soc. Am.* **58**, No. 9, 1300–1309.

Davson, H., ed. (1962). "The Eye," Vols. 1, 2, 3, and 4. Academic Press, New York.

Davson, H. (1963). "The Physiology of the Eye." Little, Brown, Boston, Massachusetts.

Deutsch, J. A. (1955). A Theory of Shape Recognition, *Brit. J. Psych.* **46**, 30–37.

Dodwell, P. C. (1970). "Visual Pattern Recognition." Holt, New York.

Dodwell, P. C. (1971). "Perceptual Processing." Appleton (Meredith), New York.

Dorn, W. S., and McCracken, D. D. (1974). "Numerical Methods with Fortran IV Case Studies," pp. 251–265. Wiley, New York.

Furman, G. G. (1965). Comprison of Models for Substructure and Shunting Lateral Inhibition in Receptor Neuron Fields. *Kybernetik* **2**, No. 6, 257–274.

Ginsburg, A. P. (1971). Psychological Correlates of a Model of the Human Visual System. M.S. thesis, Air Force Inst. of Technol., Wright-Patterson AFB, Ohio, June (AD 731 197) [*IEEE 1971 NAECON Proc., Dayton, Ohio, May* pp. 283–290].

Goodman, J. W. (1968). "Introduction to Fourier Optics." McGraw-Hill, New York.

Graham, C. H. (1965). Visual Space Perception. In "Vision and Visual Perception" (C. H. Graham, ed.), pp. 504–547. Wiley, New York.

Hall, C. F. (1978). Digital Color Image Compression in a Perceptual Space. Ph.D. Thesis, Univ. of Southern California, Los Angeles, February (unpublished).

Hall, C. F., and Hall, E. L. (1977). A Nonlinear Model for the Spatial Characteristics of the Human Visual System. *IEEE Trans. Syst., Man Cybern.* **SMC-7**, No. 3, 161–170.

Henning, G. B., Hertz, B. G., and Broadbent, D. E. (1975). Some Experiments Bearing on the Hypothesis that the Visual System Analysis Spatial Patterns in Independent Bands of Spatial Frequency. *Vision Res.* **15**, 887–897.

Heyning, J. M. (1966). The Human Observer. *Soc. Photo-Opt. Instrum. Eng. Proc., Semin. Photo-Opt. Syst. Eval., April, New York,* Ch. 3.

Hubel, D. H., and Wiesel, T. N. (1959). Receptive Fields of Single Neurons in the Cat's Striate Cortex. *J. Physiol. (London)* **148**, 574–591.

Julesz, B. (1960). Binocular Depth Perception of Computer-Generated Patterns. *Bell Syst. Techn. J.* **39**, No. 5, 1125–1163.

Julesz, B. (1962). Visual Pattern Discrimination. *IRE Trans. Inf. Theory* **IT-8**, No. 2, 84–92.

Julesz, B. (1971). "Foundations of Cyclopean Perception." Univ. of Illinois Press, Chicago, Illinois.

Kabrisky, M. (1966). "A Proposed Model for Visual Information Processing in the Human Brain." Univ. of Illinois Press, Urbana.

Klein, M. V. (1970). "Optics." Wiley, New York.

Kulikowski, J. J., and Tolhurst, D. J. (1973). Psychophysical Evidence for Sustained and Transient Channels in Human Vision. *J. Physiol. (London)* **232**, 149–163.

Land, E. (1977). Retinex Theory of Color Perception. *Sci. Am.* **232**, No. 6, 108–128.

Lowry, E. M., and DePalma, J. J. (1961). Sine Wave Response of the Visual System, II, Since Wave and Square Wave Contrast Sensitivity. **52**, No. 8, 740–746.

Mannos, J. L., and Sakrison, D. J. (1974). The Effects of a Visual Fidelity Criterion on the Encoding of Images. *IEEE Trans. Inf. Theory* **IT-20**, 525–536.

Marco, H., and Giebel, H. (1973). Recognition of Handwritten Characters with a System of Homogeneous Layers, *Nachrichtentech. Z.*, 455–459.

Munsell, A. H. (1946). "A Color Notation." Munsell Color Co., Inc., Baltimore.

Papoulis, A. (1968). "Systems and Transforms with Applications in Optics." McGraw-Hill, New York.

Pirenne, M. H. (1967). "Vision and the Eye," 2nd ed., p. 32. Chapman & Hall, London.

Rainwater, C. (1971). "Light and Color." Golden Press, New York.

Robson, J. G. (1966). Spatial and Temporal Contrast—Sensitivity Functions of the Visual System. *J. Opt. Soc. Am.* **56** (August), 1141–1142.

Rogers, D. F., and Adams, B. (1977). "Principles of Interactive Computer Graphics." McGraw-Hill, New York.

Schade, O. H. (1956). Optical and Photoelectric Analog of the Eye. *J. Opt. Soc. Am.* **46** (September), 721–739.

Sherr, S. (1970). "Fundamentals of Display System Design." Wiley, New York.

Spitzberg, R., and Richards, W. (1975). Broad Band Spatial Filters in the Human Visual System. *Vision Res.* **15**, 837–841.

Stockham, T. G., Jr. (1972). Image Processing in the Context of a Visual Model. *Proc. IEEE* **60**, No. 7, 828–842.

Tallman, O. H., II (1970). Processing of Visual Imagery by an Adaptive Model of the Visual System: Its Performance and Its Significance, Aerospace Medical Research Laboratory Rep. No AMRL-TR-70-45. Defense Documentation Center, Alexandria, Virginia.

Tolhurst, D. J. (1975a). Sustained and Transient Channels in Human Vision. *Vision Res.* **15**, 1151–1155.

Tolhurst, D. J. (1975b). Reaction Times in the Detection of Gratings by Human Observers: A Probabilistic Mechanism. *Vision Res.* **15**, 1143–1149.

Tynan, P., and Sekuler, R. (1974). Perceived Spatial Frequency Varies with Stimulus Duration. *J. Opt. Soc. Am.* **64** (September), 1251–1255.

Uhr, L. (1971). Layered "Recognition Cone" Networks that Pre-Process, Classify, and Describe, *Proc. Two Dim. Signal Processing Conf., Columbia, Missouri*, pp. 3-1-1, 3-1-12.

Vander Lugt, A. (1966). Operational Notation for the Analysis and Synthesis of Optical Data-Processing Systems. *Proc. IEEE* **54**, No. 8.

Wald, G. (1964). The Receptors for Human Color Vision. *Science* **145** (September), 1007–1017.

Wallis, R. H. (1975). Film Recording of Color Images. Ph.D. Thesis, Univ. of Southern California, Los Angeles, May (unpublished).

Westhemier, G., and Campbell, F. W. (1962). Light Distribution in the Image Formed by the Living Human Eye. *J. Opt. Soc. Am.* **52**, 1040–1045.

Wyszecky, G. W., and Stiles, W. S. (1967). "Color Science, Concepts and Methods, Quantitative Data and Formulas." Wiley, New York.

Yasuda, M., and Hiwatashi, K. (1968). A Model of Retinal Neural Networks and Its Spatio-Temporal Characteristics. NHK Lab. Note, Serial 116, January. Tokyo, Japan.

# 3 | Computer Representation of Images

## 3.1 Three-Dimensional Imaging

In analyzing pictures of natural scenes, we are often interested in inferring physical object properties such as locations, structure, color, or texture in three-dimensional space from information recorded in two-dimensional pictures. The study of the geometry of image making and viewing, as well as the transformations and invariants is therefore of fundamental concern. The topic of three-dimensional transformation discussed in this section is very familiar in the field of *computer graphics* and the discussion is based mainly upon the excellent paper by Roberts (1965) and the books by Newman and Sproull (1973), Duda and Hart (1973), and Rogers and Adams (1976). The importance of understanding the three-dimensional situation even when our data is a two-dimensional picture is now widely recognized. We will first consider a simple model for making or viewing an image.

### 3.1.1 Simple Model of Perspective Transformations

A simple model of a *pinhole camera* is shown in Fig. 3.1. An actual lens such as the one in the eye or in a photographic or television camera can always be approximated by a pinhole camera placed at a distance equal to the focal length, $f$, of the lens from the image plane. Since light rays from object points must pass through a central lens point onto the image plane, the transformation from object points to image points is called a perspective transformation. That is, object sizes, lengths, and distances appear in a similar manner to those seen by a human viewing the object. Furthermore, since the three-dimensional object points are projected onto the two-dimensional image plane, the transformation is called projective, perspective mapping. An object point maps onto a single image point; therefore, given object coordinates and a description of the lens, one should be able to determine image point coordinates. However, image points can only be back projected along a line.

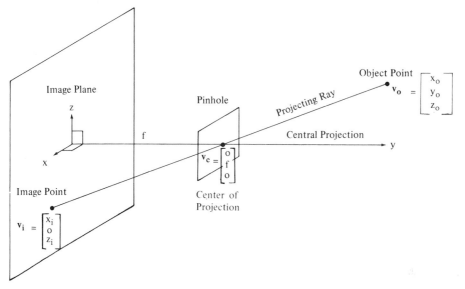

*Figure 3.1* Image recording model.

Therefore, other information such as the range or distance to the object or another intersecting line, such as in binocular vision, must be given to uniquely determine an object point. The *perspective* transformation which relates object and image points is easily derived.

Consider the image recording model shown in Fig. 3.1. A lens is placed along the central projection perpendicular to the image plane, with the coordinate system centered in the image plane. The points on the line between the image point and the lens center are described by

$$\mathbf{v} = \mathbf{v}_c + \alpha(\mathbf{v}_i - \mathbf{v}_c) = \alpha\mathbf{v}_i + (1-\alpha)\mathbf{v}_c,$$

where $\alpha$ is a parameter which is proportional to the signed distance along the line from $\mathbf{v}_c$ to $\mathbf{v}$. From the selection of the origin at the center of the image plane, we have

$$\mathbf{v}_c = \begin{bmatrix} 0 \\ f \\ 0 \end{bmatrix} \quad \text{and} \quad \mathbf{v}_i = \begin{bmatrix} x_i \\ 0 \\ z_i \end{bmatrix}.$$

Since the object point $\mathbf{v}_o$ is on this line, it must satisfy the equation

$$\mathbf{v}_o = \mathbf{v}_c + \alpha_o(\mathbf{v}_i - \mathbf{v}_c) \quad \text{or} \quad \begin{bmatrix} x_o \\ y_o \\ z_o \end{bmatrix} = \begin{bmatrix} 0 \\ f \\ 0 \end{bmatrix} + \alpha_o \begin{bmatrix} x_i \\ -f \\ z_i \end{bmatrix}.$$

This may be reduced to the coordinate equations

$$x_i = fx_o/(f - y_o), \qquad y_i = 0, \qquad z_i = fz_o/(f - y_o),$$

which clearly shows that given the object coordinates and focal length, the image coordinates may be determined. Also given the image coordinates and $f$, the line containing the object point is determined by the line which is the intersection of the planes described by each of the following equalities:

$$(f-y_o)/f = x_o/x_i = z_o/z_i.$$

The imaging model shown in Fig. 3.1 works in a manner similar to a camera or the human eye in that the object appears inverted in the image plane. To avoid this inversion, it is sometimes convenient to use the image projection model shown in Fig. 3.2. The points on the line between the lens center and the image point must again satisfy

$$\mathbf{v} = \mathbf{v}_c + \alpha(\mathbf{v}_i - \mathbf{v}_c).$$

Since the object point $\mathbf{v}_o$ is on this line,

$$\mathbf{v}_o = \mathbf{v}_c + \alpha_o(\mathbf{v}_i - \mathbf{v}_c) \qquad \text{or} \qquad \begin{bmatrix} x_o \\ y_o \\ z_o \end{bmatrix} = \begin{bmatrix} 0 \\ -f \\ 0 \end{bmatrix} + \alpha_o \begin{bmatrix} x_i \\ f \\ z_i \end{bmatrix},$$

which are easily reduced to

$$x_i = fx_o/(y_o+f), \qquad y_i = 0, \qquad z_i = fz_o/(y_o+f).$$

The direct perspective transformation which transforms object points into image points, although simple, is nonlinear. However, the transformation may be converted into a linear one by converting the vectors into *homogeneous* coordinates. The homogeneous coordinates $\hat{\mathbf{v}}$ of a point $\mathbf{v} = (x,y,z)'$ are

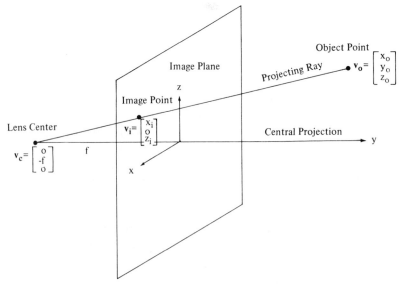

*Figure 3.2*   Image projection model.

$\hat{\mathbf{v}} = (wx, wy, wz, w)'$, where $w$ is an arbitrary constant. Clearly, given the homogeneous coordinates $\hat{\mathbf{v}}$ of a point, one may obtain the inhomogeneous coordinates by simply dividing by $w$ and selecting the first three components. Also note that the homogeneous vectors $\hat{\mathbf{v}}$ and $k\hat{\mathbf{v}}$ both correspond to the same physical point $\mathbf{v}$. The usefulness of homogeneous coordinates will now be demonstrated by showing that the direct perspective transformation in homogeneous coordinates is linear. First, consider the object point $\hat{\mathbf{v}}_{o} = (wx_{o}, wy_{o}, wz_{o}, w)'$. Then observe that a linear transformation defined by the matrix, $\mathbf{P}$, transforms object points into image points

$$\hat{\mathbf{v}}_{i} = \mathbf{P}\hat{\mathbf{v}}_{o},$$

where

$$\begin{bmatrix} x_{i} \\ y_{i} \\ z_{i} \\ w_{i} \end{bmatrix} = \begin{bmatrix} f & 0 & 0 & 0 \\ 0 & f & 0 & 0 \\ 0 & 0 & f & 0 \\ 0 & 1 & 0 & f \end{bmatrix} \begin{bmatrix} x_{o} \\ y_{o} \\ z_{o} \\ 1 \end{bmatrix}$$

Computing the matrix product gives

$$\hat{\mathbf{v}}_{i} = \begin{bmatrix} x_{i} \\ y_{i} \\ z_{i} \\ w_{i} \end{bmatrix} = \begin{bmatrix} fx_{o} \\ fy_{o} \\ fz_{o} \\ (y_{o}+f) \end{bmatrix}.$$

Converting back to physical coordinates by dividing by the fourth component gives

$$\mathbf{v}_{i} = \begin{bmatrix} x_{i} \\ y_{i} \\ z_{i} \end{bmatrix} = \begin{bmatrix} fx_{o}/(y_{o}+f) \\ fy_{o}/(y_{o}+f) \\ fz_{o}/(y_{o}+f) \end{bmatrix}.$$

Note that the $x$ and $z$ components agree exactly with the previously derived relationships. The second component cannot be taken as the physical coordinate but must be considered as a parameter and is proportional to the distance along the projecting ray from the lens center to the object point. Although computation of the second component may appear needless, its inclusion permits $\mathbf{P}$ to be a square and invertible matrix, and the value may also be used for eliminating object overlap.

Now consider the inverse perspective transformation from image points to object points. We expect that this transformation would not define a unique point but should determine the equation of the projecting ray. In homogeneous coordinates we might expect that

$$\hat{\mathbf{v}}_{o} = \mathbf{P}^{-1}\hat{\mathbf{v}}_{i}.$$

The inverse of the **P** matrix is easily determined to be

$$\mathbf{P}^{-1} = \frac{1}{f^2} \begin{bmatrix} f & 0 & 0 & 0 \\ 0 & f & 0 & 0 \\ 0 & 0 & f & 0 \\ 0 & -1 & 0 & f \end{bmatrix}.$$

The resulting calculation produces

$$\mathbf{v}_o = \begin{bmatrix} fx_i/(f-y_i) \\ fy_i/(f-y_i) \\ fz_i/(f-y_i) \end{bmatrix}.$$

The components of $\mathbf{v}_o$ are correctly specified if one considers the value $y_i$ as a parameter related to the distance along the projecting ray to the object point. Eliminating this parameter from the component equations gives the desired line:

$$x_o = (x_i/f)(y_o + f) = (x_i/z_i)z_o.$$

The use of homogeneous coordinates permits a linearization of the perspective transformation but requires a careful interpretation of the components. A plane in homogeneous coordinates always passes through the origin. A plane is represented by

$$ax + by + cz + dw = 0 \qquad \text{or} \qquad \hat{\mathbf{a}}'\hat{\mathbf{v}} = 0,$$

where the normal vector to the plane is $\hat{\mathbf{a}} = (a,b,c,d)'$, and a point on the plane in homogeneous coordinates is $\hat{\mathbf{v}} = (x,y,z,w)'$. The scale factor $w$, although arbitrary, may be used to ensure that the coordinate values stay within a given range.

The advantage of homogeneous coordinates is that a single transformation matrix **H** can accomplish a full perspective transformation involving not only perspective but also rotation, translation, and scale.

It is easily verified that the translation matrix **T** may be used to compute the new coordinates $\hat{\mathbf{v}}'$ of a point $\hat{\mathbf{v}}$ given in a reference coordinate system in which the origin has been translated to $(x_o, y_o, z_o, 1)'$:

$$\hat{\mathbf{v}}' = \mathbf{T}\hat{\mathbf{v}} = (x - x_o, y - y_o, z - z_o, 1)',$$

where

$$\mathbf{T} = \begin{bmatrix} 1 & 0 & 0 & -x_o \\ 0 & 1 & 0 & -y_o \\ 0 & 0 & 1 & -z_o \\ 0 & 0 & 0 & 1 \end{bmatrix}.$$

Similarly, if a point $\hat{\mathbf{v}}$ is translated by $(h,i,j,1)$, its new position $\hat{\mathbf{v}}'$ is described by

$$\hat{\mathbf{v}}' = \mathbf{T}\hat{\mathbf{v}} = (x + h, y + i, z + j, 1)',$$

where

$$T = \begin{bmatrix} 1 & 0 & 0 & h \\ 0 & 1 & 0 & i \\ 0 & 0 & 1 & j \\ 0 & 0 & 0 & 1 \end{bmatrix}.$$

Also, a scaling of the points may be accomplished by the matrix multiplication

$$\hat{v}' = S\hat{v} = (s_1 x, s_2 y, s_3 z, 1)',$$

where

$$S = \begin{bmatrix} s_1 & 0 & 0 & 0 \\ 0 & s_2 & 0 & 0 \\ 0 & 0 & s_3 & 0 \\ 0 & 0 & 0 & 1 \end{bmatrix}.$$

If the scale change is the same in all dimensions, the operation is particularly simple:

$$S = \begin{bmatrix} 1 & 0 & 0 & 0 \\ 0 & 1 & 0 & 0 \\ 0 & 0 & 1 & 0 \\ 0 & 0 & 0 & s \end{bmatrix} \quad \text{and} \quad \hat{v}' = S\hat{v} = (x, y, z, s)'.$$

The scaling is accomplished when the point is converted to physical coordinates.

Rotations about the coordinate axes are also easily accomplished in homogeneous coordinates. The physical rotation matrices are shown in Fig. 3.3. The matrices $R_1$, $R_2$, and $R_3$ represent right-handed rotations about the positive coordinate axes. The inverses of these matrices are obtained simply by replacing the angle with its negative:

$$R_i^{-1}(\alpha) = R_i(-\alpha), \qquad i = 1, 2, 3.$$

The corresponding matrices $\hat{R}_i$ in homogeneous coordinates are given by

$$\hat{R}_i = \begin{bmatrix} & R_i & & 0 \\ & & & 0 \\ & & & 0 \\ 0 & 0 & 0 & 1 \end{bmatrix}, \qquad i = 1, 2, 3.$$

Successive rotations about more than a single axis may be computed simply as the product of the individual rotation matrices made in any order. For example, a rotation through an angle $\alpha$ about $x$, $\beta$ about $y$, and $\gamma$ about $z$ has the transformation matrix

$$R = R_3(\gamma) R_2(\beta) R_1(\alpha).$$

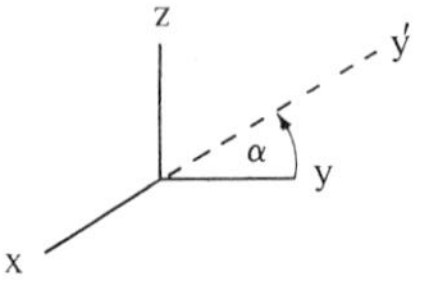

$$R_1 \; (\alpha) \; = \; \begin{bmatrix} 1 & 0 & 0 \\ 0 & \cos \alpha & -\sin \alpha \\ 0 & \sin \alpha & \cos \alpha \end{bmatrix}$$

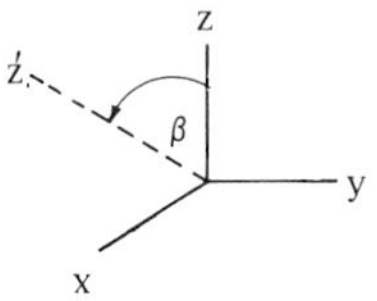

$$R_2 \; (\beta) \; = \; \begin{bmatrix} \cos \beta & 0 & \sin \beta \\ 0 & 1 & 0 \\ -\sin \beta & 0 & \cos \beta \end{bmatrix}$$

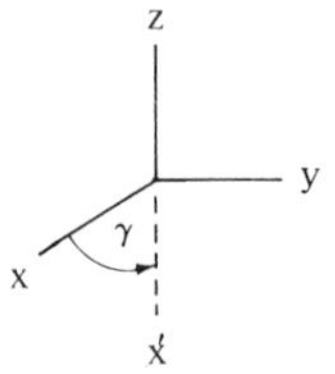

$$R_3 \; (\gamma) \; = \; \begin{bmatrix} \cos \gamma & -\sin \gamma & 0 \\ \sin \gamma & \cos \gamma & 0 \\ 0 & 0 & 1 \end{bmatrix}$$

*Figure 3.3* Rotation matrices for positive counterclockwise angles about the positive coordinate axes for a right-handed coordinate system. $R_1$, rotation through an angle $\alpha$ about the $x$ axis; $R_2$, rotation through an angle $\beta$ about the $y$ axis; $R_3$, rotation through an angle $\gamma$ about the $z$ axis.

In homogeneous coordinates, a complex succession of operations are required, such as those for the perspective transformation matrix **H**

$$\mathbf{H} = \mathbf{SPLRT}.$$

This will now be illustrated by determining a complete model for perspective transformations.

### 3.1.2 Complete Perspective Transformation Model

Ten basic elements of perspective viewing are shown in Fig. 3.4. The *object* (1) is projected onto the *image plane* (2) for viewing or recording from the *viewing point* (3). As before, the viewing point is located at a distance $f$ and is perpendicular to the image plane. For convenience in measuring points in the picture, the picture coordinate system is located at the *center of vision* (4), which is located at the intersection of the central projection ray and the image plane. As previously shown, an object point $v_o = (x_o, y_o, z_o)'$, its image point

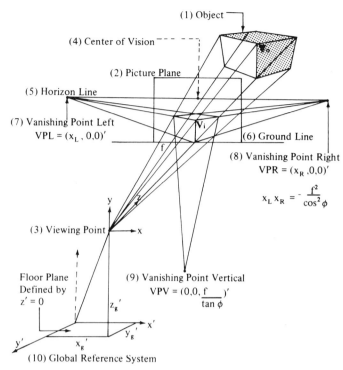

*Figure 3.4*   Elements of a perspective transformation.

$v_i = (x_i, 0, z_i)'$, and the viewing point $v_p = (0, -f, 0)'$, all lie on a line. It is also useful to establish a *horizon line* (5) along the $x$ axis in image coordinates that is the intersection of the picture plane and a plane passing through the viewing point parallel to the ground line. A *ground line* (6) may also be defined by the intersection of the plane perpendicular to the picture plane and passing through the lowest object point or support. Due to the perspective distortion, parallel horizontal lines on the object will map into intersecting lines. The *left horizontal vanishing point* (VPL) (7) is located at the intersection of the horizon line and the image line produced by a horizontal object line of negative slope. A conjugate *right horizontal vanishing point* (VPR) (8) is located at the intersection of the horizon line and the image line produced by a horizontal object line of positive slope.

Vertical lines in the object also produce image lines which intersect. The intersection point or *vertical vanishing point* (VPV) (9) is located along the image plane $z$ axis at a distance which depends upon the viewing point location.

Finally, to permit easy coordinate measurement with respect to the object, a *global* or *world coordinate system* (10) is provided. The coordinates of points measured in this reference frame will be denoted with primes. For example, $v_o'$ represents an object point in the global reference system. The same physical object point measured with respect to the image coordinate system is indicated by $v_o$.

Let us now develop an overall transformation **H** which maps points measured in the global coordinate system into points in the image reference system. In order to define this mapping, we must define the position and orientation of the picture reference system with respect to the global system. This may be done in three steps: First, we translate the global reference system to the viewing point. Then we rotate the system so that the $y'$ axis is aligned with the central projection and the $z'$ axis is aligned with the vertical picture plane $z$ axis. Since the coordinate systems are both right-handed and orthogonal, a third rotation is unnecessary. Finally, a translation from the viewing center to the image plane origin completes the transformation.

If the viewing point $v_p$ is located at $v'_p = (x'_g, y'_g, z'_g)$ in the global coordinate system, then the translation **T** required to translate the global system to this point is simply

$$\mathbf{T} = \begin{bmatrix} 1 & 0 & 0 & -x'_g \\ 0 & 1 & 0 & -y'_g \\ 0 & 0 & 1 & -z'_g \\ 0 & 0 & 0 & 1 \end{bmatrix}.$$

Let $\theta$ be the *pan angle* measured counterclockwise from the $y$ axis. Thus a rotation of $\theta$ about the $z$ axis is required to point the $y'$ axis of the system. This rotation is described by $\mathbf{R}_3(\theta)$. Let $\phi$ be the *tilt angle* of the $y'$ axis also measured positive in a counterclockwise direction. Then the rotation of $\phi$ about the $x$ axis is required to complete the alignment. This rotation is described by $\mathbf{R}_1(\phi)$. The overall rotation matrix is

$$\mathbf{R} = \mathbf{R}_1(\phi)\mathbf{R}_3(\theta)$$

or, in homogeneous coordinates,

$$\mathbf{R} = \begin{bmatrix} \cos\theta & -\sin\theta & 0 & 0 \\ \sin\theta\cos\phi & \cos\theta\cos\phi & -\sin\phi & 0 \\ \sin\theta\sin\phi & \cos\theta\sin\phi & \cos\phi & 0 \\ 0 & 0 & 0 & 1 \end{bmatrix}.$$

The final translation from the viewing center to the image plane origin is simply a translation $L$, where

$$\mathbf{L} = \begin{bmatrix} 1 & 0 & 0 & 0 \\ 0 & 1 & 0 & -f \\ 0 & 0 & 1 & 0 \\ 0 & 0 & 0 & 1 \end{bmatrix}.$$

Therefore, a complete change of reference systems from object coordinates to image coordinates is given by

$$\hat{v} = \mathbf{LRT}\hat{v}'.$$

We may now include the effect of the perspective transformation, which was developed in and is naturally expressed in terms of the picture coordinate system

$$\hat{\mathbf{v}} = \mathbf{P}\hat{\mathbf{v}}_0.$$

The last two expressions may now be combined to form the complete perspective transformation matrix **H**:

$$\mathbf{H} = \mathbf{PLRT}.$$

The mapping of object points in global coordinates into image points in picture coordinates may be expressed simply as

$$\hat{\mathbf{v}}_i = \mathbf{H}\hat{\mathbf{v}}_0.$$

Carrying out the previous operations of the equation gives

$$\hat{x}_i = f\left[ (x'_0 - x'_g)\cos\theta - (y'_0 - y'_g)\sin\theta \right],$$

$$\hat{y}_i = f\left[ (x'_0 - x'_g)\sin\theta\cos\phi + (y'_0 - y'_g)\cos\theta\cos\phi \right.$$
$$\left. - (z'_0 - z'_g)\sin\phi - f \right],$$

$$\hat{z}_i = f\left[ (x'_0 - x'_g)\sin\theta\sin\phi + (y'_0 - y'_g)\cos\theta\sin\phi \right.$$
$$\left. + (z'_0 - z'_g)\cos\phi \right],$$

$$\hat{w}_i = (x'_0 - x'_g)\sin\theta\cos\phi + (y'_0 - y'_g)\cos\theta\cos\phi$$
$$- (z'_0 - z'_g)\sin\phi.$$

To convert to physical Cartesian coordinates, we simply divide each component by $\hat{w}_i$ and recall that $\hat{y}_i$ does not correspond to the image $y$ coordinate, which is zero.

The inverse perspective transformation which back projects an image point onto the line through an object point is easily developed and may be expressed as

$$\hat{\mathbf{v}}'_0 = \mathbf{H}^{-1}\hat{\mathbf{v}}_i,$$

where $\mathbf{H}^{-1} = \mathbf{T}^{-1}\mathbf{R}^{-1}\mathbf{L}^{-1}\mathbf{P}^{-1}$. It is easily verified that the inverses exist, except for trivial cases such as with $f = 0$. They are as follows:

$$\mathbf{T}^{-1} = \begin{bmatrix} 1 & 0 & 0 & x'_g \\ 0 & 1 & 0 & y'_g \\ 0 & 0 & 1 & z'_g \\ 0 & 0 & 0 & 1 \end{bmatrix},$$

$$\hat{\mathbf{R}}^{-1} = \begin{bmatrix} \cos\theta & \sin\theta\cos\phi & \sin\theta\sin\phi & 0 \\ -\sin\theta & \cos\theta\cos\phi & \cos\theta\sin\phi & 0 \\ 0 & -\sin\phi & \cos\phi & 0 \\ 0 & 0 & 0 & 1 \end{bmatrix},$$

$$L^{-1} = \begin{bmatrix} 1 & 0 & 0 & 0 \\ 0 & 1 & 0 & f \\ 0 & 0 & 1 & 0 \\ 0 & 0 & 0 & 1 \end{bmatrix},$$

$$H^{-1} = \begin{bmatrix} f\cos\theta & -x'_g & f\sin\theta\sin\phi & f^2\sin\theta\cos\phi + fx'_g \\ -f\sin\theta & -y'_g & f\cos\theta\sin\phi & f^2\cos\theta\cos\phi + fy'_g \\ 0 & -z'_g & f\cos\phi & -f^2\sin\phi + fz'_g \\ 0 & -1 & 0 & f \end{bmatrix}.$$

The object point in global coordinates may also be expressed in the form of the projecting ray between the viewing center and object point:

$$\begin{bmatrix} x'_o \\ y'_o \\ z'_o \end{bmatrix} = \begin{bmatrix} x'_g \\ y'_g \\ z'_g \end{bmatrix} + \frac{f}{f - y_i} \begin{bmatrix} x_i\cos\theta + z_i\sin\theta\sin\phi + f\sin\theta\cos\phi \\ -x_i\sin\theta + z_i\cos\theta\sin\phi + f\cos\theta\cos\phi \\ z_i\cos\phi - f\sin\phi \end{bmatrix}.$$

### Perspective Distortion

It is interesting to determine equations for the basic elements of perspective viewing using the direct and inverse transformations.

The ground, floor, or support plane is defined by $z'_o = 0$. The intersection of this plane with the image plane determines the ground line in the image plane. To simplify the analysis, note that the intersection line is neither affected by the viewing point offset parameters $x'_g$ and $y'_g$ nor by the rotation of the ground plane about the $z'$ axis. Therefore, we may set $x'_g = y'_g = \theta = 0$.

The ground line, as shown in Fig. 3.5, is described by

$$z = (z'_g - f\sin\phi)/\cos\phi.$$

The horizon line is defined as the intersection of the image plane with a plane through the viewing center parallel to the ground plane. This line is also shown in Fig. 3.5 and is described by

$$z = f\tan\phi.$$

To determine the vanishing points, let us consider the mapping of straight lines in the object into their images in the image plane. The direct transformation with $x_g = y_g = z_g = \theta = 0$ gives

$$x_i = \frac{fx'_o}{y'_o\cos\phi - z'_o\sin\phi}, \qquad z_i = \frac{f(y'_o\sin\phi + z'_o\cos\phi)}{y'_o\cos\phi - z'_o\sin\phi}.$$

A vertical object line is traced out by a point $v'_o = (x'_1, y'_1, z')$, where $x_1$ and $y_1$ are fixed but $z$ varies over all real numbers. Substituting this point into the

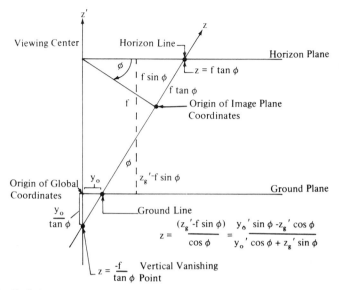

*Figure 3.5*   End view of viewing situation showing location ground plane and horizon plane.

previous expression and eliminating $z$ gives the equation of a straight line in the image plane,

$$z_i = (y_1' x_i / x_1' \sin\phi) - (f/\sin\phi).$$

Note that the $z$ intercept does not depend upon the location of the vertical line but only upon the camera and geometry. Thus, all vertical lines pass through this unique vertical vanishing point, which is shown in Fig. 3.5.

Now consider the images of a horizontal line. For simplicity, consider lines in the ground plane of the form $v' = (x_o', mx_o' + b, 0)'$, and viewing parameters $z' = x_g' = y_g' = \theta = 0$. The direct transformation gives

$$x_i = \frac{fx_o'}{y_o' \cos\phi + z_g' \sin\phi}, \qquad z_i = f\frac{(y_o' \sin\phi - z_g' \cos\phi)}{y_o' \cos\phi + z_g' \sin\phi}.$$

Substituting the point coordinates and eliminating $x_o'$ gives the equation of a straight line,

$$z_i = \left[ mz_g' x_i + f(b\sin\phi - z_g' \cos\phi) \right] / (b\cos\phi + z_g' \sin\phi).$$

Now consider the path of $z_i$ as $x_o$ approaches infinity. Taking the limit of both $x_i$ and $z_i$ in the direct transformation gives

$$\lim_{x_o' \to \infty} x_i = f/m\cos\phi = x_h,$$

$$\lim_{x_o' \to \infty} z_i = f\tan\phi = z_h.$$

Thus, $z_i$ intersects the horizon line $z_h$ with $x$ intercept $x_h$, as the object point recedes along a horizontal line, $y_o' = mx_o' + b$. The $x$ intercept may be on either side of the origin depending on whether the slope $m$ is positive or

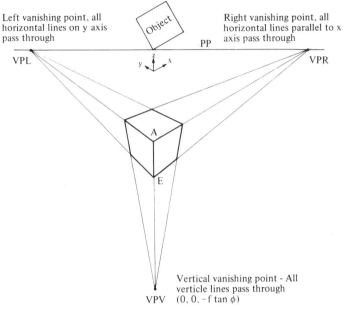

Left vanishing point, all
horizontal lines on y axis
pass through

VPL

Right vanishing point, all
horizontal lines parallel to x
axis pass through

VPR

PP

Vertical vanishing point - All
verticle lines pass through
$(0, 0, -f \tan \phi)$

VPV

*Figure 3.6*   Location of vanishing points.

negative. All lines of the same slope, i.e., parallel lines, will have exactly the same intercept point. Also, note that perpendicular lines have conjugate vanishing points which must satisfy

$$x_r x_1 = -f^2 / \cos^2 \phi$$

since $m_1 m_2 = -1$ for perpendicular lines.

The location of vanishing points is illustrated in Fig. 3.6. The measurement of vanishing points provides a simple method for determining two of the most important camera viewing parameters, $f$ and $\phi$.

## 3.2   Computer Representation of Natural Images

Scenes viewed with a camera lens and projected on a two-dimensional plane may be called *perspective, projective* images to emphasize the processes involved in their formation.

If the content of the scene is not constrained to a particular type, a general class of images of natural scenes or, simply, natural imagery must be considered. These include the total variety of images illustrating objects, concepts, and situations. The mathematical representation and the analysis of this imagery by computer is the primary concern of this section.

Several important representations of computer pictures will now be described. First, the conversion of a continuous range and tone, natural image into a numerical form is considered. In picture processing, just as in signal processing, this conversion may be considered to be two-step process involving sampling followed by quantization. The sampling theorem provides a mathematical basis for the first step. An analysis of the quantization process provides guidelines for the second.

Conversion of a computer image back into a viewable format is another basic consideration. Interpolation theory, coupled with our previous information about human viewing, gives theoretical guidelines for the display process.

In extending the analysis of functions from one to several independent variables, the two-dimensional case is particularly interesting since if a difficulty is encountered, it is the simplest case to consider; if a difficulty is not encountered, it is an interesting case to illustrate and visualize. We will see that many concepts readily extend from one- to two-dimensional analysis, for example, Fourier analysis. Yet there are other concepts which do not. For example, although a polynomial can be easily factored for a function of a single variable, no factorization theorem exists for functions of two independent variables. It may appear that extensions are either trivial or impossible; however, as we shall see, some analysis is always possible and often only a one-dimensional analysis is necessary.

It is often convenient to impose certain restrictions or constraints on the image functions considered. For example, certain constraints may be imposed upon a physical luminance image such as nonnegativity and finite extent, duration, spectral bandwidth, and energy. These constraints affect the possible solutions.

### 3.2.1   Sampling of Image Functions

An important question which we have not yet considered is: How does one get a computer image, which is a discrete valued image function, from a continuous range, continuous tone image? For example, can the continuous

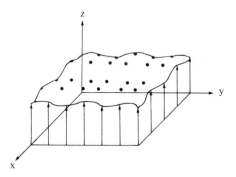

*Figure 3.7*   Sampling process. The continuous function $z = f(x, y)$ may be represented by its sampling values if the sampling is sufficiently dense.

function $f(x,y)$ be accurately represented by its values $f(mX,nY)$ at equispaced points $(mX,nY)$, as illustrated in Fig. 3.7?

The answer to part of this question is given by the sampling theorem, which easily extends from the one-dimensional case (Papoulis, 1969).

**Two-Dimensional Sampling Theorem.**  Consider a two-dimensional function $f(x,y)$ with Fourier transform $F(u,v)$, and assume that $F(u,v)$ is band limited, i.e., that $F(u,v)=0$ if $(u,v)\notin A, A\subset R$ a rectangle. Thus,

$$f(x,y)=\int\int_A F(u,v)\exp\left[\,j2\pi(ux+vy)\,\right]du\,dv.$$

Also, let $B$ be a region contained in the rectangle $R$ as shown in Fig. 3.8. Let $P_B(u,v)$ be a pulse function which is one inside $B$ and zero otherwise, i.e.,

$$P_B(u,v)=\begin{cases} 1, & (u,v)\in B, \\ 0, & (u,v)\notin B, \end{cases}$$

and let the inverse Fourier transform of $P_B$ be denoted by $p_B(x,y)$, where

$$p_B(x,y)=\int\int_B \exp\left[\,j2\pi(ux+vy)\,\right]du\,dv.$$

The two-dimensional sampling theorem states that the function $h(x,y)$ may be represented in terms of a sequence of its sampled values at equispaced values $(mX,nY)$, i.e.,

$$f(x,y)=\sum_{m=-\infty}^{\infty}\sum_{n=-\infty}^{\infty} XYf(mX,nY)p_B(x-mX,y-nY),$$

provided that the sampling is sufficiently dense.

A particular form of $p_B$ is determined when the region $B$ is specified. We will consider two cases: (1) $B=R$, a rectangle; (2) $B=C$, a circle.

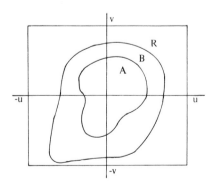

*Figure 3.8*  Bandlimited region in two dimensions. The bandlimited region $A$ is a subset of the filtering region $B$ which is a subset of the rectangular region $R$.

*Proof of Two-Dimensional Sampling Theorem.*    Consider the *periodic exten-sion* of $F(u,v)$, as illustrated in Fig. 3.9, where

$$F_p(u,v) = F(u,v) \quad \text{if} \quad (u,v) \in R$$

and

$$F_p(u+2mU, v+2nV) = F(u,v)$$

for $m, n = 0, \pm 1, \pm 2, \ldots$. Since $F_p$ is periodic, we may represent it by a two-dimensional Fourier series, i.e.,

$$F_p(u,v) = \sum_{m=-\infty}^{\infty} \sum_{n=-\infty}^{\infty} c_{mn} \exp\left[ -j2\pi(umX + vnY) \right],$$

where $X = 1/2U$ and $Y = 1/2V$ correspond to *fundamental frequencies* in the $u$ and $v$ directions, respectively, and

$$c_{mn} = \frac{1}{4UV} \int_{-V}^{V} \int_{-U}^{U} F(u,v) \exp\left[ j2\pi(umX + vnY) \right] du\, dv$$

or

$$c_{mn} = XY \int \int_{A} F(u,v) \exp\left[ j2\pi(umX + vmY) \right] du\, dv.$$

But recall that

$$f(x,y) = \int \int_{A} F(u,v) \exp\left[ j2\pi(ux + vy) \right] du\, dv.$$

From comparing the two previous equations, we see that

$$c_{mn} = XYf(mX, nY)$$

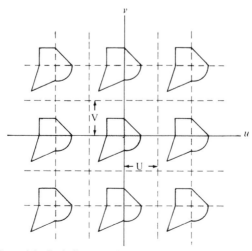

*Figure 3.9*   Periodic extension of the spatial function $F(u,v)$.

so

$$F_p(u,v) = \sum_{m=-\infty}^{\infty} \sum_{n=-\infty}^{\infty} XYf(mX,nY) \exp[-j2\pi(umX+vnY)].$$

Note that we can recover $F$ from $F_p$ by simply multiplying by the pulse $P_B(u,v)$, i.e.,

$$F(u,v) = F_p(u,v)P_B(u,v),$$

$$F(u,v) = \sum_{m=-\infty}^{\infty} \sum_{n=-\infty}^{\infty} XYf(mX,nY)P_B(u,v) \exp[-j2\pi(umX+vnY)].$$

We could select many shapes for $P_B(u,v)$. The most common are the rectangle and square.

The inverse transform of $F(u,v)$ is

$$f(x,y) = \int\int_{-\infty}^{\infty} F(u,v) \exp[j2\pi(ux+vy)]\,du\,dv$$

or

$$f(x,y) = \sum_{m=-\infty}^{\infty} \sum_{n=-\infty}^{\infty} XYf(mX,nY) \int\int_B P_B(u,v) \exp\{j2\pi[u(x-mX)$$
$$+v(y-nY)]\}\,du\,dv.$$

The calculation of the Fourier transform of a rectangular aperture gives, when $B=R$,

$$\int\int_B \exp\{j2\pi[u(x-mX)+v(y-nY)]\}\,du\,dv$$
$$=4UV\sin 2\pi U(x-mX)\sin 2\pi V(y-nY)/2\pi U(x-mX)2\pi V(y-nY),$$

which leads to the cardinal function interpolation formula

$$f(x,y) = \sum_{-\infty}^{\infty} \sum_{-\infty}^{\infty} f(mX,nY)\,\text{sinc}[2\pi U(x-mX)]\,\text{sinc}[2\pi V(y-nY)].$$

Similarly, the Fourier transform of a circular aperture, if $B$ is a circle of radius $R_0$ and $U=V=R_0$, gives

$$\int\int_B \exp\{j2\pi[u(x-mX)+v(y-nY)]\}\,du\,dv = R_0 \frac{J_1[2\pi R_0 r(x,y)]}{r(x,y)},$$

where

$$r(x,y) = [(x-mX)^2 + (y-nY)^2]^{1/2}$$

and $J_1$ is a Bessel function of the first kind. Thus, the circular region leads to the interpolation formula

$$f(x,y) = \sum_{-\infty}^{\infty} \sum_{-\infty}^{\infty} f(mx,ny) \frac{J_1[2\pi R_0 r(x,y)]}{r(x,y)}.$$

Although the two-dimensional sampling theorem is simply an extension of the one-dimensional theorem, sampling of a picture appears quite different from sampling of signal. The differences encountered are mainly due to the difference between the human auditory and visual systems. If one compares the temporal frequency response of the auditory system to the spectral frequency response of the visual system, then the only conclusion is that the eye is almost "tone deaf" since only three spectral sensors are available. A comparison of the temporal frequency bandwidth of the two systems shows a 20 Hz to 20 kHz bandwidth for the ear, but only a 0.5 to 40 Hz bandwidth for the eye. It is, of course, the ratio of the number of receptors which is about 1 to 100 million that produces the increased information capacity of the visual

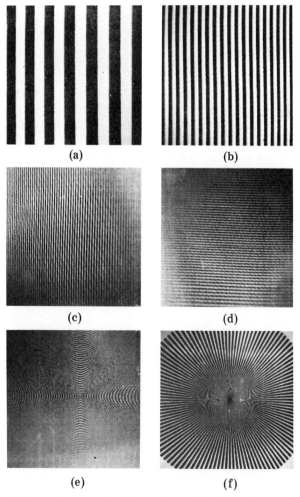

*Figure 3.10*   Aliasing patterns produced by undersampling. Each picture was sampled at 256 by 256 points: (a) course line pattern, (b) line pattern without aliasing, (c) line pattern with aliasing, (d) line pattern with aliasing, (e) rectangular grid with aliasing, and (f) star pattern with aliasing.

system. These examples illustrate that although the mathematical foundations are identical, one cannot always apply the intuition developed from signal processing directly to picture processing.

The effect of undersampling, i.e., *aliasing error*, for two-dimensional sampling is quite different from that for one-dimensional sampling. In fact, undersampling may even be used to advantage as a smoothing process. Aliasing does occur as the patterns in Fig. 3.10 illustrate. The line test images were digitized with the same resolution ($256 \times 256$ points), but as the image frequency content is increased, a diagonal aliasing pattern appears. The image in Fig. 3.10e started as a rectangular grid, while the image in Fig. 3.10f was a star test pattern. The image produced by the star test pattern may be analyzed to determine characteristics such as the width of the point spread function of the imaging system.

The perceptual effect of undersampling is shown in Fig. 3.11. The original image was scanned at a very low resolution, as shown in the block image. However, simply "defocusing", or low-pass filtering by holding the page at a distance, produces a recognizable form. This example illustrates that often high resolution is not always needed for human recognition of familiar objects or for computer recognition.

The sampling theorem may also be extended to random image functions. Let $f(x,y)$ be a stationary stochastic process with power spectrum $S_f(u,v)$. If $S_f(u,v)$ is band limited in the sense that

$$S_f(u,v) = 0, \qquad (u,v) \notin A, \quad A \subset R,$$

then the process may be represented in the mean square sense by its equispaced sampled values provided the sampling is sufficiently dense, i.e.,

$$E\left\{\left[f(x,y) - \sum_{n=-\infty}^{\infty} \sum_{m=-\infty}^{\infty} XYf(nX,mY)K_B(x-nX,y-mY)\right]^2\right\} = 0.$$

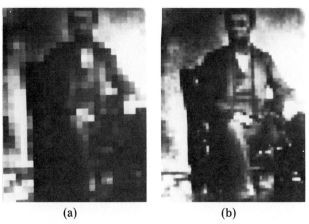

(a)                                               (b)

*Figure 3.11*   (a) $32 \times 32$ array and (b) $64 \times 64$ array of undersampled images of a familiar scene.

### 3.2.2  Quantization

The sampling theorem provides guidelines for the selection of discrete spatial values from the continuous region available. However, we must still consider the discrete representation of the magnitude of the image function so that it may be stored in a computer. Image values are almost always converted to integer or fixed point representations although some transform quantizations go directly to a floating point representation. We will only consider the quantization of spatial picture values into an integer representation at this point. Transform quantization will be considered later.

Since the quantization process is common to all signal processing by computer, the problem has been studied extensively and a theoretical solution under certain assumptions has been developed (Panter and Dite, 1951; Max, 1960; Panter, 1965; Pratt, 1970). This solution will now be presented followed by some special considerations for picture quantization.

The *quantization* of a continuous function $f$ to a finite integer representation may be considered as a decision process. Given the range of the function which without loss of generality may be written in the form

$$0 \leq f \leq A$$

and a selected number of quantization levels $L$, the problem is to determine the $L+1$ decision values

$$D_0 = 0 < D_1 < D_2 < \cdots < D_L = A$$

and the $L$ representation values, as shown in Fig. 3.12,

$$R_j, \qquad j = 1, 2, \ldots, L$$

which provides the "best" discrete representation of $f$. The optimality criterion will be taken to be the minimum mean square error, which may be expressed as

$$\overline{\varepsilon^2} = \sum_{j=0}^{L} \int_{D_{j-1}}^{D_j} (f - R_j)^2 p(f) \, df,$$

where $p(f)$ is the probability function of the amplitude of $f$ and

$$\int_0^A p(f) \, df = 1.$$

If one assumes that the number of levels is large, then the value of $p(f)$ may

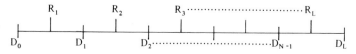

*Figure 3.12*  Decision values for representing regions in the quantization process.

be assumed constant over the interval and the error expression simplifies to

$$\overline{\varepsilon^2} = \sum_{j=1}^{L} p_j \int_{D_{j-1}}^{D_j} (f - R_j)^2 \, df,$$

where $p_j = p(D_j) - p(D_{j-1})$.

At this point, we may determine the $R_j$ by

$$\frac{\partial \varepsilon^2}{\partial R_j} = 2p_j \int_{D_{j-1}}^{D_j} (f - R_j) \, df = 0,$$

which is easily solved, giving

$$R_j = \tfrac{1}{2}(D_{j-1} + D_j).$$

Thus, the reconstruction levels should be placed at the midpoints of the decision intervals. The error expression may now be written

$$\overline{\varepsilon^2} = \sum_{j=1}^{L} p_j \frac{(D_j - D_{j-1})^3}{12}.$$

The constraint may be written

$$\int_0^A p(f) \, df = \sum_{j=1}^{L} p(f_j)(D_j - D_{j-1}) = 1$$

or

$$\int_0^A p(f)^{1/3} \, df = \sum_{j=1}^{L} p(f_j)^{1/3}(D_j - D_{j-1}) = K,$$

where $K$ is a constant. The problem of minimizing the error subject to the constraint may be solved by the method of Lagrange multipliers to obtain the result that the error is minimized when the terms

$$p_j^{1/3}(D_j - D_{j-1})$$

are equal for all $L$ quantization levels. Thus, the larger the probability, the smaller the quantization level. Furthermore, it may be shown that the optimum $D_j$ may be approximated by

$$D_j = A \int_0^{jA/L} p(f)^{-1/3} \, df \Big/ \int_0^A p(f)^{-1/3} \, df.$$

For example, if the distribution of $f$ is uniform in $(0, A)$, then

$$p(f) = 1/A, \quad 0 < f < A, \quad \text{and} \quad D_j = j/L \quad \text{for} \quad j = 0, 1, \dots, L.$$

This example shows that uniform quantization levels are optimum if the input distribution is uniform. Since the uniform quantizer is the simplest to build, a reasonable approach to the optimum quantizer for a nonuniform distribution is to first transform the signal to obtain a uniform distribution, perform uniform quantization, then implement the inverse transform to obtain the optimum quantized signal. The required transform is the cumulative distribution transformation that maps any continuous distribution into a uniform distribution. A major difficulty with this approach is that the signal distribution must be known or computed. Nevertheless, in certain situations, the distribution is known or may be approximated so that the above approach may be implemented. Devices of this type are called *compandors* and are often encountered in pulse code modulation (PCM) communication systems.

A logarithmic transformation is often performed before quantization on many image scanners. The use of a log. transformation may be justified in several ways; however, in view of the previous discussion, one should note that the cumulative distribution function may be approximated by a logarithmic curve. The inverse operation, exponentiation, may be computed, but is usually not offered as an option on the scanner and must be implemented by software.

In terms of signal and noise, it may be shown that for a Gaussian signal embedded in white noise, linear quantization produces a signal-to-noise ratio which varies with the signal amplitude, while logarithmic quantization produces a constant signal-to-noise ratio. Thus, it is preferable to insert a log amplifier before the quantizer rather than perform computer computation of the logarithm.

In the rendition of detail in a picture, the linear quantizer produces the same range for large and small values of the picture function. However, the logarithmic quantizer provides a larger number of quantization levels for small values and coarse spacing of the levels for large values of this picture function. This tends to enhance detail in the low values, but, of course, destroys the detail in the large values.

The number of quantization levels required for adequate display of a picture is an important parameter and may range from 1 to 10 bits/(picture element). In the digitization of pictures, a large number of levels may also be specified to provide immunity to round-off errors in the computations, since each computation usually decreases the number of accurate levels. However, for the display of a picture, it may be argued that it is useless to display a finer quantization than the human can perceive on the display device. The number of gray levels that may be perceived by the human depends on the particular display and viewing conditions.

In the previous chapter, we saw that the human under ideal conditions could distinguish over 500 changes in luminance and several thousand different colors. Thus a conservative estimate of the number of bits required to accurately represent luminance values would be 9 or 10. A much lower estimate is arrived at if one assumes the Weber fraction is approximate constant over the range of interest. For example, if $\Delta I / I \simeq 0.5\%$, then 200 levels or 8 bits would be required. On the other hand, most display devices cannot reproduce this large number of distinguishable levels. A color television can only display a portion of the chromaticity space and yet presents a

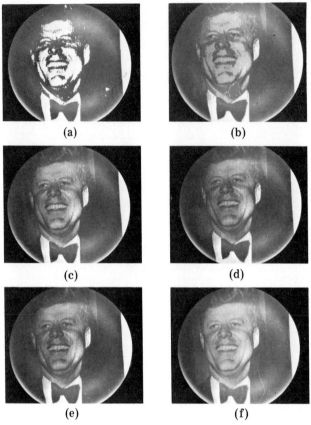

*Figure 3.13*  False contouring produced by coarse quantization: (a) 1, (b) 2, (c) 3, (d) 4, (e) 5, and (f) 6 bits/pixel.

reasonable picture. Also, the available luminance displays and photographic material usually limit the number of distinguishable levels in a displayed picture.

The effect of too few quantization levels is a characteristic pattern called "false contouring," which may be seen in the pictures in Fig. 3.13. These pictures have been sampled at 256 by 256 spatial resolution and are presented for different levels of uniform quantization. Note that when large regions are quantized to the same level, the false contouring is very objectionable. Also, the contouring effect may be produced at the initial picture quantization or by a processing step that reduces the number of levels.

An interesting method for reducing the false contouring phenomenon involves the addition of random noise to the function before quantization. This eliminates much of the contouring produced at long smooth edges.

Pseudorandom noise added before quantization and subtracted before reconstruction is a key step in the Roberts coding method (Roberts, 1962). A variety of other quantization techniques will be discussed in Chapter 5.

### 3.2.3 Interpolation and Display of Images

The illustrations presented thus far have been produced by a computer controlled multi-gray-level display and thus have been made to look very much like the original photographic images. To further emphasize the fact that computers work with numbers and also illustrate that a great deal of computer picture processing may be accomplished without access to a multi-gray-level display, several further illustrations will be presented. These illustrations were developed from low-resolution scans, i.e., $64 \times 64$ arrays. For simple computer economy, it is a wise procedure to develop computer programs on small size arrays. Successful computer programs may always be extended to large sized arrays.

The numbers representing the Lincoln image for 4 bit quantization are shown in Fig. 3.14. Note that at 4 bit quantization the image is recognizable due to the main quantization difference between the 1 and 2 digit decimal numbers. In effect, the image recognized is a 1 bit image. A successive reduction of gray levels produced the acceptable one bit quantization levels.

Another method of representing the numerical image information is by a gray level computer line printer plot. Various shades of gray are produced by

*Figure 3.14* Numerical printout of picture using 16 levels.

*Figure 3.15* Line printer gray scale pictures.

overprinting various characters as illustrated in Fig. 3.15. Also, the geometric distortion caused by the line printer writing 8 characters/in. but only 6 lines/in. was corrected by simply reducing the number of points across the page by a factor of $\frac{6}{8}$.

A final example of representing the computer image information is an isometric plot as shown in Fig. 3.16. Note that the large lesion in the upper right portion of the chest became a very visible valley in the plot.

Although there are several methods for displaying or recording a digital image, a real time cathode ray tube (CRT) may be used as an example to illustrate the interpolation and display processes. The basic problem is to display a series of images which have been sampled in three dimensions—horizontal, vertical, and temporal—so that the sampling structure is not visible, or at least not annoying, to the viewer.

The previous discussion of the sampling theorem provides the ideal interpolation method for band-limited signals. For the signal $f(x,y,t)$ band limited to a cubical region of limits $(W_x, W_y, W_t)$ and sampled at intervals $(X, Y, T)$ with

$$W_x \leq 1/2X, \qquad W_y \leq 1/2Y, \qquad W_t \leq 1/2T,$$

the *cardinal function interpolation formula* is

$$f(x,y,t) = \sum_{i=-\infty}^{\infty} \sum_{j=-\infty}^{\infty} \sum_{k=-\infty}^{\infty} f(iX, jY, kT)$$

$$\times \operatorname{sinc}\left[\left(2\pi W_x(x - iX)\right)\right] \operatorname{sinc}\left[2\pi W_y(y - jY)\right]$$

$$\times \operatorname{sinc}\left[2\pi W_t(t - kT)\right],$$

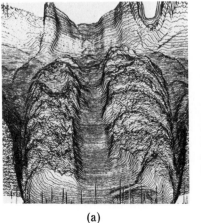
(a)

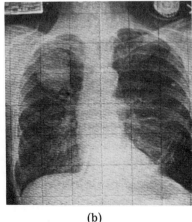
(b)

*Figure 3.16* (a) Isometric plot of chest x ray and (b) chest x ray.

where

$$\text{sinc}(x) = (\sin \pi x)/\pi x.$$

It is not possible to implement this ideal interpolation formula on a CRT display because of the negative values of luminance required. A realizable solution must therefore provide an interpolation method or the equivalent low-pass filter with a point-spread function that is nonnegative.

It is interesting to consider the solutions for the three dimensions separately and note the methods used in practical systems. A simple solution to horizontal interpolation may be found in most TV displays. The picture elements across a line are converted to the video signal and low-pass filtered with an electrical filter. In most cases, this smoothing removes the discrete structure in the horizontal dimension. Vertical interpolation is not so simple, and most TV systems have visible line structure. The low-pass filtering effect of viewing at a distance is fairly effective since the spatial frequency response of the HVS falls off sharply at high frequencies. Other methods that broaden the lines or introduce spot wobble have been implemented and are described by Pearson (1975).

Interpolation for temporal sampling is first accomplished by the exponential phosphor decay. However, the temporal cutoff characteristics of the HVS approximates that of a six to nine stage RC filter and the phosphor merely adds an additional stage, which has little effect in reducing flicker (Pearson, 1975; Sperling, 1971).

### 3.2.4  Matrix Representation

The sampled, quantized image may be naturally described as a matrix, i.e., an array of numbers, $[f]$ where

$$[f] = \{ f_{ij} : i = 1, \ldots, M; j = 1, \ldots, N \}$$

and $f_{ij}$ represents the value of the picture function in the $i$th row and $j$th column. We may sometimes assume that the matrix is square, $M = N$, and later extend the results to rectangular matrices.

The advantage of representing a picture in matrix form is that the full power of previously developed matrix theory may be applied to the analysis of picture processing operations. We shall shortly see a direct application of matrix analysis in the computation of the discrete Fourier transform of a picture. Other applications such as the computation of the singular values and rank of a picture matrix have also been developed.

One may argue that a picture is not merely an array of numbers. This argument is usually supported by citing certain properties of real pictures. Clearly, not every matrix will generate a familiar image, although every matrix does indeed represent a computer picture function. Certain constraints imposed by the physical image may also be easily imposed on the matrix. One important constraint is nonnegativeness. This constraint may be stated as

$$f_{ij} \geq 0, \qquad i = 1,\ldots,M; \quad j = 1,\ldots,N.$$

Another important constraint is boundedness, which may be expressed as

$$f_{ij} < B,$$

where $B$ is a constant. Still another is finite energy, i.e.,

$$E = \sum_{i=1}^{M} \sum_{j=1}^{N} f_{ij}^2 < E_0,$$

where $E_0$ is a constant.

One may also impose a smoothness constraint: For example, the picture function at any point does not differ from the average of its neighbors by a given amount, which may be expressed as

$$S > f_{ij} - \tfrac{1}{8}(f_{i-1,j-1} + f_{i,j-1} + f_{i+1,j-1}$$
$$+ f_{i-1,j} \qquad\qquad + f_{i+1,j}$$
$$+ f_{i-1,j+1} + f_{i,j+1} \quad + f_{i+1,j+1}),$$

where $S$ is again a constant.

In general, any constraint that may be imposed on the physical image may also be imposed on the computer matrix representation. However, note that the last two constraints, finite energy and smoothness, are difficult to express in matrix notation. These could be easily expressed for a vector, for example, the finite energy expression would simply be the vector inner product. Thus, we are motivated to develop an equivalent vector representation of the computer picture.

### 3.2.5   Vector Representation

The computer picture may be written as a vector using a column or row ordering. Again, the advantage is that vector analysis may be directly applied. The column ordering method consists of forming a column vector $\mathbf{f}$ of size $1 \times MN$:

$$\mathbf{f} = [\mathbf{f}^1, \mathbf{f}^2, \ldots, \mathbf{f}^M]',$$

where

$$\mathbf{f}^i = [f(i,1), f(i,2), \ldots, f(i,N)]'.$$

The matrix and vector representations of the computer picture are related by the expression developed by Pratt (1975),

$$\mathbf{f} = \sum_{n=1}^{N} [\mathbf{N}_n][f]\mathbf{v}_n,$$

where

$$[\mathbf{N}_n] = \begin{bmatrix} \cdots \mathbf{0} \cdots \\ \cdots \mathbf{0} \cdots \\ \cdots \vdots \cdots \\ \vdots \\ \cdots \mathbf{I} \cdots \\ \cdots \vdots \cdots \\ \cdots \mathbf{0} \cdots \end{bmatrix} \begin{matrix} 1 \\ \\ \\ n \\ \\ \\ N \end{matrix}$$

is an $MN \times M$ matrix and for which each submatrix has $M$ rows, and

$$\mathbf{v}_n = \begin{bmatrix} 0 \\ 0 \\ \vdots \\ 1 \\ \vdots \\ 0 \\ 0 \end{bmatrix} \begin{matrix} 1 \\ \\ \\ n \\ \\ \\ N \end{matrix}$$

is an $N \times 1$ vector with one nonzero element. The $\mathbf{v}_n$ multiplication extracts the $n$th column from $[f]$ and the matrix $[\mathbf{N}_n]$ places this column in the $n$th sector of the vector $\mathbf{f}$.

The inverse operation is

$$[f] = \sum_{n=1}^{N} \mathbf{N}_n' \mathbf{f} \mathbf{v}_n' .$$

Thus, given either a matrix or vector picture, we may easily convert to the other.

Let us again consider the finite energy and smoothness constraints. The energy of the vector $\mathbf{f}$ may be expressed as the inner product of the vector with itself:

$$E = \mathbf{f}'\mathbf{f} = \sum_{i=1}^{MN} f_i^2.$$

The energy is also given by the trace or sum of the diagonal terms of the outer product of the vector $\mathbf{f}$ with itself:

$$E = \mathrm{Tr}(\mathbf{f}\mathbf{f}').$$

Using the relationship between the matrix and vector representations, we may also compute the energy as

$$E = \left\{ \sum_{n=1}^{N} [\mathbf{N}_n][f]\mathbf{v}_n \right\}' \left\{ \sum_{n=1}^{N} [\mathbf{N}_n][f]\mathbf{v}_n \right\}.$$

The smoothness constraint is most easily expressed in the vector representation. Let us first consider developing a vector whose terms are

$$g_i = f_i - \tfrac{1}{8} f_{i-1}, \qquad i = 1, \ldots, N,$$

$$g_n = f_N.$$

Clearly,

$$\mathbf{g} = C\mathbf{f},$$

where $[C]$ is a bidiagonal matrix of the form

$$[C] = \begin{bmatrix} 1 & 0 & 0 & \cdots & 0 \\ -\tfrac{1}{8} & 1 & 0 & \cdots & 0 \\ 0 & -\tfrac{1}{8} & 1 & \cdots & 0 \\ \vdots & & & & \\ 0 & 0 & 0 \cdots & -\tfrac{1}{8} & 1 \end{bmatrix},$$

where the endpoint conditions are neglected. We may now extend this to the smoothness constraint given by first using the column ordering to express $\mathbf{f}$ in terms of $\{f_{ij}\}$ as

$$f_k = f_{i + M(j-1)},$$

where

$$k = 1, \ldots, MN; \qquad i = 1, \ldots, M; \qquad j = 1, \ldots, N.$$

The smoothness relation may be expressed as

$$g_k = f_k - \tfrac{1}{8}(f_{k-M-1} + f_{k-M} + f_{k-M+1}$$
$$+ \; f_{k-1} \qquad\qquad + f_{k+1}$$
$$+ \; f_{k+M-1} + f_{k+M} + f_{k+M+1}).$$

The last eight points for $g_k$ will be smoothed as much as possible with a reduced number of neighbors. Thus, the smoothing relationship may be written as

$$\mathbf{g} = [C]\mathbf{f},$$

where $[C]$ is now a banded matrix of the form given below and the edge effects at the start are ignored.

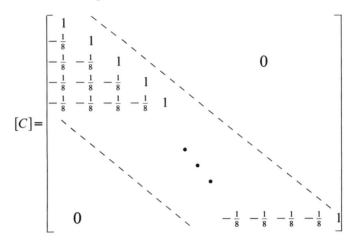

The equivalent matrix representation of $[g]$ is

$$[g] = \sum_{n=1}^{N} [N_n][C]\mathbf{fv}_n.$$

### 3.2.6  Statistical Representation

For many classes of problem, it is necessary or convenient to consider the computer image as a sample of a stochastic process, i.e., a random event. For example, film grain noise or the image on a TV screen at 4 am cannot be specified in terms of point properties as with deterministic images. However, these processes may be studied in terms of average properties. We may also find it convenient to model a set of deterministic images such as texture patterns by a stochastic process, i.e., a family of functions defined on a probability space. Rather than attempting a general development for two-dimensional random processes, we will only introduce certain definitions and notations for the sampled random image. Furthermore, since the sampled image may always be considered as a vector, we may consider a vector random variable.

Let us first consider the sampled image before quantization. The sampled image vector $\mathbf{f}$ has continuous components $f_{ij}$, where

$$\mathbf{f}' = [f_{11}, f_{12}, \dots, f_{MN}].$$

Since each component may take on a continuous set of values, the joint density function

$$p(f_{11}, f_{12}, \dots, f_{MN})$$

must be specified to describe the process. The probability that a sample picture **f** will have values which lie in the volume

$$df_{11}\, df_{12} \cdots df_{MN}$$

centered at

$$\mathbf{f'} = \left[\, f_{11}, f_{12}, \ldots, f_{MN}\,\right]$$

is

$$\text{Prob}(\mathbf{f'} < \mathbf{f} < \mathbf{f'} + d\mathbf{f}) = p(\mathbf{f})\, df_{11}\, df_{12} \cdots df_{MN},$$

where

$$\int\int\int \cdots \int_{-\infty}^{\infty} p(\mathbf{f})\, df_{11}\, df_{12} \cdots df_{MN} = 1.$$

The mean or expected value of **f** is

$$\bar{\mathbf{f}} = E\{\mathbf{f}\} = \int\int\int \cdots \int_{-\infty}^{\infty} \mathbf{f} p(\mathbf{f})\, df_{11}\, df_{12} \cdots df_{MN}.$$

We may also define a covariance matrix as

$$\mathbf{E}(\mathbf{f}) = E\left\{ (\mathbf{f} - \bar{\mathbf{f}})(\mathbf{f} - \bar{\mathbf{f}})' \right\}.$$

It is easily shown that there are only $\frac{1}{2}MN(MN+1)$ different terms in this matrix. The $MN$ diagonal elements represent variance terms $E\{f_{11}^2\}$, $E\{f_{12}^2\}, \ldots, E\{f_{MN}^2\}$. The remaining $\frac{1}{2}MN(MN-1)$ distinct terms represent covariances, such as $E\{f_{11}f_{12}\}$. If the mean value of **f** is zero, then the covariance matrix is identical to the correlation matrix which is defined as $\mathbf{R}(\mathbf{f}) = E\{\mathbf{f}\mathbf{f}^{\mathsf{T}}\}$. When the off-diagonal terms in the correlation matrix are zero, we say that the random variables in the vector **f** are uncorrelated. If the joint density function is separable, i.e.,

$$p(\mathbf{f}) = p(f_{11}) p(f_{12}) \cdots p(f_{MN}),$$

then the random variables in the vector **f** are said to be independent.

Let us now consider two random pictures **f** and **g**. The second picture **g** could be completely different from, or a subpicture of, **f**. The conditional probability density function of **f** given that **g** has taken on a given value is

$$p(\mathbf{f}/\mathbf{g}) = p(\mathbf{f}, \mathbf{g})/p(\mathbf{g}) \qquad \text{if} \quad p(\mathbf{g}) \neq 0.$$

In this development, the random event may be interpreted as randomly opening a very large book of photographs. The probability that a given point on this page has a certain brightness or that one area of the page has a given brightness given that another area has a fixed brightness could be computed from the conditional probabilities.

If we wished to extend this method to also consider a time variable, we would define a random sequence of picture vectors. A random sequence, $\mathbf{f}(K)$, is a collection of random vector variables indexed by some parameter

$$\mathbf{f}(K) = \{ \ldots, \mathbf{f}(-2), \mathbf{f}(-1), \mathbf{f}(0), \mathbf{f}(1), \mathbf{f}(2), \ldots \}.$$

This random event could be interpreted as reaching into a very large file and extracting a movie film. To characterize a random sequence completely, we must determine the joint probability function for the random variables in the sequence. Although this is generally very impractical, we may be able to estimate the first two moments, the ensemble mean vector

$$\mu(K) = E\{\mathbf{f}(K)\}$$

and the ensemble covariance matrix

$$\mathbf{\Sigma}(K) = E\{[\mathbf{f}(K) - \mu(K)][f(K) - \mu(K)]'\}.$$

It is also desirable to define the covariance matrix

$$\mathbf{\Sigma}(j,K) = E\{[\mathbf{f}(j) - \overline{\mu(j)}][f(K) - \overline{\mu(K)}]'\}$$

and the joint probability density function

$$p[\mathbf{f}(k), \mathbf{f}(j)]$$

in order to introduce the following definitions.

A random sequence is called *strict sense stationary* if the joint density function is invariant under a shift of position, i.e.,

$$p[\mathbf{f}(k), \mathbf{f}(k+1), \ldots] = p[\mathbf{f}(k+j), \mathbf{f}(k+j+1), \ldots].$$

A random process is called *wide sense stationary* if the covariance matrix is independent of a shift of stage, i.e.,

$$\mathbf{\Sigma}(j,k) = \mathbf{\Sigma}(j+i, k+i)$$

for all $i$, $j$, and $k$.

For naturally occurring pictures, the strict sense stationary process is a good model for noise-type images such as the TV image at 3 am. The wide sense stationary process is less restrictive and applies to a broader class of images. However, in most common pictures of people or manmade objects, the image function may only be reasonably assumed to have stationary statistics in local regions. Locally stationary random processes have been defined and several properties of these processes developed; however, much remains to be done in the application of these methods to image processing.

Another fundamental assumption which greatly simplifies the covariance matrix determination is the Markov assumption. A random vector process $f(t)$ is called a simple Markov process if

$$p[f(t_{k+1})/f(t_k), f(t_{k-1}), \ldots, f(t_{k-n})] = p[f(t_{k+1})/f(t_k)].$$

As a consequence of the Markov assumption, the random sequence may be completely characterized by specifying an initial density function $p[f(k_0)]$ and the conditional density function of the transition $p[f(k+1)/f(k)]$.

***Example: Markov Process.*** A large class of image processes may be modeled by a two-dimensional, stationary, first-order Markov process. If $f_{ij}$ represents the picture brightness at the point $(i,j)$, then for this process the autocorrelation function may be written

$$R(m,n) = E\{f_{ij}f_{i+m,j+n}\} = \rho^{|m-n|},$$

where $0 < \rho < 1$, and it is normally assumed that $E\{f_{ij}\} = 0$.

This assumption of zero mean is nonessential, since the mean can always be easily computed and subtracted if necessary to obtain a zero mean image.

The computation of the image statistics such as the ensemble mean and covariance is greatly simplified if the process is ergodic. A random process is called ergodic if its time (spatial) statistics are identical to its ensemble statistics. An ergodic process is stationary but a stationary process is not always ergodic. The ergodicity assumption is often made in practice. For example, whenever a histogram of the gray level values in a picture is computed and used as an estimate of the first-order probability function of the picture, this assumption has been made.

Let us now consider some specific, simple models for random picture processes. Assume that the picture brightness at any point $f$ may be represented by a stationary, ergodic process. If the picture function has been sampled but not quantized or if the quantization is sufficiently fine, then the random variable is continuous. If we further assume that the process is sufficiently characterized by its point properties, then a final assumption of a probability function will permit calculation of the picture statistics. For example, the process may have a normal of Gaussian distribution with mean $\mu$ and variance $\sigma^2$, i.e.,

$$p(f) = \frac{1}{\sqrt{2\pi}\sigma} \exp\left[(f-\mu)^2/2\sigma^2\right], \qquad -\infty < f < \infty.$$

Since this variable takes on both positive and negative values, it is inconsistent with a positivity assumption. A process which is consistent with a positivity assumption is the lognormal process. A random variable $f$ is said to have a *lognormal* distribution if $\log f$ is normally distributed. The density function for this process is

$$p(f) = (f\sqrt{2\pi}\sigma)^{-1} \exp\left[-(\ln f - \mu)^2/2\sigma^2\right]$$

with $0 < f < \infty$ and the mean and variance given by

$$E(f) = \exp\left(\mu + \tfrac{1}{2}\sigma^2\right), \qquad \mathrm{Var}(f) = (\exp\sigma^2 - 1)\exp(2\mu + \sigma^2).$$

The simplest density function which may be considered is the uniform density function with

$$p(f) = 1/(b-a), \qquad a < f < b.$$

If we wish to consider the picture function after the quantization process, a discrete probability function is appropriate. A binomial distribution is described by

$$p(f=i) = \binom{n}{i}\alpha^i(1-\alpha)^{n-i}, \qquad 0 < \alpha < 1$$

and

$$f = \sum_{i=1}^{n} X_i,$$

where $X_i = 0$ or $1$ for $i = 1, 2, \ldots, n$. If one is considering light photons, then a Poisson distribution with

$$p(f = k) = \frac{\varepsilon^{-(\lambda t)}(\lambda t)^k}{k!}, \qquad k = 0, 1, \ldots,$$

may be appropriate. Finally, the equally likely distribution with

$$p(f = i) = 1/N, \qquad i = 1, 2, \ldots, N$$

may be useful.

Let us now consider the statistical distributions of two picture points $f$ and $g$, which could be adjacent points, separated by a given distance, etc. A bivariate or second-order distribution is now appropriate. For the continuous case, the previous distribution may be extended for the normal distribution given by

$$p(f,g) = \left[ 2\pi\sqrt{(1-\rho^2)^{1/2}} \right]^{-1} \exp\left[ (f^2 - 2\rho fg + g^2)/2(1-\rho^2) \right],$$

$$-\infty < f, g < \infty,$$

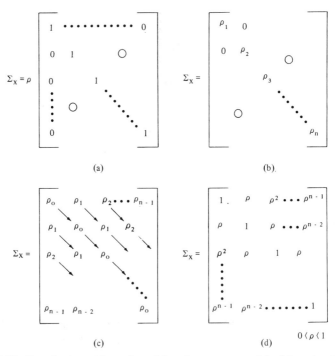

Figure 3.17  Covariance matrices of special random processes: (a) white noise, (b) uncorrelated process, (c) wide sense stationary process, and (d) wide sense stationary first-order Markov process.

while, for the discrete case, the Poisson probability function is

$$p(f=i, g=j) = \exp(-\lambda_1 - \lambda_2 + \lambda_3) \frac{a^f b^g}{f! g!} \sum_{r=0}^{S} \frac{f^r g^r \lambda_3^r}{a^r b^r r!},$$

where

$$f, g = 0, 1, \ldots, S = \min(f, g), \qquad a = \lambda_1 - \lambda_3, \qquad b = \lambda_2 - \lambda_3.$$

It is often useful to assume a certain correlation covariance or power spectral density function or to compute transitional probability matrices from the computer image.

The covariance matrices of several special processes are shown in Fig. 3.17. In Fig. 3.17a it may be seen that the covariance matrix of a white noise process is an identity matrix. An uncorrelated random process has a diagonal covariance matrix as shown in Fig. 3.17b. The covariance matrix of a wide sense stationary random process has a Toeplitz form in which all diagonal terms are equal as shown in Fig. 3.17c. Finally, the covariance matrix of a wide sense stationary, first-order Markov process has a single parameter Toeplitz form. These special cases often permit analytical solutions for image processing techniques.

## 3.3 Two-Dimensional Transforms—Image
## Representation Using Orthogonal Functions and Matrices

In signal analysis, it is often advantageous to represent a real- or complex-valued continuous signal $x(t)$ defined on the interval $(t_0, t_0 + T)$ by an expansion of the form

$$x(t) = \sum_{m=0}^{\infty} a_m \phi_m(t),$$

where $x(t)$ is assumed to be square integrable,

$$\int_T x(t) x^*(t) \, dt < \infty,$$

and the functions $\phi_m(t)$ are orthogonal,

$$\int_T \phi_m(t) \phi_n^*(t) \, dt = \begin{cases} C, & m = n \\ 0, & m \neq n. \end{cases}$$

If $C = 1$, the functions are called *orthonormal*.

The coefficients $a_m$ may be determined by multiplying both sides of the series representation by $\phi_n^*(t)$ and integrating over the interval $(t_0, t_0 + T)$ to obtain

$$a_n = \frac{1}{C} \int_T x(t) \phi_n^*(t) \, dt, \qquad n = 0, 1, 2, \ldots.$$

The set of orthogonal functions $\{\phi_n(t)\}$ with

$$\int_T \|\phi_n(t)\|^2 \, dt < \infty$$

is said to be *complete* or *closed* if there exists no square integrable function $x(t)$ for which

$$\int_T x(t)\phi_n{}^*(t) \, dt = 0, \qquad n = 0, 1, 2, \ldots,$$

or, more important, if the limit of the mean square error between the signal and a finite expansion is zero, i.e.,

$$\lim_{N \to \infty} \varepsilon_N{}^2 = 0,$$

where

$$\varepsilon_N{}^2 = \int_T \|x(t) - \sum_{n=0}^{N} a_n \phi_n(t)\|^2 \, dt.$$

When the set $\{\phi_n(t)\}$ is complete, the orthogonal functions are called an *orthogonal basis* and the signal may be represented accurately by the finite expansion

$$x(t) = \sum_{n=0}^{N} a_n \phi_n(t).$$

In the previous discussion of sampling, it was shown that a continuous picture function $f(x,y)$ could be represented by the set of orthogonal complex sinusoids, if the picture was band limited and the sampling sufficiently dense. Let us now consider another use of orthogonality theory by examining representations of the sampled image $f(m,n)$.

### Orthogonal Matrices

A sampled image $f(m,n)$ is a finite, two-dimensional array naturally represented by a matrix of sampled values

$$[f] = \{f(m,n)\} = \{f_{mn}\}.$$

The matrix $[f]$ may always be expressed in the form

$$[f] = \sum_{m=1}^{M} \sum_{n=1}^{N} f_{mn}[e_{mn}]$$

where the rank one matrices $[e_{mn}]$ have a unit term at element $(m,n)$ and zero elements otherwise, e.g.,

$$[e_{11}] = \begin{bmatrix} 1 & 0 & 0 & 0 & \cdots & 0 \\ 0 & & & & & \\ 0 & & & & & \\ \vdots & & & & & \\ 0 & & & & & 0 \end{bmatrix}.$$

$$
\begin{bmatrix} f_{11} & f_{12} & f_{13} \\ f_{21} & f_{22} & f_{23} \\ f_{31} & f_{32} & f_{33} \end{bmatrix} = f_{11} \begin{bmatrix} 1 & 0 & 0 \\ 0 & 0 & 0 \\ 0 & 0 & 0 \end{bmatrix} + f_{12} \begin{bmatrix} 0 & 1 & 0 \\ 0 & 0 & 0 \\ 0 & 0 & 0 \end{bmatrix} + \cdots + f_{33} \begin{bmatrix} 0 & 0 & 0 \\ 0 & 0 & 0 \\ 0 & 0 & 1 \end{bmatrix}
$$

*Figure 3.18* Identity expansion of picture matrix.

This *identity expansion* is illustrated in Fig. 3.18. To consider the set of $MN$ matrices $[e_{mn}]$ as an orthogonal basis set, an inner product operation between matrices could be defined. Rather than introduce a new notation for the inner product of matrices, we may without loss of generality simply use the equivalent vector notation for the image matrix. Each matrix $[e_{mn}]$ may be put into a one-to-one correspondence with a basis vector $\mathbf{e}_k$, where

$$
k = n + (m-1)N.
$$

By column stacking, as shown in Fig. 3.19, each picture matrix $[f]$ may be put into correspondence with an equivalent image vector $\mathbf{f}$. Then, clearly, the image vector may be represented by

$$
\mathbf{f} = \sum_{k=1}^{MN} f_k \mathbf{e}_k.
$$

$$
[f_{mn}] = \begin{bmatrix} f_{11} & f_{12} & f_{13} & f_{14} \\ f_{21} & f_{22} & f_{23} & f_{24} \\ f_{31} & f_{32} & f_{33} & f_{34} \\ f_{41} & f_{42} & f_{43} & f_{44} \end{bmatrix} \Rightarrow \begin{bmatrix} f_{11} \\ f_{12} \\ f_{13} \\ f_{14} \\ f_{21} \\ f_{22} \\ f_{23} \\ f_{24} \\ f_{31} \\ f_{32} \\ f_{33} \\ f_{34} \\ f_{41} \\ f_{42} \\ f_{43} \\ f_{44} \end{bmatrix} = \begin{bmatrix} f_1 \\ f_2 \\ f_3 \\ f_4 \\ f_5 \\ f_6 \\ f_7 \\ f_8 \\ f_9 \\ f_{10} \\ f_{11} \\ f_{12} \\ f_{13} \\ f_{14} \\ f_{15} \\ f_{16} \end{bmatrix} = \mathbf{f}
$$

*Figure 3.19* Column stacking of an image matrix and an equivalent vector.

The usual vector operations may now be used. For example, the orthonormality condition of the basis vectors is

$$\mathbf{e}_j'\mathbf{e}_k = \delta_{jk} = \begin{cases} 1, & k=j, \\ 0, & k \neq j. \end{cases}$$

The previous expression may also be written

$$\mathbf{f} = [\mathbf{e}_1\mathbf{e}_2 \cdots \mathbf{e}_{MN}]'\mathbf{f} = [e]'\mathbf{f},$$

where the matrix $[e]$ is formed by concatenating the columns of the basis vector of size $MN$ by $MN$. In this case, $[e]$ must be the identity matrix. However, it may often be advantageous to consider other *orthonormal matrices* $[L]$, where

$$[L][L^*]' = \mathbf{I}_{MN \times MN}.$$

Such matrices would permit an orthonormal transform representation of the form

$$\mathbf{F} = [L]\mathbf{f}$$

and the inverse transformation

$$\mathbf{f} = [L^*]'\mathbf{F}.$$

The series form of these transformations is

$$F_i = \sum_{j=1}^{(MN)^2} L_{ij}f_j \quad \text{and} \quad f_i = \sum_{j=1}^{(MN)^2} L_{ji}^* F_j.$$

Since the size of the matrix $[L]$ is $MN$ by $MN$, which is prohibitively large in most imaging applications, it is important to consider matrices that possess special structures that simplify the representation. The class of Kronecker product matrices fits this requirement and leads to the special class of *separable image transforms*.

### Separable Image Transforms

The Kronecker product of a square $M$ by $M$ matrix **A** with a square $N$ by $N$ matrix **B** is an $MN$ by $MN$ matrix **C** of the form

$$\mathbf{C} = \mathbf{A} \times \mathbf{B} = \begin{bmatrix} a_{11}\mathbf{B} & \cdots & a_{1M}\mathbf{B} \\ \vdots & & \\ a_{M1}\mathbf{B} & \cdots & a_{MM}\mathbf{B} \end{bmatrix}.$$

The important simplification introduced by the Kronecker product matrix is that the general vector transformation

$$\mathbf{F} = \mathbf{Cf}$$

may be expressed in the equivalent matrix form

$$[F] = \mathbf{A}[f]\mathbf{B}.$$

This operation is called a separable transform since the matrix **A** operates on the columns and the matrix **B** operates on the rows of the picture matrix $[f]$.

If the **A** and **B** matrices are orthonormal,

$$\mathbf{AA^{*\prime}} = \mathbf{I} \quad \text{and} \quad \mathbf{BB^{*\prime}} = \mathbf{I},$$

then an inverse transformation may be expressed as

$$[f] = \mathbf{A^{*\prime}}[F]\mathbf{B^{*\prime}}.$$

Another interesting representation may be developed by expressing the transformation in the slightly different form

$$[F] = \mathbf{A}[f]\mathbf{B^{\prime}}.$$

If the **A** and **B** matrices are now written in terms of their column basis vectors

$$\mathbf{A} = \begin{bmatrix} \mathbf{a}_1 & \mathbf{a}_2 & \cdots & \mathbf{a}_M \end{bmatrix} \quad \text{and} \quad \mathbf{B} = \begin{bmatrix} \mathbf{b}_1 & \mathbf{b}_2 & \cdots & \mathbf{b}_N \end{bmatrix},$$

then the transformation may be written in outer product form (Andrews, 1977)

$$[F] = \sum_{i=1}^{M} \sum_{j=1}^{N} f_{ij} \mathbf{a}_i \mathbf{b}_j^{\prime}.$$

Note that each outer product $\mathbf{a}_i \mathbf{b}_j^{\prime}$ forms an $M$ by $N$ rank one matrix. The total set of the $MN$ outer product matrices may be considered as a set of basis matrices, which may be used to represent the image matrix.

In the following sections, several important image transforms will be considered.

### 3.3.1 Hotelling Transform

The Hotelling transform has found wide application in both image representation and compression. This transform is based upon the statistical properties of an image and is also known as the *principal component, discrete Karhunen–Loeve,* or *eigenvector* transformation.

Several formulations of the Hotelling transformation are used in image processing. In each formulation, a random vector **X** is selected. The random vector could correspond to the equivalent image vector, to a block vector of a subset of the image, or to the multispectral coordinates of an image point.

The random vector **X** is characterized by its average properties. The mean vector $\boldsymbol{\mu}$ is given by

$$\boldsymbol{\mu} = E\{\mathbf{X}\} = \int \mathbf{x} p(\mathbf{x}) \, d\mathbf{x},$$

where the integral is calculated over the entire **x** space, $E$ denotes expected value of $p(\mathbf{x})$ is the probability density function at **X**.

The covariance matrix is defined by

$$\Sigma = E\{(\mathbf{X}-\boldsymbol{\mu})(\mathbf{X}-\boldsymbol{\mu})'\} \qquad \text{or} \qquad \Sigma = \begin{bmatrix} \sigma_{11}^2 & \sigma_{12} & \cdots & \sigma_{1n} \\ \sigma_{n1} & \sigma_{n2} & \cdots & \sigma_{nn}^2 \end{bmatrix}$$

in which the matrix components $\sigma_{ij}$ are

$$\sigma_{ij} = E\{(X_i - m_i)(X_j - m_j)\} \qquad (i,j = 1,2,\ldots,n).$$

The diagonal terms $\sigma_{ii}^2$ of the covariance matrix are the *variances* of the random components of the random vector and are always nonnegative. The off-diagonal terms $\sigma_{ij}$ are the *covariances* between components $X_i$ and $X_j$. Since $\sigma_{ij} = \sigma_{ji}$, the covariance matrix is symmetric.

It is sometimes convenient to express the covariance matrix as

$$\Sigma = E\{\mathbf{XX}'\} - \boldsymbol{\mu\mu}'.$$

The *scatter matrix* or autocorrelation matrix

$$\mathbf{S} = E\{\mathbf{XX}'\}$$

gives essentially the same information about the dispersion of the distribution but is not normalized.

To fully normalize the covariance matrix, one can convert the terms to correlation coefficients by

$$\rho_{ij} = \sigma_{ij}^2 / \sigma_{ii}\sigma_{jj}.$$

Each term $\rho_{ij}$ is then bounded by

$$-1 \le \rho_{ij} \le 1,$$

and a *correlation matrix* $\mathbf{R}$ may be formed that directly indicates the correlation between each pair of components

$$\mathbf{R} = \begin{bmatrix} 1 & \rho_{12} & \cdots & \rho_{1n} \\ \rho_{21} & 1 & & \\ \vdots & & \ddots & \\ \rho_{n1} & & & 1 \end{bmatrix}.$$

The absolute scale of variation is retained in a diagonal *standard deviation* matrix

$$\mathbf{D} = \begin{bmatrix} \sigma_{11} & 0 & \cdots & 0 \\ 0 & \sigma_{22} & & \\ \vdots & & \ddots & \vdots \\ 0 & & & \sigma_{nn} \end{bmatrix}.$$

It is now possible to express the covariance matrix in terms of the correlation and standard deviation matrices

$$\Sigma = \mathbf{DRD}.$$

It is easy to decompose the covariance matrix into the correlation and standard deviation matrices provided the individual deviations are nonzero. If one or more deviations are zero, the covariance matrix will be singular. This usually indicates a nonrandom component and requires a reduction of the dimensionality to eliminate the deterministic terms.

Two random vectors $\mathbf{X}_i$ and $\mathbf{X}_j$ are

$$\begin{aligned} \textit{uncorrelated} \text{ if} \quad & E\{\mathbf{X}_i'\mathbf{X}_j\} = E\{\mathbf{X}_i'\}E\{\mathbf{X}_j\}, \\ \textit{orthogonal} \text{ if} \quad & E\{\mathbf{X}_i'\mathbf{X}_j\} = 0, \\ \textit{independent} \text{ if} \quad & p(\mathbf{X}_i,\mathbf{X}_j) = p(\mathbf{X}_i)p(\mathbf{X}_j). \end{aligned}$$

If $\mathbf{X}_i$ and $\mathbf{X}_j$ are independent, they are uncorrelated; however, the converse is not true. If the mean vector of either $\mathbf{X}_i$ or $\mathbf{X}_j$ is equal to the zero vector, then uncorrelated and orthogonal are equivalent. If the components of a random vector $\mathbf{X}$ are mutually uncorrelated, the covariance matrix $\mathbf{\Sigma}$ is diagonal and the correlation matrix $\mathbf{R}$ is an identity matrix.

The random vector $\mathbf{X}$ can be represented without error by a certain deterministic transformation of the form

$$\mathbf{X} = \mathbf{A}\mathbf{Y} = \sum_{i=1}^{n} y_i \mathbf{A}_i,$$

where

$$\mathbf{A} = \begin{bmatrix} \mathbf{A}_1 & \mathbf{A}_2 & \cdots & \mathbf{A}_n \end{bmatrix} \quad \text{and} \quad |\mathbf{A}| \neq 0,$$

where $\mathbf{Y}$ is also an $n$-dimensional random vector and $\mathbf{A}$ is a nonsingular matrix. The matrix $\mathbf{A}$ may be considered to be made up of $n$ linearly independent column vectors called *basis vectors* which span the $n$-dimensional space containing $\mathbf{X}$. For simplicity, we may assume that $\mathbf{A}$ is orthonormal or that the columns form an orthonormal set

$$\mathbf{A}_i'\mathbf{A}_j = \delta_{ij} = \begin{cases} 1, & i=j, \\ 0, & i \neq j, \end{cases} \quad \text{or} \quad \mathbf{A}'\mathbf{A} = \mathbf{I}, \quad \mathbf{A}^{-1} = \mathbf{A}'.$$

Under these conditions, $\mathbf{Y}$ may be expressed as

$$\mathbf{Y} = \mathbf{A}'\mathbf{X} = \sum_{i=1}^{n} \mathbf{A}_i' x_i.$$

Each component of $\mathbf{Y}$ contributes to the representation of $\mathbf{X}$. Suppose we determine only $m < n$ components of $\mathbf{Y}$ and still want to represent, or estimate, $\mathbf{X}$. One method of accomplishing this is to replace the omitted components of $\mathbf{Y}$ with preselected constants and form the following estimate:

$$\hat{\mathbf{X}}(m) = \sum_{i=1}^{m} y_i \mathbf{A}_i + \sum_{i=m+1}^{n} b_i \mathbf{A}_i.$$

Since not all of the components are used, an error is introduced by the representation

$$\varepsilon = \mathbf{X} - \hat{\mathbf{X}}(m) = \sum_{i=m+1}^{n} (y_i - b_i)\mathbf{A}_i.$$

Since the components $y_i$ are random, the error vector $\varepsilon$ is also random. The mean squared value of $\varepsilon$ may be chosen to measure the effectiveness of the representation

$$\overline{\varepsilon^2}(m) = E\{\varepsilon'\varepsilon\} = E\left\{\sum_{i=1}^{n} \varepsilon_i^2\right\}$$

$$\overline{\varepsilon^2}(m) = E\left\{\sum_{i=m+1}^{n}\sum_{j=m+1}^{n}(y_i - b_i)(y_j - b_j)\mathbf{A}_i'\mathbf{A}_j\right\}$$

$$\overline{\varepsilon^2}(m) = \sum_{i=m+1}^{n} E\{(y_i - b_i)^2\}.$$

The constant $b_i$ may be selected to minimize the mean squared error

$$\frac{\partial}{\partial b_i} \overline{\varepsilon^2}(m) = -2E\{(y_i - b_i)\} = 0.$$

The optimum choice for the constants is

$$b_i = E\{y_i\} = \bar{y}_i,$$

which indicates that the omitted components should be replaced with their mean values. The optimum constants depend on the transformation as shown by

$$b_i = \bar{y}_i = \mathbf{A}_i' E\{\mathbf{X}\}.$$

The mean squared error $\overline{\varepsilon^2}(m)$ may now be expressed

$$\overline{\varepsilon^2}(m) = \sum_{i=m+1}^{n} E\{(y_i - \bar{y}_i)^2\} = \sum_{i=m+1}^{n} E\{(y_i - \bar{y}_i)(y_i - \bar{y}_i)'\}$$

$$= \sum_{i=m+1}^{n} E\{\mathbf{A}_i'(\mathbf{X} - \overline{\mathbf{X}})(\mathbf{X} - \overline{\mathbf{X}})'\mathbf{A}_i\} = \sum_{i=m+1}^{n} \mathbf{A}_i'\boldsymbol{\Sigma}_{\mathbf{X}}\mathbf{A}_i,$$

where $\boldsymbol{\Sigma}_{\mathbf{X}}$ is the covariance matrix of $\mathbf{X}$.

If the basis vectors $\mathbf{A}_i$ are selected as the eigenvectors of $\boldsymbol{\Sigma}_{\mathbf{X}}$, i.e.,

$$\boldsymbol{\Sigma}_{\mathbf{X}}\mathbf{A}_i = \lambda_i \mathbf{A}_i,$$

where $\mathbf{A}_i$ is an eigenvector and $\lambda_i$ the corresponding eigenvalue of $\boldsymbol{\Sigma}_{\mathbf{X}}$. Then, since $\mathbf{A}_i'\mathbf{A}_i = 1$,

$$\lambda_i = \mathbf{A}_i'\boldsymbol{\Sigma}_{\mathbf{X}}\mathbf{A}_i.$$

The mean squared error can be written as

$$\overline{\varepsilon^2}(m) = \sum_{i=m+1}^{n} \lambda_i.$$

The mean squared error can be minimized by ordering the eigenvalues so that

$$\lambda_1 > \lambda_2 > \cdots > \lambda_n > 0.$$

The eigenvector columns of the transformation matrix **A** are also arranged in a similar order. Now if a component $y_i$ is deleted, the mean squared error increases by $\lambda_i$. By deleting the components corresponding to the smallest eigenvalues, the mean squared error is minimized.

In summary, we may determine an orthonormal transformation **A** to represent the random vector **X**. A transform vector $\mathbf{Y} = \mathbf{A}'\mathbf{X}$ may be computed which represents **X** without error. If only $m$ of the $n$ components of **Y** are retained to represent **X** in a approximate form, then the minimum mean squared error is obtained by

(1)  deleting components corresponding to the smallest eigenvalues of $\mathbf{\Sigma_X}$,
(2)  replacing deleted components $y_i$ by their expected values, $\bar{y}_i$,
(3)  determining the transformation matrix **A** from the ordered eigenvectors of $\mathbf{\Sigma_X}$.

The resulting components $y_i$ are mutually uncorrelated, since the covariance matrix of **Y** is diagonal,

$$\mathbf{\Sigma}_y = \mathbf{A}'\mathbf{\Sigma_X}\mathbf{A} = \begin{bmatrix} \lambda_1 & & & \\ & \lambda_2 & & \\ & & \ddots & \\ & & & \lambda_n \end{bmatrix} = \Lambda.$$

Since the eigenvalues $\lambda_i$ are equal to the variances of the $y_i$ components, the rule for selecting the eigenvectors is equivalent to selecting maximum variance directions. If **X** is a Gaussian random vector, then **Y** is also a Gaussian random vector and its components $y_i$ are mutually independent.

A detailed proof is given by Fukunaga (1972), which shows that the eigenvectors of $\mathbf{\Sigma_X}$ minimize the mean squared error over all choices of orthonormal basis vectors. Transformations which are not orthonormal do not preserve distance and consequently the structure of the distribution could be arbitrarily distorted.

### *Image Rotation Example*

An orthonormal transformation may always be considered as an $n$-dimensional rotation. Selecting the orthonormal transformation which minimizes the mean squared error according to the rules developed under the Hotelling transformation is equivalent to determining a rotated set of coordinates along the axes of greatest variance of the data. If the dimensionality is reduced by omitting one of the transformed coordinates, a translation to the mean value of that coordinate should be made to prevent a bias error. The amount of mean squared error introduced by omitting a component is equal to the variance of the transformed component.

Suppose a two-element picture is observed four times to produce the data points shown in Fig. 3.20a. The first step is to compute the mean vector. Since the distribution is not known, the expected value will be estimated by the sample mean

$$\mu = E\{\mathbf{x}\} \simeq \frac{1}{4} \sum_{i=1}^{4} \mathbf{x}_i$$

$$\mu = \frac{1}{4} \left\{ \begin{bmatrix} -2 \\ 0 \end{bmatrix} + \begin{bmatrix} -1 \\ 1 \end{bmatrix} + \begin{bmatrix} 1 \\ 3 \end{bmatrix} + \begin{bmatrix} 2 \\ 4 \end{bmatrix} \right\} = \begin{bmatrix} 0 \\ 2 \end{bmatrix}.$$

Since the mean vector is not zero, the following computations may be simplified by subtracting the mean vector from each data point. This may be considered as a translation of the coordinates from the original origin to the centroid of the data. The transformation is simply

$$\begin{bmatrix} y_1 \\ y_2 \end{bmatrix} = \begin{bmatrix} x_1 \\ x_2 \end{bmatrix} - \begin{bmatrix} 0 \\ 2 \end{bmatrix}.$$

In the new coordinates the mean vector is zero and the covariance matrix is equal to the scatter matrix. The covariance matrix will be estimated by the sample covariance matrix,

$$\Sigma_Y \simeq \frac{1}{4} \sum_{i=1}^{4} \mathbf{Y}_i \mathbf{Y}_i'$$

$$= \frac{1}{4} \left\{ \begin{bmatrix} -2 \\ -2 \end{bmatrix} [-2 \ -2] + \begin{bmatrix} -1 \\ -1 \end{bmatrix} [-1 \ -1] + \begin{bmatrix} 1 \\ 1 \end{bmatrix} [1 \ 1] + \begin{bmatrix} 2 \\ 2 \end{bmatrix} [2 \ 2] \right\}$$

$$= \frac{1}{4} \left\{ \begin{bmatrix} 4 & 4 \\ 4 & 4 \end{bmatrix} + \begin{bmatrix} 1 & 1 \\ 1 & 1 \end{bmatrix} + \begin{bmatrix} 1 & 1 \\ 1 & 1 \end{bmatrix} + \begin{bmatrix} 4 & 4 \\ 4 & 4 \end{bmatrix} \right\}$$

$$= \begin{bmatrix} \frac{10}{4} & \frac{10}{4} \\ \frac{10}{4} & \frac{10}{4} \end{bmatrix}.$$

The correlated nature of the data may be observed by decomposing the covariance matrix into the product of standard deviation and correlation matrices

$$\Sigma_Y = \mathbf{DRD}$$

or

$$\Sigma_Y = \begin{bmatrix} \dfrac{\sqrt{10}}{2} & 0 \\ 0 & \dfrac{\sqrt{10}}{2} \end{bmatrix} \begin{bmatrix} 1 & 1 \\ 1 & 1 \end{bmatrix} \begin{bmatrix} \dfrac{\sqrt{10}}{2} & 0 \\ 0 & \dfrac{\sqrt{10}}{2} \end{bmatrix}.$$

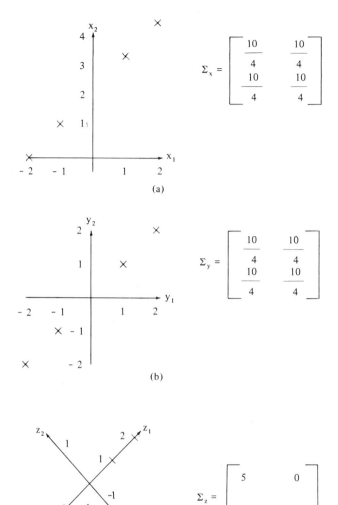

*Figure 3.20* Hotelling transformation: (a) original space, (b) translated to centroid, and (c) rotated along eigenvectors.

Since the off-diagonal correlation term $\rho_{12} = 1$, one should expect that the representation contains redundancies.

We may now determine the eigenvectors and eigenvalues of $\boldsymbol{\Sigma}_\mathbf{Y}$ by solving

$$|\boldsymbol{\Sigma}_\mathbf{Y} - \lambda \mathbf{I}| = 0 \quad \text{or} \quad \begin{vmatrix} \frac{10}{4} - \lambda & \frac{10}{4} \\ \frac{10}{4} & \frac{10}{4} - \lambda \end{vmatrix} = 0,$$

which results in the characteristic equation,

$$\lambda(\lambda - 5) = 0.$$

Therefore, the eigenvalues are $\lambda_1 = 5$, $\lambda_2 = 0$. The corresponding eigenvectors are described by the equations

$$\lambda_1 = 5, \quad y_1 - y_2 = 0,$$
$$\lambda_2 = 0, \quad y_1 + y_2 = 0.$$

To preserve structure, the eigenvectors must be normalized to obtain

$$\mathbf{A}_1 = \begin{bmatrix} \dfrac{1}{\sqrt{2}} \\ \dfrac{1}{\sqrt{2}} \end{bmatrix}, \quad \mathbf{A}_2 = \begin{bmatrix} \dfrac{1}{\sqrt{2}} \\ \dfrac{-1}{\sqrt{2}} \end{bmatrix}.$$

These basis vectors are indicated in Fig. 3.20c. Note that the direction of these vectors was selected arbitrarily and that the negative of either could have been chosen. The transformation matrix **A** may now be constructed from the eigenvectors as

$$\mathbf{A} = [\mathbf{A}_1 \ \mathbf{A}_2] = \begin{bmatrix} \dfrac{1}{\sqrt{2}} & \dfrac{1}{\sqrt{2}} \\ \dfrac{1}{\sqrt{2}} & \dfrac{-1}{\sqrt{2}} \end{bmatrix}.$$

Note that the transformed values

$$\mathbf{Z} = \mathbf{A}\mathbf{Y}'$$

have variance components of 5 along the $z_1$ axis and 0 along the $z_2$ axis. Since $\lambda_2 = 0$, the original data may be represented without error by the single component $z_1$.

### Block Transform Example

Due to the large dimensionality of most images, it is often not computationally feasible to compute the Hotelling transform of the equivalent vector image. In many cases, a block size is selected and overlapping or nonoverlapping blocks of a single image are used to estimate the covariance matrix. This procedure will be used in the following computer example in an attempt to reduce the dimensionality of the image representation.

Consider the original 8 by 8 image shown in Fig. 3.21a. A quadrant block size of 4 by 4 was selected and the nonoverlapping quadrants considered as samples. The equivalent 16-element vectors are shown in Fig. 3.21b. The mean vector and covariance matrix are shown in Fig. 3.21c, d, respectively. The eigenvalues are shown in Fig. 3.21e. Note that only three are significant and that the relative variance contained in the first principle component is 84%, and 95% in the first two components. The relative variance is computed by

$$V = \sum_{i=1}^{m} \lambda_i \Big/ \sum_{i=1}^{n} \lambda_i.$$

The corresponding eigenvectors are shown in Fig. 3.21f. The optimum mean vector is shown in Fig. 3.21g.

Using only the first two principle components to represent the data and rounding the values to the nearest integer results in the reconstructed array shown in Fig. 3.21h. Note that only six integer values are in error but that a dimensionality reduction from 16 to 2 has been obtained.

### 3.3.2 Two-Dimensional Fourier Transform

The continuous two-dimensional Fourier transform is defined by the transform pair

$$F(u,v) = \int \int_{-\infty}^{\infty} f(x,y) \exp\left[ -j2\pi(ux+vy) \right] dx\, dy,$$

$$f(x,y) = \int \int_{-\infty}^{\infty} F(u,v) \exp\left[ j2\pi(ux+vy) \right] du\, dv.$$

The significance of this transform is enhanced by the physical fact that the Fourier transform of an image field is equal to the far field or Fraunhoffer diffraction pattern of the image. For example, the twinkling of a distant light or star at night is due to the observation of its Fourier transform. This property also provides the basis for the computation of the Fourier transform with an optical system. Several properties of the continuous Fourier transform are summarized in Table 3.1. The effects of the operations of linear combination, scale change, shift of position, modulation, convolution, multiplication, correlation, rotation, differentiation, and integration are described in the table. For further information one may consult the excellent books by Goodman (1968), Bracewell (1965), and Papoulis (1969).

In both the continuous and discrete cases, the Fourier transform is a complex function, that is,

$$F(u,v) = R(u,v) + jI(u,v),$$

where $R(u,v)$ and $I(u,v)$ are the real and imaginary components and $j = \sqrt{-1}$. It is often convenient to represent $F(u,v)$ in terms of its magnitude $|F(u,v)|$ and phase spectrum $\phi(u,v)$, where

$$F(u,v) = |F(u,v)| \exp\left[ j\phi(u,v) \right]$$

or

$$|F(u,v)| = \left[ R^2(u,v) + I^2(u,v) \right]^{1/2},$$

$$\phi(u,v) = \tan^{-1}\left[ I(u,v)/R(u,v) \right].$$

(a)

(b)

(c)

(d)

MEAN
4.50
4.50
4.50
4.50
5.00
4.50
4.50
4.50
4.50
4.50
2.50
2.75

(e)

| | | | | | | | |
|---|---|---|---|---|---|---|---|
| -0.90966803D-01 | -0.92446078D-01 | -0.93925353D-01 | -0.93925353D-01 | -0.78874544D-02 | 0.00000000D 00 | 0.78874544D-02 | 0.78874544D-02 |
| 0.75191894D-01 | 0.92446078D-01 | 0.10970026D 00 | 0.10970026D 00 | 0.44497621D 00 | 0.46223039D 00 | 0.47948457D 00 | 0.53433470D 00 |
| -0.51216760D-01 | 0.32619967D-01 | 0.11645669D 00 | 0.11645669D 00 | -0.22028651D 00 | 0.00000000D 00 | 0.22028651D 00 | 0.22028651D 00 |
| -0.38935626D 00 | -0.32619967D-01 | 0.32411632D 00 | 0.32411632D 00 | -0.51983612D 00 | -0.16309983D 00 | 0.19363646D 00 | 0.35569462D 00 |
| -0.53219182D 00 | -0.15214495D-01 | 0.50176283D 00 | 0.50176283D 00 | -0.19913236D 00 | 0.00000000D 00 | 0.19913236D 00 | 0.19913236D 00 |
| 0.13392710D 00 | 0.15214495D-01 | -0.10349811D 00 | -0.10349811D 00 | 0.19478508D 00 | 0.76072477D-01 | -0.42640128D-01 | -0.94389183D-01 |
| -0.63937603D-01 | 0.16244890D-02 | 0.68002498D-01 | 0.68154521D-01 | 0.13113585D 00 | 0.00000000D 00 | -0.13391818D 00 | -0.13513521D 00 |
| 0.97135095D-01 | -0.24074959D-02 | -0.22780374D 00 | -0.23309081D 00 | -0.64652042D 00 | 0.62041157D 00 | 0.12100576D 00 | -0.62679365D-02 |
| -0.11004697D 00 | -0.57937353D-02 | 0.99981993D-01 | 0.10012780D 00 | 0.20397413D 00 | 0.00000000D 00 | -0.20746392D 00 | -0.20887548D 00 |
| 0.39845187D 00 | 0.42168101D-02 | 0.79724075D 00 | 0.18174481D-02 | -0.74829110D-01 | 0.24888829D-01 | -0.19630095D 00 | 0.20681524D-01 |
| 0.69879511D 00 | -0.62225425D-02 | 0.69889223D 00 | -0.56861609D-04 | 0.26816848D-04 | 0.00000000D 00 | -0.26840239D-04 | -0.26860658D-04 |
| 0.25570506D-01 | 0.50639571D-04 | 0.18118478D-02 | 0.18743346D-02 | 0.94518349D-01 | 0.90670420D-01 | 0.68995093D-02 | 0.73028948D-01 |
| -0.10019204D-01 | 0.99496793D 00 | -0.13026349D-02 | -0.24212751D-03 | 0.55320375D-02 | 0.00000000D 00 | -0.56674913D-02 | -0.57493538D-02 |
| 0.26128270D-01 | 0.10753794D-02 | 0.96742865D-03 | 0.10590586D-02 | 0.62827944D-01 | 0.48279179D-01 | 0.36617196D-01 | 0.39191315D-01 |
| 0.39696588D 00 | -0.44205966D-02 | -0.41324465D 00 | 0.81555318D 00 | 0.35326458D-02 | 0.00000000D 00 | -0.35717786D-02 | -0.35808879D-02 |
| 0.21521261D-01 | 0.66248926D-03 | 0.61558825D-03 | 0.73237409D-03 | 0.51043792D-01 | 0.39238240D-01 | 0.29394654D-01 | 0.31443395D-01 |
| 0.00000000D 00 | 0.00000000D 00 | 0.00000000D 00 | 0.00000000D 00 | 0.00000000D 00 | 0.10000000D 01 | 0.00000000D 00 | 0.00000000D 00 |
| 0.00000000D 00 | 0.00000000D 00 | 0.00000000D 00 | 0.00000000D 00 | 0.00000000D 00 | 0.00000000D 00 | 0.00000000D 00 | 0.00000000D 00 |
| 0.71111956D-01 | 0.23680346D-03 | -0.71988617D-01 | -0.72182327D-01 | 0.22755396D 00 | 0.00000000D 00 | -0.23445101D 00 | 0.89257077D 00 |
| 0.27633501D 00 | 0.99352703D-03 | -0.22330116D-02 | -0.22370577D-02 | -0.68838892D-01 | 0.23175055D-02 | -0.68433915D-03 | -0.17549140D-02 |
| -0.13618351D 00 | -0.20956151D-03 | 0.13798697D 00 | 0.13830841D 00 | 0.81299998D 00 | 0.00000000D 00 | -0.58384366D-02 | -0.61193445D-02 |
| -0.51619779D 00 | -0.22623500D-02 | 0.56583962D-02 | 0.60410596D-02 | 0.12516852D 00 | -0.35758672D-02 | 0.22118658D-02 | 0.50255254D-02 |
| -0.51055116D-01 | -0.21017281D-02 | 0.11763148D 00 | 0.83487054D-01 | 0.16618723D 00 | 0.00000000D 00 | -0.16860592D 00 | -C.16955415D 00 |
| 0.30586324D 00 | -0.34413116D-02 | -0.15132486D 00 | -0.15280280D 00 | -0.12714436D 00 | -0.55558060D 00 | 0.65116363D 00 | 0.55904375D-01 |
| 0.94299664D-01 | 0.23327445D-03 | -0.95491957D-01 | -0.95730926D-01 | 0.29902764D 00 | 0.00000000D 00 | 0.86218278D 00 | 0.78899560D-02 |
| 0.36320447D 00 | 0.14013826D-02 | -0.32870260D-02 | -0.34058528D-02 | -0.89593849D-01 | 0.28011708D-02 | -0.11272723D-02 | -0.27278486D-02 |
| -0.75555886D-01 | -0.40239481D-02 | 0.68484535D-01 | 0.68573419D-01 | 0.13969787D 00 | 0.00000000D 00 | -0.14223900D 00 | -0.14329081D 00 |
| 0.27274531D 00 | 0.29996187D-02 | -0.31566654D 00 | 0.85361458D 00 | -0.54419312D-01 | 0.17696803D-01 | -0.13517917D 00 | 0.18780648D-01 |
| -0.51569690D-01 | -0.18580650D-02 | 0.10299868D-01 | 0.29222164D-01 | 0.60041720D-01 | 0.00000000D 00 | -0.60904414D-01 | -0.61237310D-01 |
| 0.10727934D 00 | -0.61460060D-03 | -0.24923008D 00 | -0.25651495D 00 | -0.57258361D-01 | -0.21165458D 00 | -0.48155430D 00 | 0.75227469D 00 |
| 0.14438357D-01 | 0.88499974D-02 | 0.55805754D-02 | 0.46912196D-02 | -0.26085968D-02 | 0.00000000D 00 | 0.28222485D-02 | 0.29022782D-02 |
| -0.24954378D-01 | 0.99503956D 00 | -0.16212829D-02 | -0.17968371D-02 | -0.62546653D-01 | -0.50261651D-01 | -0.34112994D-01 | -0.36090986D-01 |

(f)

| | | | | | | | |
|---|---|---|---|---|---|---|---|
| 0.28748107D 02 | 0.36652189D 01 | 0.17741745D 01 | 0.45452561D-12 | 0.32774735D-16 | 0.27118142D-16 | 0.94213040D-17 | 0.90217720D-17 |
| 0.00000000D 00 | -0.19452463D-17 | -0.70679630D-17 | -0.11087069D-16 | -0.14602708D-16 | -0.25377543D-16 | -0.14004272D-15 | -0.81116993D-14 |

(g)

| | | | | | | | |
|---|---|---|---|---|---|---|---|
| 0.50984712D 01 | 0.48771591D 01 | 0.46558461D 01 | 0.46558456D 01 | 0.51408868D 01 | 0.50000000D 01 | 0.38591132D 01 | 0.38591127D 01 |
| 0.51833024D 01 | 0.41228414D 01 | 0.30623801D 01 | 0.30623801D 01 | 0.16746682D 01 | 0.61420715D 00 | -0.44625399D 00 | -0.91506392D 00 |

(h)

| | | | | | | | |
|---|---|---|---|---|---|---|---|
| 4 | 4 | 4 | 4 | 5 | 4 | 4 | 0 |
| 4 | 5 | 5 | 5 | 5 | 5 | 4 | 0 |
| 4 | 5 | 6 | 6 | 5 | 5 | 4 | 0 |
| 4 | 5 | 6 | 7 | 5 | 5 | 4 | 0 |
| 4 | 5 | 6 | 6 | 5 | 5 | 4 | 0 |
| 4 | 5 | 5 | 5 | 5 | 5 | 4 | 0 |
| 4 | 4 | 4 | 4 | 5 | 4 | 4 | 0 |
| 4 | 4 | 4 | 4 | 5 | 4 | 4 | 0 |

*Figure 3.21* Block transform example: (a) original data, (b) quadrant data, (c) mean vector, (d) covariance matrix, (e) eigenvalues, (f) eigenvectors, (g) optimized mean, and (h) reconstructed image.

Table 3.1

Properties of the Fourier Transform

| Spatial operations | Frequency operation | Significance |
| --- | --- | --- |
| 1. Linearity<br>$af_1(x,y) + bf_2(x,y)$,<br>where $a$ and $b$ are constants | Linearity<br>$aF_1(u,v) + bF_2(u,v)$ | Linearity and superposition apply in both domains. The spectrum of a linear sum of images is the linear sum of their spectra. Further, any function may be regarded as a sum of component parts and the spectrum is the sum of the component spectra |
| 2. Scale change<br>$f(ax, by)$ | Inverse scale change<br>$\dfrac{1}{|ab|} F\left(\dfrac{u}{a}, \dfrac{v}{b}\right)$ | Space–bandwidth invariance. Compressing a spatial function expands its spectrum in frequency and reduces its amplitude by the same factor. The amplitude reduces because the same energy is spread over a greater bandwidth. For $a = b = -1$, the spatial function is reversed. The frequency axes are also reversed, which, for real images, changes only the phase spectra |
| 3. Shift of position<br>$f(x - a, y - b)$ | Linear phase added<br>$F(u,v) \exp[-j(ua + vb)]$ | Shifting or translating a spatial function a distance $x = a$ adds a linear phase $\theta = ua$ to the original phase. Conversely, a linear phase filter produces a translation of the image. The magnitude spectrum is invariant to translation |
| 4. Modulation<br>$\exp[j(u_0 x + v_0 y)]f(x,y)$ | Shift of spectrum<br>$F(u - u_0, v - v_0)$ | Multiplying a spatial function by a complex sinusoid translates its spectrum to center at about $u_0$ rather than zero frequency |
| 5. Convolution<br>$f(x,y) * h(x,y)$<br>$= \int_{-\infty}^{\infty} \int f(\xi, \eta)$<br>$\times\, h(x - \xi, y - \eta)\, d\xi\, d\eta$ | Multiplication<br>$F(u,v)H(u,v)$ | The convolution of two spatial functions requires the reversal left to right and bottom to top of one of the functions, a translation, multiplication, and summation. Convolution occurs whenever a function is imaged or filtered by a linear, position-invariant imaging system. The frequency effect is simply to multiply the individual spectra. Note that if one of the spatial functions is a unit impulse, its spectrum remains the same |

**6. Multiplication**

$$f(x,y)g(x,y)$$

Convolution

$$F(u,v)*G(u,v)$$
$$= \int_{-\infty}^{\infty}\int F(\xi,\eta)G(u-\xi,v-\eta)\,d\xi\,d\eta$$

The product of two spatial functions occurs whenever a scene function is illuminated by another function such as in transmission or reflection. The spectrum of the product is the convolution of the spectra. If one of the spatial functions is the train of impulses, the spectrum is replicated at spacing inversely proportional to the spacing of the spatial impulse train

**7. Correlation**

$$\int_{-\infty}^{\infty}\int f(\xi,\eta)g(x+\xi,y+\eta)\,d\xi\,d\eta$$

Conjugate product

$$F(u,v)G^*(u,v)$$

The correlation of two spatial functions corresponds to the product of one spectrum and the conjugate of the other. If the two functions are identical, the spectrum is the magnitude squared or power spectral density

**8. Rotation**

$$f(x';u')$$
$$x' = x\cos\theta + y\sin\theta$$
$$y' = -x\sin\theta + y\sin\theta$$

Rotation

$$F(u',v')$$
$$u' = u\cos\theta + v\sin\theta$$
$$v' = -u\sin\theta + v\cos\theta$$

Rotating a function through an angle $\theta$ rotates the spectrum through an identical angle. Neither magnitude nor phase spectra are invariant to rotation, although invariant functions may be derived

**9. Differentiation**

$$\frac{d^n}{dx^n}f(x,y)$$

High-frequency filter

$$(ju)^n F(u,v)$$

Differentiating a spatial function in any direction corresponds to a form of high-pass filtering with the filter function shown. High-pass spatial frequency filtering characteristically "sharpens" the image

**10. Integration**

$$\int\cdots\int_{-\infty}^{x} f(\alpha,y)\,(d\alpha)^n$$
$n$-fold

Low-frequency filter

$$\frac{1}{(ju)^n}F(u,v)$$

Integrating a spatial function in any direction corresponds to a form of low-pass filtering with the filter function shown. Low-pass spatial frequency filtering characteristically "blurs" the image

Table 3.2

Properties of the Discrete Fourier Transform

| Spatial operation | Frequency operation | Significance |
|---|---|---|
| 1. Linearity $af_1(m,n) + bf_2(m,n)$, where $a$ and $b$ are constants | Linearity $aF_1(k,l) + bF_2(k,l)$ | Linearity and superposition apply in both domains. The spectrum of a linear sum of discrete images is equal to the linear sum of the individual spectra. Any function may be regarded as a sum of component parts and the spectrum as the sum of the component spectra |
| 2. Scale change $f(am, bn)$ | Dependent on $a$ and $b$ | Due to the discrete nature of the sampled signal, arbitrary scale changes involve interpolation and must be considered individually. A special case is $a = b = -1$ in which the spatial function is reversed, in which case the frequency axes are multiplied by $-1$ |
| 3. Shift of position $f(m - m_0, n - n_0)$ | Linear phase added $F(k,l) \exp \left[ \dfrac{-j(km_0 + ln_0)}{N} \right]$ | Shifting or translating the discrete spatial function in integer amounts adds a linear phase spectrum. Conversely, a linear phase addition to the spectrum produces a translation of the image. The magnitude spectrum is invariant to translation |
| 4. Modulation $\exp \left[ j \dfrac{k_0 m + l_0 n}{N} \right] f(m,n)$, where $k_0$ and $l_0$ are integer constants | $F(k - k_0, l - l_0)$ | Multiplying a spatial function by a discrete complex sinusoid translates its spectrum to center about the frequencies of the sinusoid |

5. Convolution

$$f(m,n) * h(m,n) = \sum_{i=0}^{N_1-1} \sum_{j=0}^{N_2-1} f(i,j)$$

$$h\{[m-i],[n-j]\},$$

and

$$[m-i] = (m-i) \bmod N_1,$$
$$[n-j] = (n-j) \bmod N_2$$

Multiplication

$$F(k,l) H(k,l)$$

The convolution of two discrete spatial functions corresponds to the product of the individual spectra. In the periodic convolution described, care must be taken to prevent undesirable wrap around. A periodic convolution may also be computed with an extended periodic convolution

6. Multiplication

$$f(m,n) g(m,n)$$

Convolution

$$\sum_{i=0}^{N_1-1} \sum_{j=0}^{N_2-1} F(i,j) H\{[k-i],[l-j]\}$$

The product of two discrete spatial functions corresponds to the convolution of their discrete spectra

7. Correlation

$$r(m,n)$$

$$= \sum_{i=0}^{N_1-1} \sum_{j=0}^{N_2-1} f(i,j) g([i+m],[j+n])$$

Conjugate product

$$F(k,l) G^*(k,l)$$

The periodic correlation of two discrete functions corresponds to the product of one spectrum with the conjugate spectrum of the other function. A periodic correlation and normalization may also be desirable

8. Rotation

$$f(m',n'),$$

where

$$m' = m\cos\theta + n\sin\theta,$$
$$n' = -m\sin\theta + n\cos\theta$$

Rotation

$$F(k',l')$$

$$k' = k\cos\theta + l\sin\theta,$$
$$l' = -h\sin\theta + l\cos\theta$$

Rotation of the discrete spectrum by multiples of 90° corresponds to rotation of the spectrum by exactly the same angle. Other rotations involve interpolation

9. Differentiation

$$f(m,n) - f(m-1,n)$$

$$F(k,l)[1 - \exp(-jk/N)]$$

The derivative must be approximated by an appropriate difference formula but will generally produce a form of high-pass filter

10. Integration

$$f(m,n) + f(m-1,n)$$

$$F(k,l)[1 + \exp(-jk/N)]$$

The integral must be approximated by an appropriate numerical integration formula but the result will generally be a form of low-pass filter

Since both photographic film and the human eye respond to the magnitude of an image field, it is difficult to directly observe the phase spectra. Holographic techniques are, however, based upon recording both the magnitude and phase spectra, and either can be easily computed with digital techniques. It is now known that the phase spectrum contains most of the information about the positions of edges in an image (Teschser, 1973).

The two-dimensional discrete Fourier transform (DFT) is defined by the transform pair

$$F(k,l) = \frac{1}{N} \sum_{m=0}^{N-1} \sum_{n=0}^{N-1} f(m,n) \exp\left[ \frac{-2\pi j(km+ln)}{N} \right],$$

$$f(m,n) = \frac{1}{N} \sum_{k=0}^{N-1} \sum_{l=0}^{N-1} F(k,l) \exp\left[ \frac{2\pi j(km+ln)}{N} \right].$$

A square image of size $N$ by $N$ has been assumed for simplicity and the scale factor of $1/N$ has been distributed between the forward and inverse transforms for symmetry.

The significance of the discrete Fourier transform is enhanced not only by its use to approximate the continuous transform, but also because it may be efficiently computed via the fast Fourier transform (FFT) algorithm. An important mathematical fact is that the discrete Fourier transform may be used to determine the eigenvalues of a circulant matrix.

A few of the important properties of the discrete Fourier transform are summarized in Table 3.2. The properties of linearity, shift of position, modulation, convolution, multiplication, and correlation are completely analogous to the continuous case with the main differences being due to the discrete, periodic nature of the image and its transform. The other properties, scale change, rotation, differentiation, and integration, require either interpolation or a specific numerical algorithm but can be made at least approximately analogous to the continuous case.

The kernel function of the discrete transform

$$\exp\left[ -2\pi j(km+ln)/N \right]$$

has several interesting interpretations. First, note that it is product separable into

$$\exp(-2\pi jkm/N)\exp(-2\pi jln/N).$$

Each of the factors may be described as solutions to the equation

$$W^N = 1,$$

which leads to the interpretation of

$$W_{km} = \exp(-2\pi jkm/N)$$

as a *root of unity*. Since there are only $N$ distinct roots of unity, the term $W_{km}$

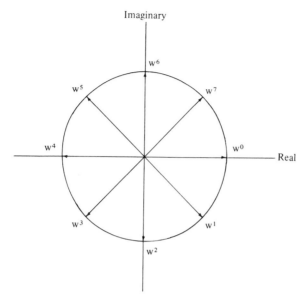

*Figure 3.22*   Phasor diagram illustrating the $N$ roots of unity for $N = 8$.

can take on only $N$ distinct values regardless of the integer values of $k$ or $m$. These $N$ values may be shown on a phasor diagram as illustrated in Fig. 3.22 (Thomas, 1971).

The values of $W_{km}$ for the principal values of $k$ and $m$, i.e.,

$$k, m = 0, 1, \ldots, N-1,$$

may be arranged in a matrix as shown in Fig. 3.23. An interesting property of the matrix is that it is *unitary*, that is, the inner product of any one column (row) with the conjugate of any other column (row) is equal to zero unless the two columns (rows) are identical.

$$\sum_{m=0}^{N-1} W_{mk} W_{ml}^* = \begin{cases} N, & k = l, \\ 0, & k \neq l. \end{cases}$$

It also follows that $\mathbf{W}^{-1} = \mathbf{W}^* / N$. Another nonobvious property was observed by Good (1958): the Fourier matrix can be factored into the product of $2n$ sparse and diagonal matrices, where $n = \log_2 N$. This observation may be considered to be the basis of the FFT algorithm.

The matrix formulation may also be used to concisely represent the two-dimensional discrete Fourier transformation. If the series representation is rewritten in the form of two consecutive one-dimensional transforms,

$$NF(k, l) = \sum_{m=0}^{N-1} H(m, l) \exp(-2\pi jkm/N),$$

$$\{w_{km}\} = \begin{bmatrix} W_0 & W_0 & W_0 & W_0 & W_0 & W_0 & W_0 & W_0 \\ W_0 & W_1 & W_2 & W_3 & W_4 & W_5 & W_6 & W_7 \\ W_0 & W_2 & W_4 & W_6 & W_0 & W_2 & W_4 & W_6 \\ W_0 & W_3 & W_6 & W_1 & W_4 & W_7 & W_2 & W_5 \\ W_0 & W_4 & W_0 & W_4 & W_0 & W_4 & W_0 & W_4 \\ W_0 & W_5 & W_2 & W_7 & W_4 & W_1 & W_6 & W_3 \\ W_0 & W_6 & W_4 & W_2 & W_0 & W_6 & W_4 & W_2 \\ W_0 & W_7 & W_6 & W_5 & W_4 & W_3 & W_2 & W_1 \end{bmatrix}$$

(a)

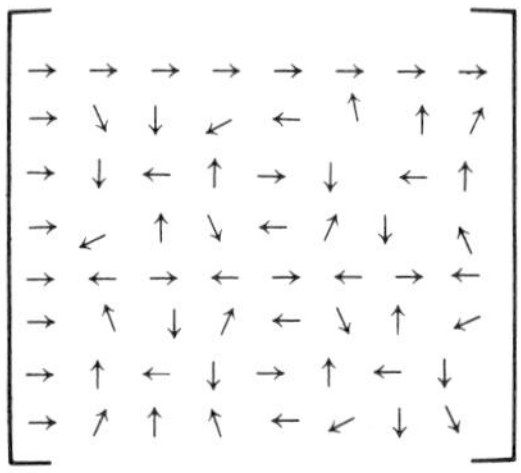

(b)

*Figure 3.23* Discrete Fourier transform for $N=8$: (a) matrix in reduced form and (b) phasor representation.

where

$$H(m,l) = \sum_{n=0}^{N-1} f(m,n)\exp(-2\pi jnl/N).$$

Note that each of the above equations can be put into the form corresponding to matrix multiplication. In general, the matrix product for square matrices subscripted from 0 to $N-1$

$$\mathbf{A} = \mathbf{BC}$$

is computed by

$$a_{ml} = \sum_{n=0}^{N-1} b_{mn} c_{nl}.$$

Letting

$$a_{ml} = H(m,l), \qquad b_{mn} = f(m,n), \qquad c_{nl} = \exp(-2\pi jnl/N),$$

and using the correspondence lead to the matrix relationship

$$[H] = [f][W],$$

where

$$[H] = [H(m,l)], \qquad [f] = [f(m,n)], \qquad [W] = [\exp(-2j\pi nl/N)].$$

Therefore, the computation of the $[H]$ matrix may be considered a postmultiplication of the picture matrix by the $[W]$ matrix or equivalently as computing the DFT of the rows of the picture matrix.

A similar correspondence may be made for the final computation of the discrete two-dimensional Fourier transform. Let

$$a_{kl} = F(k,l), \qquad b_{km} = \exp(-2\pi jkm/N), \qquad \text{and} \qquad c_{ml} = H(m,l).$$

The matrix multiplication correspondence now leads to

$$N[F] = [W][H],$$

where

$$N[F] = [F(k,l)], \qquad [W] = [\exp(-2\pi jkm/N)], \qquad [H] = [H(m,l)].$$

Therefore, the final transform operation may be considered as premultiplication by the $[W]$ matrix or equivalently as computing the DFT of the columns of the intermediate matrix $[H]$.

Combining the previous matrix equations leads to the well-known two-dimensional transform result

$$[F] = N^{-1}[W][f][W].$$

Since the matrix $[W]$ is unitary and symmetric, its inverse is equal to its conjugate and the inverse two-dimensional transform is

$$[f] = N^{-1}[W^*][F][W^*].$$

### Two-Dimensional Fourier Transform Example

The following numerical example illustrates some of the interesting numerical problems associated with the Fourier transform computation.

Consider the real valued 8 by 8 image array shown in Fig. 3.24a. The image represents a bright spot. When considered as a periodic array the North–South edge has a smooth transition but the East–West edge has a sharp change.

The first step in the two-dimensional Fourier transform computation using many of the available algorithms is to form a complex image array as shown in Fig. 3.24b. This step doubles the storage requirement and is eliminated in some algorithms. The next step is the computation of the complex Fourier transform as shown in Fig. 3.24c. Note that the DC value term of $259/8$ occurs at the origin of the array, and that the largest magnitude terms occur at the corners. To produce an origin centered Fourier transform, which corresponds to optically computed transforms and simplifies filtering operations, it is necessary to include another computation.

Using the shift property of the discrete transform, it is easily shown that if the image array is multiplied by the factor $(-1)^{i+j}$,

$$\hat{f}(i,j) = (-1)^{i+j} f(i,j).$$

As shown in Fig. 3.24d, then the new transform will be centered at $(\frac{1}{2}N, \frac{1}{2}N)$,

$$\hat{F}(k,l) = F\left(k + \tfrac{1}{2}N, l + \tfrac{1}{2}N\right).$$

**(a)**

```
4  4  4  4  4  4  4  0
4  5  5  5  4  4  4  0
4  5  6  6  5  5  4  0
4  5  6  7  6  5  4  0
4  5  6  6  6  5  4  0
4  5  5  5  5  5  4  0
4  4  4  4  4  4  4  0
4  4  4  4  4  4  4  0
```

**(b)**

| | | | | | | | |
|---|---|---|---|---|---|---|---|
| 4.000000 / 0.000000 | 4.000000 / 0.000000 | 4.000000 / 0.000000 | 4.000000 / 0.000000 | 4.000000 / 0.000000 | 4.000000 / 0.000000 | 4.000000 / 0.000000 | 0.000000 / 0.000000 |
| 4.000000 / 0.000000 | 5.000000 / 0.000000 | 5.000000 / 0.000000 | 5.000000 / 0.000000 | 5.000000 / 0.000000 | 5.000000 / 0.000000 | 4.000000 / 0.000000 | 0.000000 / 0.000000 |
| 4.000000 / 0.000000 | 5.000000 / 0.000000 | 6.000000 / 0.000000 | 6.000000 / 0.000000 | 6.000000 / 0.000000 | 5.000000 / 0.000000 | 4.000000 / 0.000000 | 0.000000 / 0.000000 |
| 4.000000 / 0.000000 | 5.000000 / 0.000000 | 6.000000 / 0.000000 | 7.000000 / 0.000000 | 6.000000 / 0.000000 | 5.000000 / 0.000000 | 4.000000 / 0.000000 | 0.000000 / 0.000000 |
| 4.000000 / 0.000000 | 5.000000 / 0.000000 | 6.000000 / 0.000000 | 6.000000 / 0.000000 | 6.000000 / 0.000000 | 5.000000 / 0.000000 | 4.000000 / 0.000000 | 0.000000 / 0.000000 |
| 4.000000 / 0.000000 | 5.000000 / 0.000000 | 5.000000 / 0.000000 | 5.000000 / 0.000000 | 5.000000 / 0.000000 | 5.000000 / 0.000000 | 4.000000 / 0.000000 | 0.000000 / 0.000000 |
| 4.000000 / 0.000000 | 4.000000 / 0.000000 | 4.000000 / 0.000000 | 4.000000 / 0.000000 | 4.000000 / 0.000000 | 4.000000 / 0.000000 | 4.000000 / 0.000000 | 0.000000 / 0.000000 |
| 4.000000 / 0.000000 | 4.000000 / 0.000000 | 4.000000 / 0.000000 | 4.000000 / 0.000000 | 4.000000 / 0.000000 | 4.000000 / 0.000000 | 4.000000 / 0.000000 | 0.000000 / 0.000000 |

*Figure 3.24* Two-dimensional Fourier transform example: (a) an 8 by 8 image array; and (b) input data, real-valued image considered as a complex array.

Figure 3.24, panels (c) and (d). The two panels contain 8 × 8 numeric arrays; each array entry lists a real part (upper number) and an imaginary part (lower number).

**(c) complex Fourier transform (not origin centered)** — each cell shown as real / imaginary

| 32.375 / 0.000 | -4.624 / -4.624 | 0.000 / -4.125 | 2.624 / -2.624 | 3.625 / 0.000 | 2.624 / 2.624 | 0.000 / 4.125 | -4.624 / 4.624 |
|---|---|---|---|---|---|---|---|
| -1.796 / -1.796 | 0.000 / 1.582 | 0.088 / -0.088 | 0.125 / 0.000 | 0.088 / 0.088 | 0.000 / 0.125 | -0.088 / 0.088 | 1.582 / 0.000 |
| 0.000 / -0.125 | 0.088 / 0.088 | -0.375 / 0.000 | 0.088 / 0.088 | 0.000 / 0.125 | -0.088 / 0.088 | 0.375 / 0.000 | -0.088 / -0.088 |
| -0.205 / 0.205 | 0.125 / 0.000 | 0.088 / 0.088 | 0.000 / -0.168 | 0.168 / 0.000 | 0.168 / 0.000 | -0.088 / -0.088 | 0.000 / -0.125 |
| -0.375 / 0.000 | 0.088 / 0.088 | 0.000 / 0.125 | 0.168 / 0.000 | 0.000 / 0.168 | 0.000 / 0.168 | 0.000 / -0.125 | 0.088 / 0.088 |
| -0.205 / -0.205 | 0.000 / 0.125 | -0.088 / 0.088 | -0.088 / 0.088 | -0.088 / -0.088 | 0.088 / -0.088 | 0.088 / -0.088 | 0.000 / 0.125 |
| 0.000 / 0.125 | -0.088 / 0.088 | 0.375 / 0.000 | -0.088 / -0.088 | 0.000 / -0.125 | 0.088 / -0.088 | -0.375 / 0.000 | 0.088 / 0.088 |
| -1.796 / 1.796 | 1.582 / 0.000 | -0.088 / -0.088 | 0.000 / -0.125 | 0.088 / -0.088 | 0.088 / 0.088 | 0.088 / 0.088 | 0.000 / -1.582 |

**(d) input data, image multiplied by $(-1)^{i+j}$** — each cell shown as real / imaginary (all imaginary parts = 0.000000)

| 4.000000 | -4.000000 | 4.000000 | -4.000000 | 4.000000 | -4.000000 | 4.000000 | -4.000000 |
|---|---|---|---|---|---|---|---|
| -4.000000 | 5.000000 | -5.000000 | 5.000000 | -5.000000 | 5.000000 | -5.000000 | 4.000000 |
| 4.000000 | -5.000000 | 6.000000 | -6.000000 | 6.000000 | -6.000000 | 5.000000 | -4.000000 |
| -4.000000 | 5.000000 | -6.000000 | 7.000000 | -7.000000 | 6.000000 | -5.000000 | 4.000000 |
| 4.000000 | -5.000000 | 6.000000 | -7.000000 | 7.000000 | -6.000000 | 5.000000 | -4.000000 |
| -4.000000 | 5.000000 | -6.000000 | 6.000000 | -6.000000 | 6.000000 | -5.000000 | 4.000000 |
| 4.000000 | -5.000000 | 5.000000 | -5.000000 | 5.000000 | -5.000000 | 5.000000 | -4.000000 |
| -4.000000 | 4.000000 | -4.000000 | 4.000000 | -4.000000 | 4.000000 | -4.000000 | 4.000000 |

*Figure 3.24* Two-dimensional Fourier transform example: (c) complex Fourier transform not origin centered; and (d) input data, image multiplied by $(-1)^{i+j}$ for each picture element $(i,j)$, $i,j=0,1,\ldots,7$.

**(e)** Fourier transform (origin centered). Each entry is shown as real part (top) over imaginary part (bottom).

| | | | | | | | |
|---|---|---|---|---|---|---|---|
| 0.375<br>0.000 | -0.088<br>-0.088 | 0.000<br>-0.125 | 0.088<br>-0.088 | -0.375<br>0.000 | 0.088<br>0.088 | 0.000<br>0.125 | -0.088<br>0.088 |
| -0.088<br>-0.088 | 0.000<br>0.168 | 0.088<br>-0.088 | 0.125<br>0.000 | -0.205<br>-0.205 | 0.000<br>0.125 | -0.088<br>0.088 | 0.168<br>0.000 |
| 0.000<br>-0.125 | 0.088<br>-0.088 | -0.375<br>0.000 | 0.088<br>0.088 | 0.000<br>0.125 | -0.088<br>0.088 | 0.375<br>0.000 | -0.088<br>-0.088 |
| 0.088<br>-0.088 | 0.125<br>0.000 | 0.088<br>0.088 | 0.000<br>-1.582 | -1.796<br>1.796 | 1.582<br>0.000 | 0.000<br>-0.088 | 0.000<br>-0.125 |
| 3.625<br>0.000 | 2.624<br>2.624 | 0.000<br>4.125 | -4.624<br>4.624 | 32.375<br>0.000 | -4.624<br>-4.624 | 0.000<br>-4.125 | 2.624<br>2.624 |
| 0.088<br>0.088 | 0.000<br>0.125 | -0.088<br>0.088 | 1.582<br>0.000 | -1.796<br>-1.796 | 0.000<br>1.582 | 0.088<br>-0.088 | 0.125<br>0.000 |
| 0.000<br>0.125 | -0.088<br>0.088 | 0.375<br>0.000 | -0.088<br>-0.088 | 0.000<br>-0.125 | 0.088<br>-0.088 | -0.375<br>0.000 | 0.088<br>0.088 |
| -0.088<br>0.088 | 0.168<br>0.000 | -0.088<br>-0.088 | 0.000<br>-0.088 | -0.205<br>0.205 | 0.125<br>0.000 | 0.088<br>0.088 | 0.000<br>-0.168 |

**(f)** Image array resulting from forward and inverse transform. Each entry is shown as real part (top) over imaginary part (bottom).

| | | | | | | | |
|---|---|---|---|---|---|---|---|
| 4.000000<br>0.000000 | 4.000000<br>0.000000 | 4.000000<br>0.000000 | 4.000000<br>0.000000 | 4.000000<br>0.000000 | 4.000000<br>0.000000 | 4.000000<br>0.000000 | 0.000000<br>0.000000 |
| 4.000000<br>0.000000 | 5.000000<br>0.000000 | 5.000000<br>0.000000 | 5.000000<br>0.000000 | 4.000000<br>0.000000 | 4.000000<br>0.000000 | 5.000000<br>0.000000 | -0.000000<br>0.000000 |
| 4.000000<br>0.000000 | 5.000000<br>0.000000 | 6.000000<br>0.000000 | 5.000000<br>0.000000 | 4.000000<br>0.000000 | 4.000000<br>0.000000 | 5.000000<br>0.000000 | -0.000000<br>0.000000 |
| 4.000000<br>0.000000 | 5.000000<br>0.000000 | 6.000000<br>0.000000 | 6.000000<br>0.000000 | 7.000000<br>0.000000 | 6.000000<br>0.000000 | 5.000000<br>0.000000 | -0.000000<br>0.000000 |
| 4.000000<br>0.000000 | 5.000000<br>0.000000 | 6.000000<br>0.000000 | 6.000000<br>0.000000 | 6.000000<br>0.000000 | 6.000000<br>0.000000 | 5.000000<br>0.000000 | -0.000000<br>0.000000 |
| 4.000000<br>0.000000 | 5.000000<br>0.000000 | 5.000000<br>0.000000 | 5.000000<br>0.000000 | 5.000000<br>0.000000 | 5.000000<br>0.000000 | 4.000000<br>0.000000 | -0.000000<br>0.000000 |
| 4.000000<br>0.000000 | 4.000000<br>0.000000 | 4.000000<br>0.000000 | 4.000000<br>0.000000 | 4.000000<br>0.000000 | 4.000000<br>0.000000 | 4.000000<br>0.000000 | 0.000000<br>0.000000 |
| 4.000000<br>0.000000 | 4.000000<br>0.000000 | 4.000000<br>0.000000 | 4.000000<br>0.000000 | 4.000000<br>0.000000 | 4.000000<br>0.000000 | 4.000000<br>0.000000 | -0.000000<br>0.000000 |

*Figure 3.24* Two-dimensional Fourier transform example: (e) Fourier transform origin centered, origin is at point (4, 4); and (f) image array resulting from forward and inverse transform.

**(g)**

| | | | | | | | |
|---|---|---|---|---|---|---|---|
| 4.0000000 / 0.0000000 | 4.0000000 / 0.0000000 | 4.0000000 / 0.0000000 | 4.0000000 / 0.0000000 | 4.0000000 / 0.0000000 | 4.0000000 / 0.0000000 | 0.0000000 / 0.0000000 | 4.0000000 / 0.0000000 |
| 5.0000000 / 0.0000000 | 5.0000000 / 0.0000000 | 5.0000000 / 0.0000000 | 5.0000000 / 0.0000000 | 5.0000000 / 0.0000000 | 4.0000000 / 0.0000000 | 0.0000000 / 0.0000000 | 4.0000000 / 0.0000000 |
| 5.0000000 / 0.0000000 | 6.0000000 / 0.0000000 | 6.0000000 / 0.0000000 | 6.0000000 / 0.0000000 | 5.0000000 / 0.0000000 | 4.0000000 / 0.0000000 | 0.0000000 / 0.0000000 | 4.0000000 / 0.0000000 |
| 5.0000000 / 0.0000000 | 6.0000000 / 0.0000000 | 7.0000000 / 0.0000000 | 6.0000000 / 0.0000000 | 6.0000000 / 0.0000000 | 4.0000000 / 0.0000000 | 0.0000000 / 0.0000000 | 4.0000000 / 0.0000000 |
| 5.0000000 / 0.0000000 | 6.0000000 / 0.0000000 | 6.0000000 / 0.0000000 | 6.0000000 / 0.0000000 | 5.0000000 / 0.0000000 | 4.0000000 / 0.0000000 | 0.0000000 / 0.0000000 | 4.0000000 / 0.0000000 |
| 5.0000000 / 0.0000000 | 5.0000000 / 0.0000000 | 5.0000000 / 0.0000000 | 5.0000000 / 0.0000000 | 5.0000000 / 0.0000000 | 4.0000000 / 0.0000000 | 0.0000000 / 0.0000000 | 4.0000000 / 0.0000000 |
| 4.0000000 / 0.0000000 | 4.0000000 / 0.0000000 | 4.0000000 / 0.0000000 | 4.0000000 / 0.0000000 | 4.0000000 / 0.0000000 | 4.0000000 / 0.0000000 | 0.0000000 / 0.0000000 | 4.0000000 / 0.0000000 |
| 4.0000000 / 0.0000000 | 4.0000000 / 0.0000000 | 4.0000000 / 0.0000000 | 4.0000000 / 0.0000000 | 4.0000000 / 0.0000000 | 4.0000000 / 0.0000000 | 0.0000000 / 0.0000000 | 4.0000000 / 0.0000000 |

**(h)**

| | | | | | | |
|---|---|---|---|---|---|---|
| 32.375 / 0.000 | 0.000 / -6.539 | 4.125 / 0.000 | 0.000 / 3.711 | 0.000 / -3.711 | 4.125 / 0.000 | 0.000 / -6.539 |
| -1.796 / -1.796 | -1.119 / 1.119 | 0.088 / 0.088 | -0.088 / 0.088 | 0.088 / -0.088 | 0.088 / 0.088 | 1.119 / -1.119 |
| 0.000 / -0.125 | 0.125 / 0.000 | 0.000 / -0.375 | -0.125 / 0.000 | 0.125 / 0.000 | 0.000 / -0.375 | -0.125 / 0.000 |
| -0.205 / 0.205 | 0.088 / 0.088 | -0.088 / 0.088 | 0.088 / -0.088 | -0.119 / -0.119 | -0.088 / 0.088 | -0.088 / -0.088 |
| -0.375 / 0.000 | 0.000 / 0.125 | -0.125 / 0.000 | 0.000 / -0.125 | 0.000 / 0.125 | -0.125 / 0.000 | 0.000 / -0.125 |
| -0.205 / -0.205 | -0.088 / 0.088 | -0.088 / -0.088 | -0.119 / 0.119 | 0.119 / -0.119 | -0.088 / -0.088 | 0.088 / -0.088 |
| 0.000 / 0.125 | -0.125 / 0.000 | 0.000 / 0.375 | 0.125 / 0.000 | -0.125 / 0.000 | 0.000 / 0.375 | 0.125 / 0.000 |
| -1.796 / 1.796 | 1.119 / 1.119 | 0.088 / -0.088 | 0.088 / 0.088 | -0.088 / -0.088 | 0.088 / -0.088 | -1.119 / -1.119 |

Figure 3.24 Two-dimensional Fourier transform example: (g) input data, image cyclically shifted one unit in the horizontal direction; and (h) Fourier transform of shifted image. (Note that the magnitude transform remains invariant.)

137

The result of this computation is shown in Fig. 3.24e. Note that the average value term is now at point (4, 4). Also note that this term is real valued. Hermitian or conjugate symmetry about the origin may also be observed. Finally, it is interesting to observe the magnitude of the terms on the horizontal and vertical axes through the origin. Since the image array had a smooth edge transition in the North–South boundaries, the North–South frequency axis values are relatively small. However, the sharp East–West boundary in the image has produced large magnitude values along the East–West frequency axis as a result of the periodicity inherent in the sampled transform.

The inverse Fourier transform of the nonorigin centered array is shown in Fig. 3.24f. This array must be converted back into a real-valued integer array to match the original image. Although the accuracy of the algorithm is on the order of $10^{-6}$, round off error occurs as indicated by the negative signs in Fig. 3.24f, which are the result of negative numbers less than $10^{-7}$.

If the original sequence is cyclically shifted one unit in the horizontal direction, as shown in Fig. 3.24g, the transform, shown in Fig. 3.24h, has the same magnitude spectrum but a different phase spectrum.

### 3.3.3   Walsh–Hadamard Transform

The matrix factorization which led to the fast Fourier transform algorithm can be extended to several other classes of matrices which may be expressed as products of Good matrices or as Kronecker products. For example the Walsh–Hadamard and the Haar transforms may be implemented via fast algorithms (Andrews and Casparie, 1970). The Walsh–Hadamard transform is characteristic of this class and will now be described. This transform has been found useful in image coding, pattern recognition, and in sequency filtering.

Fourier analysis may be described as the representation of a function by a set of orthogonal sinusoidal waveforms. The coefficients of this representation are called frequency components and the waveforms are ordered by frequency. If one chooses to represent the functions by the orthogonal Walsh functions, which are square waves, a completely analogous spectral analysis may be performed with a simple computation. The Walsh functions may be ordered by the number of zero crossings or *sequency*, and the coefficients of the representation may be called sequency components. The computational simplification arises from the fact that the Walsh functions are real rather than complex and furthermore only take on the values $\pm 1$.

Let us first briefly consider the historical foundation. Rademacher (1922) described a set of orthonormal square wave functions, which are illustrated in Fig. 3.25. Although orthonormal, this set of functions is not complete, that is, cannot be used to represent an arbitrary function. Walsh (1923) introduced a

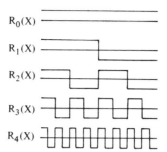

*Figure 3.25* First five Rademacher functions. [From Ahmed and Rao (1975).]

complete set of orthonormal square wave functions which are illustrated in Fig. 3.26. Walsh functions are defined on the interval $0 \le x \le 1$ by an iterative relationship

$$\phi_0(x) = 1, \qquad \phi_1(x) = \begin{cases} 1, & x < \frac{1}{2} \\ -1, & x > \frac{1}{2} \end{cases}$$

$$\phi_n(x) = \begin{cases} \phi_{[n/2]}(2x), & x < \frac{1}{2} \\ \phi_{[n/2]}(2x-1), & x \ge \frac{1}{2}, \quad n \text{ odd} \\ -\phi_{[n/2]}(2x-1), & x \ge \frac{1}{2}, \quad n \text{ even} \end{cases}$$

where $[n/2]$ is the integer part of $n/2$. The first two functions, $\phi_0(x)$ and $\phi_1(x)$, may be considered to be the fundamental even and odd functions about $x = \frac{1}{2}$. Successive functions are generated in even and odd function pairs. For $n = 2$, $\phi_2(x)$ is formed by compressing $\phi_1(x)$ into the first half of the interval and $-\phi_1(x)$ into the second half, resulting in an even function about

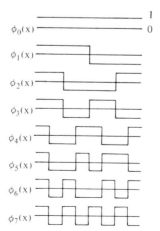

*Figure 3.26* First eight Walsh functions. [From Ahmed and Rao (1975).]

$x = \frac{1}{2}$. For $n = 3$, $\phi_3(x)$ is formed by compressing $\phi_1(x)$ into the first half and into the second half, resulting in an odd function. Harmuth (1972) has developed terminology for the Walsh functions analogous to that for the sinusoidal functions. The pairing of even and odd functions termed *cal* and *sal*, respectively, is analogous to the cosine and sine functions. The property analogous to frequency is half the number of zero crossings in the interval. The sequency of a Walsh function is sometimes defined to be one-half the number of zero crossings, which is analogous to the frequency of a sinusoidal function if the zero crossing at the left side of the interval is counted but not the one at the right side. However, an important property of the Walsh functions is that they can be ordered by the number of zero crossings in the open interval $(0, 1)$, and *sequency* will be used to describe this property.

A set of discrete Walsh functions may be developed by sampling the continuous functions at equispaced points in the interval $[0, 1]$. The number of samples should be a power of 2 to preserve the even and odd function pairing and to permit a sample in each function interval. If the Walsh functions with number of zero crossings less than or equal to $2^n - 1$ are sampled at $N = 2^n$ uniformly spaced points, a square matrix is produced as illustrated in Fig. 3.27 for $N = 8$. These matrices are orthogonal and the rows are ordered with increasing number of zero crossings. Except for the ordering of the rows, the discrete Walsh matrices are equivalent to the Hadamard matrices of rank $2^n$, (Harmuth, 1972). A Hadamard matrix $\mathbf{H}$ is an orthogonal matrix with elements of value $+1$ and $-1$ only. For $n$ greater than 2, it is known that if a Hadamard matrix of rank $n$ exists, then $n = 4m$, where $m$ is an integer. Since the Walsh–Hadamard matrix is orthogonal, the column vectors form a basis vector set. The two-dimensional Walsh–Hadamard basis matrices for $N = 8$ are shown in Fig. 3.28. Hadamard matrices are easily constructed for $N = 2^n$ by the following procedure. The order $N = 2$ Hadamard matrix is

$$\mathbf{H}_2 = \begin{bmatrix} 1 & 1 \\ 1 & -1 \end{bmatrix}.$$

The Hadamard matrix of order $2N$ may be generated by the Kronecker product operation:

$$\mathbf{H}_{2N} = \begin{bmatrix} \mathbf{H}_N & \mathbf{H}_N \\ \mathbf{H}_N & -\mathbf{H}_N \end{bmatrix}.$$

$$
\begin{bmatrix}
\text{Wal}(0,n) \\
\text{Wal}(1,n) \\
\text{Wal}(2,n) \\
\text{Wal}(3,n) \\
\text{Wal}(4,n) \\
\text{Wal}(5,n) \\
\text{Wal}(6,n) \\
\text{Wal}(7,n)
\end{bmatrix}
=
\begin{bmatrix}
1 & 1 & 1 & 1 & 1 & 1 & 1 & 1 \\
1 & 1 & 1 & 1 & -1 & -1 & -1 & -1 \\
1 & 1 & -1 & -1 & -1 & -1 & 1 & 1 \\
1 & 1 & -1 & -1 & 1 & 1 & -1 & -1 \\
1 & -1 & -1 & 1 & 1 & -1 & -1 & 1 \\
1 & -1 & -1 & 1 & -1 & 1 & 1 & -1 \\
1 & -1 & 1 & -1 & -1 & 1 & -1 & 1 \\
1 & -1 & 1 & -1 & 1 & -1 & 1 & -1
\end{bmatrix}
$$

*Figure 3.27* Discrete Walsh matrix for $N = 8$. [From Ahmed and Rao (1975).]

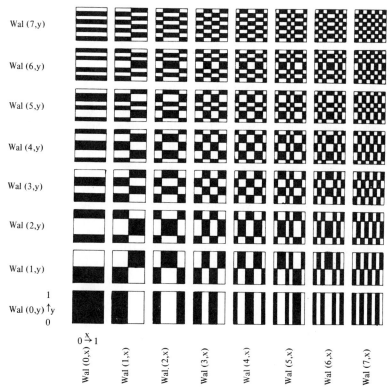

*Figure 3.28* Two-dimensional Walsh–Hadamard basis matrices for $N = 8$. Black represents $+1/N$ and white represents $-1/N$. [From Harmuth (1972).]

The sequency interpretation introduced by Harmuth for Walsh functions may be related to the elements of the Hadamard matrix by considering the rows to correspond to rectangular waves ranging between $\pm 1$ with a subperiod of $1/N$ units. In the construction given, each zero crossing number between 0 and $N - 1$ will appear in a "natural" rather than ordered manner in the rows of the matrix.

Since the Hadamard matrix of order $N = 2^n$ may be generated from the order 2 core matrix it is not necessary or desirable to store the entire matrix. Rather one may store an algorithm for generating the general element of the matrix.

The element located in row $i$ and column $j$ of $H_N$ may be expressed as the product of $n$ appropriate elements of the core matrix. Furthermore, note that if the elements of the core matrix $\mathbf{H}_2$ are addressed by a binary row index $i$ and column index $j$, then the element value is given by

$$\mathbf{H}_2(i,j) = (-1)^{ij},$$

where $i, j = 0$ or 1, and the product operation may be taken to be logical "and" or arithmetic multiplication. Similarly, if the elements of the $\mathbf{H}_4$ matrix are addressed by the binary representations

$$i = i_1 i_2, \qquad j = j_1 j_2,$$

the matrix elements are given by

$$\mathbf{H}_4(i,j) = (-1)^{i_1 j_1 \oplus i_2 j_2},$$

where $\oplus$ indicates modulo 2 addition. By induction it is easily shown that if the matrix elements are addressed by their binary representations

$$i = i_1 i_2 \cdots i_n, \qquad j = j_1 j_2 \cdots j_n,$$

then the element in row $i$ and column $j$ is given by

$$\mathbf{H}_N(i,j) = (-1)^p, \qquad \text{where} \quad p = \sum_{k=1}^{n} i_k j_k$$

and modulo 2 addition is used. Therefore any element, row, or column may be easily generated without storing the entire matrix.

The Walsh–Hadamard matrix is also easily generated in sequency ordered form. The element in row $i$ and column $j$ is given by

$$\mathbf{H}_N(i,j) = (-1)^q,$$

where

$$q = \sum_{k=1}^{n} g_k(i) j_i$$

and

$$g_1(i) = i_n,$$
$$g_2(i) = i_n \oplus i_{n-1},$$
$$g_3(i) = i_{n-1} \oplus i_{n-2},$$
$$\vdots$$
$$g_n(i) = i_2 \oplus i_1.$$

The two-dimensional Walsh–Hadamard transform of order $N = 2^n$ may therefore be based on the normalized Walsh–Hadamard matrix

$$\mathbf{H} = N^{-1/2} \mathbf{H}_N,$$

and may be defined in matrix form as

$$\mathbf{F} = \mathbf{H}[f]\mathbf{H}.$$

Since the $\mathbf{H}$ matrix is symmetric and orthogonal, the inverse transformation is simply given by

$$[f] = \mathbf{HFH}.$$

The sequency ordered Walsh–Hadamard transform of the test image is shown in Fig. 3.29. Each Walsh–Hadamard transform may be considered to

| 259 | 41 | -57 | 33 | -33 | 25 | -41 | 29 |
|---|---|---|---|---|---|---|---|
| 9 | 3 | -7 | -1 | 1 | -1 | -3 | -1 |
| -25 | -7 | 19. | 1 | -1 | 5 | 7 | 1 |
| 1 | -1 | 1 | 3 | -3 | -1 | 1 | -1 |
| -1 | 1 | -1 | -3 | 3 | 1 | -1 | 1 |
| -7 | -1 | 5 | -1 | 1 | 3 | 1 | -1 |
| -9 | -3 | 7 | 1 | -1 | 1 | 3 | 1 |
| -3 | -1 | 1 | -1 | 1 | -1 | 1 | 3 |

*Figure 3.29* Sequency ordered Walsh–Hadamard transform of test image (not normalized by $\frac{1}{8}$).

(a)

| 4.0 | 4.0 | 4.0 | 4.0 | 4.0 | 4.0 | 0.0 | 4.0 |
|---|---|---|---|---|---|---|---|
| 0.0 | 0.0 | 0.0 | 0.0 | 0.0 | 0.0 | 0.0 | 0.0 |
| 5.0 | 5.0 | 5.0 | 5.0 | 5.0 | 4.0 | 0.0 | 4.0 |
| 0.0 | 0.0 | 0.0 | 0.0 | 0.0 | 0.0 | 0.0 | 0.0 |
| 5.0 | 6.0 | 6.0 | 6.0 | 5.0 | 4.0 | 0.0 | 4.0 |
| 0.0 | 0.0 | 0.0 | 0.0 | 0.0 | 0.0 | 0.0 | 0.0 |
| 5.0 | 6.0 | 7.0 | 6.0 | 5.0 | 4.0 | -0.0 | 4.0 |
| 0.0 | 0.0 | 0.0 | 0.0 | 0.0 | 0.0 | 0.0 | 0.0 |
| 5.0 | 6.0 | 6.0 | 6.0 | 5.0 | 4.0 | 0.0 | 4.0 |
| 0.0 | 0.0 | 0.0 | 0.0 | 0.0 | 0.0 | 0.0 | 0.0 |
| 5.0 | 5.0 | 5.0 | 5.0 | 5.0 | 4.0 | 0.0 | 4.0 |
| 0.0 | 0.0 | 0.0 | 0.0 | 0.0 | 0.0 | 0.0 | 0.0 |
| 4.0 | 4.0 | 4.0 | 4.0 | 4.0 | 4.0 | 0.0 | 4.0 |
| 0.0 | 0.0 | 0.0 | 0.0 | 0.0 | 0.0 | 0.0 | 0.0 |
| 4.0 | 4.0 | 4.0 | 4.0 | 4.0 | 4.0 | 0.0 | 4.0 |
| 0.0 | 0.0 | 0.0 | 0.0 | 0.0 | 0.0 | 0.0 | 0.0 |

(b)

| 259 | 57 | -41 | 33 | 33 | -41 | 25 | -29 |
|---|---|---|---|---|---|---|---|
| 9 | 7 | -3 | -1 | -1 | -3 | -1 | 1 |
| -25 | -19 | 7 | 1 | 1 | 7 | 5 | -1 |
| 1 | -1 | 1 | 3 | 3 | 1 | -1 | 1 |
| -1 | 1 | -1 | -3 | -3 | -1 | 1 | -1 |
| -7 | -5 | 1 | -1 | -1 | 1 | 3 | 1 |
| -9 | -7 | 3 | 1 | 1 | 3 | 1 | -1 |
| -3 | -1 | 1 | -1 | -1 | 1 | -1 | -3 |

(c)

| 4 | 4 | 4 | 4 | 4 | 4 | 0 | 4 |
|---|---|---|---|---|---|---|---|
| 5 | 5 | 5. | 5 | 5 | 4 | 0 | 4 |
| 5 | 6 | 6 | 6 | 5 | 4 | 0 | 4 |
| 5 | 6 | 7 | 6 | 5 | 4 | 0 | 4 |
| 5 | 6 | 6 | 6 | 5 | 4 | 0 | 4 |
| 5 | 5 | 5 | 5 | 5 | 4 | 0 | 4 |

*Figure 3.30* (a) Shifted test images, (b) corresponding Walsh–Hadamard transform, and (c) inverse Hadamard transform.

be the summed product or nonnormalized correlation of the corresponding basis matrix and the original image. Thus, the $(1, 1)$ element may be interpreted as the average value, the $(2, 1)$ element as the correlation with a single step horizontal edge, element $(1, 2)$ as the correlation with a single step vertical edge, element $(3, 1)$ as the correlation with a horizontal line, etc. The transform of the shifted image is shown in Fig. 3.30. Note that this transform is not invariant to circular shifts. It is invariant to dyadic shifts (Ahmed and Rao, 1975).

### 3.3.4 Haar Transform

The Haar transform is based on a class of orthogonal matrices whose elements are either 1, $-1$, or 0 multiplied by powers of $\sqrt{2}$. The orthonormal Haar transform is another computationally efficient image transform. The

$$H = \frac{1}{8} \begin{bmatrix} 1 & 1 & 1 & 1 & 1 & 1 & 1 & 1 \\ 1 & 1 & 1 & 1 & -1 & -1 & -1 & -1 \\ \sqrt{2} & \sqrt{2} & -\sqrt{2} & -\sqrt{2} & 0 & 0 & 0 & 0 \\ 0 & 0 & 0 & 0 & \sqrt{2} & \sqrt{2} & -\sqrt{2} & -\sqrt{2} \\ 2 & -2 & 0 & 0 & 0 & 0 & 0 & 0 \\ 0 & 0 & 2 & -2 & 0 & 0 & 0 & 0 \\ 0 & 0 & 0 & 0 & 2 & -2 & 0 & 0 \\ 0 & 0 & 0 & 0 & 0 & 0 & 2 & -2 \end{bmatrix}$$

*Figure 3.31*   Orthonormal Haar matrix for $N = 2^3$.

transform of an $N$-point vector requires only $2(N-1)$ additions and $N$ multiplications. The Haar matrix for $N = 8$ is shown in Fig. 3.31. Note that the product of the Haar matrix with a vector results in a rough coarse-to-fine sampling. The first element gives the mean value of the components. The second results in an average difference of the first $\frac{1}{2}N$ components and the second $\frac{1}{2}N$ components. The remaining elements of the product measure adjacent differences of data elements taken four at a time or two at a time. Since these terms depend only on local properties of the data, the Haar

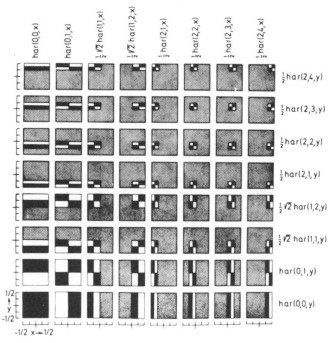

*Figure 3.32*   Two-dimensional Haar basis matrices for $N = 8$. [From Harmuth (1972).]

| 205.6 | 45.6 | 86.4 | -61.6 | 117.0 | -11.0 | -63.0 | -63.0 |
|---|---|---|---|---|---|---|---|
| 7.8 | -0.2 | 0.2 | -3.8 | 3.2 | 3.2 | -5.2 | -5.2 |
| -15.9 | 1.1 | 1.8 | 7.4 | -6.6 | -6.6 | 9.4 | 9.4 |
| 17.3 | 0.3 | 2.5 | -8.8 | 3.2 | 3.2 | -8.8 | -8.8 |
| -8.8 | -0.8 | -3.2 | 4.8 | -0.8 | -0.8 | 4.8 | 4.8 |
| 0 | 0 | 0 | 0 | -2.8 | -2.8 | 2.8 | 2.8 |
| 6.8 | -1.2 | -2.8 | -2.8 | 2.8 | 2.8 | -2.8 | -2.8 |
| 0 | 0 | 0 | 0 | 0 | 0 | 0 | 0 |

*Figure 3.33*   Haar transform of test image.

transform may be called both locally and globally sensitive in contrast to the Fourier and Walsh–Hadamard, which are globally sensitive.

The Haar transform of an image may be computed by

$$[F(u,v)] = N^{-1}\mathbf{H}[f(x,y)]\mathbf{H},$$

and the inverse transform by

$$[f(x,y)] = N^{-1}\mathbf{H}[F(u,v)]\mathbf{H},$$

where the Haar matrix is obtained by sampling the set of Haar functions.

The two-dimensional basis matrices for $N=8$ are shown in Fig. 3.32. The Haar transform of the test image is shown in Fig. 3.33. Each of the transform coefficients may be considered proportional to the correlation of the image and the corresponding basis matrix.

### 3.3.5   Slant Transform

The slant transform was introduced by Enomoto and Shibata (1971) as an orthogonal transform containing sawtooth waveforms or "slant" basic vectors. The slant matrix is constructed so that it is orthogonal, has a constant basis vector, a slant basis vector that is monotonically decreasing in constant steps from maximum to minimum, has the sequency property, and has a fast computational algorithm. Let $\mathbf{S}_N$ denote the $N$ by $N$ slant matrix with $N=2^n$. Then

$$\mathbf{S}_2 = \frac{1}{\sqrt{2}}\begin{bmatrix} 1 & 1 \\ 1 & -1 \end{bmatrix}.$$

The $\mathbf{S}_4$ matrix is obtained by the following operation:

$$\mathbf{S}_4 = \begin{bmatrix} 1 & 0 & 1 & 0 \\ a & b & -a & b \\ 0 & 1 & 0 & -1 \\ -b & a & b & a \end{bmatrix}\begin{bmatrix} \mathbf{S}_2 & 0 \\ 0 & \mathbf{S}_2 \end{bmatrix}.$$

$$S_N = \left[\begin{array}{cc|cc|cc}
\begin{matrix} 1 & 0 \\ & \\ a_N & b_N \end{matrix} & \underline{0} & \begin{matrix} 1 & 0 \\ & \\ -a_N & b_N \end{matrix} & \underline{0} & S_{N/2} & \underline{0} \\
\hline
\underline{0} & I_{(N/2)-2} & \underline{0} & I_{(N/2)-2} & & \\
\hline
\begin{matrix} 0 & 1 \\ & \\ -b_N & a_N \end{matrix} & \underline{0} & \begin{matrix} 0 & -1 \\ & \\ b_N & a_N \end{matrix} & \underline{0} & \underline{0} & S_{N/2} \\
\hline
\underline{0} & I_{(N/2)-2} & \underline{0} & -I_{(N/2)-2} & &
\end{array}\right]\left[\begin{array}{cc} & \\ & \end{array}\right]$$

*Figure 3.34*   Slant matrix recursive relationship. [From Ahmed and Rao (1975, p. 166).]

The requirement of uniform step size leads to the condition

$$a = 2b.$$

The orthogonality condition requires that

$$b = 1/\sqrt{5}\,.$$

Therefore the $S_4$ matrix becomes

$$S_4 = \begin{bmatrix}
1 & 1 & 1 & 1 \\
\dfrac{3}{\sqrt{5}} & \dfrac{1}{\sqrt{5}} & \dfrac{-1}{\sqrt{5}} & \dfrac{-3}{\sqrt{5}} \\
1 & -1 & -1 & 1 \\
\dfrac{1}{\sqrt{5}} & \dfrac{-3}{\sqrt{5}} & \dfrac{3}{\sqrt{5}} & \dfrac{-1}{\sqrt{5}}
\end{bmatrix}.$$

Note that $S_4$ does have the sequency property; the rows are ordered by the number of sign changes. The general form of the recursive relation between the slant matrix of order $N$ and $N-1$ is shown in Fig. 3.34, where $I_n$ represents a $n \times n$ identity matrix and $\underline{0}$ is a zero matrix. The constants $a_N$ and $b_N$ may be computed from the relations

$$a_{2N} = \left(\frac{3N^2}{4N^2-1}\right)^{1/2}, \qquad b_{2N} = \left(\frac{N^2-1}{4N^2-1}\right)^{1/2}, \qquad N = 2,3,4,\ldots.$$

A fast computational algorithm is a consequence of the matrix factorization in the definition of the slant matrix.

The slant transform of an image may be computed by

$$F = S[f]S'.$$

The inverse transform is given by

$$[f] = S'FS.$$

The slant matrix for $N=8$ is shown in Fig. 3.35a. The slant transform of the sampled image is shown in Fig. 3.35b.

| 1 | 1 | 1 | 1 | 1 | 1 | 1 | 1 |
|---|---|---|---|---|---|---|---|
| 1.528 | 1.091 | 0.655 | 0.218 | -0.218 | -0.655 | -1.091 | -1.528 |
| 1 | -1 | -1 | 1 | 1 | -1 | -1 | 1 |
| 0.447 | -1.342 | 1.342 | -0.447 | 0.447 | -1.342 | 1.342 | -0.447 |
| 1.342 | 0.447 | -0.447 | -1.342 | -1.342 | -0.477 | 0.477 | 1.342 |
| 0.683 | -0.098 | -0.878 | -1.659 | 1.659 | 0.878 | 0.098 | -0.683 |
| 1 | -1 | -1 | 1 | -1 | 1 | 1 | -1 |
| 0.447 | -1.342 | 1.342 | -0.447 | -0.447 | 1.342 | -1.342 | 0.447 |

(a)

| 257.8 | -40.4 | -11.2 | -21.3 | 58.0 | -40.6 | 60.1 | -6.5 |
|---|---|---|---|---|---|---|---|
| 7.4 | -2.5 | -0.1 | -3.0 | 1.9 | -4.9 | 0.6 | 0.8 |
| -1.8 | -2.3 | 1.6 | 1.0 | -1.0 | -1.6 | 2.3 | 1.8 |
| -3.2 | -0.7 | -0.3 | 2.1 | -2.1 | 0.3 | 0.7 | 3.2 |
| -25.0 | 10.6 | -0.4 | 10.1 | -5.8 | 18.6 | -3.6 | -4.3 |
| -3.7 | 4.0 | -2.1 | 1.7 | -1.0 | 4.9 | -2.9 | -0.9 |
| -7.3 | 1.5 | 1.8 | 2.6 | -1.2 | 4.0 | 0.8 | -2.0 |
| 3.3 | -0.7 | -0.8 | -1.2 | 0.5 | -1.8 | -0.3 | 0.9 |

(b)

*Figure 3.35*   (a) Slant transform matrix for $N=8$ and (b) slant transform of sample image.

### 3.3.6   Discrete Cosine Transform

The discrete cosine transform (DCT) consists of a set of basis vectors that are sampled cosine functions. The transform matrix **C** may be written

$$\mathbf{C} = N^{-1/2}[C_{kl}],$$

where

$$C_{kl} = \begin{cases} 1, & l=0 \\ \sqrt{2}\cos\left[\dfrac{(2k+1)l\pi}{2N}\right], & \begin{array}{l} k=0,1,\ldots,N-1, \\ l=1,2,\ldots,N-1. \end{array} \end{cases} \quad \text{and}$$

The basis vectors of the DCT are related to a class of discrete Chebyshev polynomials (Ahmed and Rao, 1975), which are orthogonal polynomials. Thus, the DCT is an orthogonal transform. A fundamentally important property of the DCT is that its basis vectors are directly related to the eigenvectors of a corresponding Toeplitz matrix. The DCT may also be computed with a fast algorithm. In fact, all $N$ coefficients of the DCT may be computed using a $2N$ point FFT.

The two-dimensional DCT may be defined as

$$[F(u,v)] = \mathbf{C}[f(x,y)]\mathbf{C}'$$

and the inverse transform

$$[f(x,y)] = \mathbf{C}'[F(u,v)]\mathbf{C}.$$

The DCT matrix for $N=8$ is shown in Fig. 3.36a. The discrete cosine transform of the sample image is shown in Fig. 3.36b.

$$
C(8) = \begin{bmatrix}
0.354 & 0.354 & 0.354 & 0.354 & 0.354 & 0.354 & 0.354 & 0.354 \\
0.490 & 0.416 & 0.278 & 0.098 & -0.098 & -0.278 & -0.416 & -0.490 \\
0.462 & 0.191 & -0.191 & -0.462 & -0.462 & -0.191 & 0.191 & 0.462 \\
0.416 & -0.098 & -0.490 & -0.278 & 0.278 & 0.490 & 0.098 & -0.416 \\
0.354 & -0.354 & -0.354 & 0.354 & 0.354 & -0.354 & -0.354 & 0.354 \\
0.278 & -0.490 & 0.098 & 0.416 & -0.416 & -0.098 & 0.490 & -0.278 \\
0.191 & -0.462 & 0.462 & -0.191 & -0.191 & 0.462 & -0.462 & 0.191 \\
0.098 & -0.278 & 0.416 & -0.490 & -0.490 & -0.416 & 0.278 & -0.098
\end{bmatrix}
$$

(a)

| 33.5 | -5.9 | -1.1 | 2.8 | -2.3 | 3.5 | -1.1 | -5.6 |
|------|------|------|-----|------|-----|------|------|
| -4.5 | 0.8 | 0.3 | -0.3 | 0.3 | -0.4 | 0.1 | 0.7 |
| -4.0 | 0.8 | 1.7 | 0.3 | 0.0 | -0.1 | 0.1 | 0.3 |
| 6.9 | -1.2 | 0.9 | 1.1 | -0.7 | 0.9 | -0.3 | -1.4 |
| -7.2 | 1.2 | -0.5 | -1.0 | 0.7 | -0.8 | 0.3 | 1.4 |
| 8.2 | -1.4 | 0.7 | 0.8 | -0.4 | 1.8 | -0.3 | -1.7 |
| -5.3 | 0.9 | -0.3 | -0.6 | 0.4 | -0.7 | 0.2 | 0.9 |
| -2.4 | 0.4 | -0.1 | -0.3 | 0.2 | -0.4 | 0.0 | 0.8 |

(b)

*Figure 3.36* Discrete cosine transform. (a) Matrix for $N = 8$ and (b) DCT of test image. [From Ahmed and Rao (1975, p. 178).]

### 3.3.7 Number Theoretic Transform

The discrete Fourier transform computed via the FFT has greatly expanded the applications of digital signal processing. For image processing the amount of data is so large that improvements in accuracy and speed are still welcomed. Such improvements are indeed possible as will be shown in this section concerning number theoretic transforms (NTT) or finite field transforms. This new concept that has evolved uses only integer arithmetic, promises zero round off error, has an efficient algorithm, and can compute convolutions.

Pollard (1971) defined fast Fourier transforms on a finite field and on a finite ring of integers. Rader (1972) defined transforms over residue classes of integers modulo either a Mersenne prime or Fermat prime. Agarwal and Burrus (1975) extended Rader's transform for Fermat primes for implementing fast digital convolution. Reed and Truong (1976) generalized the method to transform over the Galois field $GF(p^2)$, where $p$ is a Mersenne prime, and to complex integers. The number theoretic transform has also been applied to two-dimensional filtering by Rader (1975) and to image processing and reconstruction from projection by Kwoh (1977). Before introducing this transform, it is useful to review some basic concepts of modular arithmetic. A more detailed discussion may be found in books on number theory such as those by Szabo and Tanaka (1969) and Young and Gregory (1973).

Two integers $n$ and $m$ are said to be *congruent modulo M* if

$$n = m + kM,$$

where $k$ is some integer and $M$ is called the *modulus*. This relation is usually written

$$n \equiv m \quad (\bmod M).$$

All integers are congruent mod $M$ to some integer in the set $Z_M$,

$$Z_M = \{0, 1, 2, \ldots, M-1\}.$$

This set $Z_M$ of integers under addition and multiplication mod $M$ constitutes a finite commutative ring. If, in a ring of integers, multiplicative inverses exist for all nonzero integers, the ring becomes a field. It can be shown that $Z_M$ under addition and multiplication mod $M$ is a field if and only if $M$ is a prime. A field with a prime number of elements $p$ is called a Galois field $GF(p)$. An excellent brief history of Galois' work may be found in Kramer (1970).

Some examples of basic arithmetic operations which are permissible with modular arithmetic will now be given (Agarwal and Burrus, 1975):

(1)   Addition:   $5 + 4 = 9 \equiv 2 \ (\bmod 7)$.
(2)   Negation:   $-5 = -5 + 7 \equiv 2 \ (\bmod 7)$.
(3)   Subtraction:   $9 - 5 = 9 + (-5) = 9 + 2 \equiv 4 \ (\bmod 7)$.
(4)   Multiplication:   $9 \times 5 = 45 \equiv 3 \ (\bmod 7)$.
(5)   Multiplicative Inverse:   $3^{-1} \equiv 5 \ (\bmod 7)$. The multiplicative inverse of an integer $n$ in $Z_M$ exists if and only if $n$ and $M$ are relatively prime (have no common divisor), in that case the inverse $n^{-1}$ is an integer such that $n \times n^{-1} \equiv 1 \ (\bmod M)$.
(6)   Division:   $5/3 = 5 \times 3^{-1} = 5 \times 5 \equiv 4 \ (\bmod 7)$. The quotient $n/m$ exists if and only if $m$ has an inverse. In that case $n/m = n \times m^{-1}$.

To extend residue arithmetic to transforms and convolutions, let us briefly look at the regular DFT structure

$$F(u) = \sum_{x=0}^{N-1} f(x) r^{ux},$$

where $r$ is a root of unity of order $N$,

$$r^N = 1.$$

The corresponding root of unity concept for modular arithmetic is based upon Euler's $\phi$ function and Euler's theorem.

Euler's $\phi$ function $\phi(M)$ is defined as the number of integers in $Z_M$ that are relative prime to $M$. If $M$ is a prime, then $\phi(M) = M - 1$. If $M$ is a composite and represented in prime factored form as

$$M = p_1^{n_1} p_2^{n_2} \cdots p_l^{n_l},$$

then the general expression for $\phi(M)$ is

$$\phi(M) = M(1 - 1/p_1)(1 - 1/p_2) \cdots (1 - 1/p_l).$$

Euler's theorem states that for every number $r$ which is relatively prime to $M$

$$r^{\phi(M)} \equiv 1 \quad (\mathrm{mod}\, M).$$

If $M$ is prime, this reduces to Fermat's theorem

$$r^{M-1} \equiv 1 \quad (\mathrm{mod}\, M).$$

If $N$ is the least positive integer such that

$$r^N \equiv 1 \quad (\mathrm{mod}\, M),$$

then $r$ is called a *root of unity of order N.*

The existence of a discrete Fourier-like transform will now be given in the form of a theorem (Agarwal and Burrus, 1975).

**Theorem.** A length-$N$ transform having the DFT structure given below will implement cyclic convolution if and only if there exists an inverse of $N$ and an element $r$ which is a root of unity of order $N$. The transform pair is given by

$$F(u) = \sum_{x=0}^{N-1} f(x) r^{ux}, \qquad f(x) = N^{-1} \sum_{u=0}^{N-1} F(u) r^{-ux},$$

and

$$r^N \equiv 1 \quad (\mathrm{mod}\, M).$$

The circular convolution property ensures that the convolution of two finite sequences $f(x)$ and $h(x)$ can be obtained as the inverse transform of the product of the transforms. If

$$F(u) = \sum_{x=0}^{N-1} f(x) r^{ux}, \qquad H(u) = \sum_{x=0}^{N-1} h(x) r^{ux}, \qquad \text{and} \qquad G(u) = F(u) H(u),$$

then the inverse transform of $G(u)$,

$$g(x) = N^{-1} \sum_{u=0}^{N-1} G(u) r^{-ux},$$

is exactly equal to the convolution

$$g(x) = \sum_{y=0}^{N-1} f(y) h(x - y),$$

where $(x - y)$ denotes $(x - y) \bmod N$.

Let us now consider an example to illustrate the steps in the fast number theoretic (FNT) convolution. Consider the $N = 4$ point sequences whose circular convolution is desired:

$$\mathbf{f} = (2, -2, 1, 0), \qquad \mathbf{h} = (1, 2, 0, 0).$$

Let the prime be given as a Fermat number of the form $2^b+1$, $b=2^t$, with $t=2$, i.e., $M=2^4+1=17$. The root of unity $r=4$ will also be considered given. As a check note that

$$4^4 \equiv 1 \quad (\mathrm{mod}\, 17).$$

The transform may be written in matrix form as

$$\mathbf{F}=\mathbf{Tf},$$

where the matrix $\mathbf{T}$ is given by

$$\mathbf{T}=\begin{bmatrix} 1 & 1 & 1 & 1 \\ 1 & 4 & 4^2 & 4^3 \\ 1 & 4^2 & 4^4 & 4^6 \\ 1 & 4^3 & 4^6 & 4^9 \end{bmatrix}.$$

Using modular arithmetic $\mathbf{T}$ can be reduced to

$$\mathbf{T}\equiv\begin{bmatrix} 1 & 1 & 1 & 1 \\ 1 & 4 & 16 & 13 \\ 1 & 16 & 1 & 16 \\ 1 & 13 & 16 & 4 \end{bmatrix} \quad (\mathrm{mod}\, 17).$$

The transform of **f** can now be computed as

$$\mathbf{F}=\begin{bmatrix} 1 & 1 & 1 & 1 \\ 1 & 4 & 16 & 13 \\ 1 & 16 & 1 & 16 \\ 1 & 13 & 16 & 4 \end{bmatrix}\begin{bmatrix} 2 \\ -2 \\ 1 \\ 0 \end{bmatrix}=\begin{bmatrix} 18 \\ 78 \\ 243 \\ 213 \end{bmatrix}\equiv\begin{bmatrix} 1 \\ 10 \\ 5 \\ 9 \end{bmatrix} \quad (\mathrm{mod}\, 17).$$

Similarly, the transform of **h** can be computed to be

$$\mathbf{H}=(3,9,16,10)' \quad (\mathrm{mod}\, 17).$$

The product of the transforms is therefore

$$\mathbf{G}=(3,90,80,90)\equiv(3,5,12,5) \quad (\mathrm{mod}\, 17).$$

To compute the inverse transform matrix $\mathbf{T}^{-1}$ first note that

$$4^{-1}\equiv -4 \quad (\mathrm{mod}\, 17).$$

The inverse transformation matrix is given by

$$\mathbf{T}^{-1}=4^{-1}\begin{bmatrix} 1 & 1 & 1 & 1 \\ 1 & 4^{-1} & 4^{-2} & 4^{-3} \\ 1 & 4^{-2} & 4^{-4} & 4^{-6} \\ 1 & 4^{-3} & 4^{-6} & 4^{-9} \end{bmatrix},$$

which reduces to

$$\mathbf{T}^{-1}\equiv 13\begin{bmatrix} 1 & 1 & 1 & 1 \\ 1 & 13 & 16 & 4 \\ 1 & 16 & 1 & 16 \\ 1 & 4 & 16 & 13 \end{bmatrix} \quad (\mathrm{mod}\, 17).$$

Therefore, the inverse transform of **G** can be computed to be

$$\mathbf{g} \equiv (2, 2, 14, 2) \quad (\bmod\ 17).$$

This result may be compared with the direct computation. For example, the first convolution element is

$$\begin{bmatrix} 2 & -2 & 1 & 0 \end{bmatrix} \begin{bmatrix} 0 \\ 0 \\ 2 \\ 1 \end{bmatrix} = 2.$$

The total direct convolution vector is $(2, 2, -3, 2)$, which compares with the transform-computed **g** when we note that $14 - 17 = -3$. This example has illustrated the structure and computation of the transform matrices and integer computation. The method of scaling to and selection of primes will now be considered.

In the use of the number theoretic transform, three parameters must be chosen: $M$, $N$, and $\alpha$. Note that the prime number $M$ must, of course, be greater than the integers encountered in the data sequences **f** and **h**. For convolution, $M$ must also be greater than the convolution values **f∗h**. If the sequences **f** and **h** are in the range

$$A \leq \mathbf{f}, \mathbf{h} \leq B,$$

then it is shown by Reed *et al.* (1977) from considering the probability of overflow that to keep the circular convolution in the interval

$$-\tfrac{1}{2}(M-1) \leq g_K \leq \tfrac{1}{2}(M-1)$$

then

$$AB = 6(M-1)/\alpha\sqrt{N}\ .$$

For example, suppose that the sequence **f** is represented by a $K_1$-bit signed binary number, and sequence **h** is represented by a $K_2$-bit signed binary number. Then

$$A = 2 \cdot 2^{K_1} = 2^{K_1 + 1}, \qquad B = 2 \cdot 2^{K_2} = 2^{K_2 + 1}$$

and

$$A \cdot B = 2^{K_1 + K_2 + 2} = 6(M-1)/\alpha\sqrt{N}\ .$$

If $N = 2^8$, $\alpha = 3$, and $M = 45 \times 2^{29} + 1$, then

$$2^{K_1 + K_2} \sim 2^{30},$$

and therefore $K_1 + K_2 = 30$ bits with a small probability of overflow. The selection of the parameters therefore depends upon the binary representations of the data sequences. Further constraints are given by the existence of primes and computation word length. A table of primes and a computer program for number theoretic convolution on a PDP-10 computer may be found in Kwoh (1977).

The two-dimensional number theoretic transform is similar to the two-dimensional FFT, since the transform kernel is product separable. The transform may be written in the series form

$$F(u,v) = \sum_{x=0}^{N_1-1} \sum_{y=0}^{N_2-1} f(x,y) r_1^{ux} r_2^{vy}.$$

The inverse transform is given by

$$f(x,y) = N_1^{-1} N_2^{-1} \sum_{u=0}^{N_1-1} \sum_{v=0}^{N_2-1} F(u,v) r_1^{-ux} r_2^{-vy}.$$

Let us consider the following parameters. For image processing, the picture sizes $N_1$ and $N_2$ are generally powers of 2 such as 256, 512, or 1024. Also, the number of bits to represent gray levels in a picture $\mathbf{f}$ is 8 or less. Also, assume use of the PDP-10 computer, which has a word length of 36 bits. Furthermore, let us assume that $N_1 = N_2 = N = 2^n$ and $r_1 = r_2 = r$. Now let us consider the constraints on the parameters $N$, $r$, and the prime $M$.

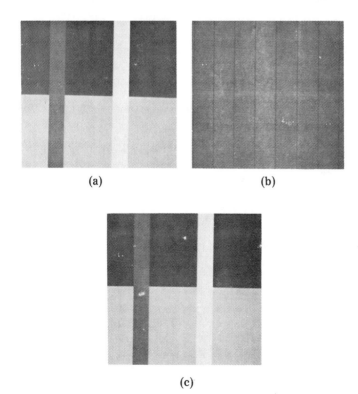

(a)                                         (b)

(c)

*Figure 3.37*   Example of the NTT: (a) original picture, (b) NTT transform, and (c) picture resulting from inverse NTT transform of (b). [From Kwoh (1977).]

The largest prime number of the form $M = K2^n + 1$ which fits the 36-bit word length excluding the sign bit is

$$M = 45 \cdot 2^{29} + 1.$$

By Fermat's theorem

$$2^{45 \cdot 2^{29}} \equiv 1 \quad (\text{mod } M),$$

since

$$\left(2^{45}\right)^{2^{29}} = 2^{45 \cdot 2^{29}}.$$

Therefore, the root of unity of order $N = 2^k$ is given by

$$r \equiv 2^{45 \cdot 2^{29-k}} \quad (\text{mod } M),$$

where $0 \leq k \leq 29$. Therefore, a variety of roots and transform sizes are available. The computation is described more completely by Kwoh (1977).

A pictorial example from Kwoh (1977) is shown in Fig. 3.37. A test image was constructed using a 512 by 512 array with 8 bits of intensity at each element as shown in Fig. 3.37a. The two-dimensional NTT of the test picture is shown in Fig. 3.37b. The original picture was then recovered exactly by taking the inverse two-dimensional NTT of Fig. 3.37b and is shown in Fig. 3.37c.

### Bibliographical Guide

Representing images and scenes mathematically is the important first step in computer analysis. Three-dimensional transformations and homogeneous coordinates are described in the paper by Roberts (1962) and the books by Newman and Sproull (1973), Duda and Hart (1973), Rogers and Adams (1976), and Giloi (1978).

The sampling theorem is described by Papoulis (1969) and Pearson (1975). Quantization methods are given by Panter and Dite (1951), Max (1960), Panter (1965), and Pratt (1970). Interpolation techniques are described by Pearson (1975) and Sperling (1971).

A computer image may be considered as a matrix or equivalent vector. The relationship between the matrix and vector representation is described by Pratt (1975). An excellent discussion of matrix representations is given by Andrews and Hunt (1977).

Orthogonal transformations are described in the books by Harmuth (1972) and Ahmed and Rao (1975). The Hotelling or Karhunen–Loeve transformation may be found in the original works of Hotelling (1933) and Karhunen (1947) and Loeve (1948). Modern treatments are given by Fukunaga (1972) and Ahmed and Rao (1975).

The continuous Fourier transform and its properties are described by Bracewell (1965), Goodman (1968), and Papoulis (1969). The discrete Fourier transform and the fast Fourier transform are described by Cooley and Tukey (1965) and Ahmed and Rao (1975).

The Walsh–Hadamard and Haar transforms were developed from considerations by Rademacher (1922) and Walsh (1923) on orthogonal square wave functions. Modern descriptions are given by Andrews and Casparie (1970), Harmuth (1972, 1977), and Ahmed and Rao (1975).

The slant transform was introduced by Enomoto and Shibata (1971) and is described by Ahmed and Rao (1975), as is the discrete cosine transform.

The number theoretic or finite field transforms are described by Pollard (1971), Rader (1972), Agarwal and Burrus (1975), Reed and Truong (1976), Kwoh (1977), and Reed *et al.* (1977). The required background may be found in the books by Szabo and Tanaka (1969) and Young and Gregory (1973).

## PROBLEMS

**3.1** Given the unit cube described by its vertex coordinates $\{000, 001, 010,$ $011, 100, 101, 110, 111\}$

(a) determine the orthographic projection on the plane described by $x = 4$.

(b) determine the perspective, projective transformation on the plane described by $z = 4$ with the camera lens center located at $(\frac{1}{2}, \frac{1}{2}, 2)$.

**3.2** For the 8 by 8 image shown below determine:

(a)  the discrete Fourier transform,

(b)  the Walsh–Hadamard transform,

(c)  the Haar transform,

(d)  the slant transform,

(e)  the discrete cosine transform,

(f)  the number theoretic transform.

Note that the example image given in Fig. 3.21a may be used as a test case. BASIC language computer programs may be used.

$$
\begin{bmatrix}
5 & 4 & 1 & 0 & 1 & 0 & 2 & 5 \\
7 & 6 & 2 & 1 & 0 & 1 & 6 & 6 \\
7 & 5 & 2 & 1 & 1 & 2 & 6 & 6 \\
7 & 4 & 1 & 0 & 1 & 0 & 5 & 6 \\
7 & 4 & 2 & 0 & 0 & 1 & 5 & 6 \\
7 & 6 & 3 & 0 & 1 & 0 & 6 & 6 \\
7 & 7 & 6 & 5 & 6 & 7 & 7 & 6 \\
7 & 7 & 2 & 1 & 2 & 2 & 7 & 4
\end{bmatrix}
$$

Test image

**3.3** An important property of the discrete cosine transform (DCT) is that its basis vectors closely approximate the eigenvectors of a class of matrices called

Toeplitz (Ahmed and Rao, 1975, p. 171). Determine the eigenvectors of the Toeplitz matrix **T** given below and compare those with the basis vectors of the $N = 8$ DCT.

$$\mathbf{T} = \begin{bmatrix} 1 & \rho & \rho^2 & \rho^3 & \rho^4 & \rho^5 & \rho^6 & \rho^7 \\ \rho & 1 & \rho & \rho^2 & \rho^3 & \rho^4 & \rho^5 & \rho^6 \\ \rho^2 & \rho & 1 & \rho & \rho^2 & \rho^3 & \rho^4 & \rho^5 \\ \rho^3 & \rho^2 & \rho & 1 & \rho & \rho^2 & \rho^3 & \rho^4 \\ \rho^4 & \rho^3 & \rho^2 & \rho & 1 & \rho & \rho^2 & \rho^3 \\ \rho^5 & \rho^4 & \rho^3 & \rho^2 & \rho & 1 & \rho & \rho^2 \\ \rho^6 & \rho^5 & \rho^4 & \rho^3 & \rho^2 & \rho & 1 & \rho \\ \rho^7 & \rho^6 & \rho^5 & \rho^4 & \rho^3 & \rho^2 & \rho & 1 \end{bmatrix}$$

$$\rho = 0.95 \text{ and } \rho = 0.90$$

# References

Agarwal, R. C., and Burrus, C. S. (1975). Number Theoretic Transforms to Implement Fast Digital Convolution. *IEEE Proc.* **63** (April), 550–560.

Ahmed, N., and Rao, K. R. (1975). "Orthogonal Transforms for Digital Signal Processing." Springer-Verlag, Berlin and New York.

Andrews, H. C., and Casparie, K. (1970). Orthogonal Transformations. In "Computer Techniques in Image Processing," pp. 73–103. Academic Press, New York.

Andrews, H. C., and Hunt, B. R. (1977). "Digital Image Restoration." Prentice-Hall, Englewood Cliffs, New Jersey.

Bracewell, R. N. (1965). "The Fourier Transform and Its Application." McGraw-Hill, New York.

Cooley, J. W., and Tukey, J. W. (1965). An Algorithm for the Machine Computation of Complex Fourier Series. *Math. Computat.* **19** (April), 297–301.

Duda, R. O., and Hart, P. E. (1973). "Pattern Classification and Scene Analysis." Wiley, New York.

Enomoto, H., and Shibata, K. (1971). Orthogonal System for Television Signals. *Proc. 1971 Symp. Appl. Walsh Functions*, Hatfield Polytechnic, Hatfield, Hertfordshire, England, pp. 11–17.

Fukunaga, K. (1972). "Introduction to Statistical Pattern Recognition." Academic Press, New York.

Giloi, W. (1978). "Interactive Computer Graphics." Prentice-Hall, Englewood Cliffs, New Jersey.

Good, I. J. (1958). The Interactions Algorithm and Practical Fourier Analyses. *J. R. Stat. Soc., Ser. B* **20**, 361.

Goodman, J. W. (1968). "Introduction to Fourier Optics." McGraw-Hill, New York.

Harmuth, H. F. (1972). "Transmission of Information by Orthogonal Functions." Springer-Verlag, Berlin and New York.

Harmuth, H. F. (1977). "Sequency Theory—Foundations and Applications." Academic Press, New York.

Hotelling, H. (1933). Analysis of a Complex of Statistical Variables into Principal Components. *J. Educ. Psychol.* **24**, 417–441, 498–520.

Karhunen, K. (1947). Uber Lineare Methoden in der Wahrscheinlich Keitsrechnung. *Ann. Acad. Sci. Fenn., Ser. A* **137**. (Engl. transl. by I. Selin. On Linear Methods in Probability Theory. T-131, RAND Corp., Santa Monica, California, 1960.)

Kramer, E. E. (1970). "The Nature and Growth of Modern Mathematics." Hawthorn, New York.

Kwoh, Y. S. (1977). "Application of Finite Field Transforms to Image Processing and X-Ray Reconstruction." Ph.D. thesis, Univ. of Southern California, Los Angeles January (unpublished).

Loeve, M. (1948). "Functions Aleatories de Seconde Ordre," Ch. 8, pp. 299–352. Hermann, Paris.

Max, J. (1960). Quantizing for Minimum Distortion. *IRE Trans. Inf. Theory* **6**, 7–12.

Newman, W. M., and Sproull, R. F. (1973). "Principles of Interactive Computer Graphics." McGraw-Hill, New York.

Panter, P. F. (1965). "Modulation Noise and Spectral Analysis Applied to Information Transmission." McGraw-Hill, New York.

Panter, P. F., and Dite, W. (1951). Quantizing Distortion in Pulse-Count Modulation with Nonuniform Spacing of Levels. *Proc. IRE* **39** (January), 44–48.

Papoulis, A. (1969). "Systems and Transforms with Applications to Optics." McGraw-Hill, New York.

Pearson, D. E. (1975). "Transmission and Display of Pictorial Information." Wiley, New York.

Pollard, J. M. (1971). The Fast Fourier Transform in a Finite Field. *Math. Computat.* **25**, No. 114, 365–374.

Pratt, W. K. (1970). Image Coding. *In* "Computer Techniques in Image Processing," pp. 135–179. Academic Press, New York.

Pratt, W. K. (1975). Vector Formulation of Two Dimensional Signal Processing Operations, *Comput. Graphics Image Processing* **4**, No. 1, 1–24.

Rademacher, H. (1922). Einige Sätze von Allegemeinen Orthogonal-Fünkeionen. *Math. Ann.* **87**, 122–138.

Rader, C. M. (1972). Discrete Convolution via Mersenne Transforms. *IEEE Trans. Comput.* **C-21**, No. 12, 1269–1273.

Rader, C. M. (1975). On the Application of Number Theoretic Transforms for High Speed Convolution to Two Dimensional Filtering, *IEEE Trans. Circuit Theory*, **CAS-22**, 575.

Reed, I. S., and Truong, T. K. (1976). Convolutions over Residue Classes of Quadratic Integers. *IEEE Trans. Inf. Theory* **IT-22**, No. 4, 468–475.

Reed, I. S., Kwoh, Y. S., Truong, T. K., and Hall, E. L. (1977). X-Ray Reconstruction by Finite Field Transforms. *IEEE Trans. Nucl. Sci.* **NS-24**, No. 1, 843–849.

Roberts, L. G. (1965). Machine Perception of Three Dimensional Solids, in "Optical and Electro-Optical Information Processing" (J. T. Tippet, ed.). MIT Press, Cambridge, Massachusetts.

Rogers, D. F., and Adams, J. A. (1976). "Mathematical Elements for Computer Graphics." McGraw-Hill, New York.

Sperling, G. (1971). Flicker in Computer-Generated Visual Displays: Selecting a CRO Phosphor and Other Problems. *Behav. Methods Instrum.* **3**, No. 3, 151–153.

Szabo, N. S., and Tanaka, R. I. (1969). "Residue Arithmetic and Its Applications to Computer Technology." McGraw-Hill, New York.

Tescher, A. G. (1973). The Role of Phase in Adaptive Image Coding. Ph.D. thesis (Electr. Eng.), Univ. of Southern California, Los Angeles, December (unpublished).

Thomas, J. C. (1971). Phasor Diagrams Simplify Fourier Transforms. *Electron. Eng.* October, pp. 54–57.

Walsh, J. L. (1923). A Closed Set of Orthogonal Functions. *Am. J. Math.* **55**, 5–24.

Young, D. M., and Gregory, R. T. (1973). "A Survey of Numerical Mathematics," Vol. 2. Addison-Wesley, Reading, Massachusetts.

# 4 | Image Enhancement and Restoration

## 4.1 Introduction

Image enhancement and restoration techniques are designed to improve the quality of an image as perceived by a human. In recent years *enhancement* has come to describe those techniques that work well as judged by human observers even though the mathematical criteria are not completely understood. For example, rectification of a geometrically distorted image produces an obvious improvement in quality to a human observer. No objective measure of this quality improvement has yet been discovered. Image *restoration* is applied to the restoration of a known distortion for which an objective criterion can be applied.

This chapter may be divided into two parts, the first concerned with enhancement techniques, while the second adds a certain mathematical rigor in the consideration of restoration methods. Contrast enhancement techniques are described in Section 4.2. Very effective techniques require only simple piecewise linear or nonlinear transformations. A novel technique, which is based upon a mapping of the gray levels to achieve a uniform distribution, is next considered. The generalization of this method to permit the specification of the enhanced image histogram is also considered. Geometric rectification methods are considered in Section 4.3. Two basic approaches to this problem are considered as well as certain problems in the subtraction of images. Edge enhancement techniques are considered next in Section 4.4. Both spatial and frequency domain methods are considered. Noise smoothing methods are described in Section 4.5. These techniques are effective for reducing the effects of random noise but also degrade the edge content of an image. The final enhancement topic, which is considered in Section 4.6, is the problem of matching the visual system response. Certain simplified models for image enhancement can be suggested by considering the known structure of the visual system.

Image restoration is the topic for the remaining sections. The basic restoration problem is considered in Section 4.7 and the space invariant and variant systems are differentiated. The restoration problem is considered in terms of estimation in Section 4.8 and several well-known matrix solutions are reviewed. The solutions given involve very large matrices that cannot be easily solved using standard methods. Several frequency domain solutions are described in Section 4.9, which are based upon circulant matrices and which may be implemented via the fast Fourier transform. Important filter solutions are described. The final restoration methods considered are directed toward motion restoration. The general problem is considered and a special technique described for restoring space variant degradation through a combination of geometric transformations and space invariant restoration.

## 4.2   Contrast Enhancement

In Chapter 2, the psychovisual properties of vision and the concepts of luminance and contrast sensitivity of the human were reviewed. We will now draw upon these relations to define contrast for an image function and to describe computer methods for contrast enhancement.

Contrast generally refers to a difference in luminance or gray level values in some particular region of an image function. If we assume that the values of $f(x,y)$ correspond to luminance values at the point $(x,y)$, then we imply that it is a real-valued, nonnegative function, since light intensity cannot be negative. The range of $f(x,y)$ will be assumed to be finite, i.e., $[m, M]$, and large values of $f(x,y)$ will correspond to bright points.

A reversed situation arises when considering photographic transparencies, if the image function values are directly related to optical density. Recall that the film intensity transmittance function $t(x,y)$ was defined as the local average of the ratio of light intensity transmitted to the intensity incident, and clearly, $0 \le t(x,y) \le 1$. In the development of a photographic transparency, the areas exposed to light turn into silver metal and thus are opaque. Since the optical density was defined as the negative logarithm of the transmittance, i.e.,

$$D(x,y) = -\log t(x,y),$$

then

$$0 \le D(x,y) < \infty.$$

Thus, density values near zero correspond to clear areas and large density values correspond to dark regions on the transparency. Since neither photographic films, the human visual system, nor image displays have perfectly symmetrical responses, the reversal of dark and light areas or *negative process* may be used for enhancing information.

For any given imaging system, only a finite luminance or optical density range will be available. An image function, $f(x,y)$, describing either situation

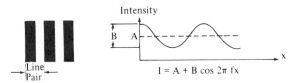

*Figure 4.1*   Spatial frequency grating bar pattern and *x* direction profile.

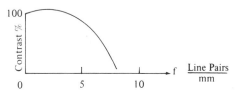

*Figure 4.2*   Intensity modulation transfer function.

would thus have a finite maximum and minimum. The difference between the maximum and minimum values of $f(x,y)$ is called the contrast range and is an important parameter for any imaging system. The ratio of the maximum to the minimum values is called the *contrast ratio* and is also a commonly used parameter. For example, film with an optical density contrast range between 0.1 and 1.0 is transparent in normal light, but one with a density range between 10 and 100 is almost totally opaque. Each has a contrast ratio of 10.

In optical systems, contrast is defined for spatial frequency gratings as

$$C = (M - m)/(M + m).$$

A spatial frequency grating might look like a bar pattern, as shown in Fig. 4.1 and may be represented in functional form as $I = A + B\cos 2\pi f x$. The *x* direction profile is also shown in Fig. 4.1. The contrast of this pattern may be defined as the ratio of the deviation in intensity to the average intensity or

$$C = B/A.$$

Note that since contrast defined in this matter is a function of spatial frequency, a set of gratings could be used to develop an intensity modulation transfer function for some system, as shown in Fig. 4.2.

### 4.2.1   Contrast Transformations

We may alter the gray level values and thus change the contrast of the information in an image by effecting a linear or nonlinear transformation, i.e., by forming a new image function, $g(x,y) = T[f(x,y)]$ for each $(x,y)$, where $T$ is some mapping. We will first consider position invariant transformations that may be characterized by a transfer function, as shown in Fig. 4.3. Furthermore, we will assume an ideal, unity transfer function scanning and display system, except when otherwise noted, for compensation transformations.

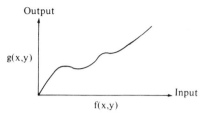

*Figure 4.3*   Position invariant transformation characterized by a transfer function.

### Scaling Transformation—Linear

A simple but useful model is the scaling transform. Suppose we have a digital image for which the contrast values do not fill the available contrast range. That is, suppose our data cover a range $(m, M)$, but that the available range is $(n, N)$. Then the linear transformation

$$g(x,y) = \{[f(x,y) - m]/(M - m)\}[N - n] + n$$

expands the values over the available range.

This transformation is often necessary when an image has been scanned, since the image scanner may not have been adjusted to use its full dynamic range.

Let us now assume that the input image function $f$ and the output image function $g$ are continuous random variables with probability distribution $p_f(x)$ and $p_g(y)$, respectively. Then, if the input probability distribution is known and zero outside the range $[m, M]$ and the system is linear, it is easy to determine the output probability distribution. Since

$$F_g(y) = \text{Prob}(g \leq y) = \text{Prob}\{f \leq [(y - n)/(N - n)](M - m) + m\}$$

$$= \begin{cases} 0, & x < m, \\ F_f(x), & m < x < M, \\ 1, & x > M, \end{cases}$$

the output distribution is of exactly the same form as in the input distribution.

Now suppose that the transfer function is linear over the range $(m, M)$, but that the input distribution is not restricted to this range, and a clipping procedure as shown in Fig. 4.4 is used.

For this situation, the output distribution is

$$F_g(y) = \begin{cases} \int_{-\infty}^{m} p_f(x)\,dx & \text{if } y = n, \\ F_f(x) & \text{if } n < y < N, \\ \int_{M}^{\infty} p_f(x)\,dx & \text{if } y = N. \end{cases}$$

Thus the output distribution is not identical to the input distribution but has

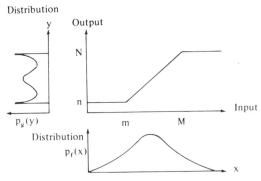

*Figure 4.4* Clipping procedure.

"spikes" at the extreme values. Since the clipping procedure is used on most image scanners, it is important to recognize the results of this procedure.

As a final example, let us consider the underuse of the dynamic range. For this situation, the same scaling transformation is used, but we assume that the maximum output used $N$ is less than that available $A$. Again, the probability distribution of the output is of exactly the same form as the input. However, if the output function is now quantized with the number of quantization levels set for the range $(n, A)$, an extremely coarse quantization and subsequent false contouring may result.

Let us now consider the mapping of output image values $g$ into digital image values $\hat{g}$ for three cases of exact over- and underuse of the input dynamic range. The quantization process for case 1 produces an ideal digital approximation to the continuous probability function. For case 2, the spikes on the continuous probability function are also mapped into spikes in the discrete probability function, while for case 3, a coarse quantization occurs. Since we have noted that the quantization noise for a linear quantization is approximately uniformly distributed, the underuse of the available dynamic range also produces a much larger level of quantization noise than is necessary.

At this point, we may say that case 1 is the ideal or preferred situation. Case 2 is undesirable because a distortion of the image information is produced. Although case 3 produces a coarser quantization and larger quantization noise than necessary, it is preferable to case 2. This analysis is based on the assumption that the input is useful information.

### Piecewise-Linear Transformations for Contrast Enhancement

A piecewise-linear transformation is often useful for contrast enhancement. For example, a three-step piecewise-linear transfer function between the input and output image function is shown in Fig. 4.5a.

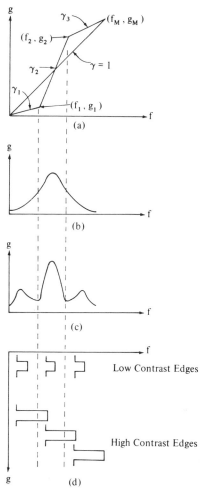

*Figure 4.5*   (a) Three-step piecewise linear transfer function between the input and output image function, (b) histogram of a Gaussian nature with low-level noise, (c) histogram showing a mixture of three components each of which has a Gaussian distribution, and (d) the effects on low- and high-contrast edge information.

As a practical example of the usefulness of this transformation, let us assume that the input image contains scratches and black emulsion blobs in addition to the useful image information. For this situation, the piecewise linear clipping transformation may be preferable to the linear transformation, since it localizes the noise information to the extreme values and provides a linear mapping of the useful information. A nonlinear digital smoothing operation that maps the extreme values into the local average of the neighboring points would now eliminate the effects of the scratches and emulsion blobs.

This function is defined by

$$
g = \begin{cases}
\gamma_1 f + b_1; & 0 \le f < f_1; \quad \gamma_1 = g_1/f_1, \quad b_1 = 0, \\
\gamma_2 f + b_2; & f_1 \le f < f_2; \quad \gamma_2 = \dfrac{g_2 - g_1}{f_2 - f_1}, \quad b_2 = g_1 - f_1 \gamma_2, \\
\gamma_3 f + b_3; & f_2 \le f \le f_M; \quad \gamma_3 = \dfrac{g_M - g_2}{f_M - f_2}, \quad b_3 = g_2 - f_2 \gamma_3.
\end{cases}
$$

Note that this mapping may also be used to compress the values below $f_1$ and linearly map the values between $f_1$ and $f_M$. Thus the range of values between $f_1$ and $f_2$ is expanded.

If a segment slope is less than unity, the image contrast information in that range is compressed; if the slope is greater than unity, the image contrast information is enhanced. The input density break points $b_1$ and $b_2$ and the segment slopes $\gamma_1$, $\gamma_2$, and $\gamma_3$ may often be selected in a reasonable manner from observation of the gray level histogram. If the histogram is of a Gaussian nature, as shown in Fig. 4.5b, with low-level noise and perhaps a dark peak due to an undesirable image edge, then the break points may be determined by selecting a probability threshold $T$. This threshold may be based on an estimate of the amount of low-level noise. If the information contained in the very low- and high-density regions is completely noise, the slopes $\gamma_1$ and $\gamma_3$ should be set equal to zero. This procedure will expand the contrast range of the information. If it is likely that some useful information may be visible in these extreme regions, then small nonzero values of $\gamma_1$ and $\gamma_3$ may be used; however, the image contrast range will not be changed. Another desirable histogram situation is shown in Fig. 4.5c. If the image information is made of a mixture of three components each of which has a Gaussian distribution, then the break points may be selected at the maximum likelihood thresholds between the distributions. If the density functions are not known, then the minimum values between the modes are reasonable breakpoints. The slopes $\gamma_1$ and $\gamma_3$ may be chosen proportional to the probability of useful information being contained in the region.

Another method for selecting the breakpoints is from consideration of the effects on low- and high-contrast edge information, as shown in Fig. 4.5d. With $\gamma_2$ greater than unity, the most likely low contrast edges are greatly enhanced while the most likely high contrast edges are only partially compressed. Finally, a practical situation for which this transformation is useful arises when a scratched transparency is scanned. The useful information may have a Gaussian distribution, but the scratch, because of its 100% transmittance, will produce a peak in the histogram on the dark side. A suitable choice of the breakpoint $b_2$ and a slope $\gamma_3$ equal to zero will move the peak to the position of $b_2$. Alternative solutions are to set these values equal to the average input value or to use a smoothing procedure.

## Logarithmic Transformation

A very useful nonlinear mapping is represented by $g(x,y) = \log f(x,y)$, $f(x,y) > 0$. A graph of this relationship is shown in Fig. 4.6.

This transformation expands the contrast of the small values of $f$ and compresses the large values. Also note that the shape of this curve can be changed considerably by changing the values of $f_{min}$ and $f_{max}$.

The log transformation has several desirable effects.

(a)  It makes the gray levels relate linearly to optical density rather than film transmittance or intensity. Suppose the film transmittance function is $t(x,y)$. Then the density function is

$$D(x,y) = -\log t(x,y).$$

If the transparency is illuminated by a unit amplitude, uniform plane wave, the transmitted luminance $l(x,y)$ is

$$l(x,y) = t(x,y).$$

If this luminance signal is linearly converted into an electronic signal $e(x,y)$, then

$$e(x,y) = kt(x,y).$$

If this electronic signal is linearly converted to a displayed luminance $d(x,y)$, then

$$d(x,y) = Ct(x,y) = C\varepsilon^{-D(x,y)}.$$

Thus, displayed luminance is related exponentially to film density. Due to this relationship, a film density wedge consisting of equal width strips of constant density may be displayed as strips of equal luminance of unequal width. If one inserts a log amplifier circuit, an electronic signal $E(x,y)$, where

$$E(x,y) = \log e(x,y)$$

will be produced. However, the displayed luminance

$$d(x,y) = C\log e(x,y) = CD(x,y)$$

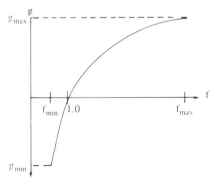

*Figure 4.6*  Graph of relationship $g(x,y) = \log f(x,y)$, $f(x,y) > 0$.

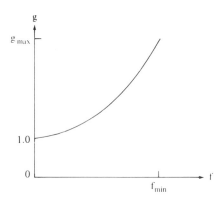

*Figure 4.7*  Exponential transformation.

is linearly related to film density and the displayed strips will have equal widths.

(b)  It makes the low-contrast detail more visible by enhancing low-contrast edges.

(c)  It provides a constant signal-to-noise ratio for quantization noise.

(d)  It somewhat matches the response of the human visual system (HVS).

(e)  It usually provides a more equal distribution of gray levels.

(f)  It transforms multiplicative noise into additive noise.

Because of these considerations, the log operations have become a standard option on most scanning equipment. They are also easily implemented in hardware with an operational amplifier with a transistor in the feedback loop.

### Exponential Transformation

Another nonlinear transformation that has been used mainly in conjunction with the logarithmic transformation in multiplicative filtering operations is the exponential transformation shown in Fig. 4.7. To obtain a unity transfer function system in which a log transformation is used in the input, an exponential operation must be used in the output (Stockham, 1972).

The effect of the exponential transfer function on edges in an image is to compress low-contrast edges while expanding high-contrast edges. This generally produces an image with less visible detail than the original and thus is not a desirable image enhancement transformation. However, an important feature of the exponential transformation is that the result is always nonnegative.

### 4.2.2  Histogram Equalization

We have seen that the logarithmic transformation generally expands a peaky input probability distribution into a broader output distribution. For example, the log normal distribution is transformed into a normal distribution. The broader output distribution results mainly from the expansion of low-contrast information, and thus the log image has enhanced low-contrast detail.

We may expand the concept of transforming the image probability function in a systematic manner by reviewing a basic theorem from probability theory.

**Theorem.**   Let $A$ be an absolutely continuous random variable with range $(a \leq X \leq b)$, and distribution function $F(x) = \text{Prob}(X \leq x)$. Then the random variable defined by

$$Y = F_X(X)$$

has a uniform $(0, 1)$ distribution, i.e.,

$$F_Y(y) = \begin{cases} 1, & 1 \leq y, \\ y, & 0 \leq y < 1, \\ 0, & y < 0. \end{cases}$$

The proof is direct: since $F_X$ is continuous,

$$F_Y(y) = \text{Prob}\left[ F_X(X) \leq y \right],$$
$$F_Y(y) = \text{Prob}\left[ X \leq F_X^{-1}(y) \right],$$
$$F_Y(y) = y, \qquad \text{for} \quad 0 \leq y \leq 1.$$

If $F_X(y)$ is not absolutely continuous, it may contain jumps. Then $\text{Prob}[F_X(X) \leq y]$ may have many values and the theorem does not hold.

The theorem states that any absolutely continuous random variable may be transformed into a uniformly distributed random variable. Since the theorem does not hold if $X$ is not absolutely continuous, the use of this transformation for digital image functions is based on the approximation of an absolutely continuous function by a finely quantized function. We may clarify this approximation by the following extension.

**Lemma.**   Let $X$ be a discrete random variable with values

$$x_i = (b - a)i/N + a, \qquad i = 0, \ldots, N,$$

i.e., $a \leq x_i \leq b$ and distribution function $F_X(x) = \text{Prob}(X \leq x)$, which is a jump function with maximum jump equal to $\varepsilon$. That is,

$$\max_i \left[ F(x_{i+1}) - F(x_i) \right] = \varepsilon, \qquad i = 1, 2, \ldots, N.$$

Then the discrete random variable $Y$, defined by $Y = F_X(X)$, has an approximately uniform distribution in the sense that

$$F_Y(y) = 1, \qquad 1 \leq y,$$
$$F_Y(y) = 0, \qquad y \leq 0,$$

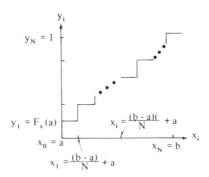

*Figure 4.8* Mapping of discrete variable $x_i$ into discrete variable $y_i$. (Note that the range of $y_i$ is $F_x(a) \leq y_i \leq 1$.)

and

$$y - F_Y(y) \leq \varepsilon, \qquad 0 < y < 1.$$

The proof is again direct. Since $X$ is discrete, $Y$ is also a discrete random variable with values $y_i = F_X(x_i)$, $i = 0, 1, \ldots, N$, e.g., $y_0 = F_X(x_0) = F(a)$ and $y_N = F_X(x_N) = F(b) = 1$. $F_X(a) \leq y_i \leq 1$, as shown in Fig. 4.8. Since the mapping is one to one, the probability function of $Y$ is equal to the probability function of $X$, which is equal to the jump at $x_i$, i.e.,

$$g_Y(y_i) = g_X(x_i) = y_i - y_{i-1}, \qquad i = 0, 1, \ldots, N-1,$$

where for convenience we define $y_{-1} = 0$.

Now the distribution function of $Y$ at any jump point of $y_i$ is simply

$$F_Y(y_i) = \sum_{j=0}^{i} g_Y(y_j) = \sum_{j=0}^{i} (y_j - y_{j-1}),$$

which is a telescoping sum and reduces to

$$F_Y(y_i) = y_i - y_{i-1} + y_{i-1} - y_{i-2} + \cdots + y_0 = y_i.$$

For any value $y_i \leq y \leq y_{i+1}$, we have

$$F_Y(y) = \sum_{j=0}^{y_i \leq y} (y_j - y_{j-1}) \leq y_i \leq y \leq y_{i+1},$$

so

$$y_i - F_Y(y) \leq y - F_Y(y) \leq y_{i+1} - F_Y(y).$$

The smallest value of $y_i - F_Y(y)$ occurs when $y = y_i$, and the largest value of $y_{i+1} - F_Y(y)$ occurs when $y = y_i$. Thus,

$$0 \leq y - F(y) \leq y_{i+1} - y_1 \leq \varepsilon$$

since the jumps are bounded by $\varepsilon$.

Since this inequality holds for any subinterval $[y_i, y_{i+1}]$ in the interval $[0, 1]$, we have $y - f(y) \leq \varepsilon$.

This theorem simply states that if we approximate a continuous distribution so that the largest jump is $\varepsilon$, then the *distribution transformation* produces a distribution which differs from a uniform distribution by at most $\varepsilon$. Clearly, as $\varepsilon$ is made smaller, the distribution more closely approximates the uniform distribution.

For many classes of images, the "ideal" distribution of gray levels is a uniform distribution, and the above theorem shows us how to obtain a uniform distribution with a gray level transformation. In general, a uniform distribution of gray levels makes equal use of each quantization level and tends to enhance low-contrast information. To use this transformation, we may

(a)   compute the histogram of the image gray level values,
(b)   add up the histogram values to obtain a distribution curve, and
(c)   use this distribution curve for the gray level transformation $g = D(f)$.

This process is illustrated in Fig. 4.9. Several other points about the computer implementation of the distribution transform should be noted. First, note that although the input random variable $X$ took on equispaced values on the interval $[a, b]$, the random variable $X$ does not in general take on equispaced values on the interval $[0, 1]$. That is, a scaling transformation must be performed to obtain the same contrast range $[a, b]$. Also, if the $X$ values are coarsely quantized before scaling, then a large reduction in the number of distinct gray levels may occur because the transformed values are closely spaced. We have found the approximation to work very well if the input image has from 512 to 1024 distinct gray level values. Also, since the human visual system can distinguish less than 64 distinct gray level values, some reduction in the number of distinct gray level values is permissible. The effect of the transformation may be increased by reducing the number of distinct levels in the distribution curve.

Finally, although this transformation is very effective for enhancing low-contrast detail, it does not discriminate between low-contrast information and noise.

A numerical example will now be given to illustrate the computational procedure for histogram equalization. Consider the simple $8 \times 8$, eight gray level image shown in Fig. 3.21a. The first step is to compute the histogram of the image. The number of occurrences may then be divided by the total number of pixels to produce relative frequency values. The gray level value, number of occurrences and relative frequencies are shown in the first three columns of Table 4.1. The next step is to compute the *empirical distribution function* by accumulating the relative frequency values. The distribution

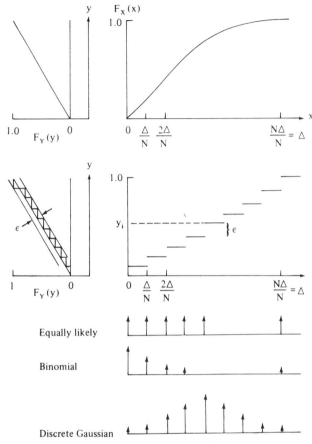

*Figure 4.9*　Histogram equalization process.

function is computed by

$$y_I = F_x(x_I) = \sum_{i=0}^{I} p(x_i).$$

The values of this function are also shown in Table 4.1. This is the desired mapping function except for scaling. Since the original gray level values are integers and the distribution function varies between 0 and 1, some method of scaling must be used. Also, for display the output gray levels should be integers. One solution to the scaling is to first convert the input gray levels to normalized values that range between 0 and 1. This may be accomplished by simply dividing by the maximum gray level value since the minimum is 0.

$$x_I = \tfrac{1}{7} I.$$

These values, which are also shown in Table 4.1, may then be mapped

*Table 4.1*

*Histogram Equalization Procedure*

| $I^a$ | $x_I = I/7^b$ | $n_i^c$ | $p(x_I) = P(I)^d$ | $y_I = F_x(x_I)^e$ | $J^f$ | $P(J)^g$ |
|---|---|---|---|---|---|---|
| 0 | 0.000 | 8 | 0.125 | 0.125 | 0 | 0.125 |
| 1 | 0.143 | 0 | 0.0 | 0.125 | 0 | 0.0 |
| 2 | 0.286 | 0 | 0.0 | 0.125 | 0 | 0.0 |
| 3 | 0.429 | 0 | 0.0 | 0.125 | 0 | 0.0 |
| 4 | 0.571 | 31 | 0.484 | 0.609 | 4 | 0.484 |
| 5 | 0.714 | 16 | 0.250 | 0.859 | 6 | 0.250 |
| 6 | 0.857 | 8 | 0.125 | 0.984 | 7 | 0.141 |
| 7 | 1.000 | 1 | 0.016 | 1.000 | 7 | — |

[a]Original gray level values.
[b]Scaled values.
[c]Number of occurrences.
[d]Relative frequencies.
[e]Distribution function.
[f]Scaled output gray levels.
[g]Relative frequencies of equalized gray levels.

directly using the distribution mapping. For example, $x_I = 0.143$ maps into $y_I = 0.125$, etc. The resulting $y_I$ values must now be scaled to the integer range of the input. This is accomplished by the *scaling* operation

$$J = \text{Int}[7(y_J - 0.125)/0.875 + 0.5],$$

where Int is the integer truncation function and 0.5 has been added to provide symmetrical rounding. The scaled output values are shown in Table 4.1.

The results of the *histogram equalization* mapping before scaling are shown in Fig. 4.10a. Note that the mapping from the original image to the equalized

| | | | | | | | |
|---|---|---|---|---|---|---|---|
| .609 | .609 | .609 | .609 | .609 | .609 | .609 | .125 |
| .609 | .859 | .859 | .859 | .859 | .859 | .609 | .125 |
| .609 | .859 | .984 | .984 | .984 | .859 | .609 | .125 |
| .609 | .859 | .984 | 1.000 | .984 | .859 | .609 | .125 |
| .609 | .859 | .984 | .984 | .984 | .859 | .609 | .125 |
| .609 | .859 | .859 | .859 | .859 | .859 | .609 | .125 |
| .609 | .609 | .609 | .609 | .609 | .609 | .609 | .125 |
| .609 | .609 | .609 | .609 | .609 | .609 | .609 | .125 |

(a)

| | | | | | | | |
|---|---|---|---|---|---|---|---|
| 4 | 4 | 4 | 4 | 4 | 4 | 4 | 0 |
| 4 | 6 | 6 | 6 | 6 | 6 | 4 | 0 |
| 4 | 6 | 7 | 7 | 7 | 6 | 4 | 0 |
| 4 | 6 | 7 | 7 | 7 | 6 | 4 | 0 |
| 4 | 6 | 7 | 7 | 7 | 6 | 4 | 0 |
| 4 | 6 | 6 | 6 | 6 | 6 | 4 | 0 |
| 4 | 4 | 4 | 4 | 4 | 4 | 4 | 0 |
| 4 | 4 | 4 | 4 | 4 | 4 | 4 | 0 |

(b)

*Figure 4.10*   Histogram equalized image: (a) before scaling and (b) after scaling.

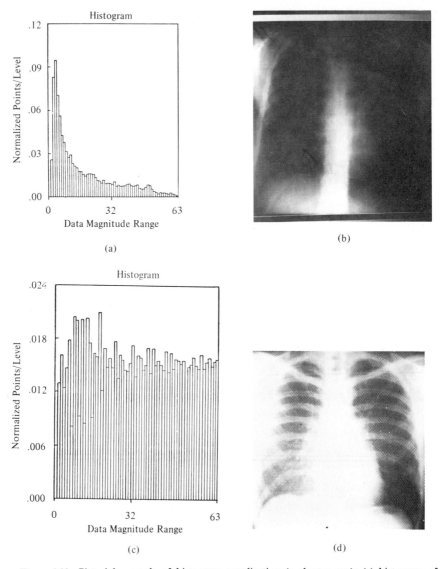

*Figure 4.11* Pictorial example of histogram equalization (a chest x ray): (a) histogram of original chest x ray, (b) original chest x ray, (c) histogram of enhanced chest x ray, and (d) enhanced chest x ray.

image can be accomplished by a table look-up, using Table 4.1. The final scaled histogram equalized image is shown in Fig. 4.10b. Note that if one were attempting to observe the circular pattern in the image, then it could be considered enhanced by the histogram equalization procedure. The equalizing effect can be observed on the output histogram and would be more pronounced for a larger sized image for which the discrete histogram more closely approximates a continuous probability function.

A pictorial example of histogram equilization is shown in Fig. 4.11. Note that the histogram of the original chest x ray is heavily biased to the dark side. The transformed image is enhanced due to the more uniform use of the gray level range.

### 4.2.3 Histogram Specification

It is often desirable to be able to effect a transformation to a specified distribution or perhaps try a variety of specified distributions for interactive image enhancement (Gonzalez and Fittes, 1975). To illustrate the histogram specification technique, consider the gray level $X$ to be a continuous random variable, normalized to the interval $(0, 1)$. Let $p_X(x)$ and $p_Y(y)$ be the original and desired probability density functions. The original image gray level variable may be transformed into a uniform distribution, using the transformation

$$s = T(x) = \int_0^x p_X(t) \, dt.$$

If the desired image gray levels were available, its distribution could also be transformed into a uniform distribution, using

$$v = U(y) = \int_0^y p_Y(t) \, dt.$$

Note that since the transformation $U(y)$ is single valued and monotonic, the inverse transformation, $U^{-1}(y)$ would reproduce the variable with $p_Y(y)$. This result would also hold for $T^{-1}(x)$. Therefore the procedure for transforming a variable $x$ to a specified distribution $p_Y(y)$ may be described as follows: First, apply the transformation

$$s = T(x)$$

to obtain a variable $s$ with a uniform distribution. Then apply the inverse of the specified distribution transformation

$$y = U^{-1}(s)$$

to obtain the variable $y$ that has the desired probability density $p_Y(y)$.

The classical mathematical problem in using this method for continuous variables is in obtaining an analytic expression for the inverse function. For example, no analytic inverse is known for the Gaussian distribution. In the discrete case this problem can be circumvented by using a discrete table of values for the inverse.

For the discrete case, the algorithm may be described as follows:

(1) Histogram-equalize the original image, $s = T(x)$.

(2) Specify the desired density function and obtain the equalizing transformation $U(y)$.

(3) Apply the inverse transformation $y = U^{-1}(s)$ to the previously equalized image $s$.

This procedure may be compressed into a single transformation of the original image by combining the transformations

$$y = U^{-1}[T(x)]$$

and thereby effecting a simple but powerful enhancement procedure. The approximation considerations for histogram equalization again apply as illustrated in the following example.

Consider the $64 \times 64$, eight level image with the histogram shown in Fig. 4.12a. It is desired to transform this histogram so that it will have the shape shown in Fig. 4.12b. The values of the specified histogram are listed in Table 4.2.

The first step in the procedure is to obtain the histogram-equalization mappings. The results are shown in Table 4.3. Next, we compute the transfor-

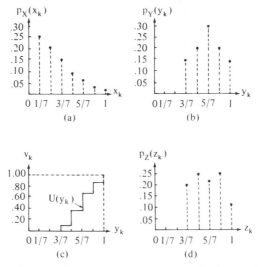

*Figure 4.12* Illustration of the histogram specification method: (a) original histogram, (b) specified histogram, (c) transformation function, and (d) resulting histogram. [Reproduced from "Digital Image Processing and Recognition," by Rafael C. Gonzalez and Paul Wintz, with permission of Addison-Wesley, Reading, Massachusetts (1977).]

Table 4.2

*Values of the Specified Histogram*[a]

| $y_k$ | $p_Y(y_k)$ | $y_k$ | $p_Y(y_k)$ |
|---|---|---|---|
| $y_0 = 0$ | 0.00 | $y_4 = \frac{4}{7}$ | 0.20 |
| $y_1 = \frac{1}{7}$ | 0.00 | $y_5 = \frac{5}{7}$ | 0.30 |
| $y_2 = \frac{2}{7}$ | 0.00 | $y_6 = \frac{6}{7}$ | 0.20 |
| $y_3 = \frac{3}{7}$ | 0.15 | $y_7 = 1$ | 0.15 |

[a]Reproduced from "Digital Image Processing and Recognition," by Raphael C. Gonzalez and Paul Wintz, with permission of Addison-Wesley, Reading, Massachusetts (1977).

Table 4.3

*Results of Histogram-Equalization Mappings*[a]

| $x_j \rightarrow s_k$ | $n_k$ | $p_S(s_k)$ |
|---|---|---|
| $x_0 \rightarrow s_0 = \frac{1}{7}$ | 790 | 0.19 |
| $x_1 \rightarrow s_1 = \frac{3}{7}$ | 1023 | 0.25 |
| $x_2 \rightarrow s_2 = \frac{5}{7}$ | 850 | 0.21 |
| $x_3, x_4 \rightarrow s_3 = \frac{6}{7}$ | 985 | 0.24 |
| $x_5, x_6, x_7 \rightarrow s_4 = 1$ | 448 | 0.11 |

[a]Reproduced from "Digital Image Processing and Recognition," by Raphael C. Gonzalez and Paul Wintz, with permission of Addison-Wesley, Reading, Massachusetts (1977).

mation function using

$$v_k = U(y_k) = \sum_{j=0}^{k} p_Y(y_j).$$

This yields the values

$$v_0 = U(y_0) = 0.00, \qquad v_1 = U(y_1) = 0.00,$$
$$v_2 = U(y_2) = 0.00, \qquad v_3 = U(y_3) = 0.15,$$
$$v_4 = U(y_4) = 0.35, \qquad v_5 = U(y_5) = 0.65,$$
$$v_6 = U(y_6) = 0.85, \qquad v_7 = U(y_7) = 1.00.$$

The transformation function is shown in Fig. 4.12c.

To obtain the $z$ levels we apply the inverse of the $U$ transformation obtained above to the histogram-equalized levels $s_k$. Since we are dealing with discrete values, an approximation must usually be made in the inverse mapping. For example, the closest match to $s_0 = \frac{1}{7} \simeq 0.14$ is $U(y_3) = 0.15$ or, using the inverse, $U^{-1}(0.15) = y_3$. Thus, $s_0$ is mapped to the level $y_3$. Using this procedure yields the following mappings:

$$s_0 = \tfrac{1}{7} \to y_3 = \tfrac{3}{7}, \qquad s_1 = \tfrac{3}{7} \to y_4 = \tfrac{4}{7},$$

$$s_2 = \tfrac{5}{7} \to y_5 = \tfrac{5}{7}, \qquad s_3 = \tfrac{6}{7} \to y_6 = \tfrac{6}{7},$$

$$s_4 = 1 \to y_7 = 1.$$

As indicated earlier, these results can be combined with those of histogram equalization to yield the following direct mappings:

$$r_0 = 0 \to y_3 = \tfrac{3}{7}, \qquad r_1 = \tfrac{1}{7} \to y_4 = \tfrac{4}{7},$$

$$r_2 = \tfrac{2}{7} \to y_5 = \tfrac{5}{7}, \qquad r_3 = \tfrac{3}{7} \to y_6 = \tfrac{6}{7},$$

$$r_4 = \tfrac{4}{7} \to y_6 = \tfrac{6}{7}, \qquad r_5 = \tfrac{5}{7} \to y_7 = 1,$$

$$r_6 = \tfrac{6}{7} \to y_7 = 1, \qquad r_7 = 1 \to y_7 = 1.$$

Redistributing the pixels according to these mappings and dividing by $n = 4096$ results in the histogram shown in Fig. 4.12d. The values are listed in Table 4.4.

Note that, although each of the specified levels was filled, the resulting histogram is not particularly close to the desired shape. As in the case of histogram equalization, this error is due to the fact that the transformation is guaranteed to yield exact results only in the continuous case. As the number of levels decreases, the error between the specified and resulting histograms tends to increase. As will be seen below, however, very useful enhancement results can be obtained even with an approximation to a desired histogram.

In practice, the inverse transformation from $s$ to $y$ is often not single-valued. This situation arises when there are unfilled levels in the specified histogram [which makes the cumulative distribution function (CDF) remain constant over the unfilled intervals], or in the process of rounding off $U^{-1}(s)$ to the nearest allowable level, as was done in the above example. Generally,

Table 4.4

Values for the Histogram shown in Fig. 4.12d[a]

| $y_k$ | $n_k$ | $p_Y(y_k)$ | $y_k$ | $n_k$ | $p_Y(y_k)$ |
|---|---|---|---|---|---|
| $z_0 = 0$ | 0 | 0.00 | $z_4 = \tfrac{4}{7}$ | 1023 | 0.25 |
| $z_1 = \tfrac{1}{7}$ | 0 | 0.00 | $z_5 = \tfrac{5}{7}$ | 850 | 0.21 |
| $z_2 = \tfrac{2}{7}$ | 0 | 0.00 | $z_6 = \tfrac{6}{7}$ | 985 | 0.24 |
| $z_3 = \tfrac{3}{7}$ | 790 | 0.19 | $z_7 = 1$ | 448 | 0.11 |

[a] Reproduced from "Digital Image Processing and Recognition," by Raphael C. Gonzalez and Paul Wintz, with permisson of Addison-Wesley, Reading, Massachusetts (1977).

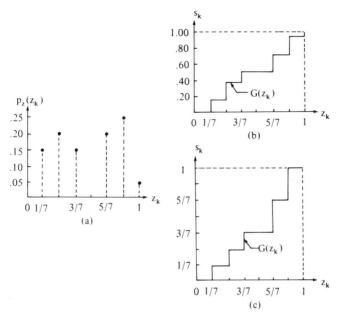

*Figure 4.13*  A nonunique transformation function: (a) specified histogram, (b) original transformation function, and (c) rounded-off function. [Reproduced from "Digital Image Processing and Recognition," by Rafael C. Gonzalez and Paul Wintz, with permission of Addison-Wesley, Reading, Massachusetts (1977).]

the easiest solution to this problem is to assign the levels in such a way as to match the given histogram as closely as possible. Consider, for example, the histogram specified in Fig. 4.13a. The original transformation function is shown in Fig. 4.13b and the rounded-off function is shown in Fig. 4.13c. It is noted that a value of $s = \frac{3}{7}$ can be mapped either to $y = \frac{3}{7}$ or $y = \frac{4}{7}$. Using the given histogram to guide the process, however, we see that the latter level was specified as being empty, so $y = \frac{3}{7}$ is used in the inverse mapping. This procedure can still be used when the ambiguous levels are filled in the specified histogram. The ambiguity arises in this case due to round-off, as illustrated in Fig. 4.13b, c. In this example, the level $s = 1$ can be assigned to either $y = \frac{6}{7}$ or $y = 1$. However, $y = \frac{6}{7}$ is chosen for the inverse mapping because the height of the specified histogram is larger at this level. It should be noted that assigning some levels to $y = \frac{6}{7}$ and some levels to $y = 1$ simply to match the histogram would make little sense since this would require dividing the pixels which have the same level into two groups. This is an artifical manipulation that could distort the appearance of the image.

The principal difficulty in applying the histogram specification method to image enhancement lies in being able to construct a meaningful histogram. Two solutions to this problem are as follows. The first is to specify a particular probability density function (i.e., Gaussian, Rayleigh, log-normal,

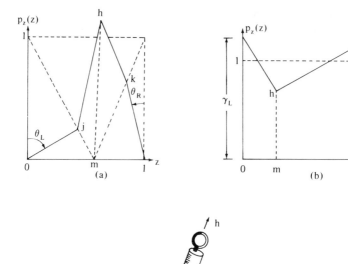

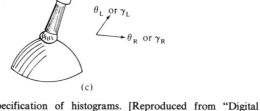

(c)

*Figure 4.14* Interactive specification of histograms. [Reproduced from "Digital Image Processing and Recognition," by Rafael C. Gonzalez and Paul Wintz, with permission of Addison-Wesley, Reading, Massachusetts (1977).]

etc.) and then form a histogram by digitizing the given function. The second approach consists of specifying a histogram of arbitrary (but controllable) shape by, for example, forming a string of connected straight line segments. After the desired shape has been obtained, the function is digitized and normalized to unit area.

Figure 4.14a illustrates this approach where the histogram is formed from line segments that are controlled by the parameters $m$, $h$, $\theta_L$, and $\theta_R$. Point $m$ can be chosen anywhere in the interval $[0, 1]$ and point $h$ can assume any nonnegative value. The parameters $\theta_L$ and $\theta_R$ specify the angle from vertical and can assume values from 0 to 90°. The inflection point $j$ moves along the line segment that connects the points $(0, 1)$ and $(m, 0)$ as $\theta_L$ varies. Similarly, point $k$ moves along the line segment that connects the points $(1, 1)$ and $(m, 0)$ as $\theta_R$ varies. The inflection points $j$ and $k$ are determined uniquely by the values of $m$, $h$, $\theta_L$, and $\theta_R$ and are therefore not specifiable. This histogram-specification method can generate a variety of useful histograms. For example, setting $m = 0.5$, $h = 1.0$, and $\theta_L = \theta_R = 0°$ yields a uniform histogram.

A second, somewhat simpler, approach is shown in Fig. 4.14b. This technique, which is also based on four parameters, is useful for contrast enhancement by emphasizing the ends of the pixel spectrum. Both of these methods depend on a small number of parameters and, therefore, lend themselves well to fast interactive image enhancement. Fig. 4.14c shows how the histogram parameters could be specified interactively by means of a joystick. This approach is particularly attractive for on-line applications in which an operator has a continuous visual output and can control the enhancement process to fit a given situation.

*Example.*   As a practical illustration of the direct histogram specification approach, consider Fig. 4.15a, which shows a semidark room viewed from a doorway. Figure 4.15b shows the histogram-equalized image, and Fig. 4.15c is the result of interactive histogram specification. The histograms are shown in Fig. 4.15d which includes, from top to bottom, the original, equalized, specified, and resulting histograms.

It is noted that histogram equalization produced an image whose contrast was somewhat high, while the result shown in Fig. 4.15c has a much more

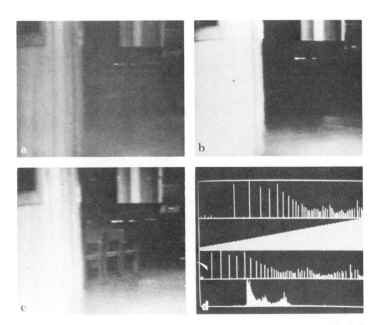

*Figure 4.15* Illustration of the histogram specification method: (a) original image, (b) histogram equalized image, (c) image enhanced by specification, and (d) histograms, from bottom to top: original image histogram, equalized histogram, desired distribution, and result of histogram specification. [Reproduced from "Digital Image Processing and Recognition," by Rafael C. Gonzalez and Paul Wintz, with permission of Addison-Wesley, Reading, Massachusetts (1977).]

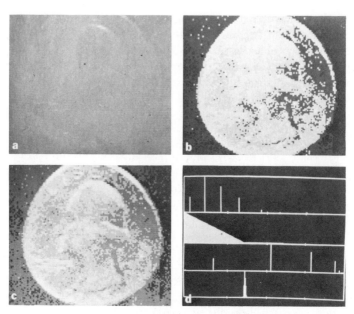

*Figure 4.16* Another illustration of the histogram specification method: (a) original image, (b) histogram equalized image, (c) image enhanced by histogram specification, and (d) histograms, from bottom to top: original image histogram, equalized histogram, desired distribution, and specified histogram. [Reproduced from "Digital Image Processing and Recognition," by Rafael C. Gonzalez and Paul Wintz, with permission of Addison-Wesley, Reading, Massachusetts (1977).]

balanced appearance. Because of its flexibility, the direct specification method can often yield results which are superior to histogram equalization. This is particularly true in cases where the image has been degraded and control of the pixel distribution becomes necessary to obtain good enhancement results. An extreme case that illustrates this point is shown in Fig. 4.16. Part (a) of this figure shows a picture of a coin taken under very poor lighting conditions. Figure 4.16b is the histogram equalized image, and Fig. 4.16c is the result of direct histogram specification. Figure 4.16d contains the histograms, which are shown in the same order as in Fig. 4.15d.

Although histogram equalization improved the image considerably, it is evident that a more balanced, clearer result was achieved by the histogram specification method. The reason for the improvement can be easily explained by examining Fig. 4.16d. The original histogram is very narrow, indicating that the pixels in the original image have little dynamic range. This is to be expected since the object of interest has such low contrast that it is barely visible. Histogram equalization improved the image by increasing the dynamic range of the pixels, thus enhancing the contrast between the coin and the background.

Since the number of different levels in this case is rather small, it is logical to deduce that having control over the shape of the final histogram should lead to some improvement over histogram equalization. In this example the improvement was realized by the specified histogram shown in Fig. 4.16d, which achieved an increase in dynamic range while biasing the histogram toward the dark side of the spectrum.

### 4.2.4  Matching Transformations

Another distribution-based technique that can provide contrast enhancement is called the sensor matching transformation. Suppose two images of the same scene were made with two different sensors such as two color image components, optical and radar, or optical and x ray sensors. In some cases, it is desirable to match the histograms of the two images. This may be accomplished by histogram equalizing each image separately or by using the histogram of one image to specify the distribution of the second. A third method may be based on the covariance matrix of the two images.

Suppose two images that are geometrically registered are given and that gross contrast reversals have been removed by inverting one image if necessary. A two-dimensional histogram of the *co-occurrent* gray levels is easily computed. Let $h(f_1, f_2) =$ (number of picture elements which have gray value $f_1$ in the first image and value $f_2$ in the second). This two-dimensional function indicates the global picture matching statistics. If the two images were identical, the nonzero histogram values would lie on a straight line. If the pictures are not identical, it may be desirable to determine the best straight line fit to the distribution and use this mapping to produce a new and hopefully enhanced image. The transformation which provides this curve fit is, of course, the Hotelling transform, which is particularly simple in this case.

To implement the Hotelling transformation, the covariance matrix of the two image variables is computed. We may first construct a vector $\mathbf{f} = (f_1, f_2)'$, then compute the 2 by 2 covariance matrix

$$\Sigma \simeq \frac{1}{N^2} \sum_{i=1}^{N^2} (\mathbf{f} - \mathbf{m})(\mathbf{f} - \mathbf{m})',$$

where

$$\mathbf{m} = \frac{1}{N^2} \sum_{i=1}^{N^2} \mathbf{f}$$

and images of size $N$ by $N$ are assumed.

This 2 by 2 covariance matrix is diagonalized, using the Hotelling transform, to obtain

$$\phi \Sigma \phi' = \Lambda,$$

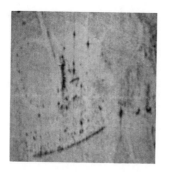

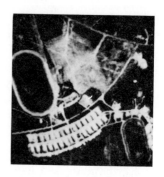

Radar Image
Intensity Reversed

Optical Image

(a)

(b)

Intensity Transformation
(Karhonen-Loeve Transform)

(c)

*Figure 4.17*  Intensity transformation.

where $\Lambda$ is a diagonal matrix with the eigenvalues of $\Sigma$ as the diagonal elements and $\phi$ is the 2 by 2 matrix whose columns correspond to the eigenvectors. The eigenvector $\phi_1$ corresponding to the largest eigenvalue will be chosen for the transformation to provide the least mean squared representation error. The final transformation to produce the enhanced image $g$ is simply

$$g = \phi_{11} f_1 + \phi_{12} f_2,$$

where $\phi_{11}$ and $\phi_{12}$ are the components of $\phi_1$.

As an example of this procedure, consider the two images shown in Fig. 4.17a, b, which were produced by optical and radar sensors from the same scene. The joint histogram function $h(f_1, f_2)$ is shown in Fig. 4.18. The idealized transformation is shown in Fig. 4.19, in which $\phi_1$ and $\phi_2$ correspond to the eigenvectors of the original distribution while $\phi_i'$ is the principal

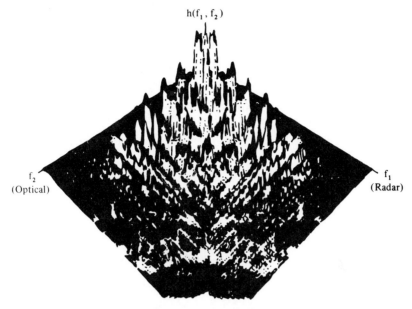

$h(f_1, f_2)$

$f_2$
(Optical)

$f_1$
(Radar)

*Figure 4.18*   Image intensity function, $h(f_1,f_2)$.

eigenvector after the transformation. The resulting transformed image is
shown in Fig. 4.17c. Note that this image has retained the primary characteristics of the radar image but is greatly enhanced.

A final example will be considered to provide a comparison between the
histogram matching and Hotelling transformation methods. The histogram
shown in Fig. 4.20 was taken as the specified histogram for a radar scene.
This histogram was computed from an optical image of the corresponding
scene. The histogram of an intensity reversed radar image is shown in
Fig. 4.21. The original and intensity reversed radar images are shown in

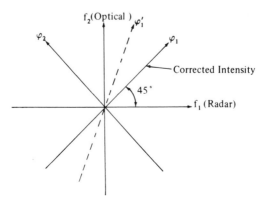

$f_2$(Optical )

$\varphi_1'$

$\varphi_2$

$\varphi_1$

Corrected Intensity

45°

$f_1$ (Radar)

*Figure 4.19*   Intensity coordinate transformation.

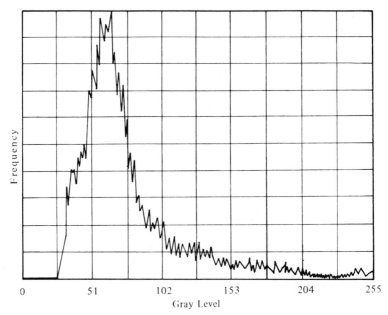

*Figure 4.20* Frequency distribution of intensity of optical image.

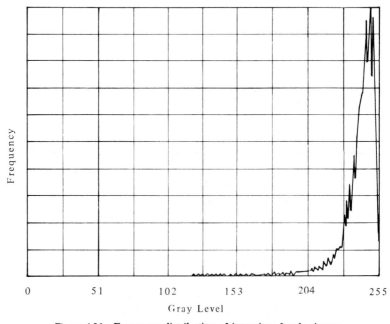

*Figure 4.21* Frequency distribution of intensity of radar image.

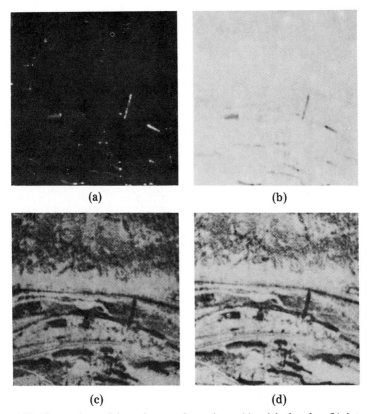

(a)                                                    (b)

(c)                                                    (d)

*Figure 4.22*   Comparison of intensity transformations: (a) original radar, (b) intensity reversed, (c) intensity transformed with sensor-dependent parameters, and (d) intensity transformed with Karhunen–Loeve transform.

Figs. 4.22a and 4.22b, respectively. Using the histogram specification technique with the specified histogram shown in Fig. 4.20, the intensity transformed image shown in Fig. 4.22c was obtained. This enhanced image may be compared to the result of the intensity transformed image shown in Fig. 4.22d, which was computed with the Hotelling or Karhunen–Loeve technique.

## 4.3   Geometric Transformations

The techniques for *geometric-based transformation* of pictures are presented in this section. The problem of rectifying the spatial coordinates of an image or of mapping one picture into another may be considered either enhancement or restoration. In either case, the problem is important and often encountered. Two basic approaches will be presented. The first may be called polynomial warping. This method can be implemented without knowledge of the image formation geometries and is based upon locating corresponding control points in a standard image and the image to be transformed. The second method will be called the perspective transformation and is based upon a knowledge of the form of the image formation processes. If the imaging parameters as well as the form are known, a good solution can be obtained. If the parameters are not known, an estimation approach is outlined.

### 4.3.1 Polynomial Warping

The spatial warping approach is illustrated in Fig. 4.23, and consists of assuming a functional form for the coordinate transformation between two images $f(x_1, x_2)$ and $f(u_1, u_2)$, and then using a series of known corresponding control points develop a transformation which will map one space into the other. This approach may be performed with manually or interactively selected control points, or automatically, using the scene matching techniques described in Chapter 8.

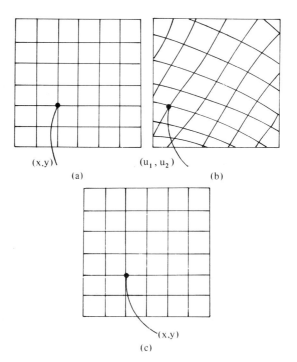

*Figure 4.23* Image geometric transformation. (a) optical image, (b) radar image, and (c) image geometric transformations.

Suppose that there is a spatial relationship between the two reference frames given by

$$x_1 = g_1(u_1, u_2), \qquad x_2 = g_2(u_1, u_2).$$

We will approximate $g_1$ and $g_2$ by polynomials in $u_1$ and $u_2$ of the form

$$x_1 = \sum_{i=0}^{N} \sum_{j=0}^{N} K_{ij}^1 u_1^i u_2^j, \qquad x_2 = \sum_{i=0}^{N} \sum_{j=0}^{N} K_{ij}^2 u_1^i u_2^j,$$

where $K^1$ and $K^2$ are the constant polynomial coefficients. For example, for $N = 1$ the polynomial relationship may be written

$$x_1 = K_{00}^1 + K_{01}^1 u_2 + K_{10}^1 u_1 + K_{11}^1 u_1 u_2.$$

Now suppose a set of control points are known, i.e.,

$$\{(x_{1i}, x_{2i}, u_{1i}, u_{2i})\} \qquad \text{for} \quad i = 1, 2, \ldots, M.$$

Using the relationship between $x_1$ and $(u_1, u_2)$, the following matrix equation may be developed:

$$
\begin{bmatrix} x_{11} \\ x_{12} \\ \vdots \\ x_{1M} \end{bmatrix}
=
\begin{bmatrix}
\mathbf{U}_{22}' & u_{11}\mathbf{U}_{21}' & u_{11}^2\mathbf{U}_{21}' & \cdots & u_{11}^N\mathbf{U}_{21}' \\
\mathbf{U}_{22}' & u_{12}\mathbf{U}_{22}' & u_{12}^2\mathbf{U}_{22}' & \cdots & u_{12}^N\mathbf{U}_{22}' \\
\mathbf{U}_{2M}' & u_{1M}\mathbf{U}_{2M}' & u_{1M}^2\mathbf{U}_{2M}' & \cdots & u_{1M}^N\mathbf{U}_{2M}'
\end{bmatrix}
\begin{bmatrix} \mathbf{K}_0 \\ \mathbf{K}_1 \\ \vdots \\ \mathbf{K}_N \end{bmatrix},
$$

where

$$\mathbf{U}_{1i}' = \left(1, u_{1i}, u_{1i}^2, \ldots, u_{1i}^N\right),$$

$$\mathbf{U}_{2i}' = \left(1, u_{2i}, u_{2i}^2, \ldots, u_{2i}^N\right),$$

and

$$\mathbf{K}_i' = \left(K_{i0}, K_{i1}, \ldots, K_{iN}\right).$$

Since there are $(N+1)^2$ unknown coefficients $K_{ij}$, the number of control points $M$ must be greater than or equal to $(N+1)^2$ to permit a unique solution. For example, the equation for $N = 1$ is

$$
\begin{bmatrix} x_{11} \\ x_{12} \\ \vdots \\ x_{1M} \end{bmatrix}
=
\begin{bmatrix}
1 & u_{21} & u_{11} & u_{11}u_{21} \\
1 & u_{22} & u_{12} & u_{12}u_{22} \\
  &  & \vdots & \\
1 & u_{2M} & u_{1M} & u_{1M}u_{2M}
\end{bmatrix}
\begin{bmatrix} k_{00} \\ k_{01} \\ k_{10} \\ k_{11} \end{bmatrix}.
$$

Since $(N+1)^2 = 4$, $M = 4$ control points would permit a unique solution. Using the previous definitions, the equation may be written simply as

$$\mathbf{x}_1 = \mathbf{U}\mathbf{K}_1.$$

A similar equation may be developed for the $x_2$ coordinates:

$$\mathbf{x}_2 = \mathbf{U}\mathbf{K}_2.$$

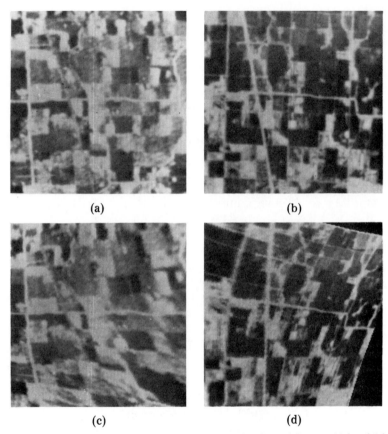

(a)                                    (b)

(c)                                    (d)

*Figure 4.24*  Results of geometric correction using $N = 2$ polynomial warp: (a) band 1 image, (b) band 2 image, (c) band 1 image transformed to band 2 image control points, and (d) band 2 image transformed to band 1 image control points.

Note that the **U** matrix is the same for both coordinates. Using linear least squares estimation for $M > (N + 1)^2$ permits the pseudoinverse solutions to be written as

$$\mathbf{K}_1 = (\mathbf{U}'\mathbf{U})^{-1}\mathbf{U}'\mathbf{x}_1 \quad \text{and} \quad \mathbf{K}_2 = (\mathbf{U}'\mathbf{U})^{-1}\mathbf{U}'\mathbf{x}_2.$$

Figure 4.24 shows the results of geometric correction using a $N = 2$ polynomial warp.

The *polynomial spatial warp* method requires no detailed knowledge of the sensor geometries. However, if information about the geometries is known, it may be used to develop a transformation that is more appropriate to the problem, as shown in the next section.

The central problem of this type is called the two image perspective, projection problem and is illustrated in Fig. 4.25 which depicts a simple scene

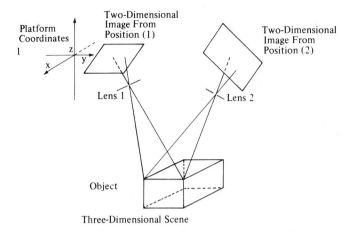

*Figure 4.25*   Two-image perspective, projection problem.

and two image planes. The hardest problem is: given image 1 and image 2 determine the transformation $T_3$ that will map one into the other. Before considering this problem, let us consider some simpler ones. First note that if the center lines are orthogonal and orthographic rather than perspective projections are recorded in the images, then for any feature that is visible in both images the three-dimensional coordinates $(x,y,z)$ are easily determined, since $(x,z)$ can be measured in one image and $(y,z)$ in the other. If a feature is not visible but perhaps obscured by some other feature, other information must be used to determine its three-dimensional coordinates. For example, hidden lines must be indicated in the plan and elevation views of a building to permit the determination of its structure.

### 4.3.2   Perspective, Projective Transformation

A problem of wide application is: given two perspective, projective images of a scene, determine the transformation that can transform common objects from one into the other. An illustration of the problem is shown in Fig. 4.26. The image planes shown are related to a controlled platform which can be moved from one position to another. The platform could correspond to an aircraft, a robot arm, or some other mounting device, and emphasizes the fact that motion would not be about a coordinate system in the image plane. For example, rotations would be about the platform origin.

The motion from platform position (1) to position (2) could take place in an infinite variety of ways. However, only the starting and ending positions are of immediate interest. Since platform position (1) may be selected as the reference coordinate system only the six spatial parameters for the translations and rotations between the platform positions $x_p$, $y_p$, $z_p$, $\theta$, $\phi$, and $\gamma$ and

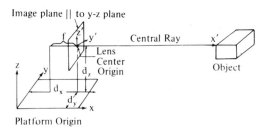

*Figure 4.26*  Relationship between platform coordinates and image coordinates.

the camera focal length $f$ need to considered, as shown in Fig. 4.25. Since the path is not important, any one of the possible combinations of translations and rotations that move the platform from position (1) to position (2) may be used. For simplicity, we will use a single translation matrix and three separate rotation matrices, followed by the perspective and projective transformations.

The transformation equations are greatly simplified by considering positions measured with respect to the image (1) coordinate system. Therefore, it is convenient to define the transformation from platform coordinates for each position as

$$\begin{bmatrix} x_i \\ y_i \\ z_i \\ w_i \end{bmatrix} = \mathbf{T} \begin{bmatrix} x \\ y \\ z \\ w \end{bmatrix}.$$

This transformation accounts for the translation and rotation differences between the platform and camera and shall be considered constant. To move the camera, we must move the platform. In practice, the camera may be aligned so that the image plane is parallel to a coordinate plane, which simplifies $\mathbf{T}$ to a translation transformation. Given the object coordinates measured with respect to the image (1) coordinate system, the image coordinates are determined by a perspective, projective transformation

$$\begin{bmatrix} 0 \\ y_i \\ z_i \\ w_i \end{bmatrix} = \begin{bmatrix} 0 & 0 & 0 & 0 \\ 0 & 1 & 0 & 0 \\ 0 & 0 & 1 & 0 \\ 0 & 0 & 0 & 1 \end{bmatrix} \begin{bmatrix} 1 & 0 & 0 & 0 \\ 0 & 1 & 0 & 0 \\ 0 & 0 & 1 & 0 \\ -1/f & 0 & 0 & 1 \end{bmatrix} \begin{bmatrix} x_0 \\ y_0 \\ z_0 \\ 1 \end{bmatrix}.$$

The problem may now be restated as follows: Suppose the controls of the platform and camera are changed, which changes the location of the image plane. How are image points before the change related to image points after the change? Since only starting and ending positions are of interest, any sequence of control changes which move the camera from one orientation to another is sufficient. For simplicity of analysis, we will consider the effects of the changes one at a time. If a linear transformation between the images can be developed for each of the control changes, then the effects of a combination of control changes may be obtained by multiplying the single effect transformation matrices. First, focal length change will be considered, then a translation of the image plane and finally each of the three possible rotations of the image plane about the platform pivot. If the focal length changes from $f$ to $f+\Delta$, the transformation to new image points is

$$\begin{bmatrix} 0 \\ y_i' \\ z_i' \\ w_i' \end{bmatrix} = \begin{bmatrix} 0 & 0 & 0 & 0 \\ 0 & 1 & 0 & 0 \\ 0 & 0 & 1 & 0 \\ -1/(f+\Delta) & 0 & 0 & 1 \end{bmatrix} \begin{bmatrix} x_0 \\ y_0 \\ z_0 \\ 1 \end{bmatrix}.$$

To determine if the old and new image points can be related, consider the homogeneous and physical coordinates

$$y_i = y_0, \qquad z_i = z_0, \qquad w_i = -(x_0/f) + 1,$$
$$y_i^P = fy_0/(f - x_0), \qquad z_i^P = fz_0/(f - x_0)$$

and

$$y_i' = y_0, \qquad z_i' = z_0, \qquad w_i = (x_0/-(f+\Delta)) + 1,$$
$$y_i'^P = (f+\Delta)y_0/(f+\Delta - x_0), \qquad z_i'^P = (f+\Delta)z_0/(f+\Delta - x_0).$$

Is there a transformation between $(y_i, z_i, w_i)$ and $(y_i', z_i', w_i')$? Note that a linear combination of $y_i$, $z_i$, and $w_i$ can be determined,

$$w_i = -(x_0/f) + 1 \qquad \text{and} \qquad w_i' = [x_0/(f+\Delta)] + 1,$$

since the two are related by

$$fw_i - 1 = -x_0 = (f+\Delta)w_i' - 1 \qquad \text{or} \qquad w_i' = w_i[f/(f+\Delta)].$$

Therefore, the required transformation is

$$\begin{bmatrix} y_i' \\ z_i' \\ w_i' \end{bmatrix} = \begin{bmatrix} 1 & 0 & 0 \\ 0 & 1 & 0 \\ 0 & 0 & f/(f+\Delta) \end{bmatrix} \begin{bmatrix} y_i \\ z_i \\ w_i \end{bmatrix}.$$

Therefore, given the original image points and using the scale of these points, $w_i = 1$, the new image points in physical coordinates are

$$y_i' = [(f+\Delta)/f]y_i, \qquad z_i' = [(f+\Delta)/f]z_i.$$

Thus only the scale is changed. The windowing due to a finite image size will not be considered.

If the image plane is translated to a new position by a movement of the platform, what is the transformation?

In the three-dimensional geometry, there is a symmetry between translating the origin and translating an object in the opposite direction.

Thus if the camera center is translated by an amount $(h,j,k)$, an equivalent image should be formed by translating the scene to $(-h, -j, -k)$ and observing with the original camera. Thus

$$
\begin{bmatrix} 0 \\ y_i' \\ z_i' \\ w_i' \end{bmatrix} = \begin{bmatrix} 0 & 0 & 0 & 0 \\ 0 & 1 & 0 & 0 \\ 0 & 0 & 1 & 0 \\ -1/f & 0 & 0 & 1 \end{bmatrix} \begin{bmatrix} x_0 - h \\ y_0 - j \\ z_0 - k \\ 1 \end{bmatrix}
$$

or

$$
y_i' = y_0 - j, \qquad z_i' = z_0 - k, \qquad w_i' = -(x_0 - h)/f + 1,
$$

so

$$
y_i' = y_0 - j, \qquad z_i' = z_0 - k, \qquad w_i' = w_i + h/f.
$$

Note that the transformation between $w_i'$ and $w_i$ is nonlinear and cannot be written in a $3 \times 3$ matrix form. However, we may introduce a new set of *superhomogeneous coordinates* and write

$$
\begin{bmatrix} y_i' \\ z_i' \\ w_i' \\ 1 \end{bmatrix} = \begin{bmatrix} 1 & 0 & 0 & -j \\ 0 & 1 & 0 & -k \\ 0 & 0 & 1 & h/f \\ 0 & 0 & 0 & 1 \end{bmatrix} \begin{bmatrix} y \\ z \\ w \\ 1 \end{bmatrix}.
$$

This transformation indicates that a translation of the camera can not only translate the image but also scale it in various perspectives. Note that the transformation from the superhomogeneous coordinates $(y_i, z_i, w_i, 1)$ to the original homogeneous coordinates is accomplished by simply omitting the last component to obtain $(y_i, z_i, w_i)$. The transformation from homogeneous coordinates $(y_i, z_i, w_i)$ to physical coordinates still involves a division by the last component, i.e., $(y_i/w_i, z_i/w_i)$.

If the image plane is rotated by a platform tilt control, what is the transformation? The main problem here is that the tilt rotation is about the platform origin rather than one of the image coordinate axes. To determine this transformation, we must use the platform coordinate–image coordinate transformation. For simplicity, assume that the image plane is parallel to the $y$-$z$ platform so that only translations are required to orient the two coordinate systems. The particular transformation is assumed to be

$$
\begin{bmatrix} \bar{x}_0 \\ \bar{y}_0 \\ \bar{z}_0 \\ 1 \end{bmatrix} = \begin{bmatrix} 1 & 0 & 0 & d_x \\ 0 & 1 & 0 & d_y \\ 0 & 0 & 1 & d_z \\ 0 & 0 & 0 & 1 \end{bmatrix} \begin{bmatrix} x_0 \\ y_0 \\ z_0 \\ 1 \end{bmatrix},
$$

where $[\bar{x}_0, \bar{y}_0, \bar{z}_0]$ are the object coordinates measured with respect to the platform coordinate system.

To determine the transformation resulting from a platform tilt or rotation about the $y$ axis, we may again use the symmetry argument. The image recorded by the camera with the platform tilted by an angle $\theta$ is the same as the image recorded in the original scene if the objects are rotated about the same point by an angle $-\theta$. To determine the object's coordinates rotated about the pivot requires a transformation from camera coordinates, the rotation, and an inverse transformation to camera coordinates. This transformation may be written as

$$\begin{bmatrix} x_0' \\ y_0' \\ z_0' \\ 1 \end{bmatrix} = \begin{bmatrix} 1 & 0 & 0 & d_x \\ 0 & 1 & 0 & d_y \\ 0 & 0 & 1 & d_z \\ 0 & 0 & 0 & 1 \end{bmatrix} \begin{bmatrix} \cos\theta & 0 & \sin\theta & 0 \\ 0 & 1 & 0 & 0 \\ -\sin\theta & 0 & \cos\theta & 0 \\ 0 & 0 & 0 & 1 \end{bmatrix} \begin{bmatrix} 1 & 0 & 0 & -d_x \\ 0 & 1 & 0 & -d_y \\ 0 & 0 & 1 & -d_z \\ 0 & 0 & 0 & 1 \end{bmatrix} \begin{bmatrix} x_0 \\ y_0 \\ z_0 \\ 1 \end{bmatrix}.$$

The result of this series of transformations is

$$x_0' = (x_0 - d_x)\cos\theta + (z_0 - d_z)\sin\theta - d_x,$$
$$y_0' = y_0,$$
$$z_0' = -(x_0 - d_x)\sin\theta + (z_0 - d_z)\cos\theta + d_z.$$

These rotated object points may now be observed on the image plane by the perspective projective transformation that results in the relations

$$y_i' = y_0,$$
$$z_i' = -(x_0 - d_x)\sin\theta + (z_0 - d_z)\cos\theta + d_z,$$
$$w_i' = -[(x_0 - d_x)\cos\theta + (z_0 - d_z)\sin\theta + d_x]/f + 1.$$

For reference, the original transformed image values are related to the object coordinates by

$$y_i = y_0, \qquad z_i = z_0, \qquad w_i = -x_0/f + 1.$$

The transformation between the two images again requires the use of super-homogeneous coordinates and is given by

$$\begin{bmatrix} y_i' \\ z_i' \\ w_i' \\ 1 \end{bmatrix} = \begin{bmatrix} 1 & 0 & 0 & 0 \\ 0 & \cos\theta & f\sin\theta & A \\ 0 & -\sin\theta/f & \cos\theta & B \\ 0 & 0 & 0 & 1 \end{bmatrix} \begin{bmatrix} y_i \\ z_i \\ w_i \\ 1 \end{bmatrix},$$

where

$$A = d_z\cos\theta + d_z + (d_x - f)\sin\theta,$$
$$B = [(d_x - f)\cos\theta + d_z\sin\theta + f - d_x]/f.$$

In a similar manner, the transformation corresponding to *panning* the camera, i.e., rotation by an angle $\phi$ about the $z$ axis, is easily shown to be

$$
\begin{bmatrix} y_i' \\ z_i' \\ w_i' \\ 1 \end{bmatrix} = \begin{bmatrix} \cos\phi & 0 & -f\sin\phi & A \\ 0 & 1 & 0 & 0 \\ (\sin\phi)/f & 0 & \cos\phi & B \\ 0 & 0 & 0 & 1 \end{bmatrix} \begin{bmatrix} y_i \\ z_i \\ w_i \\ 1 \end{bmatrix},
$$

where

$$
A = d_y \cos\phi + d_y + (d_x - f)\sin\phi
$$

$$
B = \left[ -(d_x - f)\cos\phi - d_x \sin\phi - f + d_x \right]/f.
$$

Likewise, the transformation corresponding to *platform roll*, i.e., rotation by an angle $\gamma$ about the $x$ axis, is given by

$$
\begin{bmatrix} y_i' \\ z_i' \\ w_i' \\ 1 \end{bmatrix} = \begin{bmatrix} \cos\gamma & \sin\gamma & 0 & A \\ -\sin\gamma & \cos\gamma & 0 & B \\ 0 & 0 & 1 & 0 \\ 0 & 0 & 0 & 1 \end{bmatrix} \begin{bmatrix} y_i \\ z_i \\ w_i \\ 1 \end{bmatrix},
$$

where

$$
A = d_y \cos\gamma - d_z \sin\gamma + d_y,
$$

$$
B = d_z \cos\gamma + d_y \sin\gamma + d_z.
$$

Note that the roll does not change the perspective of the two images.

In summary, for each platform control motion, there is a corresponding image-to-image transformation that is linear in superhomogeneous coordinates. The scale parameter of the original image $w_i = (x_0/f) + 1$ describes the perspective of the original image and obviously depends upon the object distance and focal length.

Since the transformations are linear in superhomogeneous coordinates, the effect of several platform control changes may be determined by multiplication of the corresponding matrices. For example, the combination of a rotation and a change of focal length may be determined by multiplying the two matrices in either order. The mathematical form of the product may, of course, be different, but the same image transformation should result.

To further illuminate this problem, let us consider the specific example shown in Fig. 4.27.

The vertex points of the cube in platform coordinates are

| Vertex | $x$ | $y$ | $z$ | Vertex | $x$ | $y$ | $z$ |
|--------|-----|-----|-----|--------|-----|-----|-----|
| 1 | 5 | 0 | 0 | 5 | 5 | 1 | 0 |
| 2 | 6 | 0 | 0 | 6 | 6 | 1 | 0 |
| 3 | 5 | 0 | 1 | 7 | 5 | 1 | 1 |
| 4 | 6 | 0 | 1 | 8 | 6 | 1 | 1 |

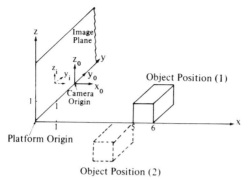

*Figure 4.27*   Platform transformation example for a cube.

The transformation from platform to camera coordinates is

$$\begin{bmatrix} x_0 \\ y_0 \\ z_0 \\ 1 \end{bmatrix} = \begin{bmatrix} 1 & 0 & 0 & -1 \\ 0 & 1 & 0 & -1 \\ 0 & 0 & 1 & -1 \\ 0 & 0 & 0 & 1 \end{bmatrix} \begin{bmatrix} x \\ y \\ z \\ 1 \end{bmatrix}.$$

Therefore, the object coordinates are now given by

| Vertex | $x_0$ | $y_0$ | $z_0$ |
|--------|-------|-------|-------|
| 1 | 4 | −1 | −1 |
| 2 | 5 | −1 | −1 |
| 3 | 4 | −1 | 0 |
| 4 | 5 | −1 | 0 |

| Vertex | $x_0$ | $y_0$ | $z_0$ |
|--------|-------|-------|-------|
| 5 | 4 | 0 | −1 |
| 6 | 5 | 0 | −1 |
| 7 | 4 | 0 | 0 |
| 8 | 5 | 0 | 0 |

These points may be transformed into image coordinates by the perspective, projective transform

$$\begin{bmatrix} 0 \\ y_i \\ z_i \\ w_i \end{bmatrix} = \begin{bmatrix} 0 & 0 & 0 & 0 \\ 0 & 1 & 0 & 0 \\ 0 & 0 & 1 & 0 \\ -1 & 0 & 0 & 1 \end{bmatrix} \begin{bmatrix} x_0 \\ y_0 \\ z_0 \\ 1 \end{bmatrix}.$$

The computed image coordinates are

| Vertex | $y_i$ | $z_i$ | $w_i$ |
|--------|-------|-------|-------|
| 1 | −1 | −1 | −3 |
| 2 | −1 | −1 | −4 |
| 3 | −1 | 0 | −3 |
| 4 | −1 | 0 | −4 |

| Vertex | $y_i$ | $z_i$ | $w_i$ |
|--------|-------|-------|-------|
| 5 | 0 | −1 | −3 |
| 6 | 0 | −1 | −4 |
| 7 | 0 | 0 | −3 |
| 8 | 0 | 0 | −4 |

The physical image coordinates are therefore

| Vertex | $y_i/w_i$ | $z_i/w_i$ | | Vertex | $y_i/w_i$ | $z_i/w_i$ |
|--------|-----------|-----------|---|--------|-----------|-----------|
| 1 | $\frac{1}{3}$ | $\frac{1}{3}$ | | 5 | 0 | $\frac{1}{3}$ |
| 2 | $\frac{1}{4}$ | $\frac{1}{4}$ | | 6 | 0 | $\frac{1}{4}$ |
| 3 | $\frac{1}{3}$ | 0 | | 7 | 0 | 0 |
| 4 | $\frac{1}{4}$ | 0 | | 8 | 0 | 0 |

Now suppose the control platform is translated 2 units in the $y$ direction, which translates the camera 2 units in the $y_0$ direction. The symmetrical object effect is to translate the object 2 units in the negative $y_0$ direction. Thus, the new vertex points of the cube are

| Vertex | $x_0$ | $y_0$ | $z_0$ | | Vertex | $x_0$ | $y_0$ | $z_0$ |
|--------|-------|-------|-------|---|--------|-------|-------|-------|
| 1 | 4 | $-3$ | $-1$ | | 5 | 4 | $-2$ | $-1$ |
| 2 | 5 | $-3$ | $-1$ | | 6 | 5 | $-2$ | $-1$ |
| 3 | 4 | $-3$ | 0 | | 7 | 4 | $-2$ | 0 |
| 4 | 5 | $-3$ | 0 | | 8 | 5 | $-2$ | 0 |

The computed image coordinates are given by

| Vertex | $y_i'$ | $z_i'$ | $w_i'$ | | Vertex | $y_i'$ | $z_i'$ | $w_i'$ |
|--------|--------|--------|--------|---|--------|--------|--------|--------|
| 1 | $-3$ | $-1$ | $-3$ | | 5 | $-2$ | $-1$ | $-3$ |
| 2 | $-3$ | $-1$ | $-4$ | | 6 | $-2$ | $-1$ | $-4$ |
| 3 | $-3$ | 0 | $-3$ | | 7 | $-2$ | 0 | $-3$ |
| 4 | $-3$ | 0 | $-4$ | | 8 | $-2$ | 0 | $-4$ |

The corresponding physical image coordinates are

| Vertex | $y_i'/w_i'$ | $z_i'/w_i'$ | | Vertex | $y_i'/w_i'$ | $z_i'/w_i'$ |
|--------|-------------|-------------|---|--------|-------------|-------------|
| 1 | 1 | $\frac{1}{3}$ | | 5 | $\frac{2}{3}$ | $\frac{1}{3}$ |
| 2 | $\frac{3}{4}$ | $\frac{1}{4}$ | | 6 | $\frac{1}{2}$ | $\frac{1}{4}$ |
| 3 | 1 | 0 | | 7 | $\frac{2}{3}$ | 0 |
| 4 | $\frac{3}{4}$ | 0 | | 8 | $\frac{1}{2}$ | 0 |

The transformation previously derived for the platform translation is

$$
\begin{bmatrix} y_i' \\ z_i' \\ w_i' \\ 1 \end{bmatrix} =
\begin{bmatrix} 1 & 0 & 0 & -2 \\ 0 & 1 & 0 & 0 \\ 0 & 0 & 1 & 0 \\ 0 & 0 & 0 & 1 \end{bmatrix}
\begin{bmatrix} y_i \\ z_i \\ w_i \\ 1 \end{bmatrix}.
$$

By comparing the two sets of coordinates, one may see that the transformation does indeed relate the two sets of image points.

An interesting variation of this example is to consider the problem with the $w_i$ values unknown. Suppose what is known is the corresponding physical coordinates for one point at each object distance. For this example, this would correspond to one point on two planes of the cube, say, vertices (1) and (2). These values are given below.

| Vertex | $y_i/w_i$ | $z_i/w_i$ | $y_i'/w_i'$ | $z_i'/w_i'$ |
|--------|-----------|-----------|-------------|-------------|
| 1 | $\frac{1}{3}$ | $\frac{1}{3}$ | 1 | $\frac{1}{3}$ |
| 2 | $\frac{1}{4}$ | $\frac{1}{4}$ | $\frac{3}{4}$ | $\frac{1}{4}$ |

For vertex point (1), the transformation may be used to derive the relationships

$$(y_i - 2)/w_i = y_i'/w_i', \qquad z_i/w_i = z_i'/w_i'.$$

Using the relationships and the known physical $y_i/w_i$ coordinate, one may determine that $w_i = -3$ and $w_i' = -3$ for vertex (1) and therefore for all points in the same $y_0 - z_0$ plane as vertex (1). A similar comparison for vertex (2) permits one to determine that $w_i = w_i' = -4$. Therefore, for this example, knowing only the transformation and a point pair correspondence for each object distance would permit the determination of the transformed points for all object points. Generalizations of this example can be very difficult. One simple case is that in which a range image is given. The $x_0$ values in this image may be used to compute $w_i = x_0/f + 1$ for each image point and the derived transformation used directly.

Since the physical image coordinates in the two images $(y_i/w_i, z_i/w_i)$ and $(y_i'/w_i', z_i'/w_i')$ are the only quantities that may be directly measured, it is interesting to reduce the linear equations to nonlinear equations that depend only upon the physical image coordinates.

Suppose the image transformation in superhomogeneous coordinates is given by

$$\begin{bmatrix} y_i' \\ z_i' \\ w_i' \\ 1 \end{bmatrix} = \begin{bmatrix} T_{11} & T_{12} & T_{13} & T_{14} \\ T_{21} & T_{22} & T_{23} & T_{24} \\ T_{31} & T_{32} & T_{33} & T_{34} \\ 0 & 0 & 0 & 0 \end{bmatrix} \begin{bmatrix} y_i \\ z_i \\ w_i \\ 1 \end{bmatrix}.$$

Factoring out the scale terms gives

$$\begin{bmatrix} y_i'/w_i' \\ z_i'/w_i' \\ 1 \\ 1/w_i' \end{bmatrix} = \frac{w_i}{w_i'} \begin{bmatrix} T_{11} & T_{12} & T_{13} & T_{14} \\ T_{21} & T_{22} & T_{23} & T_{24} \\ T_{31} & T_{32} & T_{33} & T_{34} \\ 0 & 0 & 0 & 1 \end{bmatrix} \begin{bmatrix} y_i/w_i \\ z_i/w_i \\ 1 \\ 1/w_i \end{bmatrix}.$$

For simplicity let us denote the physical image coordinates by

$$y'_{pi} = y'_i / w_i, \qquad y_{pi} = y_i / w_i,$$
$$z'_{pi} = z'_i / w'_i, \qquad z_{pi} = z_i / w_i.$$

The first two equations may now be written

$$\begin{bmatrix} y'_{pi} \\ z'_{pi} \end{bmatrix} = \frac{w_i}{w'_i} \begin{bmatrix} T_{11} & T_{12} & T_{13} & T_{14} \\ T_{21} & T_{22} & T_{23} & T_{24} \end{bmatrix} \begin{bmatrix} y_{pi} \\ z_{pi} \\ 1 \\ 1/w_i \end{bmatrix}.$$

We may now use the remaining two equations to attempt to eliminate the scale terms.

The third equation gives

$$w'_i / w_i = T_{31} y_{pi} + T_{32} z_{pi} + T_{33} + T_{34}/w_i,$$

which may be rewritten

$$w'_i = w_i \big[ T_{31} y_{pi} + T_{32} z_{pi} + T_{33} \big] + T_{34}.$$

The fourth equation gives no additional information.

Since $T_{44} = 1$, the fourth equation cannot be used. However, a solution with only a single unknown can still be determined, which will be illustrated in the following example.

Consider the translation of the camera and two perspective images of the cube given previously. The transformation is given for physical image coordinates as

$$\begin{bmatrix} y'_{pi} \\ z'_{pi} \\ 1 \\ 1/w'_i \end{bmatrix} = \frac{w_i}{w'_i} \begin{bmatrix} 1 & 0 & 0 & -2 \\ 0 & 1 & 0 & 0 \\ 0 & 0 & 1 & 0 \\ 0 & 0 & 0 & 1 \end{bmatrix} \begin{bmatrix} y_{pi} \\ z_{pi} \\ 1 \\ 1/w_i \end{bmatrix}.$$

The third equation gives the relationship $w'_i = w_i$, while the fourth equation provides no information. The first two equations may be written

$$y'_{pi} = y_{pi} - 2/w_i, \qquad z'_{pi} = z_{pi}.$$

Since we know that $w_i$ depends only upon the object distance and focal length, one method of solution is to determine a pair of corresponding coordinates $(y_{pi}, z_{pi})$ and $(y'_{pi}, z'_{pi})$ for each unique object distance. For the cube example, there are only two planes at different object distances. Suppose we determine these corresponding point pairs of physical image coordinates for vertices (1) and (2), i.e.,

| Vertex | $(y_{pi}, z_{pi})$ | $(y'_{pi}, z'_{pi})$ |
|--------|--------------------|----------------------|
| 1 | $\left(\frac{1}{3}, \frac{1}{3}\right)$ | $\left(1, \frac{1}{3}\right)$ |
| 2 | $\left(\frac{1}{4}, \frac{1}{4}\right)$ | $\left(\frac{3}{4}, \frac{1}{4}\right)$ |

This information and the previous equations may be used to determine $w_i$ for each object plane. For vertex (1)

$$w_i = -2/(y'_{pi} - y_{pi}) = -3$$

while for vertex (2)

$$w_i = -4.$$

Therefore, given the location of all vertex points in the plane of vertex (1), in particular, vertices (1), (3), (5), and (7), the transformation

$$y'_{pi} = y_{pi} + \tfrac{2}{3}$$

may be used. For the vertices in the same plane as vertex (2), namely, vertices (2), (4), (6), and (8), the transformation

$$y'_{pi} = y_{pi} + \tfrac{1}{2}$$

must be used.

The transformations previously developed may be used in two important ways to determine a vibration compensation. The simplest solution depends upon a far field assumption in which all object distances $x_p$ are replaced by an average distance $\bar{x}_0$. The solution in this case assumes

$$w_i / w'_i \simeq 1$$

and may be written

$$\begin{bmatrix} y'_{pi} \\ z'_{pi} \end{bmatrix} = \begin{bmatrix} T_{11} & T_{12} & T_{13} & T_{14} \\ T_{21} & T_{22} & T_{23} & T_{24} \end{bmatrix} \begin{bmatrix} y_{pi} \\ z_{pi} \\ 1 \\ s \end{bmatrix},$$

where

$$s = \begin{cases} (1 - T_{31} y_{pi} - T_{32} z_{pi} - T_{33})/T_{34}, & T_{34} \neq 0, \\ f_0/(1 + \bar{x}_0), & T_{34} = 0, \end{cases}$$

and $\bar{x}_0$ is an average object distance.

The near field solution applies whenever the far field solution is not valid; however, it is more complicated. The solution essentially requires a stereoscopic determination of the object distance $x_0$ for each physical image point. This involves first locating corresponding point pairs in the two physical images using a matching technique, then solving for the object distance $x_0$. This solution is also complicated by the vibration, which will produce an error in matching corresponding points.

An example of the two-image transformation procedure is shown in Fig. 4.28. An original image taken from a remotely controlled TV camera system is shown in Fig. 4.28a. Another image taken at a different control platform position is shown in Fig. 4.28b. The transformation matrix **T** given

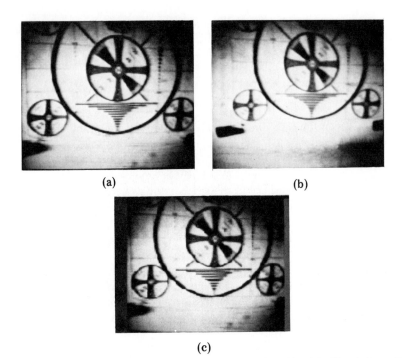

(a)                                              (b)

(c)

*Figure 4.28*  Example of two-image perspective transformation: (a) original image in position 1, (b) image at position 2, and (c) image at position 1 transformed to correspond to image at position 2.

below was used to transform the image at position (1) to approximately correspond to the image taken at position (2). Note that the correspondence is not exact and that clipping of a portion of the image was required:

$$
\mathbf{T} = \begin{bmatrix} 1 & 0 & 0 & 10 \\ 0 & 1 & 0 & 0 \\ 0 & 0 & 0 & 0 \\ 0 & 0 & 0 & 1.15 \end{bmatrix}.
$$

## 4.4  Enhancement by Spatial Filtering

Image enhancement by spatial filtering is perhaps the most powerful method for improving image quality for human viewing. The foundations of *digital filtering* of images will be presented in this section by considering linear, position invariant operations that are characterized by a convolution relationship. The input image is convolved with a point-spread function to produce an output image that is "enhanced" by the operation. Since the spatial convolution of two functions corresponds to the frequency domain

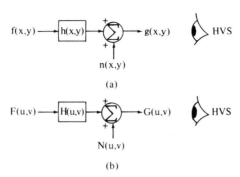

*Figure 4.29*   Image enhancement model: (a) spatial domain and (b) frequency domain.

multiplication of the transforms of the functions, design from the frequency domain is usually simpler. The basic image enhancement model is shown in Fig. 4.29 for the spatial and frequency domain methods. An input image $f(x,y)$ is operated on by a filter $h(x,y)$ then combined with a noise function $n(x,y)$ to produce the observed image. The spatial domain model corresponds to the relationship

$$g(x,y) = f(x,y) * h(x,y) + n(x,y),$$

where the two-dimensional convolution is defined by

$$f(x,y) * h(x,y) = \int \int_{-\infty}^{\infty} f(\xi,\eta) h(x-\xi, y-\eta) \, d\xi \, d\eta$$

$$= \int \int_{-\infty}^{\infty} f(x-\xi, y-\eta) h(\xi,\eta) \, d\xi \, d\eta.$$

The frequency domain relationships are given by

$$G(u,v) = F(u,v) H(u,v) + N(u,v),$$

where the frequency domain functions are equal to the two-dimensional Fourier transforms of the spatial domain functions. For example,

$$F(u,v) = \int \int_{-\infty}^{\infty} f(x,y) \exp[-j2\pi(ux+vy)] \, dx \, dy.$$

The addition of the noise term $n(x,y)$ in the image enhancement model is simply a statement of the fact that some type of "noise" is almost always present. The inclusion of the noise after the filtering operation is a mathematical convenience but may obscure the fact that both the desired signal and the noise are processed by the enhancement filter.

Certain image enhancement problems arise quite often and will be treated in detail in this section. These include edge enhancement, noise smoothing, and structured noise removal. The results of image enhancement must generally be evaluated subjectively; thus the concept of image quality is important. The theory of image enhancement is still evolving and the techniques available may be considered as tools of an artist. Much of the art of enhancement is knowing when to stop.

(a)

*Figure 4.30*   Examples of unsharp masking: (a) toy tank and (b) electron microscope image (512×512 images).

### 4.4.1   Edge Enhancement

Edge enhancement, sharpening, or crispening techniques are designed to increase the visibility of general low-contrast edges, and often lead to increased perception of detail.

One of the simplest edge enhancement techniques is the digital implementation of photographic "unsharp masking." The essence of this method is to subtract a blurred representation of an image from the image. In particular, one may calculate

$$g(x,y) = f(x,y) - \bar{f}(x,y),$$

where the function $\bar{f}(x,y)$ is a local average, for example,

$$\bar{f}(x,y) = \frac{1}{8} \sum_{i=-1}^{1} \sum_{j=-1}^{1} f(x+i, y+j), \qquad i+j \neq 0.$$

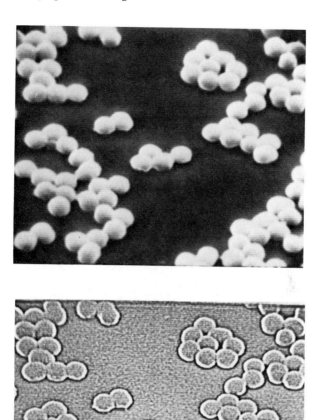

(b)

*Figure 4.30* Continued.

Two examples of digital unsharp masking using the above relationship are shown in Fig. 4.30. Note that the wheels of the toy tank are noticeably enhanced. Also, note on the electron microscope image in Fig. 4.30b that the particle edges show both dark and light bands. This effect is called *filter ringing* and may be examined more closely by considering the unsharp masking operation as a convolution.

The unsharp masking operation may be easily shown to be equivalent to a convolution of the original image with a filter whose point spread function is

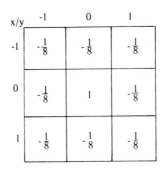

*Figure 4.31* Point-spread function $h(x,y)$ for unsharp masking operation.

shown in Fig. 4.31. The finite convolution may be written

$$g(x,y) = \sum_{i=-1}^{1} \sum_{j=-1}^{1} f(x+i,y+j)h(i,j).$$

Suppose a portion of the image containing a horizontal edge of the form $(\ldots,0,0,1,1,0,0,\ldots)$ with all 0s otherwise, passes horizontally across the center of the filter. The response to this edge function is of the form $(\ldots,0,-\frac{1}{8},\frac{7}{8},-\frac{1}{8},0,\ldots)$, which when scaled and displayed results in a gray to black to white to gray transition which is clearly visible in the electron microscope image. A careful filter design is required to eliminate the ringing response.

Edge enhancement may also be approached by considering the derivatives of the image function. Directional derivatives may be used to enhance edges with a preferential direction, such as horizontal, vertical, or diagonal. Another derivative technique requires the formation of the *digital gradient* at each point, i.e., approximating

$$\nabla f(x,y) = \frac{\partial f}{\partial x}\mathbf{i} + \frac{\partial f}{\partial y}\mathbf{j}$$

using difference equations. Since the gradient is a vector in the direction of the maximum directional derivative, both magnitude and direction may be used. For example, if first differences are used to approximate the partial derivatives

$$\frac{\partial f}{\partial x} \simeq f(x+1,y) - f(x,y) = \Delta_x f(x,y),$$

$$\frac{\partial f}{\partial y} \simeq f(x,y+1) - f(x,y) = \Delta_y f(x,y),$$

then the magnitude of the gradient is approximately

$$|\nabla f(x,y)| \simeq \left\{ \left[ \Delta_x f(x,y) \right]^2 + \left[ \Delta_y f(x,y) \right]^2 \right\}^{1/2},$$

and the direction is indicated by

$$\theta = \tan^{-1}\left[\Delta_y f(x,y)/\Delta_x f(x,y)\right].$$

This direction is generally perpendicular to the ridge of an edge; therefore an addition of 90° to $\theta$ is helpful in indicating the direction of a ridge.

Note that both the magnitude and direction angle of the gradient are nonlinear operators. A variety of other nonlinear methods may be used for edge enhancement.

### 4.4.2 Noise Smoothing

Noise smoothing for "snow" removal in TV images was one of the first applications considered for digital image processing (Graham, 1962). "Noise" is of course a relative term; however, there are some characteristic types of noise for pictures.

Noise arising from an electronic sensor generally appears as random, uncorrelated, additive errors or "snow." This type noise may give rise to extreme pixel-to-pixel changes rather than the small changes normally observed from natural scenes. In these cases, a technique which detects the extreme variations may be used, and the affected points can be replaced by a local average (Nathan, 1968). This nonlinear operation is described by

$$g(x,y) = \begin{cases} \bar{f}(x,y) & \text{if } |f(x,y)-\bar{f}(x,y)| > T \\ f(x,y) & \text{otherwise,} \end{cases}$$

where $T$ is a threshold dependent on the noise variance and the average may be computed for a local region. For the $3\times 3$ region, the average is

$$f(x,y) = \frac{1}{9} \sum_{i=-1}^{1} \sum_{j=-1}^{1} f(x+i, y+j).$$

The center point could also be omitted from the average computation.

Although nonlinear methods for noise smoothing are often more powerful than linear techniques, it is interesting to consider the linear averaging operation $\bar{f}(x,y)$ for various region sizes. If there are $N$ points in the region and a simple independent, identically distributed noise process is involved, then we may expect a $\sqrt{N}$ improvement in the signal-to-noise variance ratio by averaging. This average could be accomplished using $N$ neighbor points or a single point and $N$ repetitive frames. Therefore, the noise variance can be reduced. Let us also consider the effects of this operation on a picture without noise.

The effect of averaging is best represented in the frequency domain. Let us specifically consider an $N = n \times n$ average of the form

$$\bar{f}(x,y) = \frac{1}{N} \sum_{k=0}^{n-1} \sum_{l=0}^{n-1} f(x+k, y+l).$$

Using the properties of the discrete Fourier transform established in Chapter 3, the DFT of $\bar{f}(x,y)$ may be written

$$\bar{F}(u,v) = F(u,v)H(u,v),$$

where

$$H(u,v) = \frac{1}{N} \sum_{k=0}^{n-1} \sum_{l=0}^{n-1} \exp\left[ -j2\pi(uk + vl) \right].$$

Since the exponential term may be separated into a product, we may use the series

$$\frac{1-r^n}{1-r} = \sum_{i=0}^{n-1} r^i,$$

where $r = \exp(-j2\pi u)$ to obtain

$$H(u,v) = \frac{\phi(u,v,n)}{N} \left( \frac{\sin n\theta_x}{\sin \theta_x} \right) \left( \frac{\sin n\theta_y}{\sin \theta_y} \right),$$

where

$$\theta_x = \pi u, \qquad \theta_y = \pi v, \qquad \phi(u,v,n) = \exp\left[ j\pi(n-1)(u+v) \right].$$

The magnitude of $H(u,v)$ has a low-pass filter characteristic. The half-power width of this characteristic decreases as $N$ increases. Therefore, increasing the size of the averaging region decreases the half-power width of the filter and therefore blurs the image information.

*Frame-to-frame averaging* produces a similar function form for a low-pass filter in time. If the image stays stationary, its temporal bandwidth is very small so that temporal averaging reduces the noise variance without significantly affecting the image.

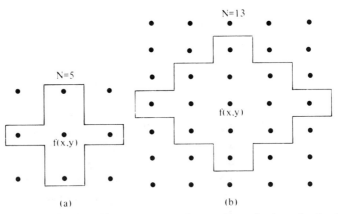

*Figure 4.32* Integer distance neighbors of a point provide a simple region for the median filter: (a) unit distance and (b) distance 2 or less.

Filtering techniques that reduce the effects of random noise without significantly affecting the useful image information are therefore desirable. The Tukey median filter (Tukey, 1971) is a nonlinear technique that is useful for reducing noise without much image degradation. The median filter is implemented by selecting a local region surrounding a center point $f(x,y)$. The filtering operation then consists of replacing the center value $f(x,y)$ by the median of the values in the region. For $N$ points with $N$ odd, $\frac{1}{2}(N-1)$ points are less than or equal and $\frac{1}{2}(N-1)$ points are greater than or equal to the median point.

Since a region with an odd number of points $N$ is desirable for a median filter, a natural choice is the integer distance neighbors of a point as shown in Fig. 4.32.

### 4.4.3 Structured Noise Removal

In some situations, structured noise rather than random noise is the predominant factor degrading image quality. One particular case is the presence of a line pattern superimposed on an image, for example, TV scan lines (Kruger *et al.*, 1970).

Consider a single line with slope $a$ represented by

$$f(x,y) = \delta(y - \alpha x).$$

The Fourier transform of $f(x,y)$ is

$$F(u,v) = \delta(u + \alpha v),$$

which is a single line of slope $-1/\alpha$. For an array of lines separated by a distance $d$, the spatial signal is given by

$$f(x,y) = \sum_{k=-\infty}^{\infty} \delta\left(y - \alpha x - kd\sqrt{1+\alpha^2}\right).$$

The Fourier transform of this array is

$$F(u,v) = \frac{1}{d\sqrt{1+\alpha^2}} \sum_{m} \sum_{n} \delta\left(u - \frac{ma}{d\sqrt{1+\alpha^2}}, v - \frac{n}{d\sqrt{i+\alpha^2}}\right).$$

Since the Fourier transform possesses this regular pattern, which is illustrated in Fig. 4.33, it is possible to design frequency filters that can eliminate the pattern. If both the slope and line spacing parameters are known, notch filters could be placed at each significant frequency point. Knowing only the slope permits one to design a band reject filter along the principal frequency line of the pattern. To illustrate this method, consider the isometric graph paper shown in Fig. 4.34a, which contains lines at 0, 30, 90, and 150°. The Fourier transform was computed for this pattern and 1° wedge-shaped frequency samples were computed to automatically detect the predominant line pattern angles. An +-shaped frequency filter was then multiplied by the image

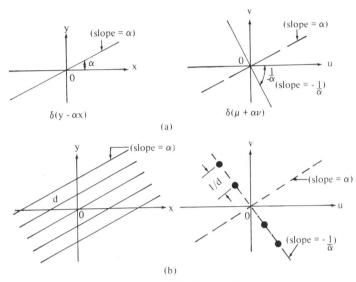

*Figure 4.33* (a) Line mass at angle $\alpha$ and Fourier transform and (b) periodic lines at angle $\alpha$ and spacing $d$ and Fourier transform.

transform to eliminate the 0 and 90° lines. The inverse Fourier transform of this filtered image is shown in Fig. 4.34b. Note that the lines at 30 and 150° are retained in the image.

### 4.4.4  High-Frequency Emphasis Filtering

One important filter for image enhancement is often called a *high-frequency emphasis filter*. This filter partially suppresses the lower-frequency components, enhances a region of high-frequency components and suppresses very-high-frequency components that may be more representative of noise than image signal. An idealized response of the high-frequency emphasis filter is shown as the square wave in Fig. 4.35. The smooth version labeled "realistic" would of course be used in practice to avoid ringing produced by Gibbs phenomenon. The ideal high-emphasis filter can be simplified in several ways. One decomposition into an all-pass filter with gain $\alpha$, an ideal high-pass filter with cutoff $f_h$ and gain $(1 - \alpha)^{1/2}$ and an ideal low-pass filter with cutoff $f_l$ is shown in Fig. 4.35b. The parallel–serial combination of these three component filters that results in the desired overall response is shown in Fig. 4.35c. This form of the ideal filter emphasizes that only three control parameters are required. The parameter $\alpha$ may be called a contrast retention factor, since it governs the amount of low-frequency contrast information

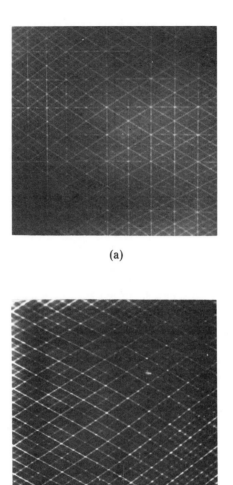

(a)

(b)

*Figure 4.34*   (a) Digitized isometric graph paper and (b) filtered image with horizontal and vertical lines removed. (From Kruger [1970].)

retained in the filtered image. The two cutoff frequency parameters $f_l$ and $f_h$ control the range of frequencies that are emphasized. Note that if $f_l \rightarrow \infty$ a high-frequency-emphasis filter without a noise frequency cutoff results. Also, if $f_h = 0$, a low-pass filter results. The parameters are assumed constrained by $0 < \alpha < 1$ and $f_l > f_h$. As $\alpha \rightarrow 0$, the filter shape approaches a band-pass filter and as $\alpha \rightarrow 1$, the filter approaches an all-pass filter.

Several examples of spatial filtering for image enhancement of a chest x ray are shown in Fig. 4.36.

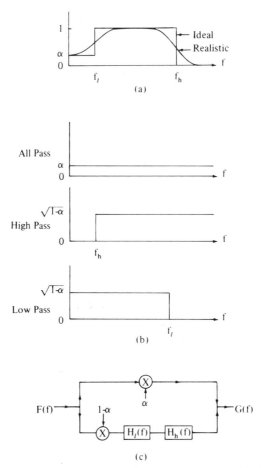

*Figure 4.35* General form of a high-frequency-emphasis filter: (a) ideal and realistic overall frequency responses, (b) components of ideal response, and (c) realization of the filter $H(f) = \alpha + (1 - \alpha)H_l(f)H_h(f)$ for $0 < \alpha < 1$.

### 4.4.5   Novel Filtering Methods

A few novel filtering methods will be described in this section to illustrate the variety of techniques which are available.

*Low sequency filtering* is shown in Fig. 4.37. The block-type error is characteristic of this type of filtering. An interesting fact is that a filter designed in one orthogonal transform domain such as the Fourier domain can be realized by filtering in another domain such as the Walsh domain (Robinson, 1972).

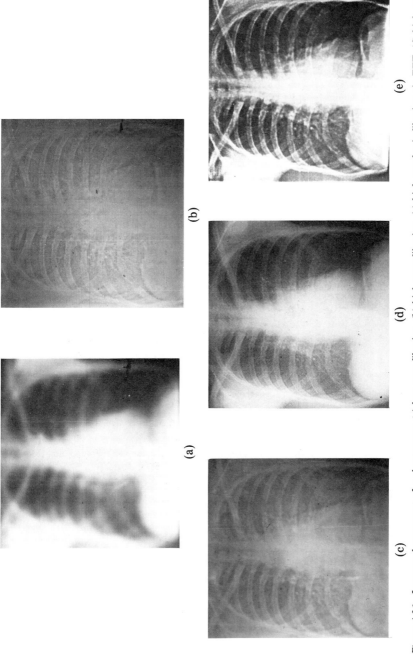

(b)

(e)

(d)

(a)

(c)

*Figure 4.36* Image enhancement of a chest x ray: (a) low-pass filtering, (b) high-pass filtering, (c) high-emphasis filtering via FFT, (d) high-emphasis filtering via a recursive filter, and (e) high-emphasis filtering via FFT and contrast enhancement. [From Hall *et al.* (1971).]

211

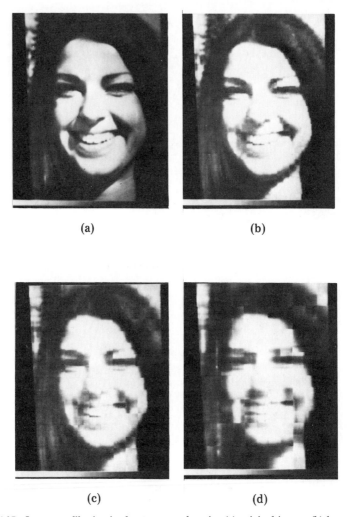

(a)                                                  (b)

(c)                                                  (d)

*Figure 4.37*  Low-pass filtering in the sequency domain: (a) original image, (b) low sequency filtered image with 60 by 60 square filtering of 256 by 256 domain, (c) result of 40 by 40 square filter, and (d) result of 20 by 20 square filter.

Filtering via the number theoretic transform offers zero round-off error and can be computed via a fast algorithm. An example is shown in Fig. 4.38.

*Image subtraction* is often proposed as a filtering technique. An example is shown in Fig. 4.39, which emphasizes the importance of accurate registration. If the transformation between the images is more than simple translation, the filtering problem becomes planar variant. Thus, an apparently simple problem may require a complex solution.

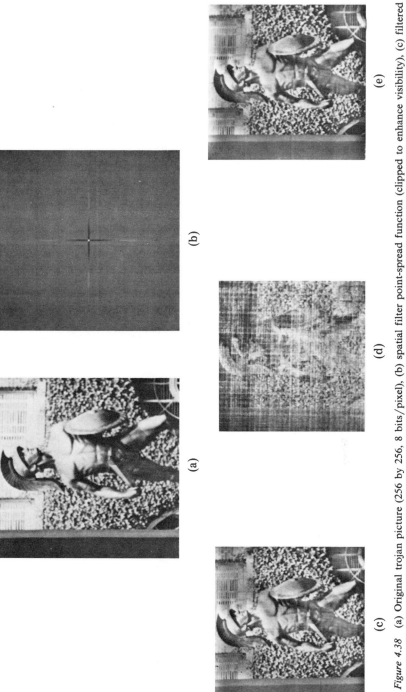

(a)

(b)

(c)

(d)

(e)

*Figure 4.38* (a) Original trojan picture (256 by 256, 8 bits/pixel), (b) spatial filter point-spread function (clipped to enhance visibility), (c) filtered image using Fourier transform, (d) roundoff error pattern of the conventional Fourier transform technique (note correlation with filtered image enhanced for visibility), and (e) filtered image using finite field or number theoretic transform (zero roundoff error). [From Kwoh (1977).]

213

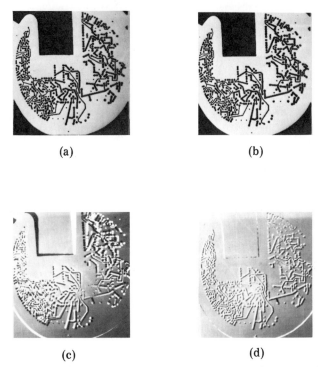

(a)                                                              (b)

(c)                                                              (d)

*Figure 4.39*   Two printed circuit board mask images that are to be compared. (a) may be considered as the master print, (b) is an item to be tested, (c) is the result of subtraction without geometric correction, and (d) is the result of subtraction with geometric correlation. (Note that dark regions appear in the test but not in the master image.)

## 4.5   Introduction to Image Restoration

In the previous section we considered a variety of image enhancement techniques. These techniques are often used for improving the quality of images for human viewing or as preprocessing techniques such as normalization of a set of images for automatic pattern recognition. For both of these applications, the effectiveness of the technique is usually determined subjectively or by evaluation experiments. In the remainder of this chapter, we will consider a related but different problem called *image restoration*. Image restoration may be defined as the improvement of image quality under an objective evaluation criterion. For example, we may use least squares or minimum mean squared error as a criterion to provide a best estimate of an original image from a blurred image. Restoration often requires measurements of the degradation phenomena and careful a priori consideration of the system that produced the degraded image. Since scientific images are produced to provide useful information about a phenomenon of interest, the restoration approach applies to the measurement sciences of imaging. In fact, the known techniques have been developed in such diverse fields as space exploration, medical and industrial radiography, atmospheric physics, geophysics, microscopy, and electron microscopy. Also, since physical imaging systems are not perfect, a recorded image will almost certainly be a degraded version of some ideal image. Fortunately, a few types of degradation phenomena appear repeatedly; thus, good system and degradation models and mathematical restoration techniques have been developed. No general theory of image restoration has yet evolved; however, some excellent specific solutions have been developed especially for linear, planar invariant systems.

The importance of mathematical models of the imaging systems and degradation phenomena cannot be overemphasized in image restoration, since objective criteria that are appropriate to the models must be specified. Two classes of imaging system models are illustrated in Fig. 4.40. The planar imaging model shown in Fig. 4.40a performs a mapping from a two-dimensional input image $f(x,y)$ to a two-dimensional output image $g(x,y)$ via a system transformation denoted by $h$. This model is the simplest form that accurately represents many physical systems, including many electronic, optical, and digital systems. If $h$ is a linear operator, then a superposition relation will relate $f$ and $g$. If $h$ is both linear and planar invariant, then $g$ will equal the convolution of $f$ and $h$. These statements shall be proven later. The second model shown, Fig. 4.40b, may be called a scene imaging model. A three-dimensional scene $f(x,y,z)$ is mapped into a two-dimensional image $g(x,y)$ by an operator $h$. Since the scene has a finite extent in the $z$ direction, the operator $h$ may appear to be scene dependent. The most common operator of this type performs a perspective, projective transformation, and is

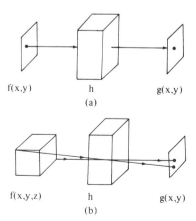

f(x,y)                 h                 g(x,y)
                      (a)

f(x,y,z)               h                 g(x,y)
                      (b)

*Figure 4.40*   Image system models: (a) planar imaging model representing a two-dimensional to two-dimensional mapping and (b) scene imaging model representing a three-dimensional to two-dimensional mapping.

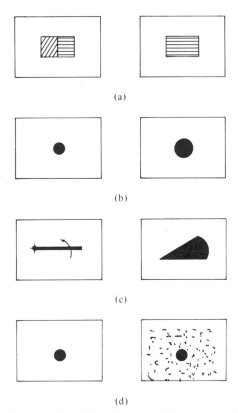

*Figure 4.41* Degradation models: (a) nonlinearities (planar invariant), (b) blurring (linear, planar invariant), (c) motion smearing (linear, planar invariant), and (d) noise (linear, planar variant).

a model for any lens imaging system. A continuous–discrete distinction for $f$, $h$, and $g$ may be made. For computer restoration, all three functions must be discrete. The natural representation for optical systems would be by continuous functions. Hybrid mixtures are of course possible. A distinction could also be made between deterministic and random systems; however, most systems are designed to be deterministic. Random phenomena are usually placed in the degradation model.

Four common degradation models are shown in Fig. 4.41. The first three are deterministic while the last is a random phenomenon. A point nonlinearity that is planar invariant is the simplest type of degradation to consider. This degradation accurately represents photographic film development and is characterized by a Hurter–Driffield (HD) curve (Dainty and Shaw, 1974). It is interesting to note that restoration by inversion of the nonlinearity of film does not have a unique solution if the HD curve saturates at the toe and shoulder of the curve. If information is lost due to saturation in the recording

*Figure 4.42*   Planar invariant image restoration model.

or development of film, an infinite variety of possible inverse solutions can be obtained. This singular or ill-conditioned nature of the image restoration occurs with most physical systems and must be accounted for in the models. A blurring degradation is illustrated in Fig. 4.41b. A model of this phenomenon can account for the finite aperture or diffraction limit of many physical imaging systems. This model must account for spatial degradations. In the simplest case of a linear, planar imaging model, a planar invariant blurring may be considered. We will see that the restoration solution to this problem is ill-conditioned, especially when the Fourier transform of the blurring function contains zero values. Another type of blurring degradation called *motion smearing* is illustrated in Fig. 4.41c. For general types of motion, for example, rotations, the planar variant model must be used to represent the physical phenomenon accurately. Finally, a random noise degration is illustrated in Fig. 4.41d. The consideration of a random degradation adds considerable realism to most models but often also requires estimation of the noise characteristics as well as estimation of the restored image. Several innovative techniques will now be considered for some of the combinations of imaging system and degradation models. Even though the restoration problem is ill-conditioned, certain constraints may be placed on the solution that often result in the optimum solution for a given criterion.

The form of the mathematical models may be easily developed for the continuous, planar imaging models shown in Fig. 4.42. In the simplest case, the output image $g(x,y)$ would exactly replicate the input image $f(x,y)$, i.e., $h$ is an identity mapping and the noise $n(x,y)=0$, so that

$$g(x,y)=f(x,y).$$

The input $f(x,y)$ may be represented by definition of the impulse function:

$$f(x,y)=\int\int_{-\infty}^{\infty}f(\zeta,\eta)\delta(x-\zeta,y-\eta)\,d\zeta\,d\eta.$$

If the input is transformed by a linear system $h$, which has the property

$$ah(f_1)+bh(f_2)=h(af_1+bf_2),$$

then

$$g(x,y)=\int\int_{-\infty}^{\infty}f(\zeta,\eta)h[\,\delta(x-\zeta,y-\eta)\,]\,d\zeta\,d\eta.$$

If we define the point-spread function $h(x,y,\zeta,\eta)$ to be the linear system response to a point source $\delta(x-\zeta,y-\eta)$,

$$h(x,y,\zeta,\eta)=h[\,\delta(x-\zeta,y-\eta)\,],$$

then the fundamental superposition equation results,

$$g(x,y) = \int\int_{-\infty}^{\infty} f(\zeta,\eta)h(x,y,\zeta,\eta)\,d\zeta\,d\eta.$$

The superposition equation is also called the *Fredholm equation* of the first kind and has been studied extensively. Note that the system defined by the superposition equation is in general planar variant, since $h(x,y,\zeta,\eta)$ depends on both the input point $(\zeta,\eta)$ and the output point $(x,y)$. The one-dimensional superposition equation

$$g(x) = \int_{-a}^{a} f(\zeta)h(x,\zeta)\,d\zeta$$

will be used to illustrate the ill-conditioned aspect of the solution. The *Riemann–Lebesgue theorem* states that if $h(x,\zeta)$ is an integrable function, then

$$\lim_{\alpha\to\infty}\int_{-a}^{a} h(x,\zeta)\sin(\alpha\zeta)\,d\zeta = 0.$$

Adding the two previous equations gives

$$g(x) = \lim_{\alpha\to\infty}\int_{-a}^{a}\left[f(\zeta)+\sin(\alpha\zeta)\right]h(x,\zeta)\,d\zeta.$$

The implications of this result (Andrews and Hunt, 1977, p. 115) for image restoration are direct. The restoration problem may be simply stated as, given $g(x)$ and $h(x,\zeta)$ determine $f(\zeta)$. However, since a high-frequency sinusoid can be added to $f(x)$ and will produce an identical $g(x)$, a unique solution cannot be expected.

Another implication of the Riemann–Lebesgue theorem is that a small change in $g(x)$ could result in a large difference between the integrands $f(\zeta)+\sin(\alpha,\zeta)$ and $f(\zeta)$. Thus, restoration may be an *ill-conditioned* problem, and oscillatory errors may occur.

If the point-spread function depends only upon the differences in position between input and output points

$$h(x,y,\zeta,\eta) = h(x-\zeta,y-\eta),$$

the superposition equation reduces to the important convolution equation

$$g(x,y) = \int\int_{-\infty}^{\infty} f(\zeta,\eta)h(x-\zeta,y-\eta)\,d\zeta\,d\eta.$$

The convolution relationship represents the planar invariant imaging system. When additive noise degradations are also present, the restoration model becomes

$$g(x,y) = \int\int_{-\infty}^{\infty} f(\zeta,\eta)h(x-\zeta,y-\eta)\,d\zeta\,d\eta + n(x,y),$$

which is the total continuous planar invariant model corresponding to Fig. 4.42.

The planar invariant imaging model will be the primary subject of the following sections, since it represents the majority of physical degradations, at least to a first-order approximation. Even for systems in which a nonlinearity or planar variant degradation is dominant, the known solutions depend heavily on reducing the problem to a planar invariant situation. The material presented must be considered of an introductory nature. For more detailed analysis the reader might start with the works of Andrews and Hunt (1977) and Huang (1975).

### 4.5.1   Restoration by Digital Computation

To introduce the problem of digital image restoration by digital filtering, we shall first consider the continuous imaging situation, then "discretize" the process for implementing via a computer. Let $f(x,y)$ represent the original undistorted picture at the point $(x,y)$, $g(x,y)$ represent the observed distorted image, $n(x,y)$ represent an additive noise process, and $h(x,y)$ represent the point-spread function that characterizes the position invariant imaging system.

If there is no noise present, the imaging system may produce a blurred image $b(x,y)$, which may be described by the convolution equation

$$b(x,y) = \int\!\!\int_{-\infty}^{\infty} h(x-x',y-y')f(x',y')\,dx'\,dy'.$$

If additive noise is present, then the distorted image is described by

$$g(x,y) = b(x,y) + n(x,y).$$

The continuous restoration problem may be stated: given $g(x,y)$, estimate $f(x,y)$. The methods for solving this problem differ mainly in their assumptions and use of a priori information about $h(x,y)$, $n(x,y)$, and $f(x,y)$.

To reformulate the problem for computer solution, two main steps are required. First, the observed image must be sampled, then the integral must be approximated by a numerical integration formula. Therefore, suppose that $g(x,y)$ and $n(x,y)$ are sampled at a finite set of points to produce $g(i,j)$ and $n(i,j)$ for $i=1,2,\ldots,I$, $j=1,2,\ldots,J$. Sampling the noise at the same points as the observed image is simply a mathematical convenience. We may now express the distorted image as

$$g(i,j) = \int\!\!\int_{-\infty}^{\infty} h(x_i-x',y_j-y')f(x',y')\,dx'\,dy' + n(i,j),$$

where the point $(x_i,y_j)$ corresponds to the sample point $(i,j)$. To complete the process, the continuous integral must be replaced by a weighted sum of the integrand values at points $(x_k',y_l')$, where $k=1,2,\ldots,K$, $l=1,2,\ldots,L$. Under these conditions, the equation becomes

$$g(i,j) = \sum_{k=1}^{K}\sum_{l=1}^{L} w_{kl}h(x_i-x_k',y_j-y_l')f(x_k',y_l') + n(i,j),$$

where $w_{kl}$ is the weighting factor at $(x'_k, y'_l)$. We may simplify the notation by letting

$$h(i-k, j-l) = h(x_i - x'_k, y_j - y'_l) \qquad \text{and} \qquad f(k,l) = f(x'_k, y'_l)$$

to obtain

$$g(i,j) = \sum_{k=1}^{K} \sum_{l=1}^{L} w_{kl} f(k,l) h(i-k, j-l) + n(i,j).$$

At this point, let us investigate a phenomenon called *edge effects*. For simplicity, let us consider the one-dimensional model with unity weighting factors and no noise. The convolution is then given by

$$g(i) = \sum_{k=1}^{K} h(i-k) f(k).$$

Now suppose $g(i)$ is known for $i = 1, 2, \ldots, I$ and that $f(k)$ is to be determined for $k = 1, 2, \ldots, K$. How many values of $h(i-k)$ must be known in order to compute the discrete convolution? Clearly, $1 \leq i - k \leq I - K$. The number of values $i - k$ is the difference between the maximum and minimum, or $I + 1$. For example, if $I = 3$ and $K = 2$, the number of required $h$ values is 4. A similar situation arises given $f(k)$ for $k = 1, 2, \ldots, K$ and $h(j)$ for $j = 1, 2, \ldots, J$. How many nonzero values of $g(i)$ can be produced? This time the answer requires consideration of the overlaps between the two sequences $f(k)$ and $h(j)$, and is $(K - J + 1) + 2(J - 1)$ if $K > J$. The term $K - J + 1$ is the number of complete overlaps of the $f$ and $k$ sequences, and the term $2(J - 1)$ is the result of the two edge overlaps. The main conclusion is simply that the lengths of the sequences $f$, $g$, and $h$ cannot be equal without loss of information. If one wishes to work with equal length sequences or square matrices, a common solution is to extend the sequences and matrices by appending zero elements to obtain the appropriate lengths. For the last example, the number of nonzero values of $g(i)$ is $K + J - 1$. We may choose any integer $M \geq K + J - 1$ and form extended sequences $g_e(i)$, $f_e(i)$, and $h_e(i)$, each containing $M$ terms, by appending zero values to the end of each original sequence. The resulting convolution equation may be written

$$g_e(i) = \sum_{i=1}^{M} h_e(i-k) f_e(k), \qquad i = 1, 2, \ldots, M.$$

Other aspects of discrete convolution are considered in the Appendix. For the remainder of this section, extended equal length sequences will be assumed, to simplify the notation. The resulting two-dimensional convolution will therefore be

$$g(i,j) = \sum_{k=0}^{M-1} \sum_{l=0}^{M-1} f(k,l) h(i-k, j-l) + n(i,j)$$

for $i,j = 0, 1, \ldots, M - 1$. Note that the numerical integration factors $w_{kl}$ have been incorporated into the $h$ values. Also, the subscripts now start at 0 for compatibility with the discrete Fourier transform notation. Finally, it should be noted that a computer solution implies the quantization of the $f$, $g$, and $h$ values.

In summary, to convert the restoration problem to a suitable form for computer solution requires sampling, numerical integration, and quantization. To conveniently work with equal length sequences and square matrices, an extending operation that consists of appending zero values to the original sequences is required.

The power of linear algebra can be brought to bear on the restoration problem by converting the convolution relationship to the equivalent vector form. The finite two-dimensional convolution summation may always be equivalently considered as a vector equation. To convert we simply perform a column scanning similar to that used when storing a two-dimensional array on a sequential data set. The following notation is necessary:

$$\mathbf{g}_d = \begin{bmatrix} \hat{\mathbf{g}}_0 \\ \hat{\mathbf{g}}_1 \\ \vdots \\ \hat{\mathbf{g}}_{M-1} \end{bmatrix} \quad \text{and} \quad \hat{\mathbf{g}}_j = \begin{bmatrix} g(0,j) \\ g(1,j) \\ \vdots \\ g(M-1,j) \end{bmatrix},$$

where $\mathbf{g}_d$ is the total picture vector and $\hat{\mathbf{g}}_j$ is the vector corresponding to one column,

$$\mathbf{n} = \begin{bmatrix} \hat{\mathbf{n}}_0 \\ \hat{\mathbf{n}}_1 \\ \vdots \\ \hat{\mathbf{n}}_{M-1} \end{bmatrix} \quad \text{and} \quad \hat{\mathbf{n}}_j = \begin{bmatrix} n(0,j) \\ n(1,j) \\ \vdots \\ n(M-1,j) \end{bmatrix},$$

$$\mathbf{f} = \begin{bmatrix} \hat{\mathbf{f}}_0 \\ \hat{\mathbf{f}}_1 \\ \vdots \\ \hat{\mathbf{f}}_{M-1} \end{bmatrix} \quad \text{and} \quad \hat{\mathbf{f}}_l = \begin{bmatrix} f(0,l) \\ f(1,l) \\ \vdots \\ f(M-1,l) \end{bmatrix},$$

and

$$[H] = \begin{bmatrix} [H_{00}] & [H_{01}] & \cdots & [H_{0,M-1}] \\ [H_{10}] & & & \\ \vdots & & & \\ [H_{M-1,0}] & & & [H_{M-1,M-1}] \end{bmatrix},$$

where each of the submatrices $[H_{ik}]$ is an $M$ by $M$ matrix of the form

$$[H_{ik}] = \begin{bmatrix} h(i-k,0) & h(i-k,-1) & \cdots & h(i-k,-M+1) \\ h(i-k,1) & h(i-k,0) & \cdots & h(i-k,-M+2) \\ \vdots & & \ddots & \vdots \\ h(i-k,M-1) & h(i-k,M-2) & & h(i-k,0) \end{bmatrix}.$$

A submatrix $[H_{ik}]$ affects observed vector $\mathbf{g}_i$, which is the $i$th row or line of the observed image, and operates on image vector $\mathbf{f}_k$, which is the $k$th row or line of the input image. A careful study of the submatrix $[H_{ik}]$ reveals that all diagonal terms are equal. Since $[H_{ij}]$ is a square matrix with equal diagonal terms we may recognize it as a Toeplitz matrix. The total matrix $[H]$ has been called *block Toeplitz* by Hunt (1972).

Further structure of the $[H]$ matrix may be realized by noting that sampled functions may without loss of generality be considered as periodic functions with periods equal to the sample interval. Thus, $f(-1)=f(M-1)$, $f(-2)=f(M-2)$, etc. Similarly, the sampled point-spread function may be considered periodic, and $h(i-k,-1)=h(i-k,M)$, $h(i-k,-2)=h(i-k,M-1)$, etc. Using this property for the submatrix $[H_{ik}]$ produces the form

$$[H_{ik}] = \begin{bmatrix} h(i-k,0) & h(i-k,M-1) & \cdots & h(i-k,1) \\ h(i-k,1) & h(i-k,0) & \cdots & h(i-k,2) \\ \vdots & & & \\ h(i-k,M-1) & h(i-k,M-2) & \cdots & h(i-k,0) \end{bmatrix}.$$

The matrix $[H_{ik}]$ now has not only equal diagonal elements, but also each row after the first is a circular right shift of the preceeding row. A matrix with this property is called a *circulant matrix*, and has the important property that it can be diagonalized by a discrete Fourier transform computation. This property is derived in Appendix 1. The total $[H]$ matrix may now be called a *block circulant* matrix (Hunt, 1972) and may be written in the form of a circulant matrix of the submatrices

$$[H] = \begin{bmatrix} [H_0] & [H_{M-1}] & \cdots & [H_1] \\ [H_1] & [H_0] & \cdots & [H_2] \\ \vdots & & & \vdots \\ [H_{M-1}] & [H_{M-2}] & & [H_0] \end{bmatrix},$$

where each submatrix has also been simplified by the periodic assumption.

The restoration equation may now be written as

$$\mathbf{g} = [H]\mathbf{f} + \mathbf{n}$$

in which the matrix $[H]$ is in block circulant form. The restoration problem may be stated as follows: given **g** and certain knowledge of $[H]$ and **n**, determine **f**. The restoration problem stated in this form is conceptually simple; however a direct solution in practice is extremely difficult. One problem is due to the large size of the matrix. For example, if a picture **g** of size 256 by 256 is observed, the $[H]$ matrix is of size 65,536 by 65,536. Direct inversion of a matrix of this size is difficult or impossible on even the largest computers. Another concern is related to the ill-conditioned nature of restoration. The matrix $[H]$ may not possess a stable inverse. Solutions to the problems will be given in the following sections.

### 4.5.2   Least Squares Restoration

In this section, a deterministic approach to the restoration problem will be considered. The restoration equation without noise may be considered as a linear relation of the form

$$\mathbf{g} = [H]\mathbf{f}$$

and can be considered in several ways. (The effects of noise **n** will be considered later in this section.) Given an observed image **g** and a knowledge of the system degradation in the form of $[H]$, we wish to approximate **f**. The criterion to be used will be to minimize the squared error between the actual observed response **g** and the observed response, using the approximate solution $\hat{\mathbf{f}}$. This approximate solution will produce, in the absence of noise, an approximate observed image $\hat{\mathbf{g}}$ given by

$$\hat{\mathbf{g}} = [H]\hat{\mathbf{f}}.$$

Therefore the error or noise in the approximation is

$$\mathbf{n} = \mathbf{g} - \hat{\mathbf{g}} = \mathbf{g} - [H]\hat{\mathbf{f}}.$$

The squared error $E$ may therefore be described by

$$E = \mathbf{n}^{*\mathrm{T}}\mathbf{n} = \|\mathbf{g} - \hat{\mathbf{g}}\|^2 = \left(\mathbf{g}^* - [H^*]\hat{\mathbf{f}}\right)\left(\mathbf{g} - [H]\hat{\mathbf{f}}\right),$$

where * indicates complex conjugate. We would now like to determine the value of $\hat{\mathbf{f}}$ that minimizes the squared error $E$. To compute the minimum, let us first expand the quadratic form, which after simplification is

$$E = \mathbf{g}^{*\prime}\mathbf{g} - \hat{\mathbf{f}}'[H^*]'\mathbf{g} - \mathbf{g}^{*\prime}[H]\,\hat{\mathbf{f}} - \hat{\mathbf{f}}'[H^*]'[H]\hat{\mathbf{f}}.$$

Two easily derived forms for the gradient or derivative with respect to a vector are

$$\frac{\partial}{\partial \mathbf{f}}(\mathbf{f}'\mathbf{w}) = \frac{\partial}{\partial \mathbf{f}}(\mathbf{w}'\mathbf{f}) = \mathbf{w},$$

where **w** is a vector, and

$$\frac{\partial}{\partial \mathbf{f}}(\mathbf{f}'\mathbf{C}\mathbf{f}) = 2\mathbf{C}\mathbf{f},$$

where **C** is a symmetric matrix.

Using these results gives

$$\frac{\partial E}{\partial \mathbf{f}} = -2[H^*]'\mathbf{g} + 2[H^*]'[H]\hat{\mathbf{f}}.$$

Equating this derivative to zero gives the least squares solution

$$\hat{\mathbf{f}} = ([H^*]'[H])^{-1}[H^*]'\mathbf{g}.$$

This result is the well-known *least squares* or *pseudoinverse* solution. Note that if $[H]$ is a square nonsingular matrix, the solution reduces to

$$\hat{\mathbf{f}} = [H]^{-1}\mathbf{g}.$$

At this point, we have arrived at a solution; however, the implementation may require the inversion of an $N^2$ by $N^2$ matrix, where $N$ may be 256, 512, or larger. Let us therefore investigate the structure of the $[H]$ matrix.

Suppose $[H]$ is a circulant matrix, as would be the case for a one-dimensional planar invariant system. It is shown in Appendix 1 that a circulant matrix can be diagonalized by a DFT operation. That is, the matrix product

$$[W]^{-1}[H][W] = [D],$$

where $[W]$ is an $N$ by $N$, DFT matrix and $[D]$ is a diagonal matrix whose elements are equal to the DFT coefficients of the first row of the circulant matrix $[H]$. We may therefore express $[H]^{-1}$ as

$$[H]^{-1} = [W][D]^{-1}[W]^{-1}.$$

The inverse matrix solution may be written

$$\hat{\mathbf{f}} = [W][D]^{-1}[W]^{-1}\mathbf{g}.$$

This sequence of matrix operations may be interpreted in a very interesting way. Let us consider separately the three main steps.

1. $[W]^{-1}\mathbf{g}$  This operation computes the DFT of **g**. We might also write this in series form as $G(u)$.

2. $[D]^{-1}$ is a diagonal matrix whose elements are the point by point inverses of the DFT coefficients of the sequence formed by the first row of the point-spread function matrix. In series form, this operator would be described as $1/H(u)$.

3. $[W]$  This operation computes the inverse DFT.

Therefore an equivalent expression for the inverse matrix solution when $[H]$ is a circulant matrix is to form the inverse filter solution

$$\hat{F}(u) = G(u)/H(u),$$

then obtain

$$\hat{f}(x) = \mathcal{F}^{-1}[F(u)].$$

This important result (Hunt, 1972) conceptually relates the algebraic and Fourier domain solutions to the restoration problem via the properties of the circulant matrix. Hunt also developed the extensions to the two-dimensional case. When $[H]$ is a block Toeplitz matrix corresponding to the planar invariant imaging system, the matrix is approximated by a block circulant matrix that again permits a DFT solution. Further details of this construction are included in Appendix 1. The corresponding image solution is

$$\hat{F}(u,v) = G(u,v)/H(u,v), \qquad \hat{f}(x,y) = \mathcal{F}^{-1}(\hat{F}(u,v)).$$

This provides a feasible solution to the computation problem.

We are still faced with the ill-conditioned nature of the restoration problem. Both the inverse matrix and the inverse filter solutions have the possibility of singular conditions. The inverse filter solution is singular if the system function $H(u,v)$ contains zero values. If the matrix $[H]$ is singular, then so is $[H]'[H]$, and both the inverse matrix and pseudoinverse matrix solutions cannot be determined. One approach to this singularity condition for the matrix solution is the singular value decomposition described by Andrews and Hunt (1977). Essentially, this approach attempts to use the linearly independent columns of a singular $[H]$ matrix. The *singular value decomposition* is an excellent method for studying the singular nature of the matrix, but is computationally difficult since it requires determining the singular values of a large matrix. Therefore, the main problems that remain with both solution methods considered are the singular conditions.

When a linear set of equations contains a singular condition, a variety of solutions may exist. It is sometimes possible to place constraints on the solution to reduce the number of possible solutions to a unique constrained solution. In fact, appropriate constraints may not only eliminate the singularity condition, but also reduce the ill-conditioned nature of the problem.

### 4.5.3   Deterministic Constrained Restoration

The least squares restoration approach is characterized by the possibility of singular solutions. A general approach to problems of this nature that permit a variety of solutions is to consider *natural* constraints that may reduce this variety, perhaps to a unique solution. The constrained restoration solution adds two important steps to the unconstrained restoration problem. First,

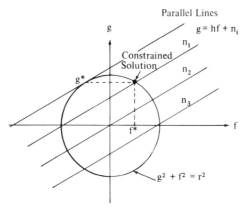

*Figure 4.43* Solution of the linear equation $g = hf + n_i$ subject to the quadratic constraint $g^2 + f^2 = r^2$ and the positivity constraint $f \geq 0$. The solution point is $(f^*, g^*)$.

natural constraints in mathematical form must be developed. Then, solutions to the restoration problem subject to these constraints must be developed. Also, these two steps must be compatible. In a practical sense, it is undesirable to have either a natural constraint with no solution or a constrained solution for an unnatural constraint.

An example of a simple constrained restoration problem will now be given to illustrate the solution method. Given values of $g$ and $h$, we wish to determine the value of $f$ that satisfies the restoration equation given below and shown in Fig. 4.43:

$$g = hf + n.$$

Since the value of $n$ is unknown, we can only determine a family of possible solutions shown as the family of parallel lines in Fig. 4.43. Suppose we are also given a constraint of the form

$$g^2 + f^2 = r^2$$

with the value of $r$ known. The simultaneous solution of the restoration equation and the quadratic constraint reduces the variety of solutions to only two possibilities. A further constraint is needed to obtain a unique solution. In this case either requiring the minimum noise $n$ or a positive value of $f$ leads to the unique solution shown in Fig. 4.43 as $(g^*, f^*)$. Note that the last constraint is an inequality constraint of the form $f \geq 0$, and that more than a single constraint was needed to obtain a unique solution.

To generalize the least squares approach to constrained restoration we may again seek the least squared error solution to

$$\mathbf{g} = [H]\mathbf{f},$$

which requires minimizing the squared error

$$E = \|\mathbf{g} - [H]\mathbf{f}\|^2 = (\mathbf{g} - [H]\mathbf{f})'(\mathbf{g} - [H]\mathbf{f})$$

subject to certain constraints. Note that the mathematical nature of $E$ is a quadratic form. The previously derived estimate which minimizes $E$ is the *pseudoinverse* solution

$$\hat{\mathbf{f}} = ([H^*]'[H])^{-1}[H^*]\mathbf{g}.$$

This solution may be used as a basis of comparison with the constrained solutions.

Several possible constraints and solutions will now be presented in order of complexity of the solutions.

One simple constraint may be interpreted as energy conservation

$$\mathbf{f}'\mathbf{f} = C = \mathbf{g}'\mathbf{g}.$$

To obtain the solution with this constraint, we form a new objective function or criterion by appending the constraint equation and a Lagrangian multiplier $\lambda$ to the previous criterion,

$$J(\mathbf{f},\lambda) = \|\mathbf{g} - [H]\mathbf{f}\| + \lambda(C - \|\mathbf{f}\|^2).$$

Equating the partial derivatives of $J(\mathbf{f},\lambda)$ to zero gives the solution

$$\hat{\mathbf{f}} = ([H^*]'[H] + \lambda\mathbf{I})^{-1}[H^*]'\mathbf{g}.$$

Another important constraint is of the form

$$\|[Q]\mathbf{f}\|^2.$$

The criterion for minimizing is

$$J(\mathbf{f},\lambda) = \|\mathbf{g} - [H]\mathbf{f}\|^2 + \lambda\|Q\mathbf{f}\|^2.$$

Solving for the minimum gives the solution

$$\hat{\mathbf{f}} = ([H^*]'[H] + \alpha[Q]'[Q])^{-1}[H^*]'\mathbf{g}.$$

Another important constraint may be developed using the statistical concept of *entropy* (Hershel, 1971; Frieden, 1972). If the image $\mathbf{f}$ is positive and normalized to unity, then each component $f_i$ can be interpreted as a probability. The entropy of the image may then be described as

$$H(\mathbf{f}) = -\sum_{i=1}^{N^2} f_i \ln f_i$$

or as

$$H(\mathbf{f}) = -\mathbf{f}'\ln\mathbf{f},$$

where $\ln\mathbf{f}$ refers to a componentwise logarithm. (A tutorial description of entropy is given in Section 6.1.) It may be desirable to maximize $H(f)$ or to minimize the negative of $H(f)$ subject to the mean squared error condition. The criterion is

$$J(\mathbf{f},\lambda) = \mathbf{f}'\ln\mathbf{f} - \lambda(\|\mathbf{g} - [H]\mathbf{f}\|^2 - \|\mathbf{n}\|^2).$$

The derivative is

$$\frac{\partial J}{\partial \mathbf{f}} = 1 + \ln \mathbf{f} + \lambda \{ 2 [H^*]'(\mathbf{g} - [H]\mathbf{f}) \}.$$

Solving for $\hat{\mathbf{f}}$ gives the solution

$$\hat{\mathbf{f}} = \exp \{ -1 - 2\lambda [H^*]'(g - [H]\hat{\mathbf{f}}) \}.$$

Although this solution is transcendental in $\mathbf{f}$, the exponential operator ensures that the solution is positive.

The constrained least squares approach to restoration performs very well for equality constraints. However, a closed form solution is not easily obtained if the constraint is in the form of an inequality, for example, positivity,

$$\mathbf{f} \geq 0.$$

Generally, numerical methods such as quadratic programming must be used, and these approaches are hampered by the large size of the image matrices.

A summary of three important constrained least square filters is shown in Table 4.5. The Fourier domain solutions are given for comparison with the algebraic solutions.

### 4.5.4  Stochastic Restoration

The class of restoration techniques considered in this section will be based upon statistical considerations of the images and noise as stochastic or random processes. The linear image restoration model

$$\mathbf{g} = [H]\mathbf{f} + \mathbf{n}$$

can be considered in several ways. If the only random process involved is the additive noise $\mathbf{n}$, then solving for the minimum mean squared estimate of $\mathbf{f}$ will be called the *regression* problem. If the image $\mathbf{f}$ is also considered as a random image with known first- and second-order moments, then the problem will be referred to as *Wiener* estimation. If the probability density function of $\mathbf{g}$ given $\mathbf{f}$ is known completely or parametrically, then *maximum likelihood* estimation may be used. Finally, if the probability density of $\mathbf{f}$ given $\mathbf{g}$ is known, the *Bayes* or *maximum a posteriori* estimation technique may be used. This overview of the stochastic restoration techniques is shown in a flowchart form in Fig. 4.44. In considering random processes in restoration one may proceed down the flowchart, which is ordered from difficult to very difficult in terms of the problem solutions. This flowchart is oversimplified and imprecise, but hopefully shows the relationships between the various well-known estimation models that have been applied to image restoration. If one arrives at the end of the flowchart without finding an appropriate model, then further research into advanced techniques is required. The given models

## Table 4.5

### Constrained Least-Squares Restoration Summary

| Filter | Criteria | Algebraic solution | Fourier solution |
|---|---|---|---|
| Inverse filter | $\|\mathbf{g}-\hat{\mathbf{g}}\|^2$ | $\hat{\mathbf{f}}=([H^*]'[H])^{-1}[H^*]'\mathbf{g}$ | $F(u,v)=G(u,v)/H(u,v)$ $-\,N(u,v)/H(u,v)$ |
| Energy constrained | $\|\mathbf{g}-[H]\mathbf{f}\|^2+\lambda(C-\|\mathbf{f}\|^2)$ | $\hat{\mathbf{f}}=\{[H^*]'[H]+\lambda\mathbf{I}\}^{-1}[H^*]'\mathbf{g}$ | $F(u,v)=\dfrac{G(u,v)H^*(u,v)}{\|H(u,v)\|^2+\lambda}$ |
| Smoothness constrained | $\|\mathbf{g}-[H]\mathbf{f}\|^2+\lambda(C-\||[Q]\mathbf{f}\||^2)$ | $\hat{\mathbf{f}}=\{[H^*]'[H]+\alpha[Q^*]'[Q]\}^{-1}[H^*]'\mathbf{g}$ | $F(u,v)=\dfrac{G(u,v)H^*(u,v)}{\|H(u,v)\|^2+\alpha\|S(u,v)\|^2}$ |
| Maximum entropy | $\|\mathbf{g}-[H]\mathbf{f}\|^2+\lambda(\mathbf{f}\ln\mathbf{f}+C)$ | $\hat{\mathbf{f}}=\exp\{-1+2\lambda[H^*]'(\mathbf{g}-[H]\hat{\mathbf{f}})\}$ | |

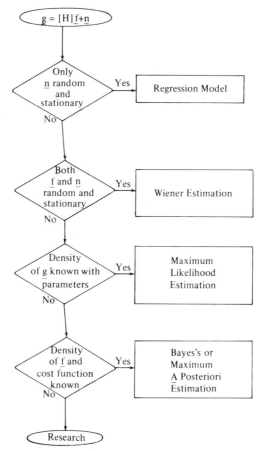

*Figure 4.44*   Stochastic restoration model decision flowchart.

are all in the form of nonrecursive estimates; however, recursive estimates or Kalman filters may also be developed for each case. We shall now consider each of the four models in detail.

### Linear Regression Model

The regression model may be considered as restoring an image **f** which has been distorted by a linear system $[H]$ and disturbed by additive random noise **n**, given an observed image **g**. Since a function of a random variable is another random variable, the random noise term in the restoration equation

$$\mathbf{g} = [H]\mathbf{f} + \mathbf{n}$$

implies that **g** is also a random process. Furthermore, any estimate $\hat{\mathbf{f}}$ of **f** that is a function of **g** will also be a random process. The regression approach attempts to approximate the *statistical* dependence of $\hat{\mathbf{f}}$ on $\mathbf{g}$ by a deterministic relation $\mathbf{T}(\mathbf{g})$. The function which minimizes the mean squared deviation

$$\varepsilon_1 = E\left[\left[\hat{\mathbf{f}} - \mathbf{T}(\mathbf{g})\right]^2\right]$$

is easily derived. Let $\varepsilon_{\mathbf{g}}$ denote the conditional expectation of $\varepsilon_1$ given the degraded image $\mathbf{g}$, i.e.,

$$\varepsilon_{\mathbf{g}} = E(\varepsilon_1/\mathbf{g}).$$

In general $\varepsilon_1$ is a random variable whose expected value over the $\{\mathbf{g}\}$ ensemble equals $\varepsilon_{\mathbf{g}}$. Therefore $\varepsilon_1$ is minimized if $\varepsilon_{\mathbf{g}}$ is minimized. It is easily verified that

$$\varepsilon_{\mathbf{g}} = E\left\{\left[\mathbf{f} - E(\mathbf{f}/\mathbf{g})\right]^2/\mathbf{g}\right\} + E\left\{\left[\hat{\mathbf{f}} - E(\mathbf{f}/\mathbf{g})\right]^2/\mathbf{g}\right\}.$$

Therefore $\varepsilon_{\mathbf{g}}$ and consequently $\varepsilon_1$ are minimized if the estimate is chosen to be

$$\hat{\mathbf{f}} = E(\mathbf{f}/\mathbf{g}).$$

Although the optimum estimator is conceptually simple, the computation of the expected value is extremely difficult for general random processes. The solution becomes tractable under certain assumptions about the form of the estimator and the random processes.

It is often sufficient to approximate the true regression by a linear function

$$\hat{\mathbf{f}} = [L]\mathbf{g}.$$

The best linear function should be *unbiased*, that is,

$$E(\hat{\mathbf{f}}) = E(\mathbf{f}),$$

and should also minimize the mean squared error

$$E(\|f - \hat{f}\|^2) = E(\|f - [L]g\|^2).$$

If the noise has zero mean and covariance matrix $\Sigma$, the optimal solution is given by the *Gauss–Markov theorem* (Lewis and Odell, 1971, p. 52) as

$$\hat{\mathbf{f}} = \left([H^*]\Sigma^{-1}[H]\right)^{-1}[H^*]\Sigma^{-1}\mathbf{g}$$

and the covariance matrix of the estimate is

$$\Sigma_{\hat{f}} = \left([H]\Sigma^{-1}[H]\right)^{-1}.$$

The optimum linear function is therefore

$$[L] = \left([H^*]'\Sigma^{-1}[H]\right)^{-1}[H^*]'\Sigma^{-1},$$

assuming the matrix inverses exist.

This solution may also be obtained by minimizing the weighted mean squared error as first developed by Gauss. We wish to determine the estimate

$\hat{\mathbf{f}}$ that minimizes the quadratic expression

$$(\mathbf{g}-\hat{\mathbf{g}})\mathbf{\Sigma}^{-1}(\mathbf{g}-\hat{\mathbf{g}})=\big(\mathbf{g}-[H]\hat{\mathbf{f}}\big)'\mathbf{\Sigma}^{-1}\big(\mathbf{g}-[H]\hat{\mathbf{f}}\big).$$

Taking derivatives and equating them to zero gives [L] as above. It is interesting to note that if the noise is white, that is, $\mathbf{\Sigma}=\mathbf{I}$, the solution reduces to the pseudoinverse solution.

### Wiener Filtering

When both the image $\mathbf{f}$ and noise $\mathbf{n}$ are considered to be random processes with known first- and second-order moments, the minimization of the mean squared error leads to the Wiener filter. This approach to image restoration was first derived by Helstrom (1967). The following presentation closely follows that of Andrews and Hunt (1977), which stresses the circulant matrix approximation and Fourier transform implementation.

Consider the linear image restoration model

$$\mathbf{g}=[H]\mathbf{f}+\mathbf{n},$$

in which both $\mathbf{f}$ and $\mathbf{n}$ are assumed to be real-valued second-order stationary random processes and statistically independent. We wish to determine a linear estimate $\hat{\mathbf{f}}$ of $\mathbf{f}$ from the observed image $\mathbf{g}$,

$$\hat{\mathbf{f}}=[L]\mathbf{g}$$

that minimizes the mean squared error

$$\varepsilon^2 = E\big(\|\mathbf{f}-\hat{\mathbf{f}}\|^2\big).$$

For the following derivation it is convenient to express the mean squared error as the trace of the outer product error matrix rather than the normal inner product

$$\varepsilon^2 = E\big\{\mathrm{Tr}\big[(\mathbf{f}-\hat{\mathbf{f}})(\mathbf{f}-\hat{\mathbf{f}})'\big]\big\} = E\big\{(f-\hat{f})'(f-\hat{f})\big\},$$

where Tr denotes the trace or sum of diagonal terms of a matrix.

The outer product form may be expanded for simplification as

$$(\mathbf{f}-\hat{\mathbf{f}})(\mathbf{f}-\hat{\mathbf{f}})' = (\mathbf{f}\mathbf{f}' - \mathbf{f}\hat{\mathbf{f}}' - \hat{\mathbf{f}}\mathbf{f}' + \hat{\mathbf{f}}\hat{\mathbf{f}}').$$

Since

$$\hat{\mathbf{f}}'=\mathbf{g}'[L]'=\big(\mathbf{f}'[H]'+\mathbf{n}'\big)[L]',$$

$$\mathbf{f}\hat{\mathbf{f}}'=\mathbf{f}\mathbf{f}'[H]'[L]'+\mathbf{f}\mathbf{n}'[L]'.$$

Also,

$$\hat{\mathbf{f}}\mathbf{f}'=[L]\mathbf{g}\mathbf{f}'=[L]\big([H]\mathbf{f}+\mathbf{n}\big)\mathbf{f}',$$

so

$$\hat{\mathbf{f}}\mathbf{f}'=[L][H]\mathbf{f}\mathbf{f}'+[L]\mathbf{n}\mathbf{f}'$$

and

$$\hat{\hat{\mathbf{f}}}' = [L]([H]\mathbf{f}\mathbf{f}'[H]' + [H]\mathbf{f}\mathbf{n}' + \mathbf{n}\mathbf{f}'[H]' + \mathbf{n}\mathbf{n}')[L]'.$$

Since the trace of a sum of two matrices is equal to the sum of the traces, the expected value and trace operators may be interchanged. Consider the expected values of the four terms separately. First,

$$E(\mathbf{f}\mathbf{f}') = \mathbf{R}_{\mathbf{f}},$$

where $\mathbf{R}_{\mathbf{f}}$ is the correlation matrix of the image process. Second,

$$E(\hat{\mathbf{f}}\mathbf{f}') = \mathbf{R}_{\mathbf{f}}[H]'[L]'$$

since the image and noise are assumed independent,

$$E(\mathbf{f}\mathbf{n}') = [0].$$

Similarly,

$$E(\hat{\mathbf{f}}\mathbf{f}') = [L][H]\mathbf{R}_{\mathbf{f}}$$

and finally,

$$E(\hat{\hat{\mathbf{f}}}\mathbf{f}') = [L][H]\mathbf{R}_{\mathbf{f}}[H]'[L]' + [L]\mathbf{R}_{\mathbf{n}}[L]',$$

where

$$E(\mathbf{n}\mathbf{n}') = \mathbf{R}_{\mathbf{n}}.$$

The mean squared error may now be expressed as

$$\varepsilon^2 = \mathrm{Tr}\big(\mathbf{R}_{\mathbf{f}} - \mathbf{R}_{\mathbf{f}}[H]'[L]' - [L][H]\mathbf{R}_{\mathbf{f}} + [L][H]\mathbf{R}_{\mathbf{f}}[H]'[L]' + [L]\mathbf{R}_{\mathbf{n}}[L]'\big).$$

Since the correlation matrix is symmetric and the trace of the transpose of a matrix is equal to the trace of the matrix, the error expression may be simplified further to

$$\varepsilon^2 = \mathrm{Tr}\big(\mathbf{R}_{\mathbf{f}} - 2\mathbf{R}_{\mathbf{f}}[H]'[L]' + [L][H]\mathbf{R}_{\mathbf{f}}[H]'[L]' + [L]\mathbf{R}_{\mathbf{n}}[L]'\big).$$

At this point the mean squared error is no longer a function of the images $\mathbf{f}$, $\mathbf{g}$, or $\mathbf{n}$, but depends only upon the statistical characteristics and transforms. The desired linear transform $[L]$ may now be determined by minimizing the scalar squared error with respect to the matrix elements. We may now differentiate $\varepsilon^2$ with respect to $[L]$.

$$\frac{\partial \varepsilon^2}{\partial [L]} = -2\mathbf{R}_{\mathbf{f}}[H]' + 2[L][H]\mathbf{R}_{\mathbf{f}}[H]' + 2[L]\mathbf{R}_{\mathbf{n}}.$$

Equating this result to zero gives the optimum filter $[L]_{\mathrm{opt}}$ which minimizes $\varepsilon^2$.

$$[L]_{\mathrm{opt}} = \mathbf{R}_{\mathbf{f}}[H]'([H]\mathbf{R}_{\mathbf{f}}[H] + \mathbf{R}_{\mathbf{n}})^{-1}.$$

Note that this solution does not depend upon the inverses of individual matrices $\mathbf{R}_{\mathbf{f}}$, $\mathbf{R}_{\mathbf{n}}$, or $[H]$. An alternative solution that does depend on the existence of the inverses may be derived, and is given by

$$[L]^* = \left([H]'\mathbf{R_n}^{-1}[H] + \mathbf{R_f}^{-1}\right)^{-1}[H]'\mathbf{R_n}^{-1}.$$

### Implementing the Wiener Filter

The form of the linear estimate that minimizes the mean squared error $[L]^*$ has now been derived. To determine the Wiener estimate, the correlation matrices of both the image $\mathbf{R_f}$ and noise processes $\mathbf{R_n}$ are required as well as the system matrix $[H]$. Even with these matrices known, the matrix inverse computation

$$\left([H]\mathbf{R_f}[H]' + \mathbf{R_n}\right)^{-1}$$

will make the computation difficult or impractical for large sized image arrays, since each of the matrices is of size $N^2$ by $N^2$. The inverse matrix could be computed easily if the matrix were a product of a diagonal matrix and a unitary matrix for which the inverse is simply the conjugate transpose. Consider the matrix $\Lambda^{\#}$

$$\Lambda^{\#} = [H]\mathbf{R_f}[H]' + \mathbf{R_n}.$$

The weakest known condition which ensures that $\Lambda^{\#-1}$ is easily computed is that $[H]$, $\mathbf{R_f}$, and $\mathbf{R_n}$ each be circulant matrices. In this case

$$[H] = [W]\Lambda_\mathbf{h}[W]^{-1}, \qquad \mathbf{R_f} = [W]\Lambda_\mathbf{f}[W]^{-1}, \qquad \mathbf{R_n} = [W]\Lambda_\mathbf{n}[W]^{-1}.$$

It is easy to show that the sum and product of two circulant matrices $\mathbf{C_1}$ and $\mathbf{C_2}$ are also circulant matrices. If

$$\Lambda_1 = [W]\mathbf{C_1}[W]^{-1}, \qquad \Lambda_2 = [W]\mathbf{C_2}[W]^{-1},$$

then

$$\Lambda_3 = [W](\mathbf{C_1} + \mathbf{C_2})[W]^{-1} \quad \text{and} \quad \Lambda_3 = [W]\mathbf{C_1}\mathbf{C_2}[W]^{-1},$$

where $\Lambda_3$ is a diagonal matrix.

Substituting the circulant correlation matrices, the estimate gives

$$\Lambda^{\#} = [W]\Lambda_\mathbf{h}\Lambda_\mathbf{f}\Lambda_\mathbf{h}^*[W]^{-1} + [W]\Lambda_\mathbf{n}[W]^{-1}.$$

Factoring out the Fourier transform matrices gives

$$\Lambda^{\#} = [W](\Lambda_\mathbf{h}\Lambda_\mathbf{f}\Lambda_\mathbf{h}^* + \Lambda_\mathbf{n})[W]^{-1}.$$

Therefore, the inverse of $\Lambda^{\#}$ can be determined by computing the inverse of a diagonal matrix and the inverse of a unitary matrix.

The Wiener estimate for this circulant matrix case reduces to

$$[L]^{\#} = [W]\Lambda_\mathbf{f}\Lambda_\mathbf{h}^*(\Lambda_\mathbf{h}\Lambda_\mathbf{f}\Lambda_\mathbf{h}^* + \Lambda_\mathbf{n})^{-1}[W]^{-1}.$$

Applying this linear estimator to an observed image **g** produces the optimum mean squared estimate $\hat{\mathbf{f}}$,

$$\hat{\mathbf{f}} = [\,W\,]\Lambda_f\Lambda_h^*(\Lambda_h\Lambda_f\Lambda_h^* + \Lambda_n)^{-1}[\,W\,]^{-1}\mathbf{g}.$$

This form of the Wiener estimate has an important interpretation in terms of Fourier implementation. Let us again consider the solution in steps from right to left.

1. $[W]^{-1}\mathbf{g}$   This is simply a computation of the discrete Fourier transform of **g**. In the two-dimensional case, the result of this operation may be written as $G(u,v)$.

2. $(\Lambda_h\Lambda_f\Lambda_h^* + \Lambda_n)^{-1}$   Since the interior matrix is a diagonal matrix, the computation of this inverse matrix gives simply the matrix of reciprocals of the diagonal elements. The diagonal matrix elements of $\Lambda_h$, $\Lambda_f$, and $\Lambda_n$ are the discrete Fourier coefficients of the corresponding circulant matrices. In the two-dimensional case, the diagonal elements of the $N^2$ by $N^2$ matrices are naturally ordered in $N$ by $N$ matrices that may be written in series form:

$$H(u,v) = \sum_{x=0}^{N-1}\sum_{y=0}^{N-1} h(x,y)\exp[-j2\pi(ux+vy)/N],$$

$$S_f(u,v) = \sum_{x=0}^{N-1}\sum_{y=0}^{N-1} R_f(x,y)\exp[-j2\pi(ux+vy)/N],$$

$$S_n(u,v) = \sum_{x=0}^{N-1}\sum_{y=0}^{N-1} R_n(x,y)\exp[-j2\pi(ux+vy)/N].$$

$H(u,v)$ is the Fourier transform of the block circulant point-spread function $h(x,y)$. $S_f(u,v)$ is called the *power spectrum* of the image and is the Fourier transform of the circulant correlation image $R_f(x,y)$. Similarly, $S_n(u,v)$ is called the power spectrum of the noise process and is the Fourier transform of the circulant noise correlation $R_n(x,y)$.

The total computation at this point may be expressed in series form as

$$\frac{G(u,v)}{H(u,v)P_f(u,v)H^*(u,v) + S_n(u,v)}.$$

3. $\Lambda_f\Lambda_h^*$   In a similar manner these terms may be related to the Fourier domain multiplication by

$$S_f(u,v)H^*(u,v).$$

The total estimate in the Fourier domain may therefore be written in the familiar Wiener filter form

$$\hat{F}(u,v) = \frac{H^*(u,v)G(u,v)}{|H(u,v)|^2 + [\,S_n(u,v)/S_f(u,v)\,]}.$$

4. [*W*]   The final operation simply transforms the estimate from the Fourier domain to the spatial domain.

An important point about the Wiener filter is that it is not often singular or ill-conditioned. If there is a lower limit $K$ of the noise-to-signal power spectral density ratio

$$S_n(u,v)/S_f(u,v) > K > 0,$$

then if $H(u,v) = 0$, the estimate at that frequency $\hat{F}(u,v)$ is also equal to zero. Also, if the noise power spectral density is much larger than the signal spectral density at some frequency, the estimate at that frequency is deemphasized. Recall that an opposite effect occurred with the inverse filter. Therefore, the Wiener filter approach permits an image restoration solution that is neither singular nor ill-conditioned. It does require estimation of the image and noise power spectral densities.

### Estimating the Statistics

A correlation matrix $\mathbf{R_f}$ is defined as

$$\mathbf{R_f} = E(\mathbf{ff}').$$

The covariance matrix was defined as

$$\Sigma_f = E\big[(\mathbf{f}-\bar{\mathbf{f}})(\mathbf{f}-\mathbf{f})'\big],$$

where $\bar{\mathbf{f}} = E(\mathbf{f})$. In Chapter 3 it was shown that the correlation and covariance matrices are related by

$$\Sigma_f = \mathbf{R_f} - \bar{\mathbf{f}}\bar{\mathbf{f}}' \qquad \text{or} \qquad \mathbf{R_f} = \Sigma_f + \bar{\mathbf{f}}\bar{\mathbf{f}}'.$$

This last expression illustrates the dependence of the correlation matrix on the mean vector $\bar{\mathbf{f}}$ of the image. If the mean vector is not constant, $\bar{\mathbf{f}} \neq \mu$, the random process is not stationary, and further simplification may not be possible. If the mean vector is constant, $\bar{\mathbf{f}} = \mu$, then the process is covariance stationary. A subtraction of $\mu$ from the estimate or of $[H]\mu$ from the observed image will therefore result in the correlation and covariance matrices being equal:

$$\mathbf{R_f} = \Sigma_f.$$

This simplification is a necessary step, and reduces our consideration to the covariance matrix only.

There are several methods that may be used to estimate the statistics required for implementing the Wiener filter. One may decide to assume a known parametric form of the statistics and then determine the parameters that best fit the experimental data. Another method is to determine a nonparametric estimate from experimental data. One must also decide upon whether a spatial or Fourier domain estimate is to be obtained. Examples of the parametric and nonparametric approaches will now be given.

To illustrate the parametric method, suppose the image statistics are separable in the horizontal and vertical directions, and that for each dimension the random variation may be modeled as a Markov process with zero mean and covariance matrix $\Sigma_f$ given by

$$
\Sigma_f =
\begin{bmatrix}
1 & \rho & \rho^2 & & & \rho^{N-1} \\
\rho & 1 & \rho & \rho^2 & \cdots & \rho^{N-2} \\
\vdots & & & & & \\
\rho^{N-1} & \cdots & & & & 1
\end{bmatrix}.
$$

This matrix depends upon a single parameter $\rho$, which is the unit distance pixel correlation coefficient. Although the covariance matrix is not a circulant, it may be approximated by a circulant matrix. It may also be noted that the Markov process matrix may be easily diagonalized in its Toeplitz form so that a circulant approximation is unnecessary (Pratt, 1972).

An easy parametric model for the noise process is white noise, which has zero mean and covariance matrix

$$
\Sigma_n =
\begin{bmatrix}
\sigma^2 & 0 & \cdots & & & 0 \\
0 & \sigma^2 & & & & \\
0 & & & \sigma^2 & & \\
\vdots & & & & \ddots & \\
0 & & & & & \sigma^2
\end{bmatrix},
$$

where $\sigma^2$ is the noise variance.

Since the point-spread function matrix $[H]$ of a planar invariant system is a circulant matrix, the optimum filter can be implemented either directly in the spatial domain or in the frequency domain without an inverse matrix computation. Only the pixel correlation coefficients, the noise variance, and the point-spread-function matrix parameters need to be determined.

Nonparametric methods may also be used for estimating the signal and noise statistics. One interesting approach involves using regions of the observed image to determine the required estimates. Suppose the observed image contains "typical" regions that characterize the effects of "signal alone" and "noise alone." Then the power spectrum estimates may be obtained by averaging the magnitude squared of the Fourier transforms of the signal or noise subregions. The magnitude squared of the Fourier transform approximates the Fourier transform of the correlation function of the region. Since the regions are normally smaller than the image, an interpolation is required to obtain the power spectrum estimates at the desired digital resolution. The point-spread function may also be estimated in this manner if one or more typical "points" or "edges" may be located. Since this procedure is implemented in the Fourier domain and various size discrete Fourier transforms may be used, careful attention must be given to the spatial frequencies involved. However, no circulant matrices need actually be constructed, and only a simple Fourier domain filter procedure with attention to convolution overlap is required.

### Advanced Stochastic Restoration

The *regression* and *Wiener estimation* models previously described are excellent restoration solutions for imaging situations that fit the underlying assumptions of constant mean and covariance stationary random processes. For imaging situations for which these assumptions are not appropriate, a variety of other restoration solutions are available that are based upon estimation and decision theory concepts. The mathematical form of the Bayesian, the mini-max, the maximum a posteriori, and the maximum likelihood estimators will now be reviewed.

Suppose that a complete statistical description is available for the measurements in the restoration model. This description could be given in the form of a joint probability density function $p(\mathbf{f}, \mathbf{g})$ from which all marginal distributions such as $p(\mathbf{f}), p(\mathbf{g}), p(\mathbf{g}, \mathbf{f}), p(\mathbf{f}/\mathbf{g})$ could be determined easily. An estimate $\hat{\mathbf{f}}$ of the ideal image $\mathbf{f}$ is desired. Suppose a nonnegative scalar cost function $C(\mathbf{f} - \hat{\mathbf{f}})$ is also given. The average cost incurred by using the estimate $\hat{\mathbf{f}}$ given the observed image $\mathbf{g}$ is given by

$$\overline{C} = E\big[ C(\mathbf{f} - \hat{\mathbf{f}})/\mathbf{g} \big] = \int_{\mathbf{f}} C(\mathbf{f} - \hat{\mathbf{f}}) p(\mathbf{f}/\mathbf{g})\, d\mathbf{f}.$$

The estimator that minimizes this average cost is called the Bayes estimator. This solution is generally the "most desirable" of the "optimum" estimators but also the most difficult to determine since both a cost function $C(\mathbf{f} - \hat{\mathbf{f}})$ and the a posteriori probability $p(\mathbf{f}/\mathbf{g})$ must be known. The cost function must be defined for all pairs of ideal and estimated images. This problem of image quality measurement is very important but is also very difficult. Methods for estimating the a posteriori density are compounded by the $N^2$ dimensionality of the image vectors. Therefore, simplifications must be made in order to realize a solution.

One approach is to minimize the maximum error (*mini-max estimate*). The mini-max estimator $\hat{\mathbf{f}}$ would minimize the expression

$$\min_{\hat{\mathbf{f}}} \int_{\mathbf{f}} \max_{\mathbf{f}} \big[ C(\mathbf{f} - \hat{\mathbf{f}}) \big] p(\mathbf{f}/\mathbf{g})\, d\mathbf{f}.$$

This may be considered to be a simplification of the Bayes estimator since only the maximum of the cost function need be specified.

Another simplification may be made by assuming a squared error cost function

$$C(\mathbf{f} - \hat{\mathbf{f}}) = \|(\mathbf{f} - \hat{\mathbf{f}})\|^2.$$

The Bayes estimate that minimizes the average cost is

$$\bar{\mathbf{f}} = E(\mathbf{f}/\mathbf{g}) = \int_{\mathbf{f}} \mathbf{f} p(\mathbf{f}/\mathbf{g}) \, d\mathbf{f}.$$

Note that this is the minimum mean squared error estimate and uses the mean of the conditional density as the estimate. Another reasonable approach is to use the mode of the conditional density as the estimate, assuming $p(\mathbf{f}/\mathbf{g})$ to be unimodal.

If the a priori probability density $p(\mathbf{f})$ is a uniform distribution, the Bayes solution simplifies to the maximum likelihood solution. That is, given an observation $\mathbf{g}^*$, the value of $\mathbf{f}$ which has maximum likelihood $p(\mathbf{g}^*/\mathbf{f})$ is chosen as the estimate $\hat{\mathbf{f}}$. For a more detailed consideration of these advanced stochastic methods the reader should consult Andrews and Hunt (1977, Chapter 9).

## 4.6  Motion Restoration

The general problem of image restoration for the deblurring of an image will now be considered. For this analysis the image recording system with origin in the center of the image plane shown in Fig. 3.1 will be used. If a point $\mathbf{v}_i$ is selected in the image plane, the instantaneous image intensity at this point will be denoted by $g(\mathbf{v}_i, t)$. The image intensity will be the result of an illumination source reflected from an object with instantaneous reflectivity $f(\mathbf{v}_o, t)$ where the object and image positions are related by

$$\mathbf{v}_o = \mathbf{v}_c + \alpha(\mathbf{v}_c - \mathbf{v}_i) \qquad \text{or} \qquad \mathbf{v}_i = \mathbf{v}_c + \beta(\mathbf{v}_c - \mathbf{v}_o),$$

where $\mathbf{v}_c$ is the position of the lens center and $\alpha$ and $\beta$ are scalar parameters related to distance from the lens center.

The recorded image over an exposure time $T$ is described by the integrated image intensity at this position,

$$g(\mathbf{v}_i) = \int_0^T g(\mathbf{v}_i, t) \, dt.$$

The medium that actually performs the integration could be photographic film or the phosphor in a camera sensor.

From physical considerations, we may assume that the power collected by the imaging system, which is radiated from a particular planar region of the object, is conserved through the imaging process at any instant of time. Therefore, the power radiated and collected by a small area in the object plane around the object point $\mathbf{v}_o = (x_o, y_o, z_o)$ is described by

$$g(x_i, 0, z_i, t) \, dx_i \, dz_i = f(x_o, y_o, z_o, t) \, dx_o \, dz_o$$

at fixed time $t$ and fixed position $y_o$. It follows that

$$g(\mathbf{v}_i, t) = f(\mathbf{v}_o, t)$$

with

$$\mathbf{v}_i = \mathbf{v}_c + \beta(\mathbf{v}_c - \mathbf{v}_o) \qquad \text{and} \qquad \beta = f/(y_o - f).$$

Therefore, the time integral may be written

$$g(\mathbf{v}_i) = \int_0^T f(\mathbf{v}_o, t) \, dt.$$

This relation may be described as the summation at each picture point $g(\mathbf{v}_i)$ over a time interval $T$ of all object points that are imaged onto the image point. If there were no motion of either object or image, this relationship would simplify to

$$g(\mathbf{v}_i) = f(\mathbf{v}_o) T.$$

If the image point is considered fixed but the object moves, then many object points such as

$$f(\mathbf{v}_o(t_1), t_1), \quad f(\mathbf{v}_o(t_2), t_2), \quad \cdots$$

would be imaged onto the same image point.

Given a description of the object motion $\mathbf{v}_o(t)$, we may convert the time integral of the moving object into a positional integral over an equivalent stationary object. Mathematically, this involves a change of variables. Assume that the instantaneous object position is described in the parametric form

$$x_o(t) = p_1(\mathbf{v}_i, t), \qquad y_o(t) = p_2(\mathbf{v}_i, t), \qquad z_o(t) = p_3(\mathbf{v}_i, t).$$

If each of the functions $p_1$, $p_2$, and $p_3$ has an inverse function $q_1$, $q_2$, and $q_3$, respectively, then

$$t = q_1(x_o, \mathbf{v}_i) = q_2(y_o, \mathbf{v}_i) = q_3(z_o, \mathbf{v}_i).$$

The object motion could therefore be described as motion along a curved path $\mathbf{r}$, where

$$\mathbf{r} = [x_o(t), y_o(t), z_o(t)],$$

and the elemental length $ds$ is given by

$$ds = \left[ \left( \frac{dx_o}{dt} \right)^2 + \left( \frac{dy_o}{dt} \right)^2 + \left( \frac{dz_o}{dt} \right)^2 \right]^{1/2} dt.$$

Therefore the time integral

$$g(\mathbf{v}_i) = \int_0^T f(\mathbf{v}_o, t) \, dt$$

may be changed to a line integral of the form

$$g(\mathbf{v}_i) = \int_{S_0}^{S_T} f(\mathbf{v}_o) \, ds \bigg/ \left[ \left( \frac{dx_o}{dt} \right)^2 + \left( \frac{dy_o}{dt} \right)^2 + \left( \frac{dz_o}{dt} \right)^2 \right]^{1/2},$$

where

$$S_0 = [x_o^2 + y_o^2 + z_o^2] | t = 0, \qquad S_T = [x_o^2 + y_o^2 + z_o^2] | t = T.$$

To "deblur" an image $g(v_i)$ degraded by motion, one must solve the previous equation for $f(v_o)$. Note that the measurements required for this solution are the positions **S** and velocities **V**. A general solution to this equation is difficult, since it involves space variant filtering; however, certain special cases are easily solved. Further information on motion restoration may be found in Sawchuk (1972).

To illustrate the steps required for solution, a linear translation case will be considered.

To determine the blurring produced by linear TV camera motion as a function of TV camera phosphor type, the integrated image $g(x,y)$ may be described as a function of the instantaneous image $f(x,y)$ as

$$g(x,y) = \int_{-T/2}^{T/2} f[x_0 - \alpha(t), y - \beta(t)] dt,$$

where $T$ is the duration of the exposure or effective integration time due to phosphor response, and $\alpha(t), \beta(t)$ describe the motion in the $x$ and $y$ directions parametrically.

The effect of the integration may be derived by Fourier analysis. Taking the Fourier transform of the above expression gives:

$$G(u,v) = \int\int_{-\infty}^{\infty} g(x,y) \exp[-j2\pi(ux + vy)] dx\, dy$$

$$= \int_{-T/2}^{T/2} \int\int_{-\infty}^{\infty} f[x - \alpha(t), \dot{y} - \beta(t)] \exp[-j2\pi(ux + vy)] dx\, dy\, dt.$$

Introduce the transformation for $\alpha(t) = V_x t$ and $\beta(t) = V_y t$,

$$\xi = x - \alpha(t) \qquad \text{or} \qquad x = \xi + \alpha(t),$$

$$\eta = y - \beta(t) \qquad \text{or} \qquad y = \eta + \beta(t).$$

Then

$$G(u,v) = \int_{-T/2}^{T/2} \int_{-\infty}^{\infty} f(\xi, \eta) \exp[-j2\pi(u\xi + v)] d\xi\, d\eta$$

$$\times \exp\{-j2\pi[u\alpha(t) + v\beta(t)]\} dt,$$

$$G(u,v) = F(u,v) \int_{-T/2}^{T/2} \exp\{-j2\pi[u\alpha(t) + v\beta(t)]\} dt.$$

Let

$$H(u,v) = \int_{-T/2}^{T/2} \exp\{-j2\pi[u\alpha(t) + v\beta(t)]\} dt.$$

Then

$$G(u,v) = F(u,v) H(u,v),$$

which shows that the effects of linear motion can be modeled as a position-invariant imaging system. If the motion is uniform in the horizontal direction, then

$$\alpha(t) = Vt, \qquad \beta(t) = 0$$

and

$$H(u,v) = \int_{-T/2}^{T/2} \exp(-j2\pi Vut)\, dt = \frac{\sin(\pi VTu)}{\pi Vu} = T \operatorname{sinc}(\pi VTu).$$

To illustrate the steps involved in experimental restoration, the following example computed by Heller (1978) will be described. The original image

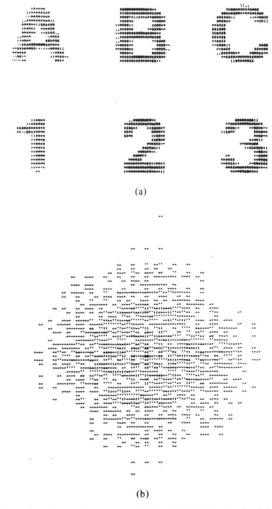

(a)

(b)

*Figure 4.45* (a) Original image and (b) Fourier transform of original image.

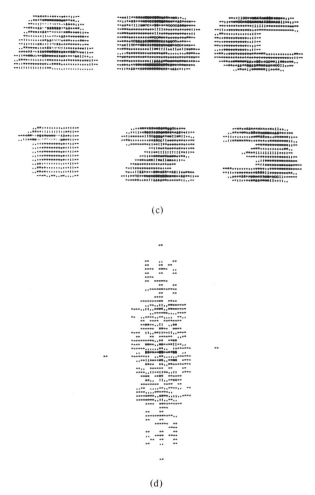

(c)

(d)

*Figure 4.45*   (c) Blurred image, distance = 4.0, noise variance = 0.33, and (d) Fourier transform of blurred image.

shown in Fig. 4.45a consists of a 64 by 64 character array. The magnitude of the Fourier transform of the original image is shown in Fig. 4.45b. To simulate motion restoration each pixel in the original image was replaced by the sum of its four horizontal neighbors with the result shown in Fig. 4.45c. The magnitude of the Fourier transform of the blurred image is shown in Fig. 4.45d. Note the severe loss of spatial frequency information in the horizontal direction. A Wiener filter was designed to restore the blurred image. The filter magnitude is shown in Fig. 4.45e, in which large values are dark and small values light. Note that the filter mainly enhances the high-frequency informa-tion but retains small values at locations in which the blur function has zeros.

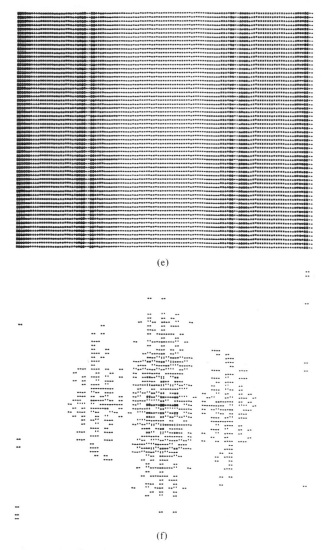

Figure 4.45   (e) Restoration filter, large values represented by black, small values by white, and (f) Fourier transform of restored image.

The result of filtering the Fourier transform of the blurred image, that is, the product of the transform shown in Fig. 4.45c and the filter shown in Fig. 4.45d, is given in Fig. 4.45f. Note that much of the high-frequency information has been restored with the exception of the zero bands. The restored image shown in Fig. 4.45g is the inverse Fourier transform of the transform of Fig. 4.45f. Note that a substantial double image remains at a decreased

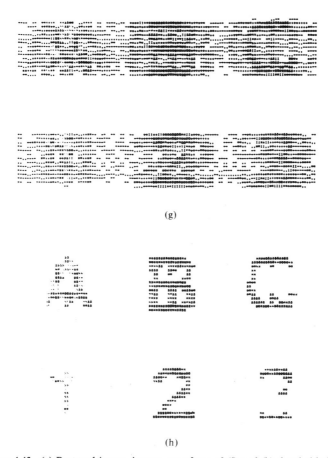

(g)

(h)

*Figure 4.45* (g) Restored image–inverse transform of (f) and (h) thresholded image.

intensity. This restored image was thresholded to produce the final restored image shown in Fig. 4.45h. Note that although the restoration is imperfect, the main characters can be determined.

## Bibliographical Guide

An introduction to digital image enhancement and restoration may be found in the general papers by Nathan (1968, 1970), Oppenheim *et al.* (1968), Prewitt (1970), Hall *et al.* (1971), Huang *et al.* (1971), Stockham (1972), Andrews *et al.* (1972), Andrews (1974), and the books by Gonzalez and Wintz (1977), Andrews and Hunt (1977), Rosenfeld and Kak (1976), Huang (1975), and Pratt (1977).

Histogram manipulations for image enhancement are described by Hall (1974), Gonzalez and Fittes (1975), Robinson and Frei (1975), and Wong (1977).

Geometric transformations are important in both digital and optical processing and are described in Nathan (1968), Billingsley (1970), Casasent (1975), and Wong and Hall (1978).

Enhancement by spatial filtering for linear spatially invariant systems is the most general and practical technique. Spatial filtering methods are described by Graham (1962), Nathan (1968), (1970), Billingsley (1970), Ekstrom and Algazi (1970), Prewitt (1970), Schreiber (1970), Eklundh (1972), Ekstrom (1973), and Hunt (1973, 1975), as well as in the books by Rosenfeld and Kak (1976), and Gonzalez and Wintz (1977).

Color techniques for image enhancement are described in Andrews et al. (1972), Fink (1976), Gazley et al. (1967), Kreins and Allison (1970), and Nichold and Lamar (1968).

Image restoration is a more satisfying formulation of the enhancement problem in that an objective criteria is imposed. A large amount of the image restoration theory was developed in optics. The methods are described in Frieden (1967, 1972, 1975), Frieden and Burke (1972), Harris (1966, 1968), Helstrom (1967), Horner (1969, 1970), Hufnagel and Stanley (1964), Jansson et al. (1968, 1970), Lohmann and Paris (1965), MacAdam (1970), McGlamery (1967), Miyamoto (1961), Mueller and Reynolds (1967), Paris (1963), Shack (1964), Slepian (1967), Som (1971), Stokseth (1969), Tsujiuchi (1963), Huang and Narendra (1975), and Huang et al. (1975).

Digital image restoration techniques are described in Andrews (1974), Andrews and Hunt (1977), Andrews and Patterson (1975), Cannon (1974), Cole (1973), Hunt (1971, 1972), Lewis and Sakrison (1975), MacAdam (1970), Mascarenhas and Pratt (1975), O'Handley and Green (1972), Pratt (1972, 1977), and Stockham et al. (1975). The entropy constrained restoration has been described by Hershel (1971), Frieden (1972, 1975, 1976), and Edward and Fitelson (1973). Space variant restoration is described by Barrett and Devich (1976), Huang (1976), Lohmann and Paris (1965), Robbins (1970), Robbins and Huang (1972), and Sawchuk (1972). Nonlinear homomorphic methods are described by Cannon (1974), Cole (1973), Stockham (1972), and Stockham et al. (1975). Stochastic restoration methods are described by Frieden (1972), Habibi (1972), Jain and Angel (1974), Mascarenhas and Pratt (1975), Nahi (1972), Pratt (1972), Pratt and Davarian (1977), Richardson (1972), Sondhi (1972), Wallis (1975), and Woods (1972). Recursive digital image filtering techniques may be computationally advantageous as shown in Hall (1972), but can also result in unstable filters. Design techniques for digital image filters are described by Anderson and Jury (1973), Dudgeon (1975), Farmer and Gooden (1971), Hu and Rabiner (1972), Huang (1972), Merserau and Dudgeon (1975), Pistor (1974), and Shanks et al. (1972).

The fundamental mathematical techniques for numerical solution of the integral equations upon which the image-restoration problem is based are described in Phillips (1962), Twomey (1963), Tikonov (1963), Albert (1972), Bellman (1970), and Rust and Burris (1972).

## COMPUTER EXPERIMENTS

**1** Select a digital image file.

(a) Write a computer program to read this file and compute a 32-level histogram. Select from this histogram break points for a contrast enhancement transformation.

(b) Write a program to effect a three-step, piecewise linear contrast enhancement transformation. Save the resulting image in a file.

(c) Use a gray scale plotting program to plot the original file and the contrast enhanced image file.

**2** Write a program to compute the discrete Fourier transform using a fast Fourier transform algorithm for the real-valued input sequence

$$A(I) = \sin(2\pi * I/8), \qquad I = 1, \ldots, 16.$$

(a) Compute the complex transform.

(b) Compute the double transform and estimate the order of magnitude of the roundoff error.

(c) Compute the log magnitude transform. Determine the half-power point.

(d) Compute the phase spectrum. Is the phase spectrum linear?

**3** Construct an image of a square, i.e., a $16 \times 16$ array of 0s with a $5 \times 5$ array of 1s in the center.

(a) Compute the forward, origin centered, two-dimensional complex transform.

(b) Compute the double transform and estimate the order of magnitude of the roundoff error.

(c) Compute the log magnitude transform.

(d) Compute the phase spectrum.

**4** Construct a square wave function, i.e., a 16-point, one-dimensional array of 0s with four 1s in the center.

(a) Compute the nonsequency ordered Hadamard transform of this array.

(b) Compute the double transform and estimate the transform roundoff error.

(c) Compute the sequency ordered transform.

(d) Low-sequency filter the transform, i.e., set the five highest sequency terms to zero and compute the inverse transform. Compare the filtered function to the original square wave.

(e) Experiment with filter shapes to determine if an edge enhancement sequency filter can be obtained.

## PROBLEMS

**4.1** A picture or image function is usually defined as a nonnegative, real-valued function of two independent variables, $X$ and $Y$. A digitized or discrete picture function $f(nX, mY)$ is nonzero only at discrete values of $x$ and $y$, i.e., $x = nX, y = mY$; $n = 0, 1, 2, \ldots, N-1, m = 0, 1, 2, \ldots, M-1$. The value of $f(nX, mY)$ at any point $(nX, mY)$ is called the *gray level* at that point.

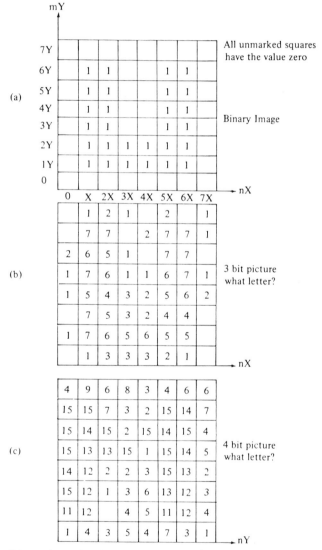

*Figure 4.46* Discrete image functions: (a) binary image of letter U, (b) 3-bit image, and (c) 4-bit image.

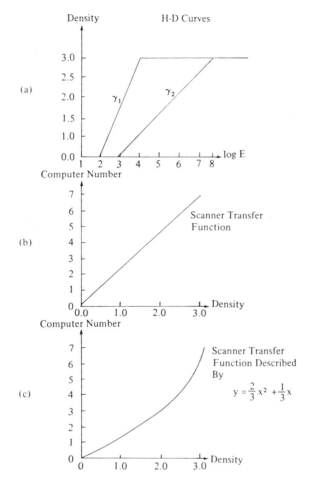

*Figure 4.47* (a) Photographic development functions. (b) Linear transfer function. (c) Nonlinear transfer function.

Usually, the largest values of $f(nX, mY)$ are called white, while the smallest values are called black.

For all computer compatible image functions, the number of distinguishable gray levels [values of $f(nX, mY)$] is discrete, i.e., there is a finite number of gray levels, usually less than 64.

A histogram of the gray levels in an image function is a function of the image that shows the relative frequency of occurrence of each gray level.

For each discrete image function given in Fig. 4.46, compute and sketch a graph of the gray level histogram.

**4.2** A photographic transmittance function is developed at $\gamma_1$ shown on the $H-D$ curve in Fig. 4.47a. This transparency is then scanned on a digitizer that has a transfer characteristic as shown in Fig. 4.47b. After this digitization

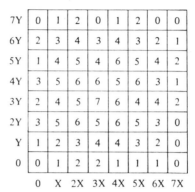

| 7Y | 0 | 1 | 2 | 0 | 1 | 2 | 0 | 0 |
|----|---|---|---|---|---|---|---|---|
| 6Y | 2 | 3 | 4 | 3 | 4 | 3 | 2 | 1 |
| 5Y | 1 | 4 | 5 | 4 | 6 | 5 | 4 | 2 |
| 4Y | 3 | 5 | 6 | 6 | 5 | 6 | 3 | 1 |
| 3Y | 2 | 4 | 5 | 7 | 6 | 4 | 4 | 2 |
| 2Y | 3 | 5 | 6 | 5 | 6 | 5 | 3 | 0 |
| Y  | 1 | 2 | 3 | 4 | 4 | 3 | 2 | 0 |
| 0  | 0 | 1 | 2 | 2 | 1 | 1 | 1 | 0 |

        0    X   2X   3X   4X   5X   6X   7X

*Figure 4.48*    Sampled digital image.

of the image shown in Fig. 4.47c, it was decided that the film did not have enough contrast and should have been developed at $\gamma_2$ as shown in Fig. 4.47a.

(a) Design a digital contrast enhancement transform to provide the effect of $\gamma_2$ development on the computer image.

(b) Compute the transformed image.

(c) Given the digital image shown in Fig. 4.48, which is assumed to have been scanned with a scanner that has a nonlinear transfer function as shown in Fig. 4.47c.

(1) Design a transformation that has the effect of linearizing this curve.

(2) Compute the resulting image.

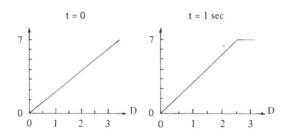

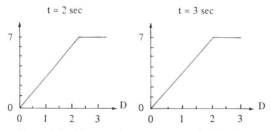

*Figure 4.49*    Time-varying scanner transfer functions.

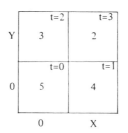

*Figure 4.50* Sampled image.

(d) Suppose the scanner amplifier gain drifts with time producing the time varying transfer function shown in Fig. 4.49. During this time we scan a small $2\times2$ image shown in Fig. 4.50. The scanner scans one point/sec.

(1) Design a compensating transformation that essentially produces the $t=0$ transfer function at all values of time.

(2) Compute the compensated image from Fig. 4.50.

**4.3** Show by direct recursion or mathematical induction that the computation

$$y(nX) = \sum_{i=0}^{L-1} a_i x(nX - iX) - \sum_{j=1}^{M-1} b_j y(nX - jX)$$

is exactly equivalent to the convolution

$$y(nX) = \sum_{i=0}^{n} h_i x(nX - iX)$$

and determine the relation between $h_i$ and $a_i$ and $b_i$.

Assume that $y(nX)$ and $x(nX)=0$ for $n<0$.

**4.4** Given the following two forms for digital convolution:

$$g(mX, nY) = \sum_{i=0}^{2M} \sum_{j=0}^{2M} h(iX, jY) f(mX - iX, nY - jY),$$

$$g(mX, nY) = \sum_{i=-M}^{M} \sum_{j=-M}^{M} h(iX, jY) f(mX - iX, nY - jY),$$

where $h(iX, jY) = 1$,

(a) compute the closed form solution for the system function for each, e.g., let $f(mX, nY) = \varepsilon^{j(\omega_x mX + \omega_y mY)}$;

(b) comment on the main difference in the magnitude and phase responses for the two systems. If each operation is used to filter an $N \times N$ image, comment on the number of computations and edge point special cases or transient effects. (*Note:* Use 50 words or less.)

**4.5** Let

$$s(x,y) = \sum_{m=-M}^{M} \sum_{n=-M}^{M} \delta(x - ma, y - nb)$$

and

$$G(x,y) = \exp\left\{-\left[(x-a)^2 + (y-b)^2\right]/2\sigma^2\right\}.$$

(a) Compute the Fourier transform of $s(x,y)$.

(b) Compute the Fourier transform of $s(x,y) ** G(x,y)$, where $**$ denotes two-dimensional convolution.

**4.6** (a) (1) Suppose the random variable $X$ has a uniform $(a,b)$ distribution. Determine a new random variable $Y = T(X)$, that has a uniform $(0,1)$ distribution.

(2) Suppose $X$ has a normal $(\mu, \sigma^2)$ distribution. Determine a new random variable $Y = T(X)$ that has a uniform $(0,1)$ distribution.

(b) A discrete variable $X$ is distributed equally likely over $N$ points, i.e., its histogram $h(x)$ is defined by

$$h(x) = \begin{cases} 1/N & \text{for } x = 0, 1/N, 2/N, \dots, (N-1)/N, \\ 0 & \text{otherwise.} \end{cases}$$

The distribution function $H(x)$ is equal to the number of $x$ values less than or equal to $x$. *Compute* this distribution function. Use the distribution function as a transformation, i.e., let $y = H(x)$. *Compute* the histogram $g(y)$ of $y$. Is this histogram still equally likely?

Now form a new distribution function $H'$ from $H$ by compressing the number of abscissa values, i.e., maintain the same range but compress every two points into one new point. This corresponds to a coarser quantization of the values.

Again *compute* the histogram $g(y)$ of $y$. Is this histogram still equally likely?

**4.7** We know that a signal $h(t)$ that is bandlimited to $f_1$ may be exactly determined from its values $h(nT)$ at equispaced points, $t = nT$, provided $T < 1/2f_1$.

State as quantitatively as possible:

(a) The effect of time limiting the signal to a duration of $NT$ seconds.

(b) For a fixed time duration of $\tau$ seconds, the effect of increasing $N$ (decreasing $T$).

(c) The effect of nonideal sampling, i.e., zero-order hold of $\tau_1$ seconds.

**4.8** Given

$$\mathbf{A} = \begin{bmatrix} a_{11} & a_{12} \\ a_{21} & a_{22} \end{bmatrix} \quad \text{and} \quad \mathbf{L} = \begin{bmatrix} L_{11} & L_{12} \\ L_{21} & L_{22} \end{bmatrix},$$

determine the following derivatives of the trace of the matrix products. Recall that

$$\text{Tr}\,\mathbf{A} = a_{11} + a_{22} \quad \text{and} \quad \frac{\partial f}{\partial \mathbf{L}} = \begin{bmatrix} \dfrac{\partial f}{\partial L_{11}} & \dfrac{\partial f}{\partial L_{12}} \\[2mm] \dfrac{\partial f}{\partial L_{21}} & \dfrac{\partial f}{\partial L_{22}} \end{bmatrix}.$$

$$(1) \quad \frac{\partial \mathrm{Tr}}{\partial \mathbf{L}}(\mathbf{AL}) \qquad (2) \quad \frac{\partial \mathrm{Tr}}{\partial \mathbf{L}}(\mathbf{LA}) \qquad (3) \quad \frac{\partial \mathrm{Tr}}{\partial \mathbf{L}}(\mathbf{L'A'})$$

$$(4) \quad \frac{\partial \mathrm{Tr}}{\partial \mathbf{L}}(\mathbf{A'L'}) \qquad (5) \quad \frac{\partial \mathrm{Tr}}{\partial \mathbf{L}}(\mathbf{LAL'}) \qquad (6) \quad \frac{\partial \mathrm{Tr}}{\partial \mathbf{L}}(\mathbf{L'AL}).$$

**4.9** Develop the $[H]$ matrix for this example.

$$(a) \quad \begin{bmatrix} f_{11} & f_{12} \\ f_{21} & f_{22} \\ f_{31} & f_{32} \end{bmatrix}, \qquad (b) \quad \mathbf{f} = \begin{bmatrix} \hat{\mathbf{f}}_1 \\ \hat{\mathbf{f}}_2 \\ \hat{\mathbf{f}}_3 \end{bmatrix} = \begin{bmatrix} f_{11} \\ f_{12} \\ f_{21} \\ f_{22} \\ f_{31} \\ f_{32} \end{bmatrix},$$

$$(c) \quad \mathbf{n} = \begin{bmatrix} n_{11} \\ n_{12} \\ n_{21} \\ n_{22} \\ n_{31} \\ n_{32} \end{bmatrix}, \qquad (d) \quad \mathbf{g} = \begin{bmatrix} g_{11} \\ g_{12} \\ g_{21} \\ g_{22} \\ g_{31} \\ g_{32} \end{bmatrix},$$

where (a) is a $3 \times 2$ image function, (b) an equivalent image vector, (c) an equivalent noise vector, and (d) an equivalent observed image vector.

## References

Albert, A. (1972). "Regression and the Moore–Penrose Pseudoinverse." Academic Press, New York.

Anderson, B. D., and Jury, E. I. (1973). Stability Test for Two-Dimensional Recursive Filters. *IEEE Trans. Audio Electroacoust.* **AU-21**, No. 4, 366–372.

Andrews, H. C. (1974). Digital Image Restoration: A Survey. *IEEE Comput.* **7**, No. 5, 36–45.

Andrews, H. C., and Hunt, B. R. (1977). "Digital Image Restoration." Prentice-Hall, Englewood Cliffs, New Jersey.

Andrews, H. C., and Patterson, C. L. (1975). Outer Product Expansions and Their Uses in Digital Image Processing. *Am. Math. Mon.* **1**, No. 82, 1–13.

Andrews, H. C., Tescher, A. G., and Kruger, R. P. (1972). Image Processing by Digital Computer. *IEEE Spectrum* **9**, No. 7, 20–32.

Barrett, E. B., and Devich, R. N. (1976). Linear Programming Compensation for Space-Variant Image Degradation. *Proc. SPIE/OSA Conf. Image Process., Pacific Grove, Calif., February* **74**, 152–158.

Bellman, R. (1970). "Introduction to Matrix Analysis" 2nd Ed. McGraw-Hill, New York.

Billingsley, F. C. (1970). Applications on Digital Image Processing. *Appl. Opt.* **9**, No. 2, 289–299.

Cannon, T. M. (1974). Digital Image Deblurring by Nonlinear Homomorphic Filtering. Ph.D. thesis, Comput. Sci. Dep., Univ. of Utah, Salt Lake City, (unpublished).

Casasent, D. P. (1975). Optical Digital Radar Signal Processing. *Int. Opt. Comput. Conf., Washington, D.C., April.*

Cole, E. R. (1973). The Removal of Unknown Image Blurs by Homomorphic Filtering. Ph.D. thesis, Dep. of Electr. Eng., Univ. of Utah, Salt Lake City, June (unpublished).

Dainty, J. C., and Shaw, R. (1974). "Image Science." Academic Press, New York.

Dudgeon, D. E. (1975). Two-Dimensional Recursive Filter Design Using Differential Correction. *IEEE Trans. Acoust., Speech, Signal Process.* **ASSP-23**, No. 3, 264–267.

Edward, J. A., and Fitelson, M. M. (1973). Notes on Maximum Entropy Processing. *IEEE Trans. Inf. Theory* **IT-19**, No. 2, 232–234.

Eklundh, J. O. (1972). A Fast Computer Method for Matrix Transposing. *IEEE Trans. Comput.* **C-21** (July), 801–803.

Ekstrom, M. P. (1973). A Numerical Algorithm for Identifying Spread Functions of Shift-Invariant Imaging Systems. *IEEE Trans. Comput.* **C-22**, No. 4, 322–328.

Ekstrom, M. P., and Algazi, V. R. (1970). Optimum Design of Two-Dimensional Nonrecursive Digital Filters. *Proc. Asilomar Conf. Circuits Syst., 4th, Pacific Grove, Calif., November.*

Farmer, C. H., and Gooden, D. S. (1971). Rotation and Stability of a Recursive Digital Filter. *Proc. Two Dimen. Signal Process. Conf., Univ. Missouri, Columbia, October.*

Fink, W. (1976). Image Coloration as an Interpretation Aid. *Proc. SPIE/OSA Conf. Image Process., Pacific Grove, Calif., February* **74**, 209–215.

Frieden, B. R. (1967). Bandlimited Reconstruction of Optical Objects and Spectra. *J. Opt. Soc. Am.* **57**, 1013–1019.

Frieden, B. R. (1972). Restoring with Maximum Likelihood and Maximum Entropy. *J. Opt. Soc. Am.* **62**, 511–518.

Frieden, B. R. (1974). Image Restoration by Discrete Deconvolution of Minimal Length. *J. Opt. Soc. Am.* **64**, 682–686.

Frieden, B. R. (1975). Image Enhancement and Restoration. *In* "Picture Processing and Digital Filtering" (T. S. Huang, ed.), pp. 179–246, Springer-Verlag, Berlin and New York.

Frieden, B. R. (1976). Maximum Entropy Restorations of Garrymede. *Proc. SPIE/OSA Conf. Image Process., Pacific Grove, Calif., February* **74**, 160–165.

Frieden, B. R., and Burke, J. J. (1972). Restoring with Maximum Entropy. II: Superresolution of Photographs of Diffraction-Blurred Images. *J. Opt. Soc. Am.* **62**, 1207–1210.

Gazley, C., Reiber, J. E., and Stratton, R. H. (1967). Computer Works a New Trick in Seeing Pseudo Color Processing. *Aeronaut. Astronaut.* **4** (April), 56.

Gonzalez, R. C., and Fittes, B. A. (1975). Gray-Level Transformations for Interactive Image Enhancement. *Mech. Mach. Theory* **12**, 111–112.

Gonzalez, R. C., and Wintz, P. (1977). "Digital Image Processing." Addison-Wesley, Reading, Massachusetts.

Graham, R. E. (1962). Snow-Removal: A Noise-Stripping Process for Picture Signals. *IRE Trans. Inf. Theory* **IT-8**, No. 1, 129–144.

Habibi, A. (1972). Two-Dimension Bayesian Estimate of Images. *IEEE Proc.* **60**, 878–883.

Hall, E. L. (1972). A Comparison of Computations for Spatial Frequency Filtering. *Proc. IEEE* **60**, No. 7, 887–891.

Hall, E. L. (1974). Almost Uniform Distribution for Computer Image Enhancement. *IEEE Trans. Comput.* **C-23**, No. 2, 207–208.

Hall, E. L., Kruger, R. P., Dwyer, S. J., McLaren, R. W., and Lodwick, G. S. (1971). A Survey of Preprocessing and Feature Extraction Techniques for Radiographic Images. *IEEE Trans. Comput.* **C-20**, No. 9, 1032–1044.

Harris, J. L., Sr. (1966). Image Evaluation and Restoration. *J. Opt. Soc. Am.* **56**, No. 5, 569–574.

Harris, J. L., Sr. (1968). Potential and Limitations of Techniques for Processing Linear Motion-Degraded Imagery. *Proc. NASA Seminar Evaluation of Motion Degraded Images*, pp. 131–138.

Heller, J. M. (1978). Image Motion Restoration. Tech. Rep., Dep. Electr. Eng., Univ. of Tennessee, Knoxville, May.

Helstrom, C. W. (1967). Image Restoration by the Method of Least Squares. *J. Opt. Soc. Am.* **57**, No. 3, 297–303.

Hershel, R. S. (1971). Unified Approach to Restoring Degraded Images in the Presence of Noise, Tech. Rep. No. 71. Optical Sciences Center, Univ. of Arizona, Tucson.

Horner, J. L. (1969). Optical Spatial Filtering with the Least-Mean-Square-Error Filter. *J. Opt. Soc. Am.* **51**, No. 5, 553–558.

Horner, J. L. (1970). Optical Restoration of Images Blurred by Atmospheric Turbulence Using Optimum Filter Theory. *Appl. Opt.* **9**, No. 1, 167–171.

Hu, J. V., and Rabiner, L. R. (1972). Design Techniques for Two-Dimensional Digital Filters. *IEEE Trans. Audio Electroacoust.* **AU-20**, No. 4, 249–257.

Huang, T. S. (1972). Stability of Two-Dimensional Recursive Filters. *IEEE Trans. Audio Electroacoust.* **AU-20**, No. 2, 158–163.

Huang, T. S. (ed.) (1975). "Picture Processing and Digital Filtering." Springer-Verlag, New York.

Huang, T. S. (1976). Restoring Images with Shift-Varying Degradations. *Proc. SPIE/OSA Conf. Image Process., Pacific Grove, Calif., February* **74**, 149–151.

Huang, T. S., and Narendra, P. M. (1975). Image Restoration by Singular Value Decomposition. *Appl. Opt.* **14**, No. 9, 2213–2216.

Huang, T. S., Schreiber, W. F., and Tretiak, O. J. (1971). Image Processing. *Proc. IEEE* **59**, 1586–1609.

Huang, T. S., Baker, D. S., and Berger, S. P. (1975). Iterative Image Restoration. *Appl. Opt.* **14**, No. 5, 1165–1168.

Hufnagel, R. E., and Stanley, N. R. (1964). Modulation Transfer Function Associated with Image Transmission Through Turbulent Media. *J. Opt. Soc. Am.* **54**, 52–61.

Hunt, B. R. (1971). A Matrix Theory Proof of the Discrete Convolution Theorem. *IEEE Trans. Audio Electroacoust.* **AU-19**, 285–288.

Hunt, B. R. (1972). Deconvolution of Linear Systems by Constrained Regression and Its Relationship to the Wiener Theory. *IEEE Trans. Autom. Control.* **AC-17**, No. 5, 703–705.

Hunt, B. R. (1973). The Application of Constrained Least Squares Estimation to Image Restoration by Digital Computer. *IEEE Trans. Comput.* **C-22**, 805–812.

Hunt, B. R. (1975). Digital Image Processing. *Proc. IEEE* **63**, No. 4, 693–708.

Jain, A. K., and Angel, E. (1974). Image Restoration, Modeling, and Reduction of Dimensionality. *IEEE Trans. Comput.* **C-23**, 470–476.

Jansson, P. A. (1970). Resolution Enhancement of Spectra. *J. Opt. Soc. Am.* **60**, 596–599.

Jansson, P. A., Hunt, R. H., and Plyler, E. K. (1968). Response Function for Spectral Resolution Enhancement. *J. Opt. Soc. Am.* **58**, 1665–1666.

Kreins, E. R., and Allison, L. J. (1970). Color Enhancement of Nimbus High Resolution Infrared Radiometer Data. *Appl. Opt.* **9**, No. 3, 681.

Kruger, R. P., Dwyer, S. J., III, and Storvick, T. S. (1970). Techniques for Automatic Isolation and Measurement of Detonation Shock Waves in Metals, Univ. of Missouri, College of Engineering Report, September. Univ. of Missouri, Columbia.

Kwoh, Y. S. (1977). Application of Finite Field Transforms to Image Processing and X-Ray Reconstruction. Ph.D. thesis, Univ. of Southern California, Los Angeles, January (unpublished).

Lewis, B. L., and Sakrison, D. J. (1975). Computer Enhancement of Scanning Electron Micrographs. *IEEE Trans. Circuits Syst.* **CAS-22**, No. 3, 267–278.

Lewis, T. O., and Odell, P. L. (1971). "Estimation in Linear Models." Prentice-Hall, Englewood Cliffs, New Jersey.

Lohmann, A. W., and Paris, D. P. (1965). Space-Variant Image Formation. *J. Opt. Soc. Am.* **55**, 1007–1013.

MacAdam, D. P. (1970). Digital Image Restoration by Constrained Deconvolution. *J. Opt. Soc. Am.* **60**, 1617–1627.

McGlamery, B. L. (1967). Restoration of Turbulence-Degraded Images. *J. Opt. Soc. Am.* **57**, No. 3, 293–297.

Mascarenhas, N. D. A., and Pratt, W. K. (1975). Digital Restoration Under a Regression Model. *IEEE Trans. Circuits Syst.* **CAS-22**, No. 3, 252–266.

Merserau, R., and Dudgeon, D. (1975). Two-Dimensional Digital Filter. *Proc. IEEE* **63** (April), 610–623.

Miyamoto, L. (1961). Wave Optics and Geometrical Optics in Optical Design. *Prog. Opt.* **1**, 1961.

Mueller, P. F., and Reynolds, G. O. (1967). Image Restoration by Removal of Random Media Degradations. *J. Opt. Soc. Am.* **57**, No. 11, 1338–1344.

Nahi, N. E. (1972). Role of Recursive Estimation in Statistical Image Enhancement. *Proc. IEEE* **60**, 872–877.

Nathan, R. (1968). Picture Enhancement for the Moon, Mars, and Man. *In* "Pictorial Pattern Recognition" (C. G. Cheng, ed.), pp. 239–266. Thompson, Washington, D.C.

Nathan, R. (1970). Spatial Frequency Filtering. *In* "Picture Processing and Psychopictorics" (B. S. Lipkin and A. Rosenfeld, eds.), pp. 151–164. Academic Press, New York.

Nichold, L. W., and Lamar, J. (1968). Conversion of Infrared Images to Visible in Color. *Appl. Opt.* **7**, No. 9, 1757.

O'Handley, D. A., and Green, W. B. (1972). Recent Developments in Digital Image Processing at the Image Laboratory at the Jet Propulsion Laboratory. *IEEE Proc.* **60**, No. 7, 821–828.

Oppenheim, A. V., Schafer, R. W., and Stockham, T. G. (1968). Nonlinear Filtering of Multiplied and Convolved Signals. *Proc. IEEE* **56**, No. 8, 1264–1292.

Paris, D. P. (1963). Influence of Image Motion on the Resolution of a Photographic System—II. *Photogr. Sci. Eng.* **7**, 233–236.

Phillips, D. L. (1962). A Technique for the Numerical Solution of Certain Integral Equations of the First Kind. *J. Assoc. Comput. Mach.* **9**, 84–97.

Pistor, P. (1974). Stability Criterion for Recursive Filters. *IBM J. Res. Dev.* **18**, No. 1, 59–71.

Pratt, W. K. (1972). Generalized Wiener Filter Computation Techniques. *IEEE Trans. Comput.* **C-21**, No. 7, 636–641.

Pratt, W. K. (1977). Pseudoinverse Image Restoration Computational Algorithms. *In* "Optical Information Processing" (G. W. Stroke, Y. Nesterikhin, and E. S. Barrekette, eds.), Vol. 2, pp. 317–328. Plenum, New York.

Pratt, W. K., and Davarian, F. (1977). Fast Computational Techniques for Pseudoinverse and Wiener Image Restoration. *IEEE Trans. Comput.* **C-26**, No. 6, 571–580.

Prewitt, J. M. S. (1970). Object Enhancement and Extraction. *In* "Picture Processing and Psychopictorics" (B. S. Lipkin and A. Rosenfeld, eds.), pp. 75–150. Academic Press, New York.

Richardson, W. H. (1972). Bayesian-Based Iterative Method of Image Restoration. *J. Opt. Soc. Am.* **62**, 55–59.

Robbins, G. M. (1970). Image Restoration for a Class of Linear Spatially-Variant Degradations. *Pattern Recognition* **2**, No. 2, 91–105.

Robbins, G. M., and Huang, T. S. (1972). Inverse Filtering for Linear Shift-Variant Imaging Systems. *Proc. IEEE* **60**, 862–872.

Robinson, G. S. (1972). Logical Convolution and Discrete Walsh Power Spectra, *IEEE Trans. Audio Electroacoust.* **AU-20**, 271–279.

Robinson, G. S., and Frei, W. (1975). Final Research Report on Computer Processing of ERTS Images. USC-IPI Rep. No. 640, Image Process. Inst., Univ. of Southern California, Los Angeles, September.

Rosenfeld, A., and Kak, A. C. (1976). "Digital Picture Processing." Academic Press, New York.

Rust, B. W., and Burrus, W. R. (1972). "Mathematical Programming and the Numerical Solution of Linear Equations." Am. Elsevier, New York.

Sawchuk, A. A. (1972). Space-Variant Image Motion Degradation and Restoration. *Proc. IEEE* **60**, 854–861.

Schreiber, W. F. (1970). Wirephoto Quality Improvement by Unsharp Masking. *Pattern Recognition* **2**, 171–121.

Shack, R. B. (1964). The Influence of Image Motion and Shutter Operation on the Photographic Transfer Function. *Appl. Opt.* **3**, 1171–1181.

Shanks, J. L., Treitel, S., and Justice, J. H. (1972). Stability and Synthesis of Two-Dimensional Recursive Filters. *IEEE Trans. Audio Electroacoust.* **AU-20**, No. 2, 115–128.

Slepian, D. (1967). Restoration of Photographs Blurred by Image Motion. *Bell Syst. Tech. J.* **40**, 2353–2362.

Som, S. C. (1971). Analysis of the Effects of Linear Smear. *J. Opt. Soc. Am.* **61**, 859–864.

Sondhi, M. M. (1972). Image Restoration: The Removal of Spatially Invariant Degradations. *Proc. IEEE* **60**, 842–853.

Stockham, T. G., Jr. (1972). Image Processing in the Context of a Visual Model. *IEEE Proc.* **60**, 828–841.

Stockham, T. G., Jr., Ingebretsen, R., and Cannon, T. M. (1975). Blind Deconvolution by Digital Signal Processing. *Proc. IEEE* **63** (April), 679–692.

Stokseth, P. A. (1969). Properties of a Defocused Optical System. *J. Opt. Soc. Am.* **59**, 1314–1321.

Tikonov, A. N. (1963). Regularization of Incorrectly Posed Problems. *Sov. Math.* **4**, No. 6, 1624–1627.

Tsujiuchi, J. (1963). Correction of Optical Images by Compensation of Aberrations and by Spatial Frequency Filtering. *Prog. Opt.* **2**, 131–180.

Tukey, J. W. (1971). "Exploratory Data Analysis." Addison-Wesley, Reading, Massachusetts.

Twomey, S. (1963). On the Numerical Solution of Fredholm Integral Equations of the First Kind by the Inversion of the Linear System Produced by Quadrature. *J. Assoc. Comput. Mach.* **10**, 97–101.

Wallis, R. (1975). Film Recording of Digital Color Images. Ph.D. thesis, Dep. Electr. Eng., Univ. of Southern California, Los Angeles, May (unpublished).

Wong, R. Y. (1977). Image Sensor Transformations, *IEEE Trans. Syst. Man Cyber.* **SMC-7**, No. 12, 836–841.

Wong, R. Y., and Hall, E. L. (1977). Sequential Hierarchical Scene Matching, *IEEE Trans. Comput.* **C-27**, No. 4, 359–365.

Woods, J. W. (1972). Two-Dimensional Discrete Markovian Fields, *IEEE Trans. Inf. Theory* **IT-18**, 232–240.

# 5 | Reconstruction from Projections

## 5.1 Introduction

An interesting and unique processing problem is the reconstruction of the image of an object from a set of transverse cross-sectional projections of the object. In several applications, an image of an interior section of an object may be produced only in this manner without physically destroying the object. The significance of this technique is shown by its wide application to areas such as medical radiology and nuclear medicine, electron microscopy, radio and radar astronomy, light microscopy and holography, and the theory of vision.

The general problem of three-dimensional reconstruction and various possible approaches to the solutions are illustrated in Fig. 5.1. Suppose the two embedded numerals can only be viewed from the sides, i.e., a transverse direction. The problem is to determine what numerical measurements would appear from the top in a thin slice. Physically slicing the object into sections is the obvious solution; however, this is impractical in many cases, such as medical diagnosis, astronomical observations, industrial nondestructive testing, photogrammetry measurements, and many others. In each of these cases, it is still possible in principle, and for some cases, in practice, to determine the cross-sectional characteristics of the objects, i.e., locate the numerals in Fig. 5.1. The three information collecting modes—transmission, emission, and reflection—are also illustrated in Fig. 5.1. The transmission mode depends upon energy transmission or absorption by the object and is most often used with x-ray beams, electron beams, light or heat, which possess a well-defined absorption law. Emission may also be used to obtain the object location, and this method has been developed for positrons, which upon disintegration emit two gamma rays in opposite directions. The time of detection of these two events may be used to locate the position of origin of the annihilation. The reflection of energy may also be used to determine the surface characteristics of the objects using, for example, light, electron beams, radar, or ultrasound

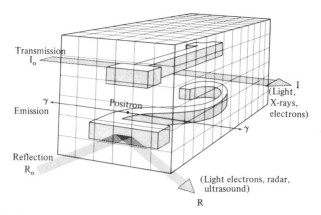

*Figure 5.1* Three-dimensional reconstruction by transmission, emission, and reflection.

as the energy source. The science of the reconstruction modes and energy sources is beyond the scope of the present discussion; however, it is interesting to note that each of the combinations for which the problem of three-dimensional reconstruction has been solved in practice has led to an exciting area of study. A final general comment is that it is interesting to differentiate between three-dimensional *reconstruction* in which the object characteristics are measured, and three-dimensional *visualization*, such as holography. Both are exciting problems; however, in this chapter we will restrict our attention to the reconstruction problem, and primarily to the transmission mode.

In this chapter a survey of the major techniques for computing a reconstruction are described. These include the direct Fourier transform method, the convolution algorithm, and the algebraic formulations. Considerations for the display of reconstructed images and objects are also given. Finally, a brief introduction is given to some important medical applications in radiography.

## 5.2  Fourier Transform Reconstruction

The Fourier transform method of reconstruction provides the simplest introduction. This method is based on the fact that the Fourier transform of a two-dimensional (one-dimensional) projection of a three-dimensional (two-dimensional) object is exactly equal to the central section of the Fourier transform of the object. By rotating the projections and thus the Fourier transform section, the entire Fourier transform plane may be first constructed and then the object reconstructed by simply taking the inverse Fourier transform.

The theory of reconstruction will now be introduced (Papoulis, 1969; Shepp and Logan. 1974).

Let $f(x,y)$ represent an image function. The two-dimensional Fourier transform is

$$F(u,v) = \int \int_{-\infty}^{\infty} f(x,y) \exp\left[-j2\pi(ux+vy)\right] dx\,dy.$$

Consider the projection of the image on the $x$ axis, i.e.,

$$g_y(x) = \int_{-\infty}^{\infty} f(x,y)\,dy.$$

An important fact is that the one-dimensional Fourier transform of the projection

$$G_y(u) = \int_{-\infty}^{\infty} g_y(x) \exp(-j2\pi ux)\,dx = \int \int_{-\infty}^{\infty} f(x,y) \exp(-j2\pi ux)\,dx\,dy$$

is exactly equal to the central profile of the two-dimensional Fourier transform

$$F(u,0) = \int \int_{-\infty}^{\infty} f(x,y) \exp(-j2\pi ux)\,dx\,dy.$$

Now suppose we project the function on a line that is rotated at an angle $\theta$ as shown in Fig. 5.2. Let us first define the rotated coordinates

$$s = x\cos\theta + y\sin\theta, \qquad t = -x\sin\theta + y\cos\theta.$$

The line upon which we project the function is chosen to be the $x$ axis. The projection points are computed by integrating the function along lines parallel to the $t$ axis at a distance $s$ from the origin. The projection may therefore be written as

$$g(s_1,\theta) = \int f(x,y)\,ds,$$

where the path is along the line

$$s_1 = x\cos\theta + y\sin\theta.$$

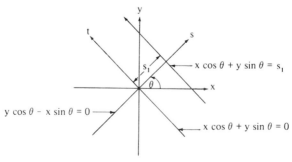

Figure 5.2 Projection geometry showing lines of projection $y = x\tan\theta$ and perpendicular direction of integration line $y = -x/\tan\theta$ at origin and at distance $s_1$ from the origin.

The one-dimensional Fourier transform of the projection is

$$G(r,\theta) = \int_{-\infty}^{\infty} g(s_1,\theta) \exp(-j2\pi r s_1) \, ds_1,$$

which may be expanded as

$$G(r,\theta) = \int\int_{-\infty}^{\infty} f(x,y) \exp\left[ -j2\pi r(x\cos\theta + y\sin\theta) \right] dx\, dy.$$

For this to equal the two-dimensional Fourier transform, we equate the terms in the exponent to arrive at

$$u = r\cos\theta, \qquad v = r\sin\theta.$$

Therefore, if the point $(u,v)$ is on a line at a fixed angle $\theta$ and variable distance $r$ from the origin, the transform of the projection will equal a line of the two-dimensional transform, $F(u,v) = G(r,\theta)$. Clearly, if the transform of the projections $G(r,\theta)$ were known for all values of $r$ and $\theta$, the two-dimensional transform would be determined. To then find the image function would require an inverse transform computation

$$f(x,y) = \int\int_{-\infty}^{\infty} F(u,v) \exp\left[ j2\pi(ux + vy) \right] dx\, dy.$$

This may be considered the basis of the reconstruction technique.

These results are easily extended to the three-dimensional situation. Let $f(x_1,x_2,x_3)$ represent an object, where $f$ may be real or complex. The three-dimensional Fourier transform is given by

$$F(u_1,u_2,u_3)$$
$$= \int\int\int_{-\infty}^{\infty} f(x_1,x_2,x_3) \exp\left[ -2\pi j(u_1 x_1 + u_2 x_2 + u_3 x_3) \right] dx_1\, dx_2\, dx_3,$$

and the central section of the transform is

$$F(u_1,u_2,0)$$
$$= \int\int_{-\infty}^{\infty} \left[ \int_{-\infty}^{\infty} f(x_1,x_2,x_3)\, dx_3 \right] \exp\left[ -2\pi j(u_1 x_1 + u_2 x_2) \right] dx_1\, dx_2.$$

By definition the profile or projection on the $x_1$, $x_2$ axes is

$$f_3(x_1,x_2) = \int_{-\infty}^{\infty} f(x_1,x_2,x_3)\, dx_3.$$

Note that the two-dimensional Fourier transform of $f_3(x_1,x_2)$ is exactly identical to the above equation for the central section of the three-dimensional transform. It may also be shown that if the projection is taken at an angle of $\theta$ with respect to the $x_1,x_2$ planes, then the corresponding transform section is at exactly the same angle $\theta$ with respect to the $u_1,u_2$ planes in the transform space. Thus, projections may be taken at different angles of orientation $\theta$ transformed and inserted into the three-dimensional transform space. An infinite number of projections are required to build up the Fourier transform space. Finally, an inverse Fourier transform reconstructs the image

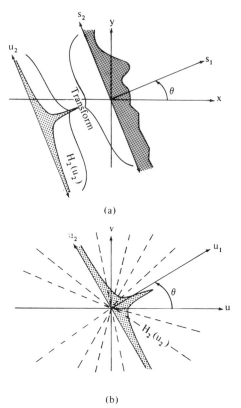

(a)

(b)

*Figure 5.3* Geometry for Fourier transform technique: (a) projection data—Fourier transform of projection and (b) assembly of Fourier transform.

$f(x_1, x_2, x_3)$. Note that a two-dimensional image $f(x_1, x_2)$, may be considered without loss of generality, since any arbitrary plane within a three-dimensional field may be reconstructed. Let us therefore rewrite the projection equations for the two-dimensional case to emphasize their dependence on $\theta$ and the projection coordinate $\rho$.

$$g(\rho, \theta) = \int_s f(x, y)\, ds,$$

where $ds$ is the differential length of the geometric ray path. The previous Fourier transform result may be given by

$$F(R, \theta) = \int_{-\infty}^{\infty} g(\rho, \theta) \exp(-j2\pi R\rho)\, d\rho$$

and

$$f(x, y) = \int\int_{-\infty}^{\infty} F(u, v) \exp[\, j2\pi(ux + vy)]\, du\, dv,$$

as shown in Fig. 5.3.

If an infinite number of projections are taken, determining the Fourier transform in rectangular coordinates $F(u,v)$ from the computed projection transform in polar coordinates $F(R,\theta)$ presents no difficulty. However, if only a finite number of projections is available, some type of interpolation in the transform may be required. Another point to note is that although only one-dimensional forward Fourier transforms of the projection data are required to construct the transform space, a two-dimensional inverse transform is required to reconstruct the image. A consequence of this is that the image cannot be partially reconstructed as the projection data is obtained, but must be delayed until all projection data are available. Before discussing other computational considerations, let us consider some other reconstruction methods. A very popular method, because of its computational simplicity, is the convolution method.

## 5.3   Convolution Algorithm

To introduce the convolution algorithm, let us first rewrite the inverse Fourier transform expression in polar coordinates with the notation illustrated in Fig. 5.4 (Bracewell and Riddle, 1967; Ramachandran and Lakshminarayanan, 1971):

$$x = r\cos\alpha, \quad u = R\cos\beta = -R\sin\theta, \quad y = r\sin\alpha, \quad v = R\sin\beta = R\cos\theta,$$

$$f(r,\alpha) = \int_0^{2\pi}\int_0^{\infty} F(R,\theta) R \exp\left[j2\pi Rr\sin(\alpha-\theta)\right] dR\, d\theta.$$

The Hermitian or symmetric conjugate property may be used to obtain

$$f(r,\alpha) = \int_{-\pi/2}^{\pi/2}\int_{-\infty}^{\infty} |R| F(R,\theta) \exp\left[j2\pi Rr\sin(\alpha-\theta)\right] dR\, d\theta.$$

Let

$$|R| \equiv R\,\text{sgn}(R) \equiv -j2\pi R\left[j\pi\,\text{sgn}(R)\right]/2\pi^2,$$

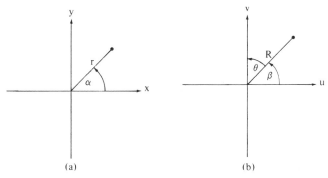

(a)                                          (b)

*Figure 5.4*   Polar coordinate representation for (a) spatial domain and (b) transform domain.

where

$$\operatorname{sgn} R = \begin{cases} 1, & R > 0, \\ 0, & R = 0, \\ -1, & R < 0, \end{cases} \qquad \text{and} \qquad z = r\sin(\alpha - \theta).$$

Then

$$f(r,\alpha) = \frac{1}{2\pi^2} \int_{-\pi/2}^{\pi/2} \int_{-\infty}^{\infty} \left[ j\pi\operatorname{sgn}(R) \right] (-j2\pi R) F(R,\theta) \exp(j2\pi Rz)\, dR\, d\theta,$$

which may be rewritten

$$f(r,\alpha) = \frac{1}{2\pi^2} \int_{-\pi/2}^{\pi/2} \left[ \frac{\partial g(z,\theta)}{\partial z} \right] * \left( \frac{1}{z} \right) d\theta,$$

where $*$ denotes convolution.

The convolution may be explicitly written

$$f(r,\alpha) = \frac{1}{2\pi^2} \int_{-\pi/2}^{\pi/2} \int_{-\infty}^{\infty} \frac{\partial g(\rho,\theta)/\partial \rho}{r\sin(\alpha - \theta) - \rho}\, d\theta.$$

The integral over $\rho$ may be interpreted as the Hilbert transform of the partial derivative of $g(\rho,\theta)$ evaluated at $r = \sin(\alpha - \theta)$. The importance of this interpretation is that one may then show that if the sample values are finite, the integral value is finite, i.e., convergent. Note that the previous integral expression with the $|R|$ term is not obviously convergent.

Another consequence of this derivation leads to a very simple method of reconstruction. Suppose one collects the projection data $g(\rho,\theta)$ and stores the data in an equivalent rectangular space. The projection data stored in this manner is sometimes called a *layergram*. Now for a fixed $\theta$ one may linearly filter that line of projection data either in the frequency domain with a rho filter, i.e., multiply by $|R|$ or in the spatial domain by convolving the projection data with a filter whose impulse response is the inverse transform of the rho frequency filter, i.e.,

$$h_\theta(\rho) = \int_{-A/2}^{A/2} |R| \exp(j2\pi R\rho)\, dR,$$

where the limit $A$ should be infinite but must be finite in practice.

The result of this process is a *rho filtered layergram*. To arrive finally at the reconstructed image, one simply integrates the rho filtered layergram over $\theta$ at a particular value of $\rho = r\cos(\alpha - \theta)$, i.e.,

$$f(r,\alpha) = \int_0^\pi g'(r\cos(\alpha - \theta),\theta)\, d\theta,$$

where $g'(\rho,\theta) = g(\rho,\theta) * h_\theta(\rho)$.

This process is illustrated in Fig. 5.5. This technique is attractive because only one dimensional filtering and integration are required. Also note that instead of producing an image $f(r,\alpha)$ in polar coordinates, appropriate rectangular values may be produced easily.

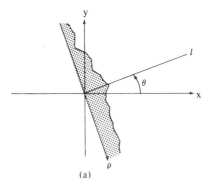

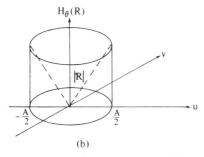

*Figure 5.5*   Geometry for convolution techniques: (a) convolution projection data and (b) rho filter for convolution restoration.

## 5.4   Algebraic Formulations

Let us now consider the numerical evaluation of the reconstruction equations. Although the equations may be computed optically, the increased dynamic range of digital sensors has led to computer implementation in most cases. The numerical evaluation of the basic projection equation

$$g(\rho,\theta) = \int_s f(x,y)\,ds$$

requires several steps. First, the projection data may only be taken at discrete values of $\rho$ and $\theta$, such as

$$g(\rho_m,\theta_n); \qquad m = 1,2,\dots,M, \quad n = 1,2,\dots,N.$$

As the next step, one may start with an integral equation solution of the form

$$f(x,y) = \int \int a(r,s) H(x,y,r,s)\,dr\,ds.$$

If this integral is evaluated numerically, we may obtain an estimate $\hat{f}(x,y)$ of the function that may be expressed as a series expansion

$$\hat{f}(x,y) = \sum_{k=0}^{K-1} \sum_{l=0}^{L-1} a_{kl} H_{kl}(x,y),$$

where the coefficients $a_{kl}$ and the generating functions $H_{kl}(x,y)$ depend upon the technique selected for the reconstruction.

The series expansion estimate should also satisfy the projection equation, which requires

$$g(\rho,\theta) = \int_s \hat{f}(x,y)\,ds = \int_s f(x,y)\,ds$$

or

$$g(\rho,\theta) = \sum_{k=0}^{K-1} \sum_{l=0}^{L-1} a_{kl} \int_s H_{kl}(x,y)\,ds.$$

The coefficients $a_{kl}$ should be selected so that this condition is satisfied; however, this may not always be possible with a finite number of terms. Finally, one may wish to estimate the image only at a discrete set of points to obtain

$$\hat{f}(x_i,y_j), \qquad i=1,2,\ldots,I, \quad j=1,2,\ldots,J.$$

Of course, quantization effects must also be considered.

At this point, one has arrived at a set of matrix equations that may be solved under certain conditions to arrive at the desired estimate. To illustrate the approach and problems encountered in numerical reconstruction, let us again consider the Fourier method.

A reasonable assumption for most reconstruction problems is that the image to be reconstructed is of finite spatial extent, i.e., $f(x,y)$ is zero outside the rectangular region defined by

$$|x| \le \tfrac{1}{2}L_x, \qquad |y| \le \tfrac{1}{2}L_y.$$

Under this assumption, we may develop a series expansion by directly applying the sampling theorem. We may first form the *periodic extension* $f_p(x,y)$ of $f(x,y)$, which is identical to $f(x,y)$ in the rectangle of interest and periodically repeated over the plane. Since $f_p(x,y)$ is periodic, we may describe it by the two-dimensional Fourier series

$$f_p(x,y) = \sum_{m=-\infty}^{\infty} \sum_{n=-\infty}^{\infty} G_{mn} \exp\left[ j2\pi(mx/L_x + ny/L_y) \right],$$

where the Fourier series coefficients are given by

$$G_{mn} = \frac{1}{L_x L_y} \int_{-L_x/2}^{L_x/2} \int_{-L_y/2}^{L_y/2} f(x,y) \exp\left[ -j2\pi(mx/L_x + ny/L_y) \right] dx\,dy.$$

This series expression may be truncated to obtain a finite series expression, but the coefficients must first be determined.

Now, by exactly the same procedure used in the proof of the sampling theorem we may represent the Fourier transform of $f(x,y)$ by cardinal function interpolation of the values on an equispaced grid in the transform domain

$$F(u,v)=L_xL_y\sum_{m=-\infty}^{\infty}\sum_{n=-\infty}^{\infty}G_{mn}\,\mathrm{sinc}\left[L_x\left(\frac{m}{L_x}-u\right)\right]\mathrm{sinc}\left[L_y\left(\frac{n}{L_y}-v\right)\right].$$

The coefficients $G_{mn}$ are equal to the sampled Fourier transform values at points

$$\left(u=m/L_x, v=n/L_y\right).$$

If one could determine these coefficients for some finite range of $m$ and $n$, say, $m=0,1,\ldots,M-1$ and $n=0,1,\ldots,N-1$, then one could use the finite Fourier series representation to compute $f(x,y)$ at any point in the region of interest. Thus, we need to determine the $G_{mn}$ given the Fourier transform points computed from the projections in polar coordinates,

$$F(u_i,v_j); \qquad u_i=R_i\cos\theta_i, \quad i=1,\ldots,I;$$

$$v_j=R_i\sin\theta_j, \quad j=1,\ldots,J.$$

One may now write the finite interpolation equation in equivalent vector form to obtain

$$\mathbf{F}=[W]\mathbf{G},$$

where $\mathbf{F}$ is a column vector with $IJ$ elements, $\mathbf{G}$ is a column vector with $MN$ elements and $[W]$ is a weighting matrix of order $IJ \times MN$. At this point, the problem has been reduced to a form that is identical to that used in linear restoration. Thus one may immediately write the least mean squared or pseudoinverse solution, if it exists, as

$$\mathbf{G}=\left([W]^{\mathrm{T}}[W]\right)^{-1}[W]^{\mathrm{T}}\mathbf{F}.$$

This solution may not be easy to compute. The order of the matrix to be inverted is equal to the number of frequency domain points to be determined. For example, to determine the points on an $80\times80$ grid in the transform domain by a direct solution requires the inversion of a $6400\times6400$ matrix. However, we may conclude that the numerical evaluation of the Fourier transform method may be considered by a series expansion approach.

Another numerical approach is based upon the assumption that $f(x,y)$ is band limited, i.e.,

$$F(u,v)=0 \qquad \text{if} \quad |u|\ge 1/2l_x, \quad |v|\ge 1/2l_y.$$

In this situation $f(x,y)$ may be represented by the cardinal function interpolation

$$f(x,y) = \sum_{m=-\infty}^{\infty} \sum_{n=-\infty}^{\infty} f(ml_x, nl_y) \operatorname{sinc}\left(\frac{x - ml_x}{l_x}\right) \operatorname{sinc}\left(\frac{y - nl_y}{l_y}\right).$$

A truncated form of this series then provides an estimate $\hat{f}(x,y)$ of the desired function,

$$\hat{f}(x,y) = \sum_{m=0}^{M-1} \sum_{n=0}^{N-1} f(ml_x, nl_y) \operatorname{sinc}\left(\frac{x - ml_x}{l_x}\right) \operatorname{sinc}\left(\frac{y - nl_y}{l_y}\right).$$

Integrating this function over a path length gives

$$\hat{g}(\rho,\theta) = \sum_{m=0}^{M-1} \sum_{n=0}^{N-1} f(ml_x, nl_y) \int_s \operatorname{sinc}\left(\frac{x - ml_x}{l_x}\right) \operatorname{sinc}\left(\frac{y - nl_y}{l_y}\right) ds.$$

To evaluate the integral over the path length, let the equation of a ray be given by

$$y = ax + b.$$

It is convenient to relocate the origin at each sample point by defining

$$x' = x - ml_x, \qquad y' = y - nl_y.$$

Then, for a given $\rho$ and $\theta$, the ray equation is

$$y' = ax' + b',$$

where

$$a' = \tan\theta, \qquad b' = \rho \sec\theta + ml_x \tan\theta - nl_y,$$

$$W_{mn}(\rho,\theta) = \int_{-\infty}^{\infty} \operatorname{sinc}\left(\frac{x - ml_x}{l_x}\right) \operatorname{sinc}\left(\frac{y - nl_y}{l_y}\right)(1 + a^2)^{1/2} dx.$$

Further evaluation of this expression leads to the equations for the weighting functions,

$$W_{mn}(\rho,\theta) = \begin{cases} (1 + a^2)^{1/2} l_x \operatorname{sinc}\left[(\rho c + ml_x a - nl_y)/l_y\right] \\ \qquad\qquad \text{for} \quad 0 \le |a| \le l_y/l_x, \\ (1 + a^2)^{1/2} (l_y/|a|) \operatorname{sinc}\left[(\rho c + ml_x a - nl_y)/l_x a\right] \\ \qquad\qquad \text{for} \quad l_y/l_x < |a| < \infty, \\ l_y \operatorname{sinc}\left[(\rho + ml_x)/l_x\right] \qquad \text{for} \quad |a| = \infty, \end{cases}$$

where $a = \tan\theta$ and $c = \sec\theta$. Thus we arrive at a set of algebraic equations

$$g(\rho,\theta) = \sum_{m=0}^{M-1} \sum_{n=0}^{N-1} W_{mn}(\rho,\theta) f(ml_x, nl_y)$$

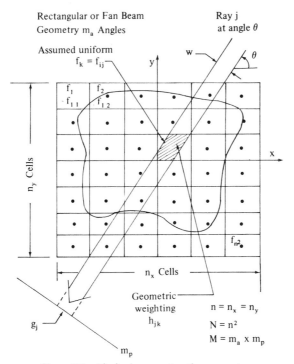

*Figure 5.6*   Algebraic reconstruction geometry.

for which the projection values and weighting functions are known at a set of discrete points and from which the image values at discrete points may be determined. In practice, the weighting function must be computed with a finite ray width, which is only slightly more complicated.

As a final illustration of the development of a series solution, let us assume that the image is represented as an array of rectangular elements as shown in Fig. 5.6 and that within each element the function has a uniform value, say, $\bar{f}(ml_x, nl_y)$. The function at any point $(x,y)$ may then be expressed as

$$\hat{f}(x,y) = \sum_{m=0}^{M-1} \sum_{n=0}^{N-1} \bar{f}(ml_x, nl_y) \operatorname{rect}\left(\frac{x - ml_x}{l_x}\right) \operatorname{rect}\left(\frac{y - nl_y}{l_y}\right).$$

In an analogous manner, this function may be integrated over a ray path, perhaps of finite width, as shown in Fig. 5.6 to determine a set of weighting functions $W'_{mn}(\rho, \theta)$ to develop the linear system of equations

$$g(\rho, \theta) = \sum_{m=0}^{M-1} \sum_{n=0}^{N-1} \bar{f}(ml_x, nl_y) W'_{mn}(\rho, \theta).$$

As illustrated, the numerical solutions of the reconstruction equations may lead to a system of linear equations that must be solved to determine the image. Without going into great detail on this well-established topic, a few points should be mentioned. First, since the sum of all the projection values at any given angle equals the integral of the image function,

$$\int g(\rho,\theta)\, d\rho = \int \int f(x,y)\, dx\, dy,$$

one may easily introduce dependent equations into the system. This situation produces a singular system of equations if the dependent equations are used. Also, it may be possible to impose an objective criterion such as minimum mean squared error and a priori constraints on the set of possible solutions. Finally, one may approach the solution to a set of linear equations with either a direct, iterative, or a combination direct–iterative algorithm (Gordon and Herman, 1973). The size of the computational facility available may make one approach advantageous.

### Reconstruction by Optimization

The reconstruction problem may also be solved by selecting a reasonable criterion function that measures the discrepancy between the actual picture and the reconstructed picture, and developing a solution method that minimizes this criterion function. This transformation of the reconstruction problem to a minimization problem was elegantly described by Kashyap and Mittal (1975) and will now be reviewed. Several *algebraic* approaches have been developed; however, the following description is illustrative.

Let us first introduce a vector notation for the reconstructed picture and projections. Let $\mathbf{f}$ represent the equivalent picture vector formed by stacking the image rows $\mathbf{f}_i$ into a column vector, i.e.,

$$\mathbf{f} = (\mathbf{f}_1, \mathbf{f}_2, \ldots, \mathbf{f}_n)'$$

as shown in Fig. 5.6. This vector is of size $n^2$.

Next, consider the projection ray at angle $\theta_k$ with respect to the horizontal as shown in Fig. 5.7. We shall consider a number $p$ of such projections at angles $\theta_1, \ldots, \theta_p$, respectively. Let $\mathbf{g}_k$ be the vector of projection values at angle $\theta_k$, which for simplicity will also be considered to have $n$ components,

$$\mathbf{g}_k = (g_{k,1}, g_{k,2}, \ldots, g_{k,n})'.$$

We may now define a vector $\mathbf{g}$ of $pn$ components by stacking the projections from each angle,

$$\mathbf{g} = (\mathbf{g}_1, \mathbf{g}_2, \ldots, \mathbf{g}_p)'.$$

The projection values are assumed to be linear combinations of the picture values that may be described by

$$g_{k,l} = \sum_{j \in D_{(k+1)n+l}} \mathbf{f}_j; \qquad k = 1, \ldots, p, \quad l = 1, \ldots, n.$$

The set $D_j$ is made up of all the cells that contribute to projection $\mathbf{g}_j$, as

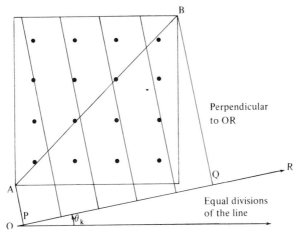

*Figure 5.7*   Projection element assignment at angle $\theta_k$.

shown in Fig. 5.6. Note that a geometrical weighting could also be computed corresponding to the fraction of intersection of the ray of width $w$ and the picture cell for element $f_{ij}$ as shown in Fig. 5.6. For simplicity, this will not be done. Rather an element will be considered to contribute to the projection value $g_{k,l}$ if a line ray at angle $\theta_k$ falls anywhere within the cell $f_{ij}$. Using this assumption, a systematic procedure for obtaining the elements of the set $D_j$ can be followed.

First, for picture element $f_{ij}$ we may consider cell $(i,j)$. This cell may also be addressed as cell $k$, where

$$k = (n-1)i + j.$$

Let the center of the cell be at $(i - \frac{1}{2}, j - \frac{1}{2})$.

Now let PQ be the line projection of the main diagonal AB of the picture as the axis OR at angle $\theta_k$ as shown in Fig. 5.7. The length of PQ is

$$\overline{PQ} = n\sqrt{2}\,\cos(45° - \theta_k).$$

Divide this length into $n$ equal parts, $P_{k,1}, P_{k,2}, \ldots, P_{k,n}$ as shown in Fig. 5.7. The rule for assigning an element to the set $D$ is: if the projection of the center of the cell $(i,j)$ falls within the corresponding increment, then consider that cell to contribute to the projection. The projection of the center of cell $(i,j)$ on the axis OR is given by

$$P_{ij} = \left[ \left(i - \tfrac{1}{2}\right)^2 + \left(j - \tfrac{1}{2}\right)^2 \right] \cos(q_{ij} - \theta_k),$$

where

$$q_{ij} = \tan^{-1}\left[\left(j - \tfrac{1}{2}\right)/\left(i - \tfrac{1}{2}\right)\right].$$

Then, if projection $P_{ij}$ for cell $(i,j)$ satisfies the condition

$$(l-1)\sqrt{2}\,\cos(45° - \theta_k) \le P_{ij} \le l\sqrt{2}\,\cos(45° - \theta_k)$$

for a given value of $l$, then that point is considered to be in the set $D_{(k-1)N+l}$. Essentially, for each projection, the center of each picture element is projected and the above condition used to determine if the picture element contributes to the projection element.

The mathematical problem can now be stated. The set of projection equations can be written

$$\mathbf{g} = \mathbf{Bf}$$

where $\mathbf{B}$ is a binary matrix of size $pn$ by $n^2$ for which

$$B_{ij} = \begin{cases} 1 & \text{if } j \in D_i \\ 0 & \text{otherwise.} \end{cases}$$

Note that this system of equations may have a unique solution if $p = n$, or be underdetermined if $p < n$, and overdetermined if $p > n$. The reconstruction problem is now reduced to the solution of simultaneous linear equations. The selection of a criterion function is the next concern.

One criterion $J_1(\mathbf{f})$ is related to the local smoothness or difference between the intensities in one cell and its local neighbors. $J_1(\mathbf{f})$ will be called a nonuniformity function and may be given by

$$J_1(\mathbf{f}) = \tfrac{1}{2}\mathbf{f'Cf},$$

where $\mathbf{C}$ is the 8-neighbor smoothing matrix described in Chapter 3. The matrix $\mathbf{C}$ is positive semidefinite by construction. Note that $J_1(\mathbf{f}) \ge 0$ and would equal zero for a uniform picture. Therefore, for minimization the solution constraint must also be imposed. This may not lead to a unique solution for the underdetermined system. Therefore, consider the criterion function $J_2(\mathbf{f})$,

$$J_2(\mathbf{f}) = \tfrac{1}{2}\mathbf{f'f} + \alpha J_1(\mathbf{f}).$$

The first term of $J_2(\mathbf{f})$ is related to the "energy" in the picture and may also be related to the sample variance $\sigma^2$, since

$$\sigma^2 = E((\mathbf{f} - \boldsymbol{\mu})'(\mathbf{f} - \boldsymbol{\mu})) = E(\mathbf{f'f}) - \boldsymbol{\mu}'\boldsymbol{\mu},$$

where $\boldsymbol{\mu} = E(\mathbf{f})$. The constant $\alpha$ may be chosen experimentally perhaps to yield the best appearing reconstruction picture.

One reconstruction problem may now be stated as minimizing

$$J_2(\mathbf{f}) = \tfrac{1}{2}\mathbf{f'f} + \alpha\tfrac{1}{2}\mathbf{f'Cf}$$

subject to the constraint $\mathbf{g} = \mathbf{Bf}$. This constrained minimization problem can be solved by introducing a Lagrange multiplier vector $\lambda$ of $np$ components. A new criterion function

$$J_3(\mathbf{f},\lambda) = J_2(\mathbf{f}) + \lambda'(\mathbf{Bf} - \mathbf{g})$$

can be directly minimized. To minimize $J_3(\mathbf{f})$ with respect to $\mathbf{f}$, we compute the derivative

$$\frac{\partial J_3}{\partial \mathbf{f}} = \mathbf{f} + \alpha \mathbf{Cf} + \mathbf{B}'\lambda.$$

Equating this derivative to zero gives

$$(\mathbf{I} + \alpha\mathbf{C})\mathbf{f} = -\mathbf{B}'\lambda.$$

Since $\mathbf{C}$ is a positive semidefinite matrix, $\mathbf{I} + \alpha\mathbf{C}$ is a nonsingular matrix, therefore we can solve for $\mathbf{f}$:

$$\mathbf{f} = -(\mathbf{I} + \alpha\mathbf{C})^{-1}\mathbf{B}'\lambda.$$

The value of $\lambda$ can be determined by first multiplying the above equation by $\mathbf{B}$,

$$\mathbf{Bf} = -\mathbf{B}(\mathbf{I} + \alpha\mathbf{C})^{-1}\mathbf{B}'\lambda,$$

and equating this result to the constraint

$$\mathbf{g} = \mathbf{Bf} = -\mathbf{B}(\mathbf{I} + \alpha\mathbf{C})^{-1}\mathbf{B}'\lambda.$$

Since $\mathbf{B}$ may be nonsingular if $p < n$, a pseudoinverse solution must be used. Thus,

$$\lambda = -\left[\mathbf{B}(\mathbf{I} + \alpha\mathbf{C})^{-1}\mathbf{B}'\right]^{\#}\mathbf{g},$$

where $\#$ indicates the pseudoinverse. The desired solution for the reconstructed picture $\mathbf{f}$ may now be written

$$\mathbf{f} = \mathbf{Fg},$$

where

$$\mathbf{F} = (\mathbf{I} + \alpha\mathbf{C})^{-1}\mathbf{B}'\left[\mathbf{B}(\mathbf{I} + \alpha\mathbf{C})^{-1}B'\right]^{\#}.$$

Note that the matrix $\mathbf{F}$ depends only upon the parameter $\alpha$, the geometry, and the constraints, and therefore could be precomputed.

Since the matrix sizes are very large and inverse or pseudoinverse computations are required, the reconstruction by optimization is not the computationally simplest solution. Various iterative solutions have been developed (Kashyap and Mittal, 1975; Gordon *et al.*, 1970) that are computationally practical. However, perhaps the most valuable aspect of the optimization-type solutions is the unified formulation, the realization that the problem can be solved using objective criteria, and the stressing of importance of the types of criteria and how they affect the solution.

## 5.5   Filter Design for Reconstruction

To illustrate the filter design problem, let us review the steps of the simplest solution, i.e., the convolution algorithm.

1. Collect projection data $g(\rho,\theta)$ and store these data in a rectangular space, i.e., layergram.

2. For a fixed $\theta$, linearly filter the layergram in the $\rho$ direction, i.e., convolve the data with the inverse transform of $|R|$ or

$$h_\theta(\rho) = \int_{-A/2}^{A/2} |R| \exp(j2\pi R\rho)\, dR.$$

3. Compute the back projection by integrating the rho-filtered layergram over $\theta$ at a particular value of $\rho$, i.e.,

$$\rho = r\cos(\alpha - \theta),$$

$$f(x,y) = f(r,\alpha) = \int_0^\pi g'(r\cos(\alpha - \theta),\theta)\, d\theta,$$

where $g'(\rho,\theta) = g(\rho,\theta) * h_\theta(\rho)$.

Therefore, there are three main steps in reconstruction. The first step, data collection, is beyond the scope of this chapter. Suffice it to say that with either a rectangular or fan beam sampling geometry, projection data can be collected and stored in a layergram. The second step, filtering, is most important in terms of the quality of the reconstructed images. The third step, back projection, is an integration that must be computed for each picture element and therefore requires a large amount of computation.

The accuracy of reconstruction depends greatly upon the filter $h_\theta(\rho)$, and several good filters have been proposed. Each approximates the $|R|$ response that is the theoretical but impractical ideal response. The main reasons that the spatial response of the $|R|$ filter is not the practical ideal filter are the lack of convergence for infinite $A$, the fact that $A$ must be finite and for finite $A$ a Gibbs phenomenon occurs, and, finally, the effects of noise are not considered. Several filter designs will now be reviewed.

The spatial impulse response of the filter defined by Ramachandran and Lakshiminarayanan (1971) is given by

$$h_1(0) = \pi/2a^2,$$

$$h_1(ka) = \begin{cases} -2/\pi k^2 a^2 & \text{for } k \text{ an odd integer} \\ 0 & \text{for } k \text{ an even integer.} \end{cases}$$

This filter was used with linear interpolation

$$h(\rho) = h(ka) + \left[(\rho - \rho_k)/a\right]\{h[(k+1)a] - h[ka]\}$$

for $ka \le \rho \le (k+1)a$ and $\rho_k = ka$. The frequency response function of this filter is

$$H(\omega) = |\omega|\operatorname{sinc}^2\left(\tfrac{1}{2}\omega a\right) \qquad \text{for} \quad |\omega| \le \pi/a,$$

where the term $\operatorname{sinc}^2(\tfrac{1}{2}\omega a)$ results from the linear interpolation between samples.

Shepp and Logan (1974) modified the above filter function to

$$h(ka) = -4/\pi a^2(4k^2 - 1) \qquad \text{for} \quad k = 0, \pm 1, \pm 2, \ldots$$

with the same linear interpolation. The corresponding frequency response is

$$H(\omega) = \left| \frac{2}{a}\sin\frac{\omega a}{2} \right| \operatorname{sinc}^2\left( \frac{\omega a}{2} \right).$$

In the presence of noise, Shepp and Logan proposed a noise smoothing filter defined by

$$\bar{h}(\rho_K) = 0.4 h(\rho_K) + 0.3 h(\rho_K + a) + 0.3 h(\rho_K - a).$$

The frequency response of this filter is

$$\overline{H}(\omega) = 0.4 H(\omega) + 0.6 H(\omega)\cos\omega a.$$

A generalized class of filters was introduced by Reed, Kwoh, *et al.* (1977) that includes the previous filters and has design parameters that can be adjusted for desired filter characteristics.

The general class of Reed–Kwoh filters has frequency response without linear interpolation of the form

$$H(\omega) = \begin{cases} a^{-1}|\omega|\exp(-\zeta|\omega|^P), & |\omega| \le \pi/a, \\ a^{-1}|2\pi/a - \omega|\exp(-\zeta|2\pi/a - \omega|^P), & \pi/a < |\omega| \le 2\pi/a, \end{cases}$$

where $\zeta$ is called the damping factor and determines the cutoff frequency, $p$ is a roll-off parameter and determines the sharpness of the filter, and because of the digital nature of the filter, $H(\omega)$ is considered periodic with period $2\pi/a$. Linear interpolation modifies the filter by a factor $a\operatorname{sinc}^2(\tfrac{1}{2}\omega a)$.

For the case $p = 1$, the impulse response of the Reed–Kwoh filter is given by

$$h(0) = (\pi\xi^2)^{-1}\left[ 1 - \exp(-\beta\zeta)(\beta\zeta + 1) \right]$$

and

$$h(ka) = \left[ \pi C_1^2(k) \right]^{-1}\left\{ (-1)^{k+1}\left[ \beta\zeta C_1(k) + C_2(k)\exp(-\beta\zeta) + C_2(k) \right] \right\}$$

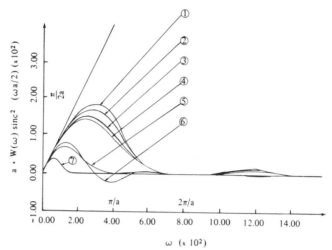

*Figure 5.8* Comparison of filter shapes including the effects of linear interpolation. 1, Ramachandran and Lakshminarayanan filter (generalized filter with $\zeta = 0$); 2, generalized filter, $p = 2, \zeta = 1.0 \times 10^{-6}$; 3, Shepp and Logan filter; 4, generalized filter, $p = 2, \zeta = 3.5 \times 10^{-6}$; 5, generalized filter $p = 2, \zeta = 3.0 \times 10^{-5}$; 6, smoothed Shepp and Logan filter; 7, generalized filter, $p = 2, \zeta = 9.6 \times 10^{-6}$. (From Kwoh, 1977.)

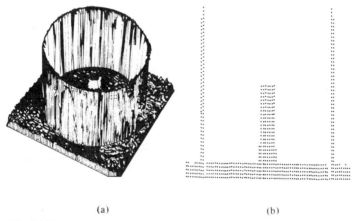

(a)                                                                                                                           (b)

*Figure 5.9* (a) Perspective view of a simulated head image and (b) cross sectional view of head image along the center line. [From Kwoh (1977).]

for $k = \pm 1, \pm 2, \ldots$, where

$$\beta = \pi / a, \qquad C_1(k) = \zeta^2 + a^2 k^2, \qquad C_2(k) = \zeta^2 - a^2 k^2.$$

For $p > 1$, numerical methods may be used to obtain the impulse response.

A comparison of the filter frequency responses is shown in Fig. 5.8.

The effects of the filters on the reconstruction of a simulated head image, shown in Fig. 5.9, using two of the filters are shown in Fig. 5.10. Further results after the addition of random noise are shown in Fig. 5.11.

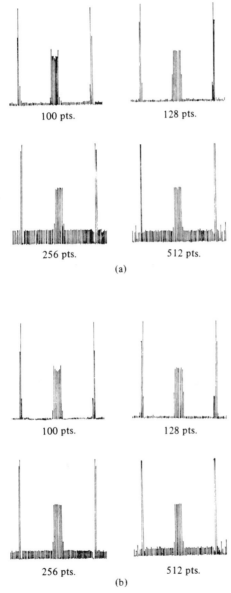

100 pts.          128 pts.

256 pts.          512 pts.

(a)

100 pts.          128 pts.

256 pts.          512 pts.

(b)

*Figure 5.10* (a) Filtering effects of the Ramachandran and Lakshminarayanan filter and (b) filtering effects of the Shepp and Logan filter. [From Kwoh (1977).]

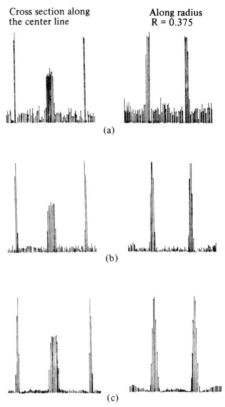

*Figure 5.11* Reconstructions with noise $\sigma = 0.01$ and 256 projection points: (a) reconstruction by Shepp and Logan filter, (b) reconstruction by smoothed Shepp and Logan filter, and (c) reconstruction by the Reed–Kwoh filter with $p = 2.6$ and $\xi = 0.96 \times 10^{-5}$. [From Kwoh (1977).]

## 5.6 Display of Reconstructed Images and Objects

Although the purpose of reconstruction is measurement rather than visualization, the large amount of information in a two-dimensional reconstructed image or a set of these matrices makes the display for visualization necessary and desirable. The nature of the problem, especially in x-ray applications, stresses current display techniques; thus, a discussion of these problems will now be given.

First, suppose one were given a three-dimensional matrix as results from the reconstruction of several slices as illustrated in Fig. 5.1. It is important to note that the instantaneous information contained in such a matrix is *beyond* the capacity of visualization of the human, since at any one instant the human is limited to stereoscopic views of surface characteristics. Since a true three-dimensional solid display device has yet to be invented, a reasonable alternative is to use *time* to represent one of the spatial dimensions, that is, produce a movie for visualization of the three-dimensional reconstructed image. Since a movie is made from a time sequence of two-dimensional images, two-dimensional display methods may be used as the starting point with three-dimensional methods added. The main problem with the movie strategy is that the total information in a single slice cannot be displayed in a single image. Thus, a sequence of images for a single slice may also be required. With these limitations in mind, we shall now first consider the display of reconstructed images, and illustrate the unique characteristics of some of these images, and then the display of reconstructed objects through movies of slices or of surface characteristics.

### 5.6.1  Display of Reconstructed Images

A primary consideration for the display of a reconstructed image is the *information density* of the image. For simplicity, one may consider a square $N \times N$ image matrix in which each element contains $2^M$ distinguishable levels. The total number of bits $T$ required to represent an image is then

$$T = N^2 M,$$

and the number of possible images $L$ that may be displayed is

$$L = 2^T.$$

For example, if $N = 160$, and $M = 10$, then $T = 327,680$ and $L \simeq 10^{100}$. The maximum information that may be contained in each picture element (pixel) is given by the maximum entropy $H$, which is

$$H = \log_2 2^M = M.$$

This, a 1024 level picture can contain 10 bits of information in each pixel. Due to element to element correlation, the actual information may be much less than this maximum. One may estimate the first-order entropy in a picture by computing a normalized histogram for each level $p_i, i = 1, 2, \ldots, 2^M$ and computing the first-order entropy

$$H = -\sum_{i=1}^{2^M} p_i \log_2 p_i.$$

This computation will be made for example images later in the chapter to illustrate the average information content of some reconstructed images.

One should also note that the tradeoff between the number of resolution elements $N$ and the number of bits/pixel $M$ is not linear; however, some psychovisual evidence indicates that for the same image quality, an inverse relationship between $M$ and $N$ must be maintained.

The limitations of the human visual system in gray scale and resolution must also be considered. Although exact values are difficult to specify, it is generally agreed that under optimum viewing conditions, the human can distinguish only a few hundred gray shades, a few thousand colors, and points spaced a few seconds of arc apart. Most viewing conditions are less than optimum.

In reconstructed images, the obtainable spatial resolution may be much less than the human resolution limit; however, the number of distinguishable levels is much greater. This necessitates an interactive capability such as windowing to permit selection of a desirable region of the gray scale for viewing and introduces the possibility of error.

### Monochrome Display

Monochrome display may be accomplished in several ways, such as with a flying spot scanner CRT, a mechanical microdensitometer, a real time TV display, or a half-tone printer. The techniques described in this section apply to all of the devices; however, a real time TV will be used as the appropriate representative display. Linearity, quantization, windowing, and enhancement techniques such as smoothing, sharpening, and high emphasis filtering shall be described.

### Linearity

Linearity is the first factor that should be considered. Given a numerical reconstructed image, what is the desired relationship between the numbers and the brightness of the display? The most fundamental desirable relationship is linear since any desirable nonlinearity is easily derived from a linear system. To obtain a linear relationship between the numbers and perceived brightness, one must consider the viewing situation and the HVS. Two common viewing situations are shown in Fig. 5.12, which illustrates both direct viewing from a CRT and photographic viewing. Because of the power law characteristic of the CRT and the HD curve of the film, the same numerical presentation will not appear linearly related to measured brightness (luminance) in both situations. Furthermore, if the characteristics of the HVS are also considered, the numbers that are linearly related to measured brightness may not be linearly related to perceived brightness.

The important feature for images is that the same numerical presentation *should not* be made for both viewing conditions. The following analysis will show that it is not possible to obtain optimum viewing of the same numerical image in both situations without a correction step. Also, the optimum correction for each case will be derived.

First consider the direct CRT veiwing situation shown in Fig. 5.12a. The light emitted $I$ at a given point on a CRT as a function of the voltage $U$ may

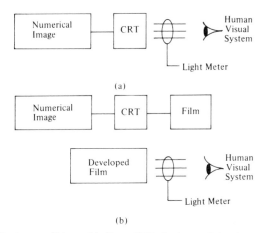

*Figure 5.12*   Viewing conditions: (a) direct CRT viewing and (b) photographic viewing.

be approximated by

$$I = U^\gamma,$$

where $\gamma$ is related to contrast and is similar to the "gamma" of a photographic film. Thus, if a voltage is produced proportional to the image numbers $N$, i.e.,

$$U = kN,$$

the light emitted is related to the numbers by an exponential function

$$I = (kN)^\gamma.$$

On the other hand, if the numbers are precorrected by the inverse power so that the numbers presented $N'$ are given by

$$N' = N^{\gamma^{-1}},$$

then a linear relationship between light emitted and image numbers may be obtained since

$$I = (kN')^\gamma = (kN^{\gamma^{-1}})^\gamma = k^\gamma N.$$

The value of $\gamma$ for a given CRT may be easily measured and a table lookup for the inverse correction developed.

The photographic viewing situation is more complicated because of the film nonlinearity.

Consider the commonly used black and white reflection print, which consists of a reflective backing coated with an emulsion of microscopic grains of silver. The image is formed by controlling the amount of silver in the emulsion and thus varying the relative light absorption of the print, within a typical dynamic range of 50 to 100 : 1. Such a photograph conveys its pictorial information to an observer irrespective of illumination variations over perhaps four to five orders of magnitude. This rather surprising phenomenon is caused by the ability of the visual system to "adapt" to ambient levels of lighting and thus to extract the reflection properties of objects. Studies of the reproduction characteristics of optimal images indicate indeed that although absolute brightness influences perceived quality, the quality criterion within the physical limitations of any given reproduction situation is greatly dependent upon its ability to reproduce relative brightness ratios. This fact is intuitively satisfying, noting that pixel brightness ratios are a property of the scene reflectances that are invariant to the absolute intensity of a uniform illumination.

Exposure of a black and white emulsion to light and subsequent development produces a light absorbing layer characterized by its optical density $D$, which is defined as the logarithm of the ratio of transmitted to incident light. With all other parameters fixed, the optical density is ideally related to the intensity of the exposing light $I$ by the function

$$D = \gamma_1 \log(It),$$

where $t$ is the duration of the exposure. This function, well known in photography, is the Hurter–Driffield or $D$–$\log E$ curve. Actual photographic materials depart from this idealized law at both ends of their useful dynamic range. The factor $\gamma_1$ describes the "contrast" of the emulsion and is positive for an ordinary negative material, negative for a reversal process. Because the unexposed emulsion and its substrate are not perfectly transparent, an additional "fog" level $D_0$ is incorporated into the above equation, yielding

$$D = D_0 + \gamma_1 \log(It).$$

The light reflected from a print $I$ is related to the incident light $I_0$ by

$$I' = I_0 10^{-D}(It) \qquad \text{or} \qquad D = -\log_{10} I/I_0,$$

where $I_0$ is the incident light.

The light produced by the CRT is related to the numbers by

$$I = N^{\gamma_1}.$$

This light, when used to expose a photographic film, produces a density $D$ by the relationship

$$D = \gamma_2 \log It = \gamma_2 \log N^{\gamma_1} t.$$

Finally, when a light of intensity $I_R$ is reflected from a film of density $D$ illuminated by the light $I_0$,

$$D = \log I_R / I_0.$$

Equating the density values gives

$$I_R = I_0 N^{\gamma_1 \gamma_2}.$$

Thus, obtaining a linear relation between reflected light requires that the

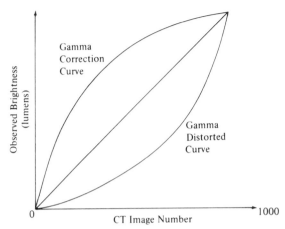

*Figure 5.13*  Compensation for CRT and/or film gamma by predisplay correction.

numbers be precorrected for both the gamma of the CRT and that of the photographic film. If

$$N' = N^{(\gamma_1 \gamma_2)^{-1}},$$

then

$$I_R = I_0 \left[ N^{(\gamma_1 \gamma_2)^{-1}} \right]^{\gamma_1 \gamma_2} \quad \text{or} \quad I_R = I_0 N.$$

Thus, only if the film gamma, $\gamma_2$, is unity can the same correction suffice for both CRT and photographic viewing. The general shapes of the distortion and compensation curves are shown in Fig. 5.13.

### Window Techniques

Since the number of quantization levels retained in a reconstructed image permits the possible visualization of 0.1% detail (1000 levels), and available image viewing devices permit direct viewing of only about 50 levels, a window technique must be used. Since fixing an information window of 50 levels is difficult, an interactive window is required. The amount of information that would be lost is not as large as might be expected, as shown by the histogram in Fig. 5.14. The histogram is an accumulation of image values in each gray level. Note that most of the picture values are clustered about the central region (CR).

Prospective window centers and widths can be selected from the histogram of the original 1000 quantized level computerized tomography (CT) picture. As observed, most of the useful information lies slightly above the central region. For instance, a width of about 30 gray levels extends on either side of the peak. Thus, a window center (CR) of 20 and a total width (WD) of 40 is selected to scale the original CT picture, and the result is demonstrated in Fig. 5.15a.

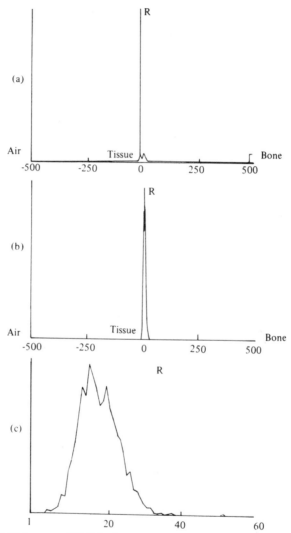

*Figure 5.14* (a) Histogram of EMI brain scan. (b) Histogram of brain area determined by restricting the computation to the region internal to the brain. [Note that this information corresponds to the small peak in the histogram (a).] (c) Internal brain histogram scaled from 1 to 60. The tumor gray level range lies in the region of the peak above 20.

In order to confirm whether or not a good selection of window values was chosen, the following method has experimentally proven to be very effective. After the histogram method gives a good notion about the selection of window center and width, the original CT picture is requantized and displayed. At this juncture, candidate(s) for suspected and identifiable tumor regions can be spotted. An interactive cursor is then moved to this suspected

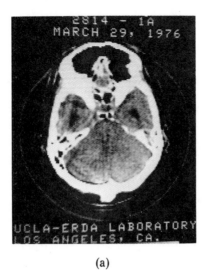

(a)

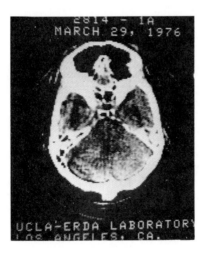

(b)

*Figure 5.15*  Effect of window center and width as determined from the histogram: (a) center of 20, width of 40 and (b) center of 20, width of 20.

location and its gray level (GL) is read. Gray level differentiation of the neighborhood picture elements gives another measure on the new width (GD). The GL is then used as the new window center and the width is 20(2·GD). Again, a requantized picture is displayed as shown in Fig. 5.15b. It can be concluded that the combination of the histogram and interactive utilizations definitely results in a better presentation of the reconstructed image especially for low-contrast detail.

## Pseudocolor Enhancement

The purpose of pseudocolor enhancement is to present the original numerical image in a chromatic manner that increases the perceptibility of important details or increases the speed of object recognition. The technqiue for mapping a single spectral representation into a three-color representation is called "pseudocolor," and has been considered by many authors in several applications. The fundamental reason for considering pseudocolor is that the human visual system contains three color sensitive receptors and consequently has more capacity for distinguishing color information. For example, consider the

*Table 5-1*
*Luminance to Chromaticity Mapping*

| Gray level | Blue | Red | Green | Gray level | Blue | Red | Green |
|---|---|---|---|---|---|---|---|
| 0 | 0 | 0 | 0 | 32 | 0 | 0 | 15 |
| 1 | 15 | 0 | 15 | 33 | 0 | 1 | 15 |
| 2 | 15 | 0 | 14 | 34 | 0 | 2 | 15 |
| 3 | 15 | 0 | 13 | 35 | 0 | 3 | 15 |
| 4 | 15 | 0 | 12 | 36 | 0 | 4 | 15 |
| 5 | 15 | 0 | 11 | 37 | 0 | 5 | 15 |
| 6 | 15 | 0 | 10 | 38 | 0 | 6 | 15 |
| 7 | 15 | 0 | 9 | 39 | 0 | 7 | 15 |
| 8 | 15 | 0 | 8 | 40 | 0 | 8 | 15 |
| 9 | 15 | 0 | 7 | 41 | 0 | 9 | 15 |
| 10 | 15 | 0 | 6 | 42 | 0 | 10 | 15 |
| 11 | 15 | 0 | 5 | 43 | 0 | 11 | 15 |
| 12 | 15 | 0 | 4 | 44 | 0 | 12 | 15 |
| 13 | 15 | 0 | 3 | 45 | 0 | 13 | 15 |
| 14 | 15 | 0 | 2 | 46 | 0 | 14 | 15 |
| 15 | 15 | 0 | 1 | 47 | 0 | 15 | 15 |
| 16 | 15 | 0 | 0 | 48 | 0 | 15 | 14 |
| 17 | 14 | 0 | 1 | 49 | 0 | 15 | 13 |
| 18 | 14 | 0 | 2 | 50 | 0 | 15 | 12 |
| 19 | 13 | 0 | 3 | 51 | 0 | 15 | 11 |
| 20 | 12 | 0 | 4 | 52 | 0 | 15 | 10 |
| 21 | 11 | 0 | 5 | 53 | 0 | 15 | 9 |
| 22 | 10 | 0 | 6 | 54 | 0 | 15 | 8 |
| 23 | 9 | 0 | 7 | 55 | 0 | 15 | 7 |
| 24 | 8 | 0 | 8 | 56 | 0 | 15 | 6 |
| 25 | 7 | 0 | 9 | 57 | 0 | 15 | 5 |
| 26 | 6 | 0 | 10 | 58 | 0 | 15 | 4 |
| 27 | 5 | 0 | 11 | 59 | 0 | 15 | 3 |
| 28 | 4 | 0 | 12 | 60 | 0 | 15 | 2 |
| 29 | 3 | 0 | 13 | 61 | 0 | 15 | 1 |
| 30 | 2 | 0 | 14 | 62 | 0 | 15 | 0 |
| 31 | 1 | 0 | 15 | 63 | 0 | 15 | 0 |

images in Plate VII. A pseudocolor enhancement of the black and white image involves the assignment of a particular chromaticity–luminance combination in the pseudocolor image to correspond to each particular shade of gray in the original black and white image. By using all three visual coordinates, one can transform a small gray scale difference in the black and white original into an easily distinguished color difference in the pseudocolor enhancement. The specified chromaticities may be produced from the luminance values by a mapping procedure such as that shown in Table 5.1.

The pseudocolor mapping from luminosity (one-dimensional space) to chromacity (three-dimensional space) is difficult to define in an optimum manner, not only because of the dimensionality increase but also because of the effects of color perception. For each luminosity value $Y$ a set of three numbers $(r,g,b)$ must be specified to combine the spectral responses and produce a color sensation. The arbitrariness permitted by forcing a mapping from one to three dimensions is both the beauty and the curse of pseudocolor.

Conceptually, a dimensional increase may be made in a completely specified manner if constraints are imposed on the mapping. One such constraint is a curve path. For example, a line may be uniquely mapped into a three-dimensional spiral if distance along the line is made to correspond to distance along the spiral. The most familiar pseudocolor mapping occurring in nature is the rainbow. The curve path of the rainbow shown on the chromaticity diagram is similar to a parabola and is shown in Fig. 5.16. Note that not all colors are visible in the rainbow (purple is absent) but that a large number are present. Another feature of the rainbow mapping is that, since humans are familiar with it, a minimal amount of training is required to specify a scale (e.g., blue–cold, green–medium, red–hot).

Another important constraint that can be imposed on the mapping is related to perceptual distance. A perceptual distance is easily assumed for luminance perception: black is farther from white than from gray. Furthermore, just perceptible distances may be determined along the range. With pseudocolor, the distance must be defined and learned by the observer. The spectral distance is one familiar concept. However, the just perceptible

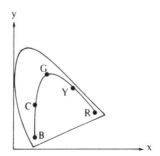

*Figure 5.16*   Curve path specified by the rainbow pseudocolor mapping.

difference can be used to define a chromaticity mapping in which equal lengths along the curve correspond approximately to equal perceptual distances. This concept has lead to the $(u^*, v^*, w^*)$ coordinate system described in Chapter 2.

At the present state of development, the theoretical constraints on curve shape and distance are not sufficient to uniquely define an optimum pseudocolor mapping. Thus, empirical methods must be used to select an appropriate mapping for each application. Fortunately, the mappings require only a table lookup procedure for implementation and several may be stored easily on a computer for experimental use. In fact, each mapping can be rotated in time during display to produce an effective search method.

An experimental procedure that could test the effectiveness of a pseudocolor mapping may be outlined as follows: A set of borderline images should be selected. The larger the number the more accurate the resulting statistics. These images should be mixed with normal images in a random manner. Then the complete set may be viewed by one or more observers and a receiver operating characteristic (ROC) curve developed. The ROC curves for black and white may then be compared to those obtained with pseudocolor and an objective comparison made.

### 5.6.2 Display of Reconstructed Objects

The display of a three-dimensional matrix of reconstructed information is in itself a challenging problem. Several alternatives are available to the designer. One fundamental decision that must be made is to display density or surface information. For most applications, the density information that is produced by the reconstruction algorithm is displayed directly in the slice geometry in which the projection data were collected. The third dimension is displayed by simply presenting a collection of two-dimensional pictures.

It is also possible to display a series of views different from those collected in projection. Since the thickness of slices is usually much greater than the picture element width, a blurred image would result from a direct display; for example, from the sagittal plane. A unique solution to the sagittal plane problem has been developed by Glenn *et al.* (1975) that consists of first deblurring, then displaying the sagittal cross section images.

A different approach is to first detect the surfaces of objects represented by the density information, then display a perspective view of this surface with hidden lines removed and perhaps shaded. This method has been used by Herman and Liu (1977). Due to the development of shaded graphics by Christenson and Stephenson (1976) and others, the display of shaded perspective images can be easily accomplished. An example of a shaded object is shown in Fig. 5.17. Therefore, object detection is the main difficult step. A complete discussion of this topic will be given in Chapter 7.

(a)

(b)

*Figure 5.17*   (a) Real apple before slicing and (b) view of shaded apple. (From Jolley, 1977.)

## 5.7   Application to Medical Radiography

Within the past few years, an entirely new concept in imaging has been developed that is now being generically called computed tomography (Huth and Hall, 1977). It was first applied to diagnostic radiology using transmitted x rays as the information gathering source. Using a transverse-axial motion, this development has been rather ponderously termed computerized transverse axial tomography—CTAT or CAT for short. We will use the simpler term *Transmission Computed Tomography* to encompass any situation in which transmitted radiation is used as the information gathering source. Computed tomography has more recently been applied in nuclear medicine to image positron-emitting isotopes. We will term this emission computed tomography (ECT).

The aim of this section is to give a brief introduction to computed tomography systems and their fundamental principles of operation. It is hoped that this brief review will provide a greater appreciation of the advantages and limitations of computed tomography and help in the proper application and interpretation of this exciting new tool.

Computed tomography began with the introduction to diagnostic radiology of the now famous x-ray transmission scanning systems for the brain, developed by EMI Ltd. in England (Hounsfield, 1971). Subsequently, Ter-Pogossian and Phelps *et al.* (1975) have applied the principles of computed tomography to nuclear medical imaging with their development of tomographic systems for use with isotopes that decay by positron emission. More recently, Budinger *et al.* (1975) have used the principles of computed tomography in conjunction with very high energy, accelerator-produced, heavy ions in reconstructing tomographic images of the ventricles. Each of these three concepts are discussed in more detail below.

### The Basic Aspects of Computed Tomography

Figure 5.18 indicates the basic elements of an x-ray transmission system. In conventional radiography, as illustrated here with a cerebral angiogram, information from the three-dimensional object is superimposed in two dimensional form on film. In some instances, contrast enhancing agents (generally iodinated) are required to regain and obtain diagnostically useful information. Computed tomography is "tomography" in the sense that the image is obtained from a transverse section, or cut, of the object. Within this section, an imaginary image matrix is constructed that is composed of square ele-

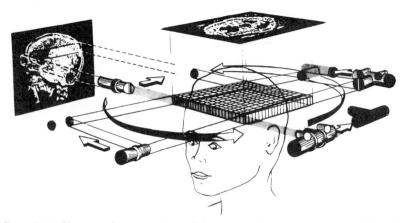

*Figure 5.18* Elements of an x-ray transmission computed tomography system. [From G. C. Huth and E. L. Hall (1977). Computed Tomography and Its Application to Nuclear Medical Imaging. *In* "Current Concepts in Radiology" (E. J. Potchem, ed.), Vol. 3, p. 57. Mosby, St. Louis, Missouri.]

ments of predetermined size. In the x-ray brain imaging system the element size is 1–3 mm square. The matrix must be made sufficiently large to generally encompass the target to be imaged. In the case of the head, therefore, typically 148 elements are used to span approximately 25 cm.

To obtain the transmission data from which to form the image, an aligned x-ray source–detector arrangement is used. This assembly is highly collimated to the width of one image element. The initial transverse part of the scanning motion is a linear motion in which each of the 148 or more rows (each containing 148 or more elements) is interrogated. Thus 148 or more separate data points are entered into the computer during this traverse. After each single traverse, the entire source-detector geometry is rotated a preset angular increment (1° for purposes of discussion here) and the linear, transverse scan motion commences again. If 180 angular projections are used, the result is 180 times 148 or 26,640 separate projections having been entered into the computer. Data in digital form are stored in the computer as the scan is in progress. A very important aspect of x-ray transmission computed tomography is that the information density of the final image is controllable at this point. The EMI scanner has aptly demonstrated that variations in x-ray absorption coefficients can be measured to the order of a few percent. This corresponds to the small differences between fat, muscle, and other tissue. But radiation statistics must be satisfied to achieve this, which means that each data point must contain 40,000–50,000 counts. This number of x-ray photons must therefore exit from the object being scanned and be counted by the x-ray detector.

Radiographic film has a very high information density, while the gamma camera or scanner in nuclear medicine produces an image that has a relatively low value. This is readily apparent to the eye. The computed tomographic image lies midway between the two but additionally possesses a peculiar character. It is composed of relatively coarse blocks (the "elements" discussed above), but the information contained in each block is very well determined as indicated by the statistical argument used above. This information is related to the concept of a "gray scale" and can be displayed in discrete steps of gray. The initial computed reconstruction does not allow the blocks to interact, so there is no "spillover" from one block to another as is introduced to the gamma camera image through the mechanism of photon scatter. The image, therefore, is at first glance deceptive because the unparalleled level of determinations within each block is masked by a coarse overall appearance. It should be noted that this coarseness can be artificially obscured in some image displays.

While the scan is underway, the digital data (representing projections) is being entered into the computer for processing into the final image. We will attempt to give some insight into how the computer accomplishes this task. Comparison with traditional radiographic film tomography is helpful. In that technique, the transverse motion and the x-ray detector of the computed

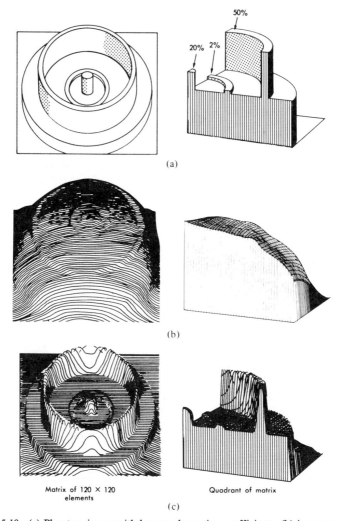

Matrix of 120 × 120          Quadrant of matrix
elements
(c)

*Figure 5.19* (a) Phantom image with known absorption coefficients, (b) image resulting from superposition, and (c) image resulting from deblurring. [From G. C. Huth and E. L. Hall (1977). Computed Tomography and Its Application to Nuclear Medical Imaging. *In* "Current Concepts in Radiology" (E. J. Potchem, ed.), Vol. 3, p. 61. Mosby, St. Louis, Missouri.]

tomography system are replaced by a film cassette. The film must obviously be placed at an angle through the object to accept the x-ray beam, while computed tomography is directly "end-on," but this makes no difference in our argument here. The point is that film possesses only the ability to integrate what it sees. Thus, adjacent components of the image are superimposed (a principle of "superposition"), resulting in the blurred quality characteristic of film tomography.

The following example illustrates the reconstruction process for transmission radiography. A phantom object image was designed with varying absorption coefficient values as shown in Fig. 5.19a (Cho *et al.*, 1975). The concentric ring values were selected to be 20, 2, and 50% above the background. A computer program was used to generate the projections—these projections would in an x-ray transmission system be experimentally generated. A set of such projections was produced for each angle from 0 to 180°. The results were first superimposed to simulate integration. The resulting image is shown in the isometric plot of Fig. 5.19b. Note that the 50% variation and background are visible but the 2 and 20% rings are barely distinguishable. The simple addition of the ray lines produces a barely recognizable and inaccurate representation of the cross section image.

The result of the corrections on this phantom image is shown in Fig. 5.19c. Both weighting factors and the general deblurring function were implemented. Note the increased sharpness of the edges and the ease of visualization of not only the 20% but also the 2% object.

The previous techniques are applicable to transmission systems with rectangular geometries or fan beam geometries, or emission systems with rectangular or ring detectors. It is easily shown that a projection data point collected on one system could also have been collected on a system with a different geometry.

### The Application of Computed Tomography to Imaging Positron Emitting Isotopes

The potential usefulness of positron emitting isotopes in nuclear medicine has been recognized for many years. Much has been written concerning the significance of carbon-11, oxygen-15, and nitrogen-13 (half-lives of 20, 2, and 10 min, respectively) to medical and physiological measurements. However, except for the limited use of fluorine-18 in bone scanning, positron emitting isotopes have not so far been utilized extensively in nuclear medicine.

Positron decay results in the annihilation of the positron and emission of two 0.511-MeV gamma rays at a 180° angle to each other (back-to-back). Gamma ray scatter and absorption is minimized at this high energy. One obstacle to positron emitter usage has been the lack of an imaging system with adequate sensitivity at the required 0.511-MeV energy level. The gamma ray camera system of Anger concept is only a few percent efficient at this energy, losing precious photons and negating the potential advantage of the high specific activity of these isotopes. Brownell *et al.* (1968) have perhaps contributed most to the area of positron imaging with development of a conceptually unique but complex camera system. Ter-Pogossian and Phelps have more recently applied the principles of computed tomography to imaging positron-emitting isotopes in a progression of systems to which they give the name PETT, or "positron emission transaxial tomography."

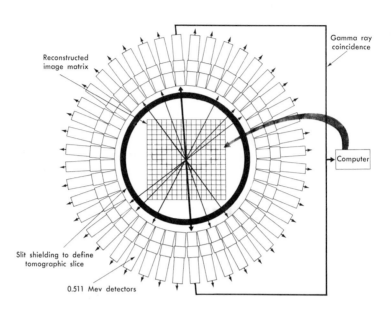

*Figure 5.20*   PETT gamma ray detectors. [From G. C. Huth and E. L. Hall (1977). Computed Tomography and Its Application to Nuclear Medical Imaging. *In* "Current Concepts in Radiology" (E. J. Potchem, ed.), Vol. 3, p. 63. Mosby, St. Louis, Missouri.]

In the PETT concept, gamma ray detectors are arranged around the object to be measured as shown in Fig. 5.20. The shape of the array could take the form of a circle, as proposed by Eriksson *et al.* (1975) and Thompson *et al.* (1975) or a hexagonal shape as preferred by Ter-Pogossian *et al.* (1975). In both configurations, opposing detectors are placed in coincidence to detect the back-to-back gamma rays from positron annihilation. With the detection of each coincidence event, a "line" or projection is effectively drawn to be applied to the eventual image reconstructed by the computer. It is possible to increase the sensitivity of the system by placing each detector in coincidence with a number of opposing detectors, in effect, increasing the number of projections.

Figure 5.21 presents a side-looking view of two opposing detectors of the array. Shielding can be used to define the tomographic plane to be imaged and helps to reject detection of unwanted annihilation events that occur out of this plane. As indicated in Fig. 5.20, coincidence events representing projections enter and are stored in the computer and the image reconstruction proceeds. The computer solution or algorithm used for reconstruction can be very similar to the one used in the x-ray transmission systems. One such approach rearranges the projections to simulate the transverse-axial geometry of the x-ray systems. The solution is then identical.

An important aspect of positron emission tomography derives from the penetration and lower scatter of the 0.511-MeV annihilation gamma rays as

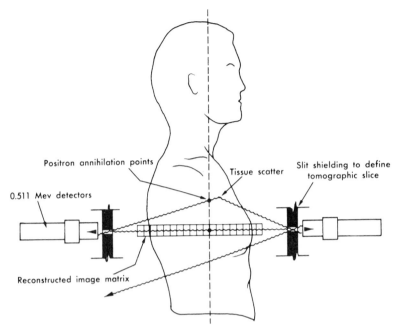

*Figure 5.21* Side view of positron detectors. [From G. C. Huth and E. L. Hall (1977). Computed Tomography and Its Application to Nuclear Medical Imaging. *In* "Current Concepts in Radiology" (E. J. Potchem, ed.), Vol. 3, p. 64. Mosby, St. Louis, Missouri.]

compared to the 0.140-MeV energy of technetium, which is so widely used in nuclear medicine. The higher energy, as alluded to earlier, has in the past been a disadvantage primarily because of the insensitivity of the imaging systems available for use with it. In positron emission tomography the total distance traveled by both gamma rays is independent of the location of the annihilation. This, in conjunction with lower scatter and absorption, results in a "purer" image, i.e., one more independent of depth, as shown in Fig. 5.22. Images of the same deep water phantom are compared using a traditional gamma camera and a 0.140-MeV gamma emitter and the PETT emission tomography system in connection with a 0.511-MeV positron emitter. The depth-independent quality of the 0.511-MeV image is obvious.

The most fully engineered emission tomography system at this time is the PETT III system developed and operated by Ter-Pogossian and his group at Washington University in St. Louis. A view of this system and its control console is shown in Fig. 5.23. The computational facility for reconstructing the image and the image display unit are located in an adjacent room. The PETT III system utilizes 48 sodium iodide gamma ray detectors solving 192 coincidence pairs (each detector is in coincidence with eight detectors opposing it). Some transverse motion is used with each bank of eight detectors being shifted through 3° increments 20 times to obtain more projections. The

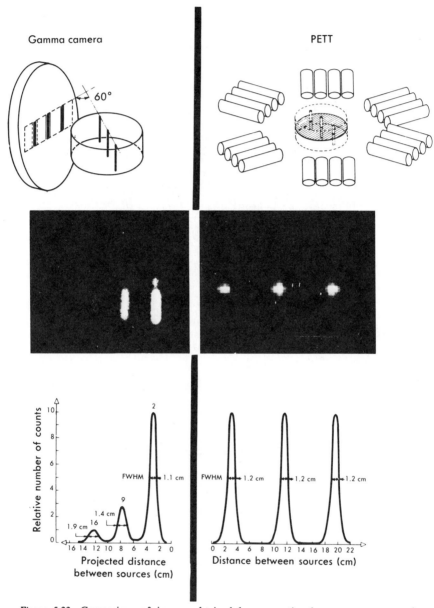

*Figure 5.22* Comparison of images obtained by conventional gamma camera and new positron camera. (From Ter-Pogossian *et al.*, 1975.)

*Figure 5.23*   Positron emission transaxial tomography system. [From Phelps *et al.* (1976).]

system produces high quality images with integration times of 2 to 3 min in the human head, thorax and abdomen with 10–20 mCi of positron activity introduced.

The image reconstructed in the PETT III system is a $50 \times 50$ matrix. Nominal resolution of this system, and most other emission tomography systems under development, is in the 0.8–1.0-cm range. Although the image is somewhat coarser than the x-ray transmission images ($148 \times 148$), this element size is consistent with 0.511-MeV scatter and detection considerations. They have the same quality discussed earlier in regard to the x-ray images, wherein each picture element, although coarse, is statistically well determined. It is simple, as illustrated by the PETT III scans shown below, to eliminate the coarse steps in the image display. A gamma camera image possessing an equal number of events appears smeared by comparison because of scatter— in this case into adjacent, hypothetical picture elements. This tends to mask the fact that the resolutions for both are roughly comparable.

A group of representative images obtained with the PETT III system is shown in Fig. 5.24. The series of tomographic cuts in the top row was taken at the level of the fourth intercostal space. In the "transmission" image (the method of obtaining this is discussed below), the myocardium and rib structure is imaged. The bottom series was taken 2–3 cm lower, imaging a part of the liver in a transmission view. Fifteen mCi ($\pm 20\%$) activity levels were utilized to obtain the emission images. The lower series is particularly interesting. In the $^{13}NH_3$ experiment the myocardial wall is well defined on the right while with liver uptake appears at the left in the image. In the $^{11}CO$ image, the four chambers of the heart are shown—the left ventricle just appearing on the lower right.

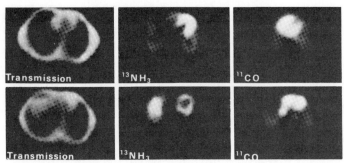

Figure 5.24   Images obtained by positron emission and transmission. [From Phelps *et al.* (1976).]

What are termed "transmission" images are also shown in Fig. 5.24. They are obtained by placing a uniform (solution) ring source inside of the detector array and around the patient to be imaged. Operation of the system is identical to that described earlier except that one of the coincidence gammas travels the entire distance through the patient (thus "transmission") with the other being detected very close to the point of annihilation. In addition to having a potential diagnostic significance, this mode of operation is used to correct the image for differential absorption within the object to be imaged and in standardizing the entire detector array.

### Computed Tomographic Imaging with Heavy Charged Particles

Budinger and his colleagues (1975) have used the principles of computed tomography in connection with beams of heavy charged particles obtained from the Berkeley 184-in. cyclotron to obtain reconstructed images of sections of the brain. These images have an interesting and potentially important

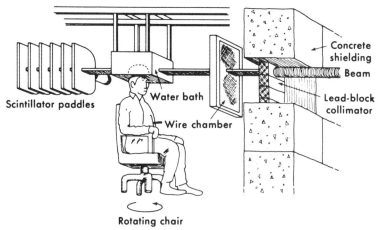

Figure 5.25   Heavy ion tomography system. [From Budinger (1975).]

character. The use of such singular or esoteric sources of radiation should not deter one from gaining an understanding of the potential of the new capability that this experiment could portend.

Heavy charged particles such as protons, helium ions, and even heavier nuclei can be accelerated so that they penetrate through the body. In Budinger's experiment 910-MeV helium ions with a range of 30 cm of tissue were used in transmission through the human head. The fascinating properties of heavy ions that make them attractive as information gathering probes are (a) minimum scatter, and (b) an extremely well defined stopping point in tissue. These properties allow highly accurate determinations of the thickness and electron density of the tissue traversed. In heavy ion "radiography" or computed tomography, this is translated into high contrast and high depth resolution. Information is thus obtained by measurement of the stopping point or residual energy of each ion rather than the attenuation or elimination of photons from a beam characteristic of x-ray tomography.

The heavy ion experimental situation is depicted in Fig. 5.25. The patient is rotated to obtain the axial motion. The residual range of each can be determined by a succession of scintillator "paddles" with the x-ray position (or specific projection coordinate) determined by a multiwire, proportional gas counting chamber. This information, along with the angular coordinate, is fed into the computer and image reconstruction performed.

Figure 5.26 is the helium ion transmission image obtained of a section of the brain. One hundred twenty-eight discrete angular projections of 5000 counts each were determined. In the figure, the resultant image is compared with an EMI x-ray transmission image of the same subject. The interesting point is that, even in this preliminary helium ion image, the expected high definition of the ventricular structure is readily apparent.

In summary, some of the medical applications of computed tomography systems have been presented. While technology in this field is rapidly evolving, these systems have the common feature that the resultant images are entirely a product of the computer. In three systems considered here, images are reconstructed from multiple data points obtained by a radiation probe

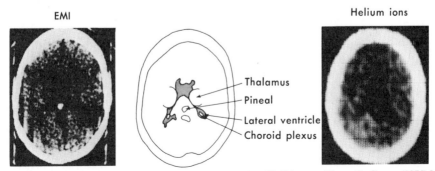

*Figure 5.26*   Helium ion transmission image compared to EMI image. [From Budinger (1975).]

*Table 5.2*

*Characteristics of Computed Tomography Systems[a]*

| Radiation, source energy energy | Mode of operation | Information gathering principle | Primary attributes |
|---|---|---|---|
| X-rays 0.04–0.10 MeV | Transmission | Removal of x-ray photons from transmitted beam by photoelectric and Compton absorption | Noninvasive organ visualization of unparalleled sensitivity |
| Positrons, radioisotopes  Cyclotron—0.511 MeV | Emission | Detection of the back-to-back annihilation gamma rays from positron decay | Ability to utilize the organ specificity of agents labeled with positron-emitting isotopes |
| Heavy ions, accelerator  Cyclotron—910 MeV | Transmission | Measurement of stopping point or residual energy of each ion | Potential ability to visualize abrupt density differences such as tissue–air interfaces; large dose reduction |

[a]From G. C. Huth and E. L. Hall (1977). Computed Tomography and Its Application to Nuclear Medical Imaging. *In* "Current Concepts in Radiology" (E. J. Potchem, ed.), Vol. 3, p. 69. Mosby, St. Louis, Missouri.

unique to the system: (1) transmitted x rays, (2) positron-emitting isotopes, or (3) heavy ions. The bases of operation and primary attributes of each system are summarized in Table 5.2.

There is one profound lesson, we believe, in the discovery and evolution of computed tomography—specifically as applied to x rays and diagnostic radiology. In that field it was very easy to think that the technology was essentially complete and that the probability of any new "turning point" developments was remote indeed. And yet they happened, and an entirely new dimension in extracting information from the x ray was brought about. It seems that the lesson might be that we should search for more such developments in what might have been assumed to be mature technological areas.

## Bibliographical Guide

The earliest consideration of the reconstruction problem is described in a paper by Radon (1917), in which the Radon inversion formula is derived and proved. The simplest theorem of modern reconstruction is that the Fourier transform of a projection of an object is equal to a slice of the Fourier transform of the object. This theorem was first applied to the reconstruction problem by Bracewell (1956, 1960). The x-ray applications were considered by Cormack (1963).

The back projection technique and its application to nuclear medicine are described by Kuhl and Edwards (1968).

The Fourier technique is also described in the text by Papoulis (1969). An optical implementation is described by Garrison *et al.* (1969).

The simpler solution based on the convolution algorithm was described by Bracewell and Riddle (1967). The convolution algorithm was also developed for different fields by DeRosier and Klug (1968), Vainshtein (1971), and Ramachandran and Lakshminarayanan (1971).

The reconstruction problem, with emphasis on Fourier and convolution techniques and applications, is described by Garrison *et al.* (1969), Berry and Gibbs (1970), Tretiak *et al.* (1971), Crowther and Klug (1971), Muehllehner and Wetzel (1972), Smith *et al.* (1973), Mersereau (1973), Mersereau and Oppenhiem (1974), Shepp and Logan (1974), Cho *et al.* (1975), Reed *et al.* (1977), Kwoh (1977), Huth and Hall (1977), and Horn (1978).

The algebraic approach is described by Gordon *et al.* (1970), Bender *et al.* (1970), Frieder and Herman (1971), Bellman *et al.* (1971), Goitein (1972), Herman (1973), Krishnan *et al.* (1973), Gordon and Herman (1973), and Kashyap and Mittal (1975).

Emission techniques are described by Phelps *et al.* (1976), Budinger *et al.* (1975), Eriksson *et al.* (1975), and Thompson *et al.* (1975).

The significance of the machines which followed the patent of Hounsfield (1971, 1973) can be counted in terms of reduced suffering and prolonged human life.

## PROBLEMS

**5.1.** Fan Beam Reconstruction. Consider the reconstruction of an object $f(x, y)$ when the radiation source $S$ emits a diverging fan beam rather than a parallel beam as shown in Fig. 5.27. The projection function $h(s, \theta)$ is the integral of $f(x, y)$ along a ray at angle $\theta$ to the source and at distance $s$ from the origin as shown in Fig. 5.27. Note that a particular ray path of the fan beam source is exactly the ray path of some parallel beam source. Show that the projection function of a parallel beam source $g(s, \theta)$ may be obtained from the projection function of the fan beam source $h(s, \theta)$ by the transformation

$$g(l, \theta) = h\left\{ l / \left[ 1 - (l/D)^2 \right]^{1/2}, \theta - \sin^{-1}(l/D) \right\},$$

where $D$ is the radius of the circle which contains the source $S$.

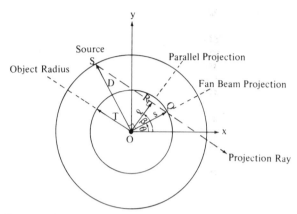

*Figure 5.27* Geometry for comparison of fan beam and parallel beam reconstruction. The object is enclosed in the circle of radius $T$. The source rotates about circle of radius $D$.

**5.2.** Compute the projections of the sampled image given in Fig. 3.22a and perform reconstruction using

(a)  the direct Fourier transform method,
(b)  the convolution method,
(c)  a fan beam source and the convolution method,
(d)  the algebraic method.

**5.3.** Circularly Symmetric and Separable Functions.

(a)  Show that if $f(x,y)$ is circularly symmetric, it may be reconstructed exactly using a single projection.

(b)  Show that if $f(x,y) = g(x)h(y)$, i.e., $f(x,y)$ is product separable, it may be reconstructed exactly using only two projections taken orthogonal to the coordinate axes.

## References

Bellman, S. H., Bender, R., Gordon, R., and Rowe, J. E. (1971). ART is Science, Being a Defense of Algebraic Reconstruction Techniques for Three-Dimensional Electron Microscopy. *J. Theor. Biol.* **32**, 205–216.

Bender, R., Bellman, S. H., and Gordon, R. (1970). ART and the Ribosome: A Preliminary Report on the Three-Dimensional Structure of Individual Ribosomes Determined by an Algebraic Reconstruction Technique. *J. Theor. Biol.* **29**, 483–487.

Berry, M. V., and Gibbs, D. F. (1970). The Interpretation of Optical Projections. *Proc. R. Soc. London, Ser. A* **314**, 143–152.

Bracewell, R. N. (1956). Strip Integration in Radio Astronomy. *Aust. J. Phys.* **9**, 198–217.

Bracewell, R. N. (1960). Two-Dimensional Aerial Smoothing in Radio Astronomy. *Aust. J. Phys.* **9**, No. 3, 297–314.

Bracewell, R. N., and Riddle, A. C. (1967). Inversion of Fan Beams in Radio Astronomy. *Astron. J.* **150** (November), 427–434.

Brownell, G. L., Burnham, C. A., and Silensky, S. (1968). *IAEA Med. Isot. Scintigr.* **96/97**, 163, 174.

Budinger, T. F. (1975). Axial Scanning with 900 Neoalpha Particles. *IEEE Trans. Nucl. Sci.* **22**, 1952.

Budinger, T. F., Crowe, K. M., Cahoon, J. L., Elisher, V. P., Huesman, R. H., and Kanstein, L. L. (1975). Transverse Section Imaging with Heavy Charged Particles: Theory and Application. *Proc. Conf. Image Process. 2-D 3-D Reconstr. Project., Stanford Univ., August.*

Cho, Z. H., Chan, J. K., Hall, E. L., Kruger, R. P., and McCaughey, D. P. (1975). A Comparison of 3-D Reconstruction Algorithms with Reference to Number of Projections and Noise Filtering. *IEEE Trans. Nucl. Sci.* **21**, No. 1, 335–358.

Christenson, H., and Stephenson, M. (1976). MOVIE, BYU: A General Purpose Computer Graphics Display System, Brigham Young Univ. Tech. Rep., December, 1976.

Cormack, A. M. (1963). Representation of a Function By Its Line Integrals, with Some Radiological Applications. *J. Appl. Phys.* **34**, 2722–2727.

Crowther, R. A., and Klug, A. (1971). Art and Science or Conditions for Three-Dimensional Reconstruction from Electron Micrograph Images. *J. Theor. Biol.* **32**, 199–203.

DeRosier, D. J., and Klug, A. (1968). Reconstruction of Three-Dimensional Structures from Electron Micrographs. *Nature (London)* **217**, 130–134.

Eriksson, L., Chan, J. K., and Cho, Z. H. (1975). Emission Tomography with a Circular Ring Transverse Axial Positron Camera. *Proc. Conf. Image Process. 2-D 3-D Reconstr. Project., Stanford Univ., August.*

Frieder, G., and Herman, G. T. (1971). Resolution in Reconstructing Objects from Electron Micrographs. *J. Theor. Biol.* **33**, 189–211.

Garrison, J. B., Grant, D. G., Guier, W. H., and Johns, R. J. (1969). Three-Dimensional Roentgenography. *Am. J. Roentgenol., Radium Ther. Nucl. Med.* **60**, No. 4, 903–908.

Glenn, W. V., Taveras, J. M., Morton, P. E., and Dwyer, S. J., III (1975). Image and Display Techniques for CT Scan Data: Thin Transverse and Reconstructed Coronal and Sagittal Planes, *Invest. Radiol.* **10**, No. 4, 403–416.

Goitein, M. (1972). Three-Dimensional Density Reconstruction from a Series of Two-Dimensional Projections. *Nucl. Instrum. Methods* **101**, 509–518.

Gordon, R., and Herman, G. T. (1971). Reconstruction of Pictures from Their Projections. *Commun. ACM* **14**, 759–768.

Gordon, R., Bender, R., and Herman, G. T. (1970). Algebraic Reconstruction Technique (ART) for Three-Dimensional Electron Microscopy and X-Ray Photography. *J. Theor. Biol.* **29**, 471–481.

Herman, G. T. (1973). Two Direct Methods for Reconstructing Pictures from Their Projections: A Comparative Study. *Comput. Graphics Image Proc.* **1**, 123–144.

Herman, G. T., and Liu, H. K. (1977). Dynamic Boundary Surface Detection, *Proc. Symp. Comput. Aided Diag. Med. Images, San Diego, November, 1976*, pp. 27–32.

Horn, B. K. P. (1978). Density Reconstruction Using Arbitrary Ray-Sampling Schemes. *IEEE Proc.* **66**, No. 5, 551–562.

Hounsfield, G. N. (1971). A Method of and Apparatus for Examination of a Body by Radiation Such as X-Ray or Gamma Radiation. London Patent Specif. No. 1,283,915.

Hounsfield, G. N. (1973). Method and Apparatus for Measuring X or γ-Radiation Absorption or Transmission at Plural Angles and Analyzing the Data. U.S. Patent No. 3,778,614, December 11.

Huth, G. C., and Hall, E. L. (1977). Computed Tomography and Its Application to Nuclear Medical Imaging. *In* "Current Concepts in Radiology" (E. J. Potchem, ed.), Vol. 3, pp. 55–70. Mosby, St. Louis, Missouri.

Jolley, S. P. (1977). An Input System for the Creation of Three-Dimensional Shaded Graphics Pictures. M.S. thesis, Univ. of Tennessee, Knoxville, December (unpublished).

Kashyap, R. L., and Mittal, M. C. (1975). Picture Reconstruction for Projections. *IEEE Trans. Comput.* **24**, No. 9, 915–923.

Krishnan, S., Prabhu, S. S., and Krishnamurthy, E. V. (1973). Probabilistic Reinforcement Algorithms for the Reconstruction of Pictures from Their Projections. *Int. J. Syst. Sci.* **4**, 661–670.

Kuhl, D. E., and Edwards, R. Q. (1968). Reorganizing Data from Transverse Section Scans of the Brain Using Digital Processing. *Radiology* **91**, 975–983.

Kwoh, Y. S. (1977). Application of Finite Field Transforms to Image Processing and X-Ray Reconstruction. Ph.D. thesis, Univ. of Southern California, Los Angeles, January (unpublished).

Mersereau, R. M. (1973). The Digital Reconstruction of Multi-Dimensional Signals from Their Projections. Sc.D. thesis, MIT, Cambridge, Massachusetts, (unpublished).

Mersereau, R. M., and Oppenheim, A. V. (1974). Digital Reconstruction of Multi-Dimensional Signals from Their Projections. *IEEE Proc.* **62**, No. 10, 1313–1338.

Muehllehner, G., and Wetzel, R. A. (1972). Section Imaging by Computer Calculation. *J. Nucl. Med.* **12**, 76–84.

Papoulis, A. (1969). "Systems and Transforms with Applications to Optics." McGraw-Hill, New York.

Phelps, M. E., Hoffman, E. J., Mullani, N. A., Higgins, C. S., and Ter Pogossian, M. M. (1976). Design Considerations for a Positron Emission Transaxial Tomograph (PETT III), *IEEE Trans. Nucl. Sci.* **23**, 516–522.

Radon, J. (1917). Uber die Bestimmung von Funktionen durch ihre Integralwerte langs gewisser Mannigfaltigkelten. *Ber. Verh. K. Saechs. Ges. Wiss., Math.-Phys. Kl.* **69**, 262–279.

Ramachandran, G. N., and Lakshminarayanan, A. V. (1971). Three-Dimensional Reconstruction from Radiographs and Electron Micrographs: Application of Convolutions Instead of Fourier Transforms. *Proc. Natl. Acad. Sci. U.S.A.* **68**, 2236–2240.

Reed, I. S., Kwoh, Y. S., Truong, T. K., and Hall, E. L. (1977). X-Ray Reconstruction by Finite Field Transforms. *IEEE Trans. Nucl. Sci.* **24**, No. 1, 843–849.

Shepp, L. A., and Logan, B. F. (1974). The Fourier Reconstruction of a Head Section. *IEEE Trans. Nucl. Sci.* **21**, 21–43.

Smith, P. R., Peters, T. M., and Bates, R. H. T. (1973). Image Reconstruction from Finite Numbers of Projections. *J. Phys. A* **6**, 361–382.

Ter-Pogossian, M. M., Phelps, M. E., Hoffman, E. J., and Mullani, N. A. (1975). A Positron-Emission Transaxial Tomography for Nuclear Imaging (PETT). *Radiology* **114**, 89.

Thompson, C. J., Yamamoto, Y. L., and Meyer, E. (1975). Reconstruction of Images from a Multiple Detector Ring. *Proc. Conf. Image Process. 2-D 3-D Reconstr. Project., Stanford Univ., August.*

Tretiak, O., Ozonoff, D., Klopping, J., and Eden, M. (1971). Calculation of Internal Structure from Multiple Radiograms. *Proc. Two-Dimensional Digital Signal Process. Conf., Univ. Missouri, Columbia* pp. 6-2-1–6-2-3.

Vainshtein, B. K. (1971). Finding the Structure of Objects from Projections. *Sov. Phys.—Crystallogr.* **15**, No. 5, 781–787.

# 6 | Digital Television, Encoding, and Data Compression

## 6.1 Introduction

The digital transmission of pictorial information was first accomplished with telegraph equipment over 50 years ago. The current importance of and interest in this area are reflected in the large number of publications and special issues of journals directed to image coding. Furthermore, with the advent of microcomputers, high-speed A/D and D/A converters, and decreasing memory cost, more complex image compression techniques can now be considered feasible.

The transmission and storage of pictorial information is of practical importance in diverse application areas, including digital television, picturephones, space exploration, aerial reconnaissance, remote sensing, and biomedical and industrial quality control. In certain applications, particularly television in the US, the amount of existing analog equipment is so enormous that an immediate conversion to digital transmission is unlikely. For other applications, such as space exploration, digital image transmission is mandatory because of the desired error rates and limited power capability of a spacecraft. For the majority of applications a combination of digital methods with analog and optical methods may be employed to achieve efficient, high-quality image communications. However, we shall concentrate on digital coding techniques in this chapter, mainly because the trend in modern communications systems is toward the more flexible and controllable digital system.

The central problem in image communication is channel capacity or achieving storage reduction while maintaining an acceptable fidelity or image quality. The possibility of a bandwidth reduction is indicated by two observations. First, there is a large amount of statistical redundancy or correlation in normal images. For example, two points that are spatially close together tend to have nearly the same brightness level. Second, there is a large amount of psychovisual redundancy in most images. That is, a certain amount of

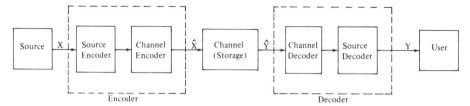

*Figure 6.1*   Communication system model.

information is irrelevant and may be eliminated without causing a loss in subjective image quality, and a large amount may be removed without causing a complete loss of detail. Many methods have been devised that reduce the channel capacity or storage requirement by removing the redundant and irrelevant information from the images, and several of these will now be considered.

The general digital communications system model shown in Fig. 6.1 consists of an information source, encoder, channel, decoder, and an information user. It is assumed that the information source generates a sequence of discrete symbols taken from a finite set. That is, the images must be sampled and quantized for transmission. The encoder is a device that transforms the source output into a form that may be transmitted over the channel. Errors occur during transmission over the channel. Then the decoder converts the channel output into a form suitable for interpretation by the user. The usual communication problem assumes that the source, channel, and user are given and cannot be altered, so the communication system designer must construct the encoder and decoder to satisfy the rate and error criteria desired of the system.

A physical implementation of a digital–analog communication might consist of the following steps. The information source may first be filtered to ensure that the signal is band limited and to reduce high-frequency noise. Next, the signal is sampled and quantized to provide a pulse code modulated signal. Certain encoding may then be performed to reduce the psychovisual or statistical redundancy. The signal is then encoded for reduction of channel errors and transformed by the modulator into a form suitable for transmission over the channel. The signal received from the channel is demodulated into base band signals, decoded for channel errors and the psychovisual and statistical redundancy is restored. Finally, an equalizing or enhancing two-dimensional filtering may be performed before the signal is received by the user.

A simple but useful model for the digital image communication system is shown in Fig. 6.2. The source symbols $M$ may be thought of as sampled and quantized picture elements that require $l$ bit representations. The source encoder making use of a psychovisual or statistical scheme may reduce the number of bits required to represent each picture element and thus produce $k$-bit binary sequences. The channel encoder may add $n - k$ check bits to

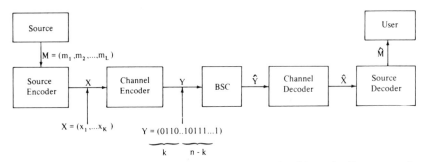

*Figure 6.2* Simple model for communication system using binary $(n, K)$ group code and binary symmetric channel.

each code word in order to implement an error detection or correction at the decoder. The binary symmetric channel adds noise independently to each bit of the transmitted code world $Y$. The channel decoder uses the redundant $n - k$ check bits to implement an error detection or correction and produces a $k$-bit estimated received symbol $\hat{X}$. The source decoder attempts to compensate for the removed psychovisual or statistical redundancy and produces a $j$-bit reconstructed picture element $\hat{M}$.

### 6.1.1 Pulse Code Modulation for Digital TV

Digital TV is an exciting application for pictorial data compression and several major research groups are now engaged in this challenging area (Goldberg, 1976). This application will be used to illustrate the trade-off between analog and digital transmission.

In digital television, the analog picture and sound signals are sampled, quantized, and transmitted using pulse code modulation (PCM). The three major steps—sample and hold, quantize, and encode—are shown in Fig. 6.3. The sampling frequency $F_S$ must be greater than twice the bandwidth $W$ of the signal to prevent aliasing.

Also, $F_S$ should be at least 20% high to compensate for the responses of the low-pass filters before and after the codex (coder/decoder) units. Thus, if $W = 4.3$ MHz for the National Television Standards Committee (NTSC) signals, then $F_S = 10.3$ MHz. Furthermore, to prevent a visible beat frequency with the color subcarrier frequency, it is desirable to have $F_S$ harmonically related to the carrier frequency $F_C = 3.58$ MHz. The third and fourth harmonics provide rates of 10.74 and 14.32 MHz, which are above the Nyquist rate. The broadcasting industry is currently examining whether the $3F_C$ or $4F_C$ encoding rate is better.

It is interesting to consider the rates required for digital television using the NTSC rates. A similar analysis could be made using the parameters of the Phase Alternating Line (PAL) system in the United Kingdom, Western

Table 6.1

International Commercial Television System Parameters[a]

| System | Region or country | Scan lines/frame (Picture elements/frame) | Frame rate | Nominal picture bandwidth | Color transformation and subcarrier frequency |
|---|---|---|---|---|---|
| National Television System Committee (NTSC) | North America South America Japan | 525 (130,000) | Full frame at 30/sec with interlaced frames at 60/sec. | 4.3 MHz | 3.58 |
| Phase Alternating Line (PAL) | United Kingdom Western Europe South America Africa Australia | 405–625 (130,000–210,000) | Full frame at 25/sec with interlaced frames at 50/sec. (South America is 60/sec.) | 3–6 MHz | 4.3 |
| Sequential Couleur à Memoire (SECAM) | France Eastern Europe Russia | 625–819 (210,000–440,000) | Full frame at 25/sec with interlaced frames at 50/sec. | 6–10.4 MHz | 4.3 |

[a]Bandwidth and color subcarrier frequencies are approximate.

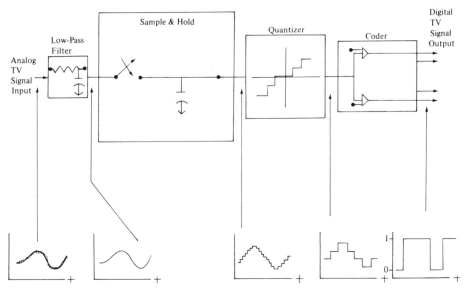

*Figure 6.3* Conversion of analog TV to digital PCM.

Europe, South America, and Africa or the Sequential Couleur à Memoire (SECAM) system in France, Eastern Europe, and Russia with parameters as shown in Table 6.1.

Suppose the NTSC signal bandwidth of 4.3 MHz is A/D converted into 8-bit or 256-level signals. The sampling rate will be chosen to be the third harmonic of the color subcarrier frequency, which is 10.74 MHz. This gives a binary rate of 86 Mbits/sec. Using the correspondence between sequency and frequency, the binary rate of 86 Mbits/sec corresponds to the base bandwidth of the analog signal of 4.3 MHz and indicates that a data compression of 20:1 is required for competitive digital transmission. Circuit costs also depend on signal-to-noise ratio or equivalently the cost of channel information capacity in bits/second. Therefore, the error rate must also be considered. The possibility of modifying the binary codes to reduce the bandwidth at the expense of the bit error rate will now be explored. The processing that is currently used in digital TV consists of an analog-to-digital and digital-to-analog combination performed by a codex circuit. Thus, although 6- or 7-bit PCM encoding may produce acceptable quality initial pictures, 8-bit PCM encoding is recommended to ensure an acceptable signal-to-error ratio (SER) if three or four codex units are used in the circuit.

The full tradeoff between analog and digital TV depends on both bandwidth and signal-to-noise ratio. An analog TV signal requires one 4.3-MHz channel with a SER of 55 dB. The digital TV signal requires one wide band 43-MHz channel or ten multiplexed narrow band channels of 4.3 MHz each to accommodate the 86-Mbit/sec rate. However, as noted, error rate is a

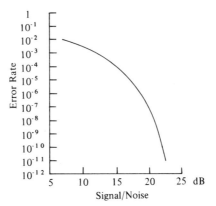

**Figure 6.4**   Error rate versus signal-to-noise ratio for digital transmission. [From Goldberg, (1976). Reprinted from *Electronics*, Feb. 5, 1976; Copyright © McGraw-Hill, Inc., 1976.]

fundamental factor. An error rate of $10^{-8}$ or one bit error in 100 million transmitted bits is practically undetectable in a digital picture. Channel error rates of $10^{-5}$ still permit acceptable pictures, especially if error correction techniques are used.

The selection of sampling word size and rate are a primary consideration in digital TV. The code word size must ensure not only undetectable initial quantization errors, but also provide a "guard band" to further processing errors. An 8-bit PCM television signal differs from the original analog signal by a maximum error of $\pm\frac{1}{2}$ the least significant bit (LSB), or $\pm 0.2\%$. This "fine" quantization would appear as white noise if viewed as a picture. The root mean squared value of this quantization noise is approximately equal to $(Q/12)^{1/2}$, where $Q = 256$ is the number of levels, and a uniformly distributed error is assumed. The equivalent SER of a PCM signal is the ratio of the peak-to-peak value of the signal to the RMS quantization error. For codes with length $n$ of 4 or more bits, the SER is given by

$$SER = 6.02n + 18.8 \quad \text{dB}.$$

Theoretically, the SER with 8 bits is 59 dB and for each 1 bit reduction in quantization, the SER is reduced 6 dB. The actual SER of a composite PCM color TV signal is about 4 dB less because of bandwidth limiting of the quantization error, safety factors in the quantization range, and the way the NTSC signals are measured. Thus, 8-bit PCM encoding of a noise-free NTSC composite color signal yields a SER of 55 dB. A bit error rate of $10^{-8}$ is practically undetectable. From Fig. 6.4, this requires a SER of only 21 dB. If the rate of $10^{-5}$ and error correction bits are added, a SER of 18 dB may be sufficient. Therefore, for comparable picture quality, the SER need not exceed 18 dB. This requires less than 1 bit/picture element. The essential problem in digital TV coding is therefore to reduce the bandwidth at the expense of the bit error rate and retain acceptable picture quality.

A comparable trade-off for the experimental US Picturephone and UK Viewphone would be more favorable for digital techniques since the nominal bandwidth on these systems is only 1.0 MHz (Pearson, 1975).

The techniques for digital image coding may be classified in several ways. In terms of photometric content and complexity, the image could contain half-tone or binary graphics, multi-gray-level monochrome, color, or multi-spectral data. The techniques may be quite different for each type. Processing methods for PCM information may also be divided into reversible and nonreversible techniques. Reversible techniques may permit a data compression of 2:1, but often require a variable code word length. If a certain amount of error is permitted, nonreversible methods may be used and higher compression rates of 10:1 are possible. Implementation also provides another classification basis, since the techniques may be based on line, frame, or interframe processing.

In this chapter, we shall assume that a pulse code modulation system is used for sampling and quantizing the picture information and consider the digital gray level values as messages to be transmitted over a communication system.

The desire to reduce storage or channel capacity for digital image information has led to several approaches that reduce the redundancy in these images. Two distinct image communication problems are considered. For reversible coding, methods are sought that reproduce the sampled, quantized image information exactly. The noiseless coding theorem indicates that the amount of redundancy reduction that can be achieved is bounded by the source entropy. Thus, methods that reduce the source entropy and codes that have rates close to the lower bound are described.

The second problem considered assumes that a certain amount of distortion due to coding is permissible. The statistical rate distortion theory provides guidelines for transmitting images under this constraint. However, difficulties with the practical implementation of this theory are encountered because no known objective distortion measure has been found that always agrees with human subjective judgment of image quality. For example, noise may either decrease or increase the subjective quality. If a human observer must judge the quality of a received image, then properties of the human visual system may be used to develop more efficient coding methods. A wide variety of techniques are available that exploit statistical and psychovisual redundancies to provide efficient communications. Techniques for the coding of movies and color information will also be considered.

## 6.1.2 Information Theory Guidelines

A picture is worth a thousand words. Information theory provides a mathematical method for investigating this adage as well as important insight

*Figure 6.5*   Single resolution cell picture.

into the performance that can be expected when transmitting pictorial information. Several concepts of the Shannon theory (Shannon and Weaver, 1949) will now be reviewed in the context of transmitting pictures by pulse code modulation. An excellent introduction to the general information theory may be found in Abramson (1963). The concepts of a measure of information, entropy, information rate, channel capacity, and the rate distortion function for picture transmission and compression will now be described.

### Measure of Information

The fundamental concept of Shannon's theory is that statistical information may be quantitatively measured in a manner that agrees with intuition. Only statistical uncertainty rather than semantic or structural content is considered. However, in viewing an image or receiving a message, the more the uncertainty is reduced, the greater the amount of information received by the observer. Consider the simplest picture that contains a single resolution cell, as shown in Fig. 6.5. If it is certain that the gray level of the cell is always black, then no information can be conveyed. However, suppose that the gray level could take on one of two possible values such as $m_0 = 0$, or black, and $m_1 = 1$, or white, and that the probability of each level is

$$\text{Prob}(0) = p, \qquad \text{Prob}(1) = 1 - p.$$

With this introduction of randomness, one may ask the question: How much information is obtained by observing the picture? Suppose $p = 0.99$ or that we know a priori that the likelihood of observing a black picture is 99%. If one now observes the picture and it is black, one has obtained some information. If one observes the picture and it is white, one has obtained a greater amount of information, since this event was more unlikely. To measure the amount of information, first generalize the situation so that any one of $M$ gray levels $m_1, m_2, \ldots, m_M$ could occur with a priori probabilities $\text{Prob}(m_i) = p_i$ and

$$\sum_{i=1}^{M} p_i = 1.$$

The information measure should be a function of the probability of occurrence of the event.

One may now postulate that an information measuring function $h$ should satisfy the following conditions:

(1)  $h(p)$ is continuous for $0 < p < 1$;
(2)  $h(p) = \infty$ if $p = 0$;
(3)  $h(p) = 0$ if $p = 1$;
(4)  $h(p_2) > h(p_1)$ if $p_1 > p_2$.

If another picture independent of the first is jointly considered with gray level values $n_1, n_2, \ldots, n_N$ and probabilities $q_1, q_2, \ldots, q_N$, then an additional property is additivity.

(5)  $h(p_i, q_j) = h(p_i) + h(q_j)$.

It may be shown that these postulates can be satisfied if and only if

$$h(p) = -\log_b p.$$

The logarithm base $b$ is chosen so that one unit of information is generated by an elementary choice, or for convenience. The most common elementary choice involves picking one of two equally likely states, i.e., $b = 2$. The unit of information for this base is called the *bit*. For a set of $2^n$ equally likely levels, the information may be transmitted as an $n$-digit binary number. The amount of self-information contained in each message is then:

$$h(p) = -\log_2 p = -\log_2(1/2^n) = n \quad \text{bits.}$$

### Entropy of a Source

If there are several possible gray levels, one may measure the *entropy* or average information per level of an element by the average or expected value of the information contained in each possible level. If the picture has $M$ levels $S_i$ each with probability of occurrence $p_i$, then

$$H(M) = \sum_{i=1}^{M} p_i h(p_i) = -\sum_{i=1}^{M} p_i \log_2 p_i.$$

The maximum possible entropy occurs when the levels are equally likely, i.e.,

$$H_{max} = -\sum_{i=1}^{M} \frac{1}{M} \log_2 \frac{1}{M} = \log_2 M.$$

If there are two picture elements $X$ with gray levels $\{x_i\}$, and $Y$ with gray levels $\{y_i\}$, then the average joint information or joint entropy is

$$H(X, Y) = -\sum_{i=1}^{n} \sum_{j=1}^{m} p(x_i, y_j) \log_2 p(x_i, y_j).$$

It is easily shown that this joint information is always less than or equal to the sum of the entropies of the two pictures elements and equal only when the two gray levels are statistically independent.

If one is interested in the average information that may be obtained from viewing one picture element given that he has observed another element, then the conditional entropy is appropriate. The *average conditional entropy* of the element $X$ with respect to $Y$ is defined as

$$H(X/Y) = - \sum_{i=1}^{n} \sum_{j=1}^{m} p(y_j)p(x_i/y_j)\log_2 p(x_i/y_j).$$

Several properties of entropy will now be stated for reference.

1. Entropy is a nonnegative number, i.e., $H(p) \geq 0$ with $H(p) = 0$ only for $p = 0$ or $p = 1$.
2. $H(p_1, p_2, \ldots, p_n) \leq \log_b n$ with equality if and only if $p_i = 1/n$ for $i = 1, \ldots, n$.
3. $H(X, Y) \leq H(X) + H(Y)$ with equality if and only if $X$ and $Y$ are statistically independent.
4. $H(X/Y) = H(X, Y) - H(Y) \leq H(X)$ with equality if and only if $X$ and $Y$ are statistically independent.

As an example, let us consider a binary picture element $X$ with levels $x_1 = 0$ or $x_2 = 1$ and probabilities $p$ and $q = 1 - p$.

The entropy or average information that may be conveyed by this source is

$$H(X) = -p\log_2 p - q\log_2 q$$

This entropy function is shown in Fig. 6.6. It is easily shown that $H(p)$ attains its maximum value of one bit whenever the levels are equally likely.

The average information in a picture can be increased in two ways, by increasing either the spatial resolution or the gray scale resolution. If the number of gray levels is increased from 2 to $2^n$, the maximum entropy of the element increases from 1 to $n$ bits. If the number of resolution elements is increased from 1 to $N$, then the maximum entropy for the binary picture increases from 1 to $N$. Since this maximum entropy occurs if and only if the values are statistically independent, the actual entropy may be considerably less than the maximum.

The entropy of a picture is important because it gives the lower bound on the noise-free transmission rate. The picture elements may be considered as symbols produced by a discrete information source with the gray levels as the states. If successive pixels are independent, the entropy $H$ is simply

$$H = - \sum P_i \log_2 P_i,$$

where $P_i$ is the probability of gray level $i$. Since the pixels are not usually independent, one must consider sequences of pixels to estimate the entropy. The following theorem from Shannon's classic paper (Shannon and Weaver, 1949, p. 55) may be used as the basis of an entropy estimation procedure.

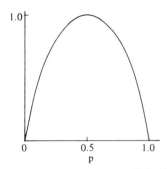

*Figure 6.6* Entropy function for binary source. (From C. E. Shannon and W. Weaver, "The Mathematical Theory of Communication," by permission of the University of Illinois Press, Urbana, and Copyright © 1949 by the University of Illinois.)

**Theorem.** Let $p(B_i)$ be the probability of a sequence $B_i$ of picture elements. Let

$$H_N = -\frac{1}{N}\sum_i p(B_i)\log_2 p(B_i)$$

when the sum is over all sequences $B_i$ containing $N$ picture elements. Then $H_N$ is a monotonic decreasing function of $N$ and

$$\lim_{N\to\infty} H_N = H.$$

As an example of the sequence estimation of entropy, consider the digital image shown in Fig. 2.24. The eight possible gray levels, the relative frequencies, and the empirical entropy estimate are shown in Table 6.2, which shows that $H_1 = 1.853$. To estimate the second-order entropy, the picture is assumed to be raster scanned with a wraparound from line to line and from the end to the beginning. The pairs of gray levels, their relative frequencies and the corresponding estimate of $H_2 = 1.533$ are shown in Table 6.2. The results of this procedure for several other sequence lengths are shown in Fig. 6.7. Note that the sequence of estimates is monotonic decreasing and that the value of $H_8 = 0.659$ is considerably smaller than $H_1$.

A measure that is important in transmitting pictures over a communication system is *mutual information*, which is defined by

$$I(X, Y) = H(X) - H(X/Y) = H(Y) - H(Y/X).$$

The mutual information may be interpreted as the information transmitted over a communication channel, since $H(X)$ is the information at the input of the channel and equivocation $H(X/Y)$ is the information about the input given that the transmitted picture $Y$ is known. Note that if $Y$ is totally correlated with $X$, then $H(X/Y) = 0$ and $I(X/Y) = H(X)$. However, if $Y$ is independent of $X$, then $H(X/Y) = H(X)$ and $I(X/Y) = 0$.

Suppose one is interested in transmitting a picture over a binary communication channel. Some method must be used to encode the gray level

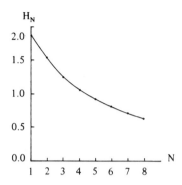

*Figure 6.7* Sequential estimation of picture entropy for picture in Fig. 3.21a.

*Table 6.2*

*Entropy Sequences Computed for the Digital Image of Fig. 3.21*

| First-Order Estimate | | | |
|---|---|---|---|
| Gray level | Frequency | Probability | Entropy |
| 0 | 8 | 0.125 | 0.375 |
| 1 | 0 | 0 | 0 |
| 2 | 0 | 0 | 0 |
| 3 | 0 | 0 | 0 |
| 4 | 31 | 0.484 | 0.507 |
| 5 | 16 | 0.250 | 0.500 |
| 6 | 8 | 0.125 | 0.375 |
| 7 | 1 | 0.016 | 0.096 |
| Total | 64 | 1.000 | 1.853 |

$H_1 = 1.853$

Second-Order Estimate

| Gray level Pairs | Frequency | Probability | Entropy |
|---|---|---|---|
| (4,4) | 18 | 0.2813 | 0.515 |
| (4,0) | 8 | 0.1250 | 0.375 |
| (0,4) | 8 | 0.1250 | 0.375 |
| (4,5) | 5 | 0.0781 | 0.287 |
| (5,4) | 5 | 0.0781 | 0.287 |
| (5,6) | 3 | 0.0469 | 0.207 |
| (6,5) | 3 | 0.0469 | 0.207 |
| (6,7) | 1 | 0.0156 | 0.094 |
| (7,6) | 1 | 0.0156 | 0.094 |
| (6,6) | 4 | 0.0625 | 0.250 |
| (5,5) | 8 | 0.1250 | 0.375 |
| Total | 64 | 1.0000 | 3.066 |

$H_2 = \frac{1}{2}(3.066) = 1.533$

information into a binary code suited to the channel. For a digital PCM picture that has been sampled into an $L$ by $L$ array of picture elements with $B$ bits/element, a total of $L^2B$ bits must be transmitted. Typical sizes of $L = 512$ and $B = 8$ require about 4 million bits. This large number of bits motivates one to consider data compression techniques and wonder if it is possible to reduce this requirement and by how much. The answer to this question is given by Shannon's first theorem. Before introducing the theorem, it is helpful to introduce some concepts about coding.

A *code* is mapping of words from a source alphabet into the words of a code alphabet. Coding thus consists of mapping source-symbol sequences into code-symbol sequences. A *word* is a finite sequence of symbols from an alphabet. For example, if the alphabet is $A = 0, 1$, then $0, 001, 1011, 11111$, etc., are words. The decoding of an original message is not ensured by the code definition, i.e., two source symbols could be mapped into the same code word. A code is called *distinct* if each code word is distinguishable from the other code words. Even with a distinct code, a problem of recovering the original message can occur when the words are placed together without commas. For example, the code words $1, 01, 0$ cannot be distinguished in the sequence $1101001$. Since commas are simply symbols, only comma-free codes are usually considered. A distinct code is *uniquely decodable* if every code word is identifiable when immersed in a sequence of code words.

Another desirable property of a *uniquely* decodable code is that it be decodable on a *word-to-word* basis. This property is ensured if no code word may be formed by adding code symbols to another code word; that is, no code word is a prefix of another. A uniquely decodable code with this prefix property is called *instantaneously* decodable. A code is said to be *optimal* if it is instantaneously decodable and if it has the minimum average length $L$ for a given source and a given probability distribution for the source symbols. The optimal code thus guarantees both the exact recovery of the original information and the minimum average length.

### *Noiseless Coding Theorem for Binary Transmission*

Given a code with an alphabet of 2 symbols and a source $A$ with an alphabet of 2 symbols, the average length of the code words per source symbols may be made arbitrarily close to the lower bound $H(A)$, as desired, by *encoding sequences* of the source symbols rather than individual symbols.

In particular, the average length $L(n)$ of encoded $n$ sequences is bounded by

$$H(A) \leq L(n)/n \leq H(A) + 1/n.$$

This remarkable theorem provides both an upper and lower bound on the average length for any finite value of $n$. The source entropy $H(A)$ may be considerably less than the maximum. Furthermore, the sequence limit of the average length is $H(A)$.

If the entropy of the PCM image source is sufficiently small, it is then possible by suitable encoding to obtain a significant data compression. The entropy of the PCM source is extremely difficult to determine, since the picture elements are statistically dependent, especially point-to-point, line-to-line, and frame-to-frame. Computation of the source entropy requires the symbols to be considered in blocks over which the statistical dependence is negligible. Suppose the dependencies exist over $T$ sec in time, with $F$ frames/sec. Then the sequence length that must be considered is

$$S = TFBL^2.$$

For digital PCM with $T=5$, $F=30$, $B=8$, and $L=512$, $S \simeq 3 \times 10^8$. A joint probability function for a vector length of 300 million is prohibitively difficult to determine experimentally. Thus, our approach must be limited to a single pixel or small blocks of pixels and to overestimates of $H(A)$. If the blocks of pixels are chosen so that the sequence entropy estimates converge rapidly to the limit, then block coding methods may be close to the minimum length. Estimates of entropy have been made that indicate that for most natural scenes the source entropy is less than 1 bit per picture element. Thus, a bit compression ratio from 8 bits/picture element to 0.8 bits/picture element may be possible by using optimal codes and long block sequences. Shannon binary code construction, the Shannon–Fano encoding method, and the Huffman minimum redundancy code procedure may be used to construct optimal noiseless codes.

### Rate Distortion Theory

An information theory guideline for the nonreversible image communication problem is rate-distortion theory. The foundations of this theory were also introduced in Shannon's classic 1949 paper. The book by Berger (1971) is exclusively devoted to this topic. This theory applies to the case in which distortion or noise is introduced by the channel or encoding process.

In the design of a communication system, it is desired to transmit the maximum amount of mutual information; however, this amount is limited by the channel noise. The *channel capacity* $C$ is defined as the maximum over all possible sources of the mutual information and is given by

$$C = \max_{\{p(x)\}} I(X, Y),$$

where $\{p(x)\}$ indicates the set of possible a priori probability distributions.

To illustrate the computation of channel capacity, the simplest binary channel to consider is the binary symmetric channel.

The binary symmetric channel (BSC) is characterized by a single parameter, the bit error probability $p$. The flow diagram for the first-order BSC is shown in Fig. 6.8, in which the conditional probability of an input digit being transformed into an output digit is indicated by the letter above the corresponding flow branch. Since the transition probabilities $p(0/1)=p(1/0)$ are equal, the channel is called symmetric. The channel is characterized equally well by the transition matrix $\pi_2$

$$\pi_2 = \left[ p(x_i/y_j) \right] = \begin{bmatrix} 1-p & p \\ p & 1-p \end{bmatrix},$$

where $x_i$ and $y_j = 0$ or 1.

Since for the BSC it is assumed that the bit errors are independent, the transition matrix for higher-order BSC may be developed using the Kronecker product operation

$$\pi_n = \begin{bmatrix} (1-p)\pi_{n-1} & p\pi_{n-1} \\ p\pi_{n-1} & (1-p)\pi_{n-1} \end{bmatrix}.$$

The flow diagram of $\pi_4$ is also shown in Fig. 6.8.

We may easily calculate the channel capacity of the BSC. The mutual information is

$$I(A,B) = H(B) - H(B/A).$$

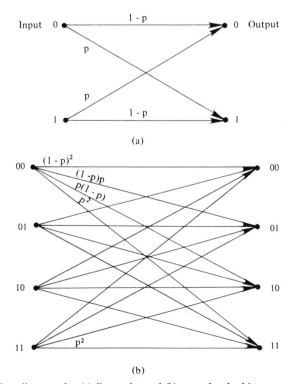

*Figure 6.8*   Flow diagrams for (a) first-order and (b) second-order binary symmetric channel.

The maximum of $H(B)$, as previously developed, is 1 bit/picture element and occurs when the output symbols are equally likely, which also occurs for the BSC when the input symbols are equally likely. The channel equivocation $H(B/A)$ is independent of the input probability distribution and is given by

$$H(B/A) = -p\log p - q\log q, \qquad p+q=1.$$

Therefore, the channel capacity $C$ of the BSC is

$$C = 1 + p\log p + q\log q.$$

We shall now present an introduction to rate distortion theory for the simple situation of a binary memoryless picture source and channel. The main theorem of rate distortion theory states that the suitably coded output of a picture source with entropy $H$ may be transmitted over a channel of capacity $C$, with average distortion less than $D$, provided that the rate $R(D)$, which is a function of the distortion, is less than the channel capacity $C$.

There are several ways to apply the rate distortion theory to the image communication problem. One could encode each digital picture into a single symbol. For an $L$ by $L$ picture array with $B$ bits/picture element, the set of symbols would contain the large but finite set of $S = 2^{L^2 B}$ symbols. Also, a distortion matrix of size $S$ by $S$ that specifies the distortion cost for each pair of transmitted and received pictures must be specified. This approach is not computationally practical; however, smaller block sizes may be feasible. A particularly simple case arises from considering a single picture element as a symbol. If $B$ bits are used to represent gray levels, only $S = 2^B$ symbols must be considered. Furthermore, the distortion cost matrix of size $S$ by $S$ should be reasonable. Some intermediate block size between these extremes may be appropriate for a particular imaging situation.

To illustrate the rate distortion theory, consider a binary picture source with symbol set $A = \{m_1, m_2\}$, where $m_1 = 0$ and $m_2 = 1$, with probabilities $p_1$ and $p_2$ for each symbol, respectively. We shall only consider a special case of a binary memoryless source—the $n$th extension source. The words of the $n$th extension of this source $A^n$ are $n$-bit binary sequences in which successive letters are independent and identically distributed. For every source word $\mathbf{x} = (x_1, x_2, \ldots, x_n)'$,

$$p(\mathbf{x}) = \prod_{i=1}^{n} p(x_i).$$

Now consider the problem of reconstructing the output of the source $A_n$ within a certain prescribed fidelity at the receiver. To determine the prescribed accuracy, we must specify a distortion value for every possible approximation of a source word. Assume that a nonnegative cost function $\rho_n(\mathbf{x}, \mathbf{y})$, called the *word distortion function*, is given that specifies the cost for reproducing the source word $\mathbf{x}$ by the vector $\mathbf{y}$. For any fixed $n$, the word distortion function could be stored in a $2^n$ by $2^n$ matrix.

A sequence of word distortion measures

$$F = \{\rho_n(\mathbf{x}, \mathbf{y}), \ 1 \leq n < \infty\}$$

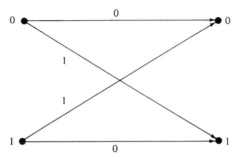

*Figure 6.9*   Single letter distortion measure for binary source and probability of error used as the distortion measure.

is called a fidelity criterion. A fidelity criterion composed of word distortion measures of the form

$$\rho_n(\mathbf{x}, \mathbf{y}) = \frac{1}{n} \sum_{j=1}^{n} \rho(x_j, y_j)$$

is called a single-letter fidelity criterion $F_j$. With a single-letter fidelity criterion, the distortion between a source word and the received word is simply the arithmetic average of the distortion between corresponding letters as specified by a single-letter distortion measure $p$. The single-letter distortion measure $p$ may be represented by an $n \times n$ distortion matrix or as a flow diagram analogous to that used to represent a discrete, memoryless channel. For example, the diagram in Fig. 6.9 illustrates the single letter distortion measure for a binary source when probability of error is used for the error criterion. The lines in Fig. 6.9 are labeled with single letter cost values rather than conditional probabilities of output letters given input letters.

Note that the mean squared error (MSE) between two binary picture elements may be represented by a single letter fidelity criterion

$$\rho_n(\mathbf{x}, \mathbf{y}) = \frac{1}{n} \sum_{j=1}^{n} E\big[(x_j - y_j)^2\big] 2^{j-1}.$$

A least squares single-letter fidelity criterion would be

$$\rho_n(\mathbf{x}, \mathbf{y}) = \frac{1}{n} \sum_{j=1}^{n} (x_j - y_j)^2 2^{j-1}.$$

The Hamming distance could also be used as a single-letter fidelity criterion that would equally weight the location of a bit error

$$\rho_n(\mathbf{x}, \mathbf{y}) = \frac{1}{n} \sum_{j=1}^{n} (x_j - y_j)^2.$$

We will now define the rate distortion function for the binary memoryless source $A^n$, with respect to a single-letter fidelity criterion $F_p$. Any channel may be characterized by conditional probabilities $Q(k/j)$. We may compute the average distortion $d(Q)$ associated with the conditional distribution

$Q(k/j)$ as

$$d(Q) = \sum_{j=1}^{2^n} \sum_{k=1}^{2^n} p(x_j) Q(k/j) p(j,k).$$

A conditional probability assignment $Q(k/j)$ defining a channel is said to be *D-admissable* if and only if $d(Q) \le D$. The set of all $D$-admissable channels or conditional probability assignments is denoted $Q_D$ and defined by

$$Q_D = \{ Q(k/j), d(Q) \le D \}.$$

For each channel or conditional probability assignment, we may also compute the average mutual information $I(Q)$,

$$I(Q) = \sum_{j=1}^{2^n} \sum_{k=1}^{2^n} P(x_j) Q(k/j) \log_2 \frac{Q(k/J)}{Q(k)},$$

where

$$Q(k) = \sum_{j=1}^{2^n} Q(k/j) P(x_j)$$

is the received symbol probability. For $D$ fixed, we define the rate distortion function $R(D)$ by

$$R(D) = \min_{Q \in Q_D} I(Q).$$

The rate distortion function $R(D)$ is thus the minimum over all $D$-admissable channels of the average mutual information. Since the source rather than the channel is given, the minimization is over the admissable channels. The minimization provides the minimum mutual information rate that must be conveyed to the user in order to achieve a prescribed fidelity.

The following theorem is useful for computing the rate distortion function for the discrete memoryless channel.

**Theorem.** The rate distortion function for the source $U = \{0, 1, \ldots, K-1\}$ with a priori probabilities $P(U=j) = P(j)$ and the distortion measure

$$\rho(j,k) = \rho_{jk}$$

is given by

$$R(D) = H(U) - H(D) - D \log(K-1)$$

for

$$D \le (K-1) Q_{min}, \qquad Q_{min} = \min_k Q(k),$$

where

$$Q(k) = \sum_{j=0}^{K-1} Q(k/j) P(j), \qquad H(D) = -D \log D - (1-D) \log(1-D),$$

$$H(U) = -\sum Q(k) \log Q(k).$$

As an example, let us compute $R(D)$ for the binary source and probability of error distortion measure $\rho(j/k) = 1 - \rho_{jk}$.

The minimum possible value that the average distortion can assume is found by letting

$$\rho = \min_k \rho(j/k),$$

then for each $j$ setting $Q(k/j) = 1$ for that $k$ (or one of those $k$) that realizes the minimum. This results in the minimum average distortion value

$$D_{\min} = \sum_{j=1}^{2^n} P(x_j)p_j$$

with the error distortion measure assumed $p_j = 0$, and thus $D_{\min} = 0$.

Let $D_{\max}$ represent the least average distortion that may be achieved when the average mutual information is zero. Since $I(X, Y) = 0$ if and only if $X$ and $Y$ are independent, then $R(D) = 0$ if and only if

$$Q(k/J) = Q(k).$$

The average distortion for this situation is

$$d(Q) = \sum_k Q_k \sum_j P(x_j)p(j/k).$$

The minimum value of $d(Q)$ may thus be obtained by setting $Q_k = 1$ for one of the values for which

$$\sum_j P(x_j)p(j/k)$$

attains its minimum value and $Q_k = 0$ for all others. The value of average distortion is then

$$D_{\max} = \min \sum_j P(x_j)p(j/k).$$

If

$$P(x_1) \leq \tfrac{1}{2},$$

then clearly

$$D_{\max} = P(x_1).$$

Now, since

$$0 \leq I(X, Y) \leq H(x) \leq \log_2 M,$$

then

$$0 \leq R(D) \leq \log_2 M = 1.$$

For $D < 0$, the set $Q_D$ is empty and $R(D)$ is not defined. Thus, $R(D)$ is positive for $0 \leq D < D_{\max}$, and 0 for $D \geq D_{\max}$. Furthermore, for $D = 0$ the channel must be noiseless with $Q(k/j) = \rho_{kj}$ and $Q(k) = P(m_k)$, so

$$R(0) = -P(m_1)\log P(m_1) - P(m_2)\log P(m_2) = H[P(m_1)].$$

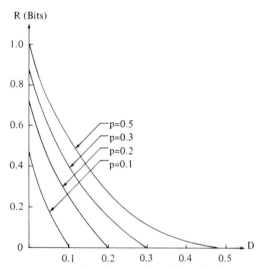

*Figure 6.10* Binary source rate distortion function. (From Toby Berger, "Rate Distortion Theory: A Mathematical Basis for Data Compression" © 1971, pp. 7, 48, 100. Reprinted by permission of Prentice-Hall, Inc., Englewood Cliffs, New Jersey.)

Finally, using the theorem we see that

$$R(D) = H[P(m_1)] - H(D).$$

The rate distortion curves are shown in Fig. 6.10 for several values of $P(m_1)$.

Finally, the important theorem of rate distortion theory specialized to the discrete memoryless channel will now be stated.

***Theorem.*** Given a discrete memoryless source $U$ of alphabet size $K$, entropy $H(U)$, and minimum letter probability $P_{\min}$ that is connected to a destination by a discrete memoryless channel of capacity $C$ bits per source symbol and $D \leq (K-1)Q_{\min}$, then an error probability per source digit of $D$ may always be achieved through appropriate coding if

$$C \geq R(D) = H(U) - H(D) - D\log(K-1),$$

and can never be achieved if $C < R(D)$.

In general, $R(D)$ can be shown to possess the following properties: $R(D)$ is a positive, convex, strictly decreasing function for $0 < D < D_{\max}$. The maximum value is $R(0)$. There exists a value $D_{\max}$ such that $R(D) = 0$ for $D \geq D_{\max}$.

In summary, for a given source and a given distortion measure $D$, we can find a function $R(D)$, such that one may transmit the picture to the user with a distortion as close to $D$ as desired as long as $R(D)$ is less than the channel capacity $C$. As with the other capacity theorems, the rate distortion theory tells one what can be accomplished but does not describe how to accomplish it. Thus, there are several difficulties in trying to apply the rate distortion theory. In the use of a channel or storage, errors are generated. In fact, most channels are characterized to some extent by the probabilities of errors occurring. However, a more complete characterization is given by the channel capacity $C$, which is the maximum error-free transmission rate.

The capacity concept is directly related to the noisy coding theorem, which states that if $H > C$, then any communication system designed to transmit the source information over the channel must lose at least $H - C$ bits of information per source letter. Is it thus possible to reproduce the source at the channel output with fidelity $D$? For at least $R(D)$ bits of information per letter to be received with distortion less than $D$, at most $H - R(D)$ bits may be lost. Thus, it is impossible to obtain an average distortion of $D$ unless $H - R(D)$ exceeds $H - C$, or $R(D) \leq C$. This heuristic argument indicates the intimate link between rate distortion theory and the noisy coding theorem.

Two important points are:

(a)   In order to calculate the rate $R$, one must have a mathematical model (statistical) of the source.

(b)   One also needs a mathematical expression or description of the distortion measure $D$ that agrees reasonably well with subjective human judgments.

## 6.2   Reversible Information Preserving Picture Coding

The noiseless coding theorem indicates that in a binary code for a source with an alphabet of $q$ symbols, the average bit length $b$ may be made arbitrarily close to the lower bound $H(A)$ by encoding sequences of source symbols rather than individual symbols. The digital PCM image that has been sampled into an $L \times L$ array with $B$ bits/pixel may be considered as a sequence of source messages. The messages may be the brightness levels of individual picture elements. Alternatively, the messages may be pairs of neighboring pixels or four-element arrays or first differences of adjacent pixels along each horizontal line. Of the variety of ways in which messages can be chosen, the main requirement in this section is for methods that are reversible. The digital picture should be coded and decoded exactly. Two techniques, Huffman and run-length coding will now be described.

### 6.2.1   Huffman Coding

Consider a particular set of $n$ messages derived from the digital picture $(m_1, m_2, \ldots, m_n)$, and let the probability function of these messages over the appropriate ensemble of digitized pictures be $p_1, p_2, \ldots, p_n$ where

$$p_i \geq 0, \quad i = 1, 2, \ldots, n, \quad \text{and} \quad \sum_{i=1}^{n} p_i = 1.$$

The intuitive idea is to assign short code words to the most probable messages and the longer code words to the least likely messages to obtain a code with a small average length.

The elegantly simple procedure of Huffman (1952) guarantees that one may obtain a uniquely decodable code with the minimum average number of bits per message $R$ with

$$H \leq R \leq H + 1, \qquad \text{where} \quad H = -\sum_{i=1}^{n} p_i \log_2 p_i.$$

The average length, of course, varies with the probability distribution. Fortunately, the differences of adjacent pixels often have nonuniform first-order probability distributions for which the Huffman coding procedure is highly efficient.

As an example of the Huffman coding procedure, consider the simple 8 by 8 image given in Fig. 3.24. Since the gray levels range between 0 and 7, 3 bits/pixel would be required for direct PCM. The gray levels will be considered as messages, and the normalized image histogram will be used to estimate the probabilities. A tabulation of the possible gray levels, the number of occurrences, and the normalized histogram values are given in Table 6.3. Note that levels 1, 2, and 3 do not occur in the image and therefore the normalized histogram values are 0 for those levels. The construction procedure of the Huffman code may be considered in two steps. First, the probabilities are ordered by magnitude, then the smallest two are combined repeatedly until only two remain. The code word assignment is then made by reversing this procedure. At the first step the input probabilities are ordered by magnitude. The zero probability elements may be ignored. The resulting list is

$$0.484, \quad 0.250, \quad 0.125, \quad 0.125, \quad 0.016.$$

Table 6.3

Huffman Code Construction for Image Given in Fig. 3.21a

| Gray level | Number of occurrences | Normalized histogram, $p_i$ | Code word | $-p_i \log_2 p_i$ | $p_i l_i$ |
|---|---|---|---|---|---|
| 0 | 8 | 0.125 | 0000 | 0.375 | 0.5 |
| 1 | 0 | 0.0 | — | — | — |
| 2 | 0 | 0.0 | — | — | — |
| 3 | 0 | 0.0 | — | — | — |
| 4 | 31 | 0.484 | 1 | 0.507 | 0.484 |
| 5 | 16 | 0.250 | 01 | 0.5 | 0.5 |
| 6 | 8 | 0.125 | 001 | 0.375 | 0.375 |
| 7 | 1 | 0.016 | 0001 | 0.095 | 0.064 |
| Total | 64 | 1.000 | | 1.852 | 1.923 |

Combining the two smallest probabilities and recording the list gives

$$0.484, \quad 0.250, \quad 0.141, \quad 0.125.$$

Note that equal probabilities may be combined or ordered arbitrarily. The two smallest probabilities are again combined and reordered, resulting in

$$0.484, \quad 0.266, \quad 0.250.$$

Again the elements are combined, giving only two probabilities

$$0.516, \quad 0.484.$$

The combining procedure is now stopped.

The code word assignment may now be started. First, assign code bits 0 and 1 to the last two element probabilities. Next, reverse the combination procedure assigning a code bit 0 or 1 to each branching path. Note that this procedure corresponds to the binary tree shown in Fig. 6.11. When all combinations have been exhausted, the final code word is obtained by simply concatenating the code bits assigned in traversing the tree, as shown in Fig. 6.11. It is also interesting to compare the entropy $H(X)$ and average length $L$. From Table 6.3,

$$H(X) = - \sum_{i=1}^{5} p_i \log_2 p_i = 1.852 \quad \text{bits,}$$

$$L = \sum_{i=1}^{5} p_i l_i = 1.923 \quad \text{bits,}$$

where $l_i$ is the number of bits in code word $i$. Note that $L$ is indeed very close to $H(X)$ and indicates that an average length of less than 2 bits/pixel rather than 3 can be obtained without error.

The variable length of the Huffman code makes it more difficult to implement than a fixed length code. Also, note that a change of the source probabilities would require a new code mapping to ensure minimal length.

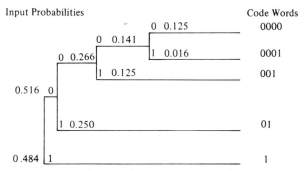

*Figure 6.11*   Code word assignments for Huffman code.

## 6.2.2 Run Length Encoding

Another statistical characteristic of digital pictures, picture element to element correlation, may be exploited by a simple reversible coding technique. In the digital image, a run may be defined as a sequence of consecutive pixels of identical values along a specified direction. For simplicity, the horizontal scan line may be used as the direction. A reduction over PCM in the average bit rate may be achieved if long runs occur by simply transmitting the start and length of the run rather than the individual pixels. Note that in a highly textured region or a region with a large high-frequency content, the run lengths could be very short and more bits could be required for the specification of the start and length than for the pixel.

As an example of run length coding, again consider the image of Fig. 3.24. The sequence of gray level and horizontal run length pairs for the image is

$$(4,7)(0,1),$$
$$(4,1)(5,5)(4,1)(0,1),$$
$$(4,1)(5,1)(6,3)(5,1)(4,1)(0,1),$$
$$(4,1)(5,1)(6,1)(7,1)(6,1)(5,1)(4,1)(0,1),$$
$$(4,1)(5,1)(6,3)(5,1)(4,1)(0,1),$$
$$(4,7)(0,1),$$
$$(4,7)(0,1).$$

Note that 30 pairs or 60 numbers are used in this representation compared to 64 pixels in the original image.

The sequence of gray levels and vertical run lengths are

$$(4,9)(5,5)(4,3)(5,1)(6,3),$$
$$(5,1)(4,3)(5,1)(6,1)(7,1),$$
$$(6,1)(5,1)(4,3)(5,1)(6,3),$$
$$(5,1)(4,3)(5,5)(4,10)(0,8).$$

In this direction, only 20 pairs or 40 numbers are required.

The efficiency of run length coding depends on the number of gray level transitions or edges and could therefore be expected to be most efficient for images with a small number of edges and gray levels. The method is most suitable for black and white facsimile images with little edge and texture content. The method is also sensitive to errors, since a single bit error could change the length of a run and thus offset the entire image. This can be reduced to single line errors by line synchronization and greatly alleviated by using only three or four standard run lengths. An example of run length coding of picture intensities in which an average bit rate of 3 bits/pixel was achieved for TV quality pictures is given in Cherry *et al.* (1963).

Bit plane images rather than intensity values have also been run length coded. An *n*-bit PCM picture may be divided into *n* bit planes in which each element is equal to the binary value of the corresponding element of the original picture. Run length coding of bit planes was used by Spencer and Huang (1969) to achieve an average of 2 to 4 bits/pixel on $256 \times 256$ 6-bit images. The rate varied with picture complexity but exact reconstruction was achieved.

Area coding may be considered as an extension of run length coding to two dimensions. An area may be defined as a connected, continuous group of picture elements that have identical values. The area may then be specified by a set of boundary points and the intensity value. Only these values would need to be transmitted. Unfortunately, finely quantized pictures do not contain many such areas. A study of 16-level pictures was made by Schreiber *et al.* (1972), who found that an average bit rate of about 2.5 bits/pixel would be required using area coding. For 6 or 8 bits/pixel, no redundancy reduction could be expected.

## 6.3   Minimum Distortion Encoding

In the previous section we have considered methods for distortion-free encoding. The information theory guideline indicated that the maximum distortion-free transmission rate is bounded by the source entropy. For example, for a given ensemble of images with $H = 1.0$ bits/pixel, a distortion-free rate $R = 0.1$ bits/pixel is not possible. Since a certain amount of distortion may be tolerable in some image transmission applications, it is interesting to consider the maximum distortion that might be expected if a rate greater than the distortion-free rate is used. The information theory guideline from rate distortion theory indicates that for a given average distortion $D$, the maximum distortion-$D$ transmission rate is $R(D)$. A communication channel with capacity only slightly greater than $R(D)$ is required to transmit the information with distortion less than or equal to $D$.

In applying rate distortion theory to picture transmission, both the picture source probability distribution and the distortion criteria must be specified. An upper bound on the achievable performance may be obtained by considering the picture vector source components to be independent Gaussian random variables and the distortion measure to be mean squared error (Davisson, 1972). Let

$$X = \{ x_i, i = 1, 2, \ldots, N \}$$

represent the source samples, which are Gaussian with zero mean and common variance $\sigma^2$, and

$$Y = \{ y_i, i = 1, 2, \ldots, N \}$$

represent the output of the decoder. Then the average mean squared error or

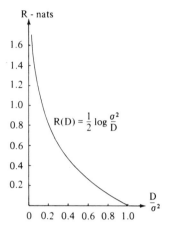

*Figure 6.12*   Rate distortion function for Gaussian source.

distortion is

$$D = \frac{1}{N} \sum_{i=1}^{N} E\left[(x_i - y_i)^2\right].$$

The corresponding rate distortion function (Berger, 1971) is

$$R(D) = \begin{cases} \frac{1}{2}\log_2(\sigma^2/D), & 0 \le D \le \sigma^2, \\ 0, & D > \sigma^2. \end{cases}$$

This function is shown in Fig. 6.12. This performance could theoretically be achieved if the encoder were designed so that the output error vector was an independent Gaussian random vector with zero mean and common variance $D$.

The rate is the number of bits/pixel required for transmission. Therefore, the number of levels $N$ is

$$N = 2^{R(D)}.$$

This may be simplified to

$$N = \left(\sigma^2/D\right)^{1/2}.$$

Therefore, for a given $D$, which may be considered as the quantization error variance, and a given pixel variance $\sigma^2 > D$, the number of quantization levels $N$ may be determined. The optimum placement of the levels along the pixel amplitude range to provide minimum distortion $D$ is the *Max quantizer* described by Max (1960), who also gives an empirical data for the functional dependence between the distortion $D$ and the number of levels $N$. If the distortion $D$ is greater than the pixel variance, the transmission rate is zero, indicating that no information need be transmitted.

In summary, this analysis indicates that a distortion-$D$ transmission rate $R(D)$ in bits/pixel may be determined given the input and output pixel variances. Also, the number of output quantization levels is proportional to the square root of the variance ratio. We have assumed that the method of encoding will ensure an independent Gaussian error with variance $D$. To apply this procedure to image information, we must verify this assumption. The first problem encountered is that the equivalent vector of an image does not have independent components; in fact, pixels that are spatially adjacent may have correlation coefficients as large as 0.95. Therefore, we shall now consider methods that will at least decorrelate the picture components, which for Gaussian variables will ensure statistical independence.

### 6.3.1   Hotelling Transformation and Block Quantization

The PCM picture may be considered as either an array of size $m \times n$ or as an equivalent vector of size $N = m \times n$. The next step in pictorial data compression is obtaining statistically independent components that may be coded separately. Knowledge of the factorability of the joint probability density of the vector components is required to ensure independence. Such detailed knowledge can rarely be determined. However, we have seen that the Hotelling transformation produces uncorrelated components. For the Gaussian distribution, uncorrelatedness implies statistical independence.

Consider the problem of determining a linear transformation $[T]$ of the form

$$\mathbf{Y} = [T]\mathbf{X}$$

where $\mathbf{X}$ is the original image vector, $\mathbf{Y}$ is the transformed image vector and $[T]$ is a linear transformation selected to decorrelate the components of $\mathbf{Y}$. It is also natural to restrict $[T]$ to be an orthonormal transformation to avoid the situation in which the total error is dominated by a single transformed component, e.g., $y_1 = 10^6 x_1$. In Chapter 3 the Hotelling transformation, which may be defined by the condition

$$\Sigma_{\mathbf{Y}} = [T]\Sigma_{\mathbf{X}}[T]' = \begin{bmatrix} \sigma_1^{\,2} & & & \\ & \sigma_2^{\,2} & & \mathbf{0} \\ & & \ddots & \\ \mathbf{0} & & & \sigma_{N^2} \end{bmatrix},$$

where $\Sigma_{\mathbf{X}}$ and $\Sigma_{\mathbf{Y}}$ are the covariance matrices of $\mathbf{X}$ and $\mathbf{Y}$, respectively, was shown to produce the desired decorrelation of the $\mathbf{Y}$ components. Furthermore, it was shown that if the components of $\mathbf{Y}$ were rearranged so that the variances were ordered from largest to smallest,

$$\sigma_1^2 \geq \sigma_2^2 \geq \cdots \geq \sigma_m^2 \geq \sigma_{m+1}^2 \geq \sigma_n^2 \geq 0,$$

and if only the first $m$ components of $\mathbf{Y}$ were retained to represent $\mathbf{X}$, with unused components replaced by their mean values, then the mean squared error is

$$\overline{\varepsilon^2}(m) = E(\|\mathbf{Y} - \hat{\mathbf{Y}}\|) = \sum_{i=m+1}^{N} \sigma_i^2,$$

where $\hat{\mathbf{Y}} = [y_1, y_2, \ldots, y_m, \mu_{m+1}, \ldots, \mu_N]'$. If the variance of any component is zero, this component can be replaced by its mean value without contributing to the error. It is also interesting to note that due to the orthonormal nature of the transformation, the mean squared error is invariant under any other linear, orthonormal transformation.

### Block Quantization

The Hotelling transformation produces uncorrelated principal components to use in the image representation. The next step in encoding is assigning bit representations to these components. If the $\mathbf{Y}$ components were of equal importance in the representation and if the variances of these components were equal, then there would be no reason for not using an equal bit length code. However, the variances of the principal components vary widely.

One strategy that might be considered for the unequal variances is normalizing each component by its standard deviation,

$$y_i' = y_i / \sigma_i, \qquad i = 1, 2, \ldots, N.$$

The normalized components $y_i'$ have unity variance and might be quantized to the same number of bits. The quantization error would be equal for the normalized components $y_i'$. However, the principal component $y_i$ quantization error would be greatest for the components with largest variance. Since these components are most important in terms of contributing to the mean squared error of the representation, this quantization strategy is not optimal.

The optimal assignment of $M$ bits to $N$ components that have Gaussian distributions with unequal variances was first considered by Huang and Schultheiss (1963). They found that the number of bits $m_K$ used to code coefficient $y_K$ should be proportional to $\log_2 \sigma_K$. Algorithms for assigning the $M$ bits to minimize the mean squared error are often called block quantization algorithms. Wintz and Kurtenbach (1968) developed the following rule, which was successfully used by Habibi and Wintz (1971), for image coding with several linear transformations. The possible assignment numbers $\hat{m}_k$ are first computed by

$$\hat{m}_k = \frac{M}{N} + 2\log_2 \sigma_k^2 - \frac{2}{N} \sum_{i=1}^{N} \log_2 \sigma_i^2.$$

Each of the $\hat{m}_k$ numbers is then rounded to the nearest integer and the total arbitrarily adjusted so that

$$M = \sum_{i=1}^{N} m_k.$$

A similar rule was obtained by observing that minimal mean squared error quantization of Gaussian variables $y_i$ using a quantizer with $2^{b_i}$ levels introduces a quantization error $e_i$ whose variance is

$$E\left(e_i^2\right) \simeq \sigma_i (10)^{-b_i/2}.$$

The bit assignment rule

$$\hat{m}_i = (M/N) + 2\left[\log_2 \sigma_i / |\Sigma_X|^{1/M}\right]/\log_2 10$$

gives

$$E\left(e_i^2\right) \simeq |\Sigma_X|^{1/M}(10)^{-b/2} = \Delta.$$

This indicates that the block quantization of $Y$ results in an equal quantization error $\Delta$ for all components of $Y$. Habibi (1974) points out that $\Delta$ should be considered as a lower bound on the quantization error since inaccuracies in the integer representations for the $b_i$ are not considered.

Since quantization error may be considered as independent, additive noise, the quantized error $Y^*$ is

$$Y^* = AY + Q \qquad \text{and} \qquad E(QQ^T) = \Delta I,$$

where $I$ is the identity matrix.

The total square error is equal to the trace of the covariance matrix of the error vector $X - X^*$. Since $A$ is unitary,

$$\overline{\varepsilon^2} = N^{-1} \text{Tr}(\Delta I) = \Delta.$$

This coding system not only minimizes the coding error but also gives an uncorrelated error in the signal domain. The error reduction over simply coding all components of $Y$ with $b$ bits may be determined by noting that the average error would then be

$$\overline{\varepsilon_2^2} = N^{-1} \text{Tr}\left[A^{-1}E(\hat{Q}\hat{Q}^T)(A^{-1})^T\right],$$

where the components of the quantization error $Q_2$ are uncorrelated with variances $E(y_r^2)10^{-b/2}$. Thus,

$$\varepsilon_2^2 = N^{-1}10^{-b/2}\text{Tr}\left[A^{-1}C_Y(A^{-1})^T\right] = 10^{-b/2}\sigma_X^2,$$

where $\sigma_X^2$ is the common variance of the components of the $X$ vector.

## 6.3.2 Coding Examples

Since the bit assignment using block quantization assigns a number of bits proportional to the logarithms of the component variances, the first concern in implementation is to estimate the variances. Since the equivalent image vector has $N^2$ components, the covariance matrix is of size $N^2$ by $N^2$ which is usually computationally impractical. Two simplifications that are often used by researchers are the block transform and Markov process assumptions. The block transform method consists of selecting a subimage size of $M$ by $M$, where $M^2 < N^2$ is selected to be a computationally feasible size. An example of the covariance matrix computation for the block transform method is given in Sec. 3.3.1. If the covariance matrix $\Sigma_x$ is estimated from subimages of the original image and a linear transformation $\mathbf{T}$ used for coding, the transform domain covariance

$$\Sigma_Y = \mathbf{T}\Sigma_X\mathbf{T}'$$

must be used for the block quantization bit assignment. Given the $M^2 \times M^2$ diagonal covariance matrix, an $M \times M$ bit mapping matrix may be formed

(a)

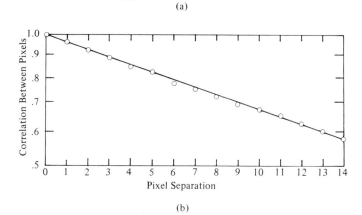

(b)

*Figure 6.13*   (a) Kodak girl original image and (b) correlation between pixels of (a).

Hue    Saturation    Brightness

The three types of color sensations

Brightness

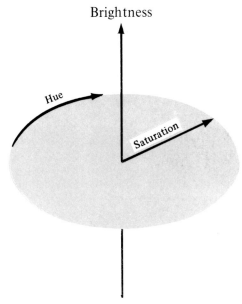

*Plate I* Psychological color sensations of hue, saturation, and brightness.

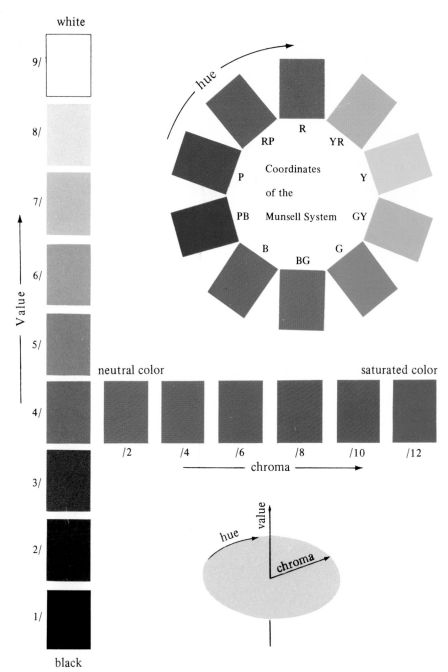

*Plate II* Munsell color system variables of hue, chroma, and value.

y
1.0

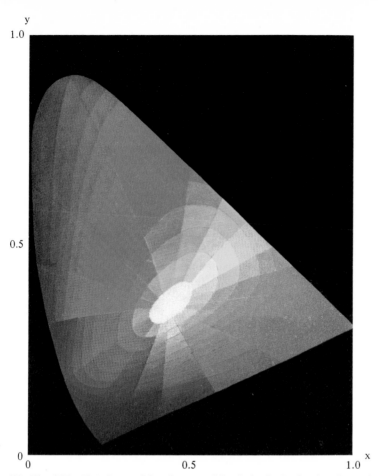

0.5

0
0                    0.5                    1.0
                                                    x

*Plate III*   The CIE (1931) chromaticity diagram. (Oil painting by L. Condax. Reproduced by permission of the Optical Society of America and Eastman Kodak Corporation.)

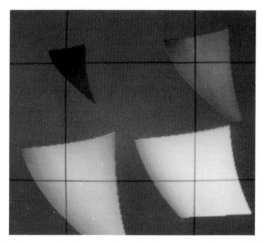

*Plate IV*   Constant luminance slices in *Lab* space. (From Wallis, 1975.)

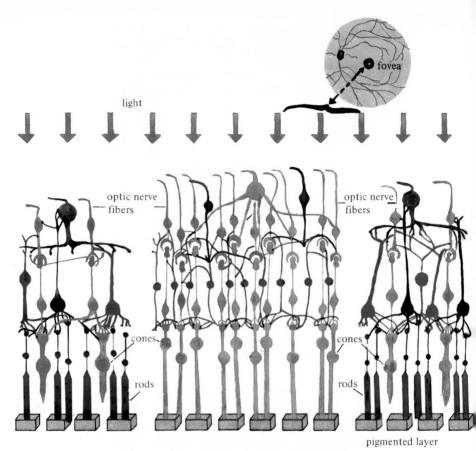

light

optic nerve
fibers

optic nerve
fibers

fovea

cones

cones

rods

rods

pigmented layer

*Plate V*   Structure of the retina. (Adapted from Rainwater, 1971.)

*Plate VI*   Color Mach bands. (From Wallis, 1975.)

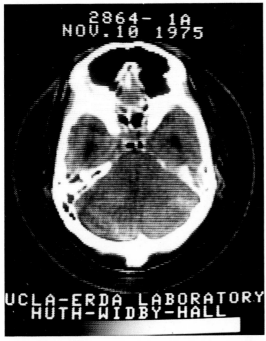

(a)

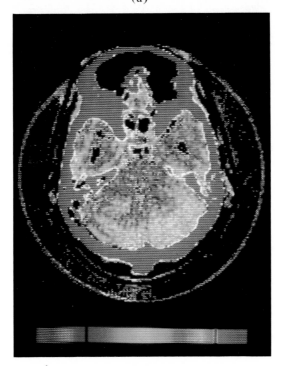

(b)

*Plate VII*   Comparison of (a) monochrome and (b) pseudocolor display.

*Plate VIII* Cosine transform coded images with various transformations: (a) original image; (b) YIQ, NMSE, red = 0.58%, green = 0.98%, blue = 1.69%; (c) *Lab*, NMSE, red = 0.49%, green = 0.76%, blue = 0.03%; (d) perceptual, NMSE, red = 0.58%, green = 0.79%, blue = 1.03%. (From Hall, 1978.)

*Plate IX*   Cosine and Fourier coded images at 256 × 256 and 1 bit/pixel: (a) cosine coded, YIP, NMSE, red = 0.42%, green = 0.73%, blue = 1.25%; (b) cosine coded, perceptual, NMSE, red = 0.36%, green = 0.52%, blue = 0.85%; (c) Fourier coded, YIP, NMSE, red = 0.42%, green = 0.76%, blue = 1.59%; (d) Fourier coded, perceptual, NMSE, red = 0.39%, green = 0.52%, blue = 0.86%. (From Hall, 1978.)

*Plate X* Perceptual power spectrum coded images at 256 × 256 and various rates: (a) original; (b) 2 bits/pixel, NMSE, red = 0.10%, green = 0.18%, blue = 0.70%; (c) 1 bit/pixel, NMSE, red = 0.20%, green = 0.33%, blue = 0.84%; (d) 0.5 bit/pixel, NMSE, red = 0.43%, green = 0.66%, blue = 1.1%. (From Hall, 1978.)

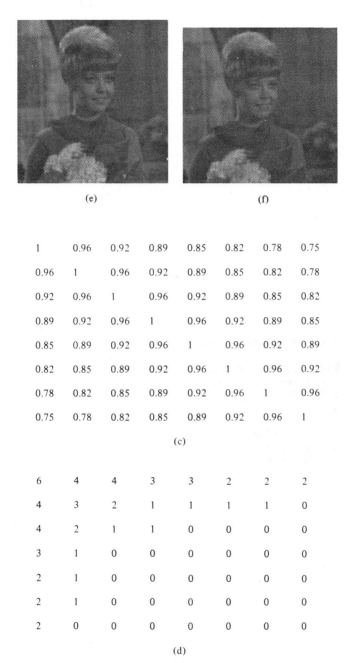

(e)    (f)

| 1 | 0.96 | 0.92 | 0.89 | 0.85 | 0.82 | 0.78 | 0.75 |
|---|---|---|---|---|---|---|---|
| 0.96 | 1 | 0.96 | 0.92 | 0.89 | 0.85 | 0.82 | 0.78 |
| 0.92 | 0.96 | 1 | 0.96 | 0.92 | 0.89 | 0.85 | 0.82 |
| 0.89 | 0.92 | 0.96 | 1 | 0.96 | 0.92 | 0.89 | 0.85 |
| 0.85 | 0.89 | 0.92 | 0.96 | 1 | 0.96 | 0.92 | 0.89 |
| 0.82 | 0.85 | 0.89 | 0.92 | 0.96 | 1 | 0.96 | 0.92 |
| 0.78 | 0.82 | 0.85 | 0.89 | 0.92 | 0.96 | 1 | 0.96 |
| 0.75 | 0.78 | 0.82 | 0.85 | 0.89 | 0.92 | 0.96 | 1 |

(c)

| 6 | 4 | 4 | 3 | 3 | 2 | 2 | 2 |
|---|---|---|---|---|---|---|---|
| 4 | 3 | 2 | 1 | 1 | 1 | 1 | 0 |
| 4 | 2 | 1 | 1 | 0 | 0 | 0 | 0 |
| 3 | 1 | 0 | 0 | 0 | 0 | 0 | 0 |
| 2 | 1 | 0 | 0 | 0 | 0 | 0 | 0 |
| 2 | 1 | 0 | 0 | 0 | 0 | 0 | 0 |
| 2 | 0 | 0 | 0 | 0 | 0 | 0 | 0 |

(d)

*Figure 6.13* (c) Block Toeplitz matrix of Kodak girl picture for $M=8$ and $\rho=0.96$, and $8\times8$ block size, (d) bit mapping matrix for 1 bit/pixel rate $\rho=0.96$, and $8\times8$ block size, (e) coded image using $8\times8$ block size, 1 bit/pixel, NMSE$=0.39\%$, and (f) coded image using $16\times16$ block size, 1 bit/pixel, NMSE$=0.36\%$. [From Hall (1978).]

using the vector to image transformation. The elements of this matrix are the integer number of bits to assign to each transform coefficient. The previously described bit assignment rule may be used to form the bit mapping matrix and is easily computed for large matrix sizes. For smaller matrix sizes, an iterative procedure developed by Pratt (1978) may be used that optimizes the assignment for Gaussian data. The algorithm uses the Gaussian error function to decrement the largest variance in the covariance matrix one bit at a time until the total number of desired bits have been used. Each time a variance is decremented the corresponding bit mapping matrix element is incremented. If the desired average bit rate is $B$ and the subpicture size is $M \times M$, this procedure requires $BM^2$ passes through the $M \times M$ matrix elements. Therefore, the computation grows rapidly with $M$ and for $M > 32$ the cost versus error should be carefully considered.

An example from Hall (1978) of the block transformation and block quantization will now be considered. The original image of the Kodak girl is shown in Fig. 6.13a. The spatial correlation function of this image is shown in Fig. 6.13b. The slope of the correlation curve is 0.96 and is constant, therefore indicating that a first-order Markov process with $\rho = 0.96$ is appropriate. The vertical and horizontal correlations were nearly equal and both modeled accurately by the value of $\rho = 0.96$. A transform block size $M = 8$ was now selected. The block Toeplitz correlation matrix of the form shown in Problem 3.3 was computed with the result shown in Fig. 6.13c. A transform method must now be selected. The discrete cosine transform will be used. The next step is to estimate the covariance matrix in the transform domain. This is accomplished by computing the two-dimensional discrete cosine transform of the correlation matrix to form

$$\Sigma_Y = [C] \Sigma_X [C]'.$$

Applying the discrete cosine transform matrix described in Sec. 3.3.6 to the correlation matrix given in Fig. 6.13c results in the transform covariance matrix.

From the variance matrix, the bit mapping matrix shown in Fig. 6.13d was formed using the iterative procedure for a 1 bit/pixel rate. Note that the elements of the bit mapping matrix sum to 64. The results of coding the Kodak girl with this procedure are shown in Fig. 6.13e. The normalized mean squared error was 0.39%. A similar coding result with a transform block size $M = 16$ is shown in Fig. 6.13f and resulted in a mean squared error of 0.36%. A close inspection of the coded images, especially around the flowers, indicates one problem with this type of coder. The coarseness of quantization especially of the dc term produces a visible block error. Channel errors in the dc term produce similar errors.

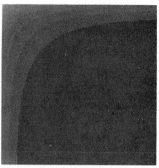

*Figure 6.14* A 256×256 cosine transform domain bit mapping matrix, $\rho = 0.96\%$. [From Hall (1978).]

## Full Image Transform Coding

Due to the advent of high-speed hardware transform processors, it is important to consider full image transform coding techniques for which $M = N$, the digital image size. This also alleviates entirely the block error pattern previously mentioned with $M < N$. This method was developed by Hall (1978). Again, first-order Markov stationary statistics will be assumed. The method requires the determination of an $N \times N$ block Toeplitz matrix for the horizontal and vertical correlation, the transformation of this matrix to estimate the variance matrix and the formation of an $N \times N$ bit mapping matrix.

An example for $N = 256$ will now be described. The variance matrix of size $256 \times 256$ is computed as previously described. The bit mapping matrix is formed using the relation

$$b_{ij} = \text{Int}\left[ -B + 2\log_{10}\sigma_{ij}^2 - \frac{2}{N}\sum_{k=1}^{N}\sum_{l=1}^{N}\log_{10}\sigma_{kl}^2 + 0.5 \right],$$

where $b_{ij}$ is the $(i,j)$th element of the bit mapping matrix, $B$ is the desired average bit rate, and $\sigma_{ij}^2$ is the variance of the $(i,j)$th transform coefficient. The $b_{ij}$ values are rounded to the nearest integer, which introduces a slight error over the previous assignment procedure. A bit matrix for the discrete cosine transform is shown in Fig. 6.14. The white region near the origin (upper left) represents the value 9 and the dark region (lower right) has value 0 as a result of the small variances for large $i$ and $j$ values. Note that the equal bit contours in Fig. 6.14 are hyperbolic in shape. An image coded to 1 bit/pixel with the discrete cosine transform and this method is shown in Fig. 6.15. An important variation from the bit coding rule was used on the dc term. A full 36 bits was used to code the dc term to provide stability in the image contrast and therefore eliminate the need for rescaling before display.

*Figure 6.15*   A 256×256 cosine transform coded image, 1 bit/pixel, NMSE=0.24%. [From Hall (1978).]

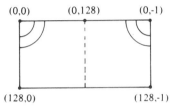

*Figure 6.16*   Fourier transform domain frequency locations for 256×256 image. [From Hall (1978).]

*Figure 6.17*   A 256×129 Fourier transform domain bit mapping matrix. [From Hall (1978).]

These extra 27 bits out of the total 65,536 represent an insignificant increase (0.000412) in the average bit rate.

The full image transform coding method may be used with any of the image transforms. With the Fourier transform, a difference is encountered since the transform values are complex valued. Although it may appear that twice the number of coefficients must be coded, the transform of a real-valued image is conjugate symmetric; therefore, only half of the complex transform coefficients are unique. For image coding, it is convenient to work with a non-origin-centered Fourier transform domain as shown in Fig. 6.16

*Figure 6.18*   A 256×256 Fourier transform coded image, 1 bit/pixel, NMSE=0.23%. [From Hall (1978.)]

for a 256×256 image. The semicircles in this illustration represent contours of constant radial frequency. A bit mapping matrix of size 256×129 was generated from the variance matrix and is shown in Fig. 6.17. Note that the hyperbolic contours are still present in this mapping matrix. An image coded with the Fourier transform using Max quantization and the above bit mapping matrix for both the real and imaginary components is shown in Fig. 6.18 (Hall, 1978). The bit rate is 1 bit/pixel.

### Coding in the Perceptual Domain

In the previous coding techniques, no special consideration was given to the spatial frequency response of the HVS. A coding technique that includes a visual system model was developed by Hall (1978). The method uses the visual spatial frequency model as a preprocessor before the quantization step. The first-order Markov assumption for the input image statistics is made. However, a new bit mapping procedure is used that is based not on a variance matrix, but rather on the power spectral density in the preprocessed transform or perceptual domain shown in Fig. 6.19. The results of the HVS preprocessing are shown in Fig. 6.20. The power spectrum equation derived

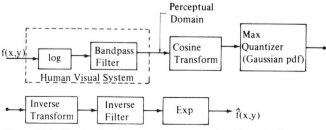

*Figure 6.19*   Psychovisual cosine transform coder. [From Hall (1978).]

*Figure 6.20*   A HVS preprocessed image. [From Hall, (1978).]

by Hall is

$$S(\omega_r) = \left[ 2\alpha\sigma_Y^2 / (\alpha^2 + \omega_r^2) + 2\pi\mu_Y^2 \delta(\omega_r) \right] |H(\omega_r)|^2,$$

where $\omega_r$ is radial spatial frequency and

$$H(\omega_r) = 2.6(0.0192 + 0.018\,\omega_r) \exp(-0.018\omega_r^{1.1})$$

and $\alpha$ is the correlation parameter of the input image whose correlation is assumed to be of the Markov form

$$\rho_x(\tau) = \exp(-\alpha|\tau|)$$

with a lognormal distribution. The parameters $\mu_Y$ and $\sigma_Y^2$ are the mean and variance of the image after the log conversion. If one chooses not to code the dc term to eliminate contrast errors, one need only consider

$$S(\omega_r) = \left[ 2\alpha\sigma_Y^2 / (\alpha^2 + \omega_r^2) \right] |H(\omega_r)|^2, \qquad \omega_r > 0.$$

Since this equation defines the power spectrum for any $\omega_r > 0$, it may be used to determine the transform component variance and therefore the bit allocation mapping. Thus, the generation of a variance matrix and bit mapping matrix is not required.

To obtain a bit assignment, one merely solves the above equation at each $\omega_r$

$$\omega_r = \omega_s \sqrt{i^2 + j^2} \,,$$

where $i$ and $j$ are the Fourier coefficient indices and $\omega_s$ is the scale factor for conversion from linear spatial frequency to radians/degree. Let the variance of element $(i,j)$ be $\sigma_{ij}^2$. Then

$$\sigma_{ij}^2 = S\left(\omega_s \sqrt{i^2 + j^2}\right).$$

The bit allocation for this transform coefficient may then be described as

$$b_{ij} = \text{Int}\left[ \log_2 \gamma\sigma_{ij}^2 + 0.5 \right].$$

The factor $\gamma$ is selected to yield the desired bit rate. Experimental values used by Hall for average bit rates of 0.1–1.0 bits/pixel were $8.9 \times 10^4$–$9 \times 10^5$. A

*Figure 6.21*   Perceptual power spectrum bit mapping matrix. [From Hall (1978).]

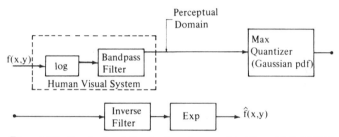

*Figure 6.22*   Psychovisual Fourier transform coder. [From Hall (1978).]

*Figure 6.23*   A $256 \times 256$ psychovisual coded image, 1 bit/pixel, NMSE = 0.26%. [From Hall (1978).]

bit mapping using perceptual coding is shown in Fig. 6.21. Note that the equal bit contours are now circles of constant radial frequency.

The psychovisual Fourier coding system shown in Fig. 6.22 was used to code the image shown in Fig. 6.23 at 1 bit/pixel. The effects of variation in bit rate are shown in the series of perceptually coded images shown in Fig. 6.24. The Fourier coefficients were normalized by $\sigma_{ij}^2$ and Max quantized to $b_{ij}$ bits for the real and $b_{ij}$ bits for the imaginary parts. Since it was not necessary to compute or store variance and bit mapping matrices, it was

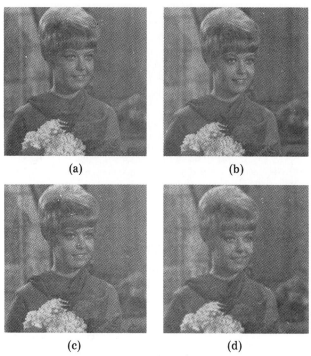

(a)                                        (b)

(c)                                        (d)

*Figure 6.24*  Perceptual power spectrum coded images (512×512): (a) original image; (b) 0.5 bit/pixel, NMSE = 0.28%; (c) 0.35 bit/pixel, NMSE = 0.50%; and (d) 0.1 bit/pixel, NMSE = 0.72%. [From Hall (1978).]

possible to code any size image for which a transform may be computed. The images shown were coded by Hall at 512 by 512 (a resolution comparable with standard TV). The bit rates are on the order of $\frac{1}{10}$ the rates achieved with the previous methods. Furthermore, the degradation with decreasing rates is "graceful." Errors appear in the form of "snow" distributed throughout the image.

### 6.4   Predictive and Interpolative Coding

Predictive coding techniques including differential pulse code modulation (DPCM) will now be considered. To introduce this method it is interesting to consider a special class of lower triangular linear transformations.

### 6.4.1   Lower Triangular Transformation

A unified theory of a class of lower-triangular transformations has been developed by Habibi and Hershel (1974), which shows the relationship between general transform coding and DPCM. This approach provides a fresh outlook on DPCM by showing that it is a special case of a more general coding system that uses a lower triangular matrix to transform the data to a set of uncorrelated signals.

Consider again the PCM picture as an $N$-dimensional data vector $\mathbf{X}$, where $\mathbf{X}$ represent a sample vector from an ensemble of zero mean random variables. A vector $\mathbf{Y}$ may always be generated from a linear combination of $x_i$s as

$$y_1 = x_1, \qquad y_j = x_j - \sum_{k=1}^{j-1} l_{jk} x_k, \quad \text{for } j = 2, 3, \ldots, N$$

or, in vector form,

$$\mathbf{Y} = \mathbf{LX},$$

where $\mathbf{L}$ is a unit lower triangular matrix

$$\mathbf{L} = \begin{bmatrix} 1 & 0 & 0 & 0 & \cdots & 0 & 0 \\ -l_{21} & 1 & 0 & 0 & \cdots & 0 & 0 \\ -l_{31} & -l_{32} & 1 & 0 & \cdots & 0 & 0 \\ \vdots & & & & & \vdots & \vdots \\ -l_{N1} & -l_{N2} & & & & -l_{N,N-1} & 1 \end{bmatrix}.$$

If the covariance matrix of $\mathbf{X}$ is $\boldsymbol{\Sigma}_{\mathbf{X}}$, then the covariance matrix of $\mathbf{Y}$, $\boldsymbol{\Sigma}_{\mathbf{Y}}$, is given by

$$\boldsymbol{\Sigma}_{\mathbf{Y}} = \mathbf{L}\boldsymbol{\Sigma}_{\mathbf{X}}\mathbf{L}^{\mathrm{T}}.$$

It is known that for every symmetric positive definite matrix $\boldsymbol{\Sigma}_{\mathbf{X}}$, there exists a real nonsingular lower triangular matrix $\mathbf{L}$ such that the matrix $\mathbf{L}\boldsymbol{\Sigma}_{\mathbf{X}}\mathbf{L}^{\mathrm{T}}$ is diagonal (Fox, 1954). Martin and Wilkinson (1965) have considered numerical algorithms for finding $\mathbf{L}$ and $\boldsymbol{\Sigma}_{\mathbf{Y}}$ and have developed efficient techniques requiring only $\frac{1}{6}N^3$ multiplications.

The fact that the transformation diagonalizes the covariance matrix of $\mathbf{X}$ indicates that the transformed elements of $\mathbf{Y}$ are uncorrelated. In general, the components would also have unequal variances. However, Habibi and Hershel (1974) have shown that for $n$th-order Markov processes, the variances of the $\mathbf{Y}$ components after the first $n$ are all equal. The implementation of a lower triangular transformation is closely related to the important method called predictive coding, which includes differential pulse code modulation (DPCM) and delta modulation.

## 6.4.2   Predictive and Interpolative Coding

The lower triangular transformation illustrates that one may achieve uncorrelated pixels by transformations other than the Hotelling. It will now be shown that predictive and interpolative techniques may also be used. Since picture elements in a local region are highly correlated, one might expect that a currently considered pixel might be accurately predicted from a knowledge of previous pixels (prediction) or that a current pixel might be inferred from a knowledge of some past and some future elements (interpolation). If accurate prediction or interpolation is possible, then it should not be necessary to transmit the pixel but rather a predictor or interpolator could be placed at the receiver. Of course, some starting value might be required or actual pixel values periodically transmitted to avoid large errors.

There is a trade-off in the complexity of the predictor and the transmission bit rate. A highly accurate predictor should require only a starting value and therefore a very low bit rate. However, a predictor of such high accuracy may require a large amount of a priori knowledge about the image characteristics or require a complex nonlinear computation. A simple predictor such as a linear predictor may produce a large prediction error if only a starting value is transmitted. A compromise solution is to use a linear predictor at the transmitter but transmit not only a starting value but also an estimate of the prediction error, quantized for digital transmission. (The transmitter would estimate the pixel value from its predicted value plus the received error value.) If the prediction error is small, a low transmission bit rate should result.

To introduce the elements of predictive coding, consider forming a linear estimate for a picture element $f(m,n)$ based upon the previous horizontal element $f(m-1,n)$, i.e.,

$$\hat{f}(m,n) = \alpha f(m-1,n).$$

The mean squared error between the original and estimated values may be expressed as

$$\overline{\varepsilon^2} = E\left\{\left[f(m,n) - \hat{f}(m,n)\right]^2\right\}.$$

The error may be expanded to obtain

$$\overline{\varepsilon^2} = E\left[f^2(m,n)\right] - 2\alpha E\left[f(m,n)f(m-1,n)\right] + \alpha^2 E\left[f^2(m-1,n)\right].$$

Since the autocorrelation function of a random picture element is defined as

$$R(i,j,k,l) = E(f(i,j)f(k,l)),$$

the expected values of the products of the shifted image functions may be expressed in terms of the autocorrelation function values. For simplicity, we may assume that the image is stationary so that

$$R(i-k,j-l) = R(i,j,k,l).$$

The mean squared error may then be expressed as

$$\varepsilon^2 = R(0,0) - 2\alpha R(1,0) + \alpha^2 R(0,0).$$

Differentiating $\varepsilon^2$ with respect to $\alpha$ and equating the result to 0 gives

$$\alpha = R(1,0)/R(0,0).$$

Therefore, the optimum previous element predictor is

$$\hat{f}(m,n) = R(1,0)f(m-1,n)/R(0,0).$$

If the autocorrelation is normalized, $R(0,0)=1$. Also, let $R(1,0)=\rho$. The prediction equation then simplifies to

$$\hat{f}(m,n) = \rho f(m-1,n).$$

### Predictive Coding

In a general predictive coding system, an equation of prediction between the present picture value $x_n$ and $n-1$ past picture values $x_1,\ldots,x_{n-1}$ is first assumed:

$$\hat{x}_n = f(x_1, x_2, x_3, \ldots, x_{n-1}),$$

where $\hat{x}_n$ is the predicted value. The mean squared error between the actual

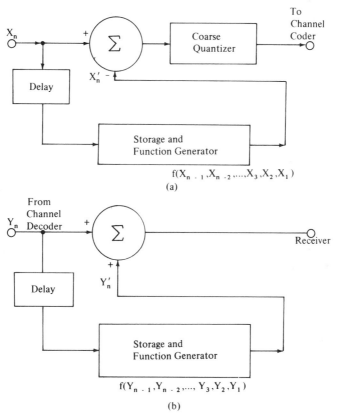

*Figure 6.25*   Predictive image coding system: (a) encoder system and (b) decoder system.

and predicted values is

$$\overline{\varepsilon_n^{\,2}} = E\Big[(x_n - \hat{x}_n)^2\Big].$$

Minimizing the mean squared error gives an optimum prediction method. Other criteria could also be used. Given the predicted value $\hat{x}_n$, which may be computed at the receiver, and the received error signal $\varepsilon_n$, the picture value may be computed by

$$x_n = \hat{x}_n + \varepsilon_n.$$

One form of the predictive image coding system is shown in Fig. 6.25. The encoder shown in Fig. 6.25a computes the error between the current value and a predicted value. This error signal is quantized and transmitted through the channel. The decoder system shown in Fig. 6.25b forms the received signal by adding the quantized error to the computed predicted value. Several variations of this basic predictive system are also used. One variation is to place the quantizer before the summing element in the encoder. This has the effect of combining both prediction and quantization effects in the transmitted error signal. An important modern form of the predictive coding system is differential pulse code modulation (DPCM), which will now be considered.

### 6.4.3  Differential Pulse Code Modulation

The block diagram of a differential pulse code modulation system using previous element prediction is shown in Fig. 6.26 and exemplifies the simplicity of DPCM. The transmitter predictor computes the error between the actual picture value and the predicted value. This error signal is then quantized in a coarse, nonuniform manner and transmitted. The received error value is added to the previous predicted value, delayed and possibly attenuated to form the next predicted value in the feedback loop. The attenuation factor or "leak" can be adjusted, depending on the correlation properties of the scenes to be transmitted. The quantized output may then be binary coded for error correction and transmitted over the channel.

The received difference values are added to the predicted value by the receiver predictor to produce the received picture value. The predicted value duplicates the operation of the source prediction. To provide tracking and eliminate long error bursts, the receiver predictor may be periodically updated. For example, the source and receiver predictors may be reset to zero at the beginning of each line to ensure only within line errors.

The performance of DPCM is dependent upon the method used to set the quantizer levels and attenuation thresholds. Quantization error, the difference between the input and output of the quantizer, produces a characteristic pictorial error called granularity error that forms a texture pattern in regions

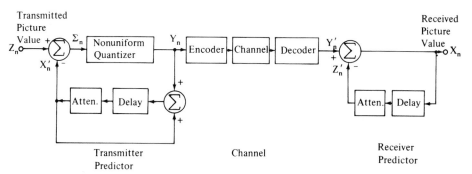

*Figure 6.26* Differential pulse code modulation using previous element prediction.

of constant brightness. Quantization step size also limits the slew rate of the coder and leads to an error called slope overload. The combination of these errors also occurs in natural images and appears as a mixture of a form of slope overload for a larger signal and granularity error of the larger value.

### 6.4.4   Delta Modulation System

Delta modulation $\Delta M$ is the simplest form of DPCM in which a 1-bit quantizer is used (De Jager, 1952). A block diagram of the $\Delta M$ system is shown in Fig. 6.27. The difference between the incoming video signal and its predicted value is quantized to 1 bit for transmission. The feedback loop in the encoder may be described by the transfer function

$$H_1(Z) = \beta Z / (1 - \beta Z) = \beta Z + \beta^2 Z^2 + \beta^3 Z^3 + \cdots,$$

where $Z$ represents the unit delay operator $Z = \exp(-ST)$ and $T$ is the sampling interval. Therefore, the output of the feedback loop is the weighted sum of all past inputs by $T$.

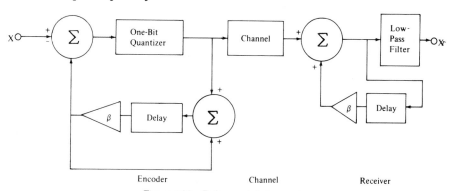

*Figure 6.27* Delta modulation system.

The transfer function of the decoder loop is

$$H_2(Z) = (1 - \beta Z)^{-1} = \beta Z + \beta^2 Z^2 + \beta^3 Z^3 + \cdots .$$

Thus, the decoder output is the sum of all past points. This signal can be low-pass filtered or integrated to estimate the input signal.

The most fundamental parameters of the $\Delta M$ system are the sampling rate $f_s = 1/T$ and the quantization step size $k$ or output levels $\pm k$. For a given sampling rate, slope overload noise can be minimized by increasing the step size $k$. We might like to have $kf_s$ greater than the maximum slope of the signal. However, if the step size is increased, the granularity noise increases. Although one might attempt to determine the optimum step size for subjective image quality for the simplest $\Delta M$ system, another approach is to adaptively adjust the step size dependent on the signal characteristic.

Several forms of adaptive $\Delta M$ systems have been proposed. In the high-information delta modulation (HIDM) system (Winkler, 1965) the step size is adjusted at the transmitter and receiver as a function of the slope of the signal as estimated from the three past values. The effect of this variable step size is greatly reduced slope overload as illustrated in Fig. 6.28. Other forms of adaptive $\Delta M$ have also been developed. A study by O'Neal (1966) has shown analytically that for good quality pictures, multibit differential coders have larger signal-to-noise ratios than equivalent bit rate delta modulators. Thus, to obtain equivalent picture quality a higher bit rate would be required. Nevertheless the simple circuitry of the delta modulator makes it attractive in some applications.

### 6.4.5 Interpolative Coding

The basic principle of interpolative coding is that the receiver may fill in missing samples that have not been transmitted by interpolation. A straight-

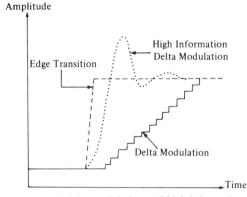

*Figure 6.28*   Edge response of delta modulation and high information delta modulation.

forward example that achieves a 2 to 1 compression is to transmit every other scan line and at the receiver fill in the missing lines by interpolation.

The general interpolation equation is

$$x'_n = f(\ldots, x_{n+2}, x_{n+1}, \ldots, x_{n-1}, x_{n-2}, \ldots).$$

The error $e_n$ between the interpolated values is

$$e_n = x_n - x'_n$$

and may be used to determine the effectiveness of the method. If the interpolation technique is carefully chosen, the error may be made very small. For example, by using the cardinal function interpolation formula developed in the sampling theorem, it is possible to represent the band limited signal exactly by interpolation. Unfortunately, the amount of computation is also large so that simpler methods have been considered.

### The Basic Principle

Judiciously chosen samples in the picture are omitted in transmission. At the receiver, these missing samples are filled in by interpolation from the transmitted samples. Sometimes corrective values are also transmitted for the interpolated samples. In a good scheme, these corrective values should have a very peaky distribution so that efficient statistical coding can be used.

### Piecewise-Linear Approximation

Youngblood (1968) was among the first to investigate interpolative coding. His method was piecewise-linear approximation. For each scan line, consider the intensity $z$ as a function of position $n$,

$$z_n = f(n).$$

He approximated $f(n)$ by a (sampled) piecewise-linear curve $g(n)$ such that

$$|g(n) - f(n)| \le \varepsilon,$$

where $\varepsilon$ was a preselected threshold. At the transmitting end, only the locations and amplitudes of the joint points of the straight line segments were sent. From these, the receiver could reconstruct $g(n)$ by linear interpolation. Using this scheme, Youngblood was able to reconstruct good-quality pictures at an average bit rate of about 1 bit/sample for a girl's face, and 3 bits/sample for a crowd scene. The pictures contained about $240 \times 240$ samples.

### Lowpass and Corrective Signals

In this scheme, according to Cunningham (1970), a picture was divided into small square blocks, and the average intensity of each block transmitted to represent the intensity of the central sample of the block. At the receiver, linear interpolation was used to obtain intervening intensities. Coarsely quantized correction signals had to be sent for each example to make the resulting picture acceptable.

With this scheme, good-quality pictures could be obtained at an average bit rate of about 0.9 bit per sample for a picture of a girl's face, and 1.7 bits/sample for a crowd scene. These were the same pictures used by Youngblood.

### 6.5  TV and Color Image Coding

The number of television sets in the United States is on the order of 100 million with the greater percentage color receivers. The rapid change from monochrome to color was due in large part to the compatibility of the color system with the monochrome systems. This compatibility is illustrated in Fig. 6.29 (Pearson, 1975). The bandwidth allocation is the same, because the chrominance signals are interleaved in the zero regions of the luminance signal. When compatible color transmissions are received on a monochrome receiver, the precise relationship between the line frequency and color subcarrier frequency permits the color signal to fall in a zero region of the luminance signal and minimize interference. To further reduce the monochrome interference the chrominance signal is transmitted inverted in amplitude for each line period. The integration of the inverted signals in the display phosphors and the HVS virtually eliminates the interference. Thus, the monochrome receiver is effectively immune to the chrominance signal.

To permit compatibility the red, green, and blue signals from a TV camera are transformed to luminance and chrominance signals before transmission. The transformation for the NTSC system is

$$\begin{bmatrix} Y \\ I \\ Q \end{bmatrix} = \begin{bmatrix} 0.299 & 0.587 & 0.114 \\ 0.596 & -0.275 & -0.321 \\ 0.212 & -0.523 & 0.311 \end{bmatrix} \begin{bmatrix} R \\ G \\ B \end{bmatrix}.$$

The $I$ and $Q$ signals are filtered and transmitted in quadrature in the chrominance. Quadrature transmission consists of multiplying one signal by a cosine and the other by a sine wave, which permits the receiver to separate the two components.

At the receiver, after demodulation, the inverse transformation given below re-forms $R, G, B$ signals from the received $\hat{Y}, \hat{I}, \hat{Q}$ signals.

$$\begin{bmatrix} R \\ G \\ B \end{bmatrix} = \begin{bmatrix} 1 & 0.956 & 0.620 \\ 1 & -0.272 & -0.647 \\ 1 & -1.108 & 1.705 \end{bmatrix} \begin{bmatrix} \hat{Y} \\ \hat{I} \\ \hat{Q} \end{bmatrix}.$$

The development of digital TV may await the discovery of a compatible

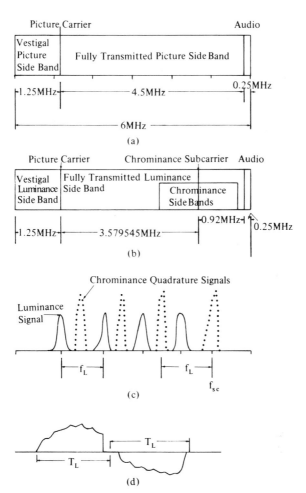

*Figure 6.29* Compatible color TV: (a) bandwidth allocation for monochrome broadcasting in the US. (b) Bandwidth allocation for compatible color broadcasting in the US. (c) Relationship between the line frequency $f_L = 15.734264$ kHz and the color subcarrier frequency $f_{SC} = 3.579545$ MHz. (d) The chrominance signal is inverted in amplitude each line period to eliminate an interference signal on monochrome receivers. [From Pearson (1975).]

digital system either for the camera and receiver or the transmitter and receiver. The bandwidth requirement for compatible color transmission is exactly the same as for digital monochrome transmission. Therefore, bit rates of a fraction of a bit for each picture element are required.

### Digital Color Coding Examples

Several color image coding experiments will be described to illustrate digital color coding techniques. For efficient coding, it is preferable to

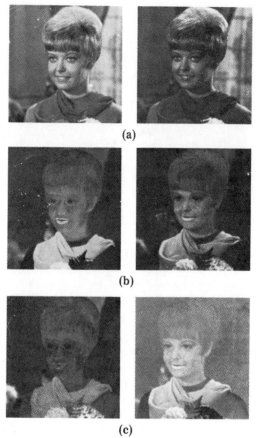

(a)

(b)

(c)

*Figure 6.30*  Kodak girl image components using *Lab* and *YIQ* transformations: (a) *L* and *Y* components. (b) *a* and *I* components. (c) *b* and *Q* components. [From Hall (1978).]

transmit statistically independent samples. If the class of images satisfied the stationary and Gaussian assumptions, this could be achieved by a $3N^2$-dimensional Hotelling transformation. If point to point independence was also assumed, a three-dimensional Hotelling transformation could produce three uncorrelated component images. For compatibility with standard color TV, the *YIQ* transformation is preferable. Fortunately, it has been indicated that the *YIQ* transformation is a close approximation to the Hotelling transformation. For comparison, the *Lab* and *YIQ* component images are shown in Fig. 6.30. The correlations between these component images are compared to the *RGB* component images in Table 6.4. Note that the *YIQ* components are much less correlated than the *RGB* components.

The techniques of transformation and quantization described for monochrome image coding such as DPCM, transform, and perceptual coding may be extended to color image coding; however, three component images must now be considered. To compare the effectiveness of different coding methods,

*Table 6.4*
*Correlation Values for RGB, YIQ, and Lab Transformations of Kodak Girl Image*

| | Correlations | | |
| --- | --- | --- | --- |
| System | Components (1) and (2) | Components (1) and (3) | Components (2) and (3) |
| *RGB* | 0.771 | 0.682 | 0.913 |
| *YIQ* | 0.018 | −0.315 | 0.527 |
| *Lab* | −0.353 | −0.270 | 0.646 |

an *overall* bit rate $R$ will be used that is the sum of the component bit rates $R_Y$, $R_I$, and $R_Q$,

$$R = R_Y + R_I + R_Q.$$

Given a desired overall bit rate $R$, one must determine the component rates $R_Y$, $R_I$, and $R_Q$. The optimum approach would involve specifying a distortion $D$ for the color image and deriving the corresponding component rates. Experimental results from monochrome coding may be extended to the luminance component $Y$ including the rate distortion theory criteria. The chrominance signals in color TV are band limited to a fraction on the order of $\frac{1}{3}-\frac{1}{9}$ the luminance bandwidth; therefore, it should be possible to reduce the number of samples by corresponding fractions. This approach has been considered by Limb *et al.* (1971) for DPCM color coding. The bandwidth energy compaction and transformations were also considered by Pratt (1971).

The following experimental results developed by Hall (1978) illustrate the possibilities of efficient color image coding at fractional overall bit rates. Proportional values for the *YIQ* component rates of 0.625/0.275/0.1 were used. Given the desired component bit rate, the component image may be coded as a monochrome image. To aid in comparing the image quality, the original image and three processed images have been displayed and processed as an entity. Therefore, any difference between quadrants ia a result of coding and not the reproduction process.

The first series of images shown in Plate VIII illustrate the block transform method with $16 \times 16$ block size. The original image shown in the upper left is the Kodak girl at $256 \times 256$. The three processed images were coded with the cosine transform at a rate of 1 bit/pixel in the *YIQ*, *Lab*, and a perceptual space. Note that block-type errors are apparent in the color images.

The second series of images, shown in Plate IX, was coded at 1 bit/pixel with a full image block size of $256 \times 256$ both with the cosine transform and with the Fourier transform.

The final example, shown in Plate X, illustrates the variation in bit rate. The perceptual coding method was used in which bits are allocated in proportion to the estimate of the perceptual power spectrum. Note the change in image quality as the bit compression is changed from 1 to 1, to 12 to 1, 24 to 1, and 48 to 1.

## 6.6   Interframe Image Coding

The reproduction of natural motion scenes by television is possible mainly because of the persistance of the retinal receptors. Since the brain retains the image of an illumination pattern for about 0.1 sec after the source of illumination has been removed from the eye, a minimum of about 10 frames/sec can simulate natural motion. To depict rapid motion smoothly, about 25–30 frames/sec are required. In many cases involving rapid motion, a large fraction of the picture elements correspond to stationary background. In extensive statistical measurements, Seyler (1965) showed that only about 10% of the picture elements change more than 8% in brightness from frame to frame. Thus, both psychovisual properties and statistical correlations may be exploited to develop interframe coding techniques. Several combinations of predictive and transform coding methods that have been applied to interframe coding will now be reviewed.

### *Interframe Transform Coding*

Consider a three-dimensional digital image $f(x,y,t)$ of gray level values, where

$$x=0,1,\dots,N_1-1, \qquad y=0,1,\dots,N_2-1, \qquad t=0,1,\dots,N_3-1.$$

The three-dimensional transform is given by

$$F(u,v,w)=\sum_{x=0}^{N_1-1}\sum_{y=0}^{N_2-1}\sum_{t=0}^{N_3-1}f(x,y,t)K(x,u,y,v,t,w),$$

where the kernel functions $K(x,u,y,v,t,w)$ represent the basis functions of the expansion. For example, the kernel function for the discrete Fourier transform is

$$K(x,u,y,v,t,w)=(N_1N_2N_3)^{-1}\exp\left[-2\pi j(ux/N_1+vy/N_2+wt/N_3)\right].$$

The Fourier kernel function is product separable; therefore, the three-dimensional transform may be computed using only three one-dimensional transforms. For the Fourier transform, the one-dimensional transform is of the form

$$G(u)=\frac{1}{N}\sum_{x=0}^{N-1}g(x)\exp\left(\frac{-2\pi jux}{N}\right),$$

where $g(x)$ is the complex function resulting from the intermediate transforms. For the discrete cosine transform the one-dimensional transform is of the form

$$G(u)=\frac{1}{N}\sum_{x=0}^{N-1}g(x)\cos\left[(2x+1)\frac{u\pi}{2N}\right],$$

and $g(x)$ and $G(u)$ are real-valued functions.

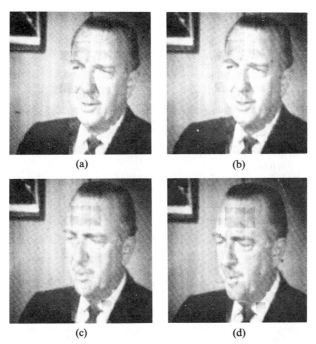

(a)    (b)

(c)    (d)

*Figure 6.31* Three-dimensional cosine transform coding results of a 16-frame data base at 1 bit/pixel using $16 \times 16 \times 16$ blocksize: (a) frame 1, (b) frame 2, (c) frame 8, and (d) frame 16. [From Robinson and Roese (1975).]

An example of three-dimensional transform coding using the cosine transform is shown in Fig. 6.31 (Robinson and Roese, 1975). Four frames out of a 16-frame sequence are shown. The three-dimensional transform was computed on blocks of size $16 \times 16 \times 16$ on 16 frames of size $256 \times 256$. The average bit rate used was 1.0 bits/pixel.

### Interframe DPCM Coding

Predictive coding techniques may also be used to decorrelate the samples for interframe coding. The DPCM implementation requires the prediction of each picture element using a set of previously scanned elements. One implementation of the DPCM predictor may be written

$$\hat{f}(x,y,t) = \alpha f(x-1,y,t) + \beta f(x,y-1,t) + \gamma f(x,y,t-1).$$

The prediction parameters may be selected to minimize the mean squared prediction error

$$\overline{\varepsilon^2} = E\left\{ \left[ f(x,y,t) - \hat{f}(x,y,t) \right]^2 \right\}.$$

Assuming stationarity, the mean squared error may be expressed in terms of the correlation values

$$R(i,j,k) = E\left[ f(x,y,t)f(x+i,y+j,t+k) \right],$$

$$\overline{\varepsilon^2} = (1 + \alpha^2 + \beta^2 + \gamma^2)R(0,0,0) - 2\alpha R(1,0,0) - 2\beta R(0,1,0)$$
$$- 2\gamma R(0,0,1) + 2\alpha\beta R(1,1,0) + 2\alpha\gamma R(1,0,1) + 2\beta\gamma R(0,1,1).$$

This error may be minimized by differentiating with respect to the parameters, which results in the following algebraic equations:

$$\alpha R(0,0,0) + \beta R(1,1,0) + \gamma R(1,0,1) = R(1,0,0),$$
$$\alpha R(1,1,0) + \beta R(0,0,0) + \gamma R(0,1,1) = R(0,1,0),$$
$$\alpha R(1,0,1) + \beta R(0,1,1) + \gamma R(0,0,0) = R(0,0,1).$$

Given the correlation values, the parameters $\alpha$, $\beta$, and $\gamma$ may be determined. Note that the correlation values may be estimated from the correlation coefficients from sample messages.

An example of interframe DPCM coding is shown in Fig. 6.32 for frames 1, 2, 8, and 16 of a 16-frame sequence (Robinson and Roese, 1975). For the first frame only spatial prediction was used. For the remaining frames the three

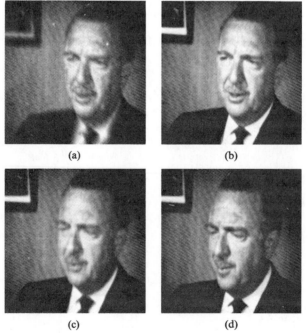

(a)　　　　　　　　　　　　(b)

(c)　　　　　　　　　　　　(d)

*Figure 6.32*　Three-dimensional DPCM coding results at 1.0 bit/pixel: (a) frame 1, (b) frame 2, (c) frame 8, and (d) frame 16. [From Robinson and Roese (1975).]

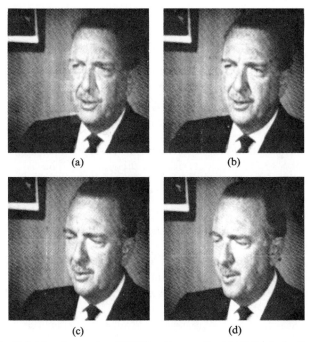

(a)                              (b)

(c)                              (d)

*Figure 6.33*   Hybrid cosine–cosine–DPCM coding results at 1.0 bit/pixel with 16×16 block-size: (a) frame 1, (b) frame 2, (c) frame 8, and (d) frame 16. [From Robinson and Roese (1975).]

spatial and temporal elements were used to predict the present value. The coding rate is 1.0 bits/pixel.

### Hybrid Interframe Coding

A variety of combinations of coding techniques may be used for each of the three dimensions of $f(x,y,t)$. Hybrid techniques offer the capability of matching the image characteristics in each dimension. For example, one may consider transform techniques for the spatial dimensions and DPCM for the temporal dimension. Only a single frame of storage would be required for this combination.

An example of hybrid interframe coding using the two-dimensional cosine transform for each frame and DPCM for the temporal dimension (Robinson and Roese, 1975) is shown in Fig. 6.33. Again, frames 1, 2, 8, and 16 from a 16-frame sequence are shown. The average bit rate used was 1.0 bits/pixel. The normalized mean squared error was 0.017%. The storage requirement may be estimated as the product of the number of frames times the number of pixels/frame times the average number of bits/pixel. For this hybrid technique, the total storage requirement is only $1 \times 8 \times 256^2 = 512$ Kbits.

**Bibliographical Guide**

The theoretical foundations of image encoding and compression are described in the classical book by Shannon and Weaver (1949). The book by Abramson (1963) is an excellent introduction. More detailed descriptions may be found in the books by Gallager (1965), Ash (1968), and Berger (1971), and the papers by Kolmogorov (1956), Peterson and Middleton (1963), Kortman (1967a, b), Hayes *et al.* (1970). Tasto and Wintz (1971), and Davisson (1972).

Digital transmission of television is an exciting application of the encoding and compression techniques described in this chapter. The astute reader will also observe that the analog–digital trade-off appears favorable to digital techniques. The possibility of digital television and several general methods are described by Mertz and Grey (1934), Lewis (1954), Laemmel (1951), Schreiber and Knapp (1958), Julesz (1959), Gabor and Hill (1961), Seyler (1962), Roberts (1962), Cherry *et al.* (1963), Deutsch (1965), Schwartz and Barker (1966), Hochman *et al.* (1967), Thompson and Sparkes (1967), Robinson and Cherry (1967), Richards and Bisignani (1967), Ehrman (1967), Graham (1967), Seitzer *et al.* (1969), Brainard and Candy (1969), Limb (1967, 1969, 1970), Huang and Tretiak (1968, 1972), Landau and Slepian (1971), Class (1971), Stucki (1971), Millard and Maunsell (1971), Thompson (1971), Beaudette (1972), and May and Spencer (1972), Maier and Gardenhire (1972), Bauch *et al.* (1974), Pearson (1975), Goldberg (1976), and Pratt (1978).

The theory of reversible encoding techniques for minimum redundancy is described by Huffman (1952). Run length encoding is described by Capon (1959) and Goloumb (1966). The use of reversible encoding for pictures is described by Happ (1969), Wintz and Wilkins (1970), White *et al.* (1972), and Huang (1972).

Quantization techniques are described by Max (1960), Huang and Schultheiss (1963), Kimme and Kuo (1963), Roe (1964), Bruce (1965), Wood (1969), Curry (1970), Lippel *et al.* (1971), Woods and Huang (1972), Kretz (1975), Huhns (1975), and Pratt (1978).

Transform coding techniques are described by Huang (1966), Anderson and Huang (1969, 1971, 1972), Wintz and Kurtenbach (1968), Wilkins and Wintz (1969, 1970), Andrews *et al.* (1969), Andrews and Pratt (1969), Gattis and Wintz (1971), Habibi and Wintz (1971), Habibi (1971, 1974), Enomoto and Shibata (1971), Pearl (1971), Jain (1974), Pratt *et al.* (1974), Habibi and Hershel (1974), Ahmed *et al.* (1974), Habibi and Robinson (1974), Rao *et al.* (1975), Reis *et al.* (1976), Andrews and Patterson (1976b), and Hall (1978).

Predictive coding techniques including pulse code modulation and differential pulse code modulation are described by Goodall (1951), De Jager (1952), Harrison (1952), Kramer and Mathews (1956), Graham (1958), Cabrey (1960), Huang (1965), Bisignani *et al.* (1966), Harper (1966), Protonotarios (1966), Limb and Mounts (1969), Aughenbough *et al.* (1970), Abbott (1971),

Chow (1971), Kutz and Davisson (1971), Connor *et al.* (1971), Brown (1972), and Kaul and Golding (1972).

Delta modulation techniques are described by De Jager (1952), Winkler (1963, 1965), Inose and Yasuda (1963), Balder and Kramer (1962), Young and Mott-Smith (1965), O'Neal (1966, 1967, 1971), Bosworth and Candy (1969), Schindler (1970), and Candy (1971). Interpolation techniques are described by Andrews (1976) and Andrews and Patterson (1976a).

Adaptive coding techniques are described by Cutler (1952), Gardenhire (1964), Andrews *et al.* (1967), Katz and Sciulli (1968), Jayant (1970), Tasto and Wintz (1971), Kamunski and Brown (1971), Rice and Plaunt (1971), Frei *et al.* (1971), Limb (1972), Andrews and Tescher (1974), and Cox and Tescher (1976).

Interframe coding techniques are described by Seyler and Budrikis (1965), Mounts (1969, 1972), Cunningham (1970), Pease and Limb (1971), Limb and Pease (1971), Limb (1972), Haskell *et al.* (1972), Stockham (1972), Robinson and Roese (1975), Roese and Pratt (1976), and Roese *et al.* (1977).

Color and multispectral coding techniques are described by Limb *et al.* (1971), Golding and Garlow (1971), Pratt (1971), Bhushan (1972), Habibi (1972), Jain (1972), Jain and Pratt (1972), Chen (1976), and Hall (1978).

Studies on the statistics of images, effects of errors and image quality are described by Cowan (1928), Mertz *et al.* (1950), Kretzmer (1952), Farnsworth (1958), Linfoot (1958), Fredenhall and Behrend (1960), Dean (1960), Knight (1962), Cherry *et al.* (1963), Prosser *et al.* (1964), Schade (1964), Seyler (1965), Scoville and Huang (1965), Nishikawa *et al.* (1965), McLean (1966), Harris (1966), Frieden (1966), Brainard (1967), Brainard *et al.* (1967), Graham (1967), Huang *et al.* (1967, 1971), Erdmann and Neal (1968), Weaver (1968), Arps *et al.* (1969), Boyce (1976), Arguello *et al.* (1971), Lippel and Kurland (1971), Barnard (1971), Sakrison and Algazi (1971), Wilder (1972), Schreiber *et al.* (1972), Pearson (1972), Snyder (1973), Dainty and Shaw (1974), and Preuss (1975).


## PROBLEMS

**6.1**  Differential pulse code modulation.   Use single element prediction to develop a DPCM coding system for the sample image given in Fig. 3.21a. First, compute the horizontal difference image using line to line and top to bottom wraparound. Then, compute the histogram of the difference image.

(a)  Using the histogram as an estimate of the first difference probability density, determine a Huffman code for the image. What is the compression ratio?

(b)  Estimate the single element correlation and implement the minimum mean squared error encoding and decoding systems.

**6.2**  Bit plane encoding.   Investigate the efficiency of bit plane encoding for

the image given in Fig. 3.21a. First, divide the image into 3 bit planes, corresponding to the 3 bit or 8 gray level representation. Then use run length encoding for each line in the bit plane image. What is the compression ratio?
**6.3**  Transform coding.  Use the full image transform coding method described in Sec. 6.3.2 to encode the image of Fig. 3.21a. Use the single element horizontal pixel correlation computed on a line to line and bottom to top wraparound basis for the model.
(a)   Use the Fourier transform.
(b)   Use the cosine transform.
(c)   Use the Hadamard transform.
What are the mean squared errors for (a), (b), and (c) for a rate of 1 bit/pixel?

# References

Abbott, R. P. (1971). A Differential Pulse-Code-Modulation Coder for Videotelephony Using Four Bits per Sample. *IEEE Trans. Commun. Technol.* **19**, No. 6, 907–913.
Abramson, N. (1963). "Information Theory and Coding." McGraw-Hill, New York.
Ahmed, N., Natarajan, T., and Rao, K. R. (1974). On Image Processing and a Discrete Cosine Transform. *IEEE Trans. Comput.* **23** (January), 90–93.
Anderson, G. B., and Huang, T. S. (1969). Errors in Frequency-Domain Processing of Images. *Proc. Spring Jt. Comput. Conf., Boston, Mass.* **34**, 173–185.
Anderson, G. B., and Huang, T. S. (1971). Picture Bandwidth Compression by Piecewise Fourier Transformation. *IEEE Trans. Commun. Technol.* **19**, 133–140.
Anderson, G. B., and Huang, T. S. (1972). Piecewise Fourier Transformation for Picture Bandwidth Compression. *IEEE Trans. Commun.* **20**, No. 3, 488–491.
Andrews, C. A., Davies, J. M., and Schwartz, G. R. (1967). Adaptive Data Compression. *Proc. IEEE* **55**, No. 3, 267–277.
Andrews, H. C. (1970). "Computer Techniques in Image Processing." Academic Press, New York.
Andrews, H. C. (1976). Digital Interpolation of Discrete Images. *IEEE Trans. Comput.* **25**, No. 2, 196–202.
Andrews, H. C., and Patterson, C. L. (1976a). Digital Interpolation of Discrete Images. *IEEE Trans. Comput.* **25**, No. 2, 196–202.
Andrews, H. C., and Patterson, C. L. (1976b). Singular Value Decomposition (SVD) Image Coding. *IEEE Trans. Commun.* **24**, No. 4, 425–432.
Andrews, H. C., and Pratt, W. K. (1969). Transform Image Coding. *Proc. Symp. Comput. Process. Commun., Polytechnic Inst. Brooklyn, New York, April* **19**, 63–84.
Andrews, H. C., and Tescher, A. G. (1974). The Role of Adaptive Phase Coding in Two and Three Dimensional Fourier and Walsh Image Compression. *Proc. Walsh Funct. Symp., Washington, D.C., March.*
Andrews, H. C., Kane, J., and Pratt, W. K. (1969). Hadamard Transform Image Coding. *Proc. IEEE* **57**, No. 1, 58–68.
Arguello, R. J., Sellner, H. R., and Stuller, J. A. (1971). The Effect of Channel Errors in the Differential Pulse-Code Modulation Transmission of Sampled Imagery. *IEEE Trans. Commun. Technol.* **19**, No. 6, 926–933.
Arps, R. B., Erdmann, R. L., Neal, A. S., and Schlaepfer, C. E. (1969). Character Legibility Versus Resolution in Image Processing of Printed Matter. *IEEE Trans. Man-Mach. Syst.* **10**, No. 3, 66–71.

Ash, R. B. (1968). "Information Theory and Reliable Communication." Wiley, New York.

Aughenbough, G. W., Irwin, J. D., and O'Neal, J. B. (1970). Delayed Differential Pulse Code Modulation. *Proc. Annu. Princeton Conf. 2nd, October* pp. 125–130.

Balder, J. C., and Kramer, C. (1962). Video Transmission by Delta Modulation Using Tunnel Diodes. *Proc. IRE* **50**, No. 4, 428–431.

Barnard, T. W. (1971). An Image Evaluation Method. *Symp. Sampled Images, Perkin-Elmer Corp., Norwalk, Conn., June*, pp. 5-1–5-7.

Bauch, H. H., Musmann, H. G., Ohnsorge, H., and Wengenroth, G. A. (1974). Picture Coding. *IEEE Trans. Commun.* **20**, No. 9, 1158–1167.

Beaudette, C. G. (1972). An Efficient Facsimile System for Weather Graphics. *In* "Picture Bandwidth Compression" (T. S. Huang and O. J. Tretiak, eds.), pp. 217–220. Gordon & Breach, New York.

Berger, T. (1971). "Rate Distortion Theory: A Mathematical Basis for Data Compression" Prentice-Hall, Englewood Cliffs, New Jersey.

Bhushan, A. K. (1972). Transmission and Coding of Color Pictures. *In* "Picture Bandwidth Compression" (T. S. Huang and O. J. Tretiak, eds.), pp. 697–725. Gordon & Breach, New York.

Bisignani, W. T., Richards, G. P., and Whelan, J. W. (1966). The Improved Grey Scale and the Coarse–Fine PCM Systems: Two New Digital TV Bandwidth Reduction Techniques. *Proc. IEEE* **54**, No. 3, 376–390.

Bosworth, R. H., and Candy, J. C. (1969). A Companded One-Bit Coder for Television Transmission. *Bell Syst. Tech. J.* **48**, No. 5, 1459–1479.

Boyce, B. (1977). A Summary Measure of Image Quality. *Proc. Symp. Curr. Math. Probl. Image Sci., Nav. Postgrad. Sch., Monterey, Calif., November*.

Brainard, R. C. (1967). Subjective Evaluation of PCM Noise-Feedback Coder for Television. *Proc. IEEE* **55**, No. 3, 346–353.

Brainard, R. C., and Candy, J. C. (1969). Direct-Feedback Coders: Design and Performance with Television Signals. *Proc. IEEE* **57**, No. 5, 776–786.

Brainard, R. C., Mounts, F. W., and Prasada, B. (1967). Low Resolution TV: Subjective Effects of Frame Repetition and Picture Replenishment. *Bell Syst. Tech. J.* **46**, No. 1, 261–271.

Brown, E. F. (1972). Sliding-Scale Operation of Differential Type PCM Codes for Television. *In* "Picture Bandwidth Compression" (T. S. Huang and O. J. Tretiak, eds.), pp. 303–322. Gordon & Breach, New York.

Bruce, J. D. (1965). Optimum Quantization. MIT Res. Lab. Electron., Tech. Rep. No. 429, March.

Cabrey, R. L. (1960). Video Transmission over Telephone Cable Pairs by Pulse Code Modulation. *Proc. IRE* **48**, No. 9, 1546–1551.

Candy, J. C. (1971). Refinement of a Delta Modulator. *In* "Picture Bandwidth Compression" (T. S. Huang and O. J. Tretiak, eds.), pp. 323–339. Gordon & Breach, New York.

Capon, J. (1959). A Probabilistic Model for Run-Length Coding of Pictures. *IRE Trans. Inf. Theory* **5**, No. 4, 157–163.

Chen, W. H. (1976). Adaptive Coding of Color Images Using Cosine Transform. *Proc. Int. Commun. Conf., Philadelphia, Pa., June*, pp. 47-7–47-13.

Cherry, C., Kubba, M. H., Pearson, D. E., and Barton, M. P. (1963). An Experimental Study of the Possible Bandwidth Compression of Visual Image Signals. *Proc. IEEE* **51**, 1507–1517.

Chow, M. C. (1971). Variable-Length Redundancy Removal Coders for Differentially Coded Video Telephone Signals. *IEEE Trans. Commun. Technol.* **19**, No. 6, 922–926.

Class, F. (1971). Video Bandwidth Compression by Multiplexing: Experimental Results. *IEEE Trans. Commun. Technol.* **19**, No. 3, 371–372.

Connor, D. J., Pease, R. F. W., and Scholes, W. G. (1971). Television Coding Using Two-Dimensional Spatial Prediction. *Bell Syst. Tech. J.* **50** (March), 1049–1063.

Cowan, A. (1928). Test Cards for Determination of Visual Acuity. *Arch. Ophtholmol.* **57**, 283–295.

Cox, R. B., and Tescher, A. G. (1976). Channel Rate Equalization Techniques for Adaptive Transform Coders. *Proc. SPIE Conf. Adv. Image Transmission Tech.*, *San Diego, Calif.*, *August*, Vol.187.

Cunningham, J. E. (1970). Frame Correction Coding. *In* "Picture Bandwidth Compression" (T. S. Huang and O. J. Tretiak, eds.), pp. 623–652. Gordon & Breach, New York.

Curry, R. (1970). "Estimation and Control with Quantized Measurements." MIT Press, Cambridge, Massachusetts.

Cutler, C. C. (1952). Differential Quantization of Communication Signals. U.S. Patent No. 2,605,361, July 29.

Dainty, J. C., and Shaw, R. (1974). "Image Science." Academic Press, New York.

Davisson, L. D. (1972). Rate Distortion Theory and Applications. *Proc. IEEE* **60**, No. 7, 800–808.

Dean, C. E. (1960). Measurements of the Subjective Effects of Interference in Television Reception. *Proc. IRE* **48**, No. 6, 1035–1050.

De Jager, F. (1952). Deltamodulation: A Method of PCM Transmission Using a One-Unit Code. *Philips Res. Rep.* **7**, 442–466.

Deutsch, S. (1965). Pseudo-Random Dot Scan Television Systems. *IEEE Trans. Broadcast.* **11**, No. 1, 11–21.

Ehrman, L. (1967). Analysis of Some Redundancy Removal Bandwidth Compression Techniques. *Proc. IEEE* **55**, No. 3, 278–287.

Enomoto, H., and Shibata, K. (1971). Orthogonal Transform Coding System for Television Signals. *Appl. Walsh Funct. Symp.*, *Washington, D.C.* pp. 11–17.

Erdmann, R. L., and Neal, A. S. (1968), Word Legibility as a Function of Letter Legibility with Word Size, Word Familiarity, and Resolution as Parameters. *J. Appl. Psychol.* **52**, No. 5, 403–409.

Farnsworth, D. (1958). A Temporal Factor in Colour Discrimination. *In* Visual Problems in Colour, II. *Natl. Phys. Lab.* (*U.K.*), *Symp.* No. 8, p. 429.

Fox, L. (1954). Practical Solution of Linear Equations and Inversion of Matrices. *Natl. Bur. Stand.* (*U.S.*), *Appl. Math. Ser.* **39**, 1–54.

Fredenhall, G. L., and Behrend, W. L. (1960). Picture Quality-Procedures for Evaluating Subjective Effects of Interference. *Proc. IRE* **48**, No. 6, 1030–1034.

Frei, A. H., Schindler, H. R., and Vettiger, P. (1971). An Adaptive Dual-Mode Coder/Decoder for Television Signals. *IEEE Trans. Commun. Technol.* **19**, No. 6, 933–944.

Frieden, B. R. (1966). Image Evaluation by Use of the Sampling Theorem. *J. Opt. Soc. Am.* **56**, No. 10, 1355–1362.

Gabor, D., and Hill, P. C. J. (1961). Television Bandwidth Compression by Contour Interpolation. *IEE Conf. Proc.* (*London*) **108B**, No. 39, 303–315, 634.

Gallagher, R. B. (1965). "Information Theory." Wiley (Interscience), New York.

Gardenhire, L. W. (1964). Redundancy Reduction, the Key to Adaptive Telemetry. *Proc. Natl. Telemeter. Conf.*, *Los Angeles, Calif.*, pp. 1-5-1, 1-5-16.

Gattis, J. L., and Wintz, P. A. (1971). Automated Techniques for Data Analysis and Transmission. Rep. TR-EE-37, Sch. Eng., Purdue Univ., Lafayette, Indiana, August.

Goldberg, A. A. (1976). Digital Techniques Promise to Clarify the Television Picture. *Electronics* February 5, 94–100.

Golding, L. S., and Garlow, R. (1971). Frequency Interleaved Sampling of a Color Television Signal. *IEEE Trans. Commun. Technol.* **19**, 972–979.

Golomb, S. W. (1966). Run Length Encodings. *IEEE Trans. Inf. Theory* **12**, No. 3, 399–401.

Goodall, W. M. (1951). Television Transmission by Pulse Code Modulation. *Bell Syst. Tech. J.* **30** (January), 33–49.

Graham, D. N. (1967). Image Transmission by Two-Dimensional Contour Coding. *Proc. IEEE* **55**, No. 3, 336–346.

Graham, R. E. (1958). Predictive Quantizing of Television Signals. *IRE WESCON Conv. Rec.* Part 4, pp. 142–157.

Habibi, A. (1971). Comparison of *n*th-order DPCM Encoder with Linear Transformation and Block Quantization Techniques. *IEEE Trans. Commun. Technol.* **19**, 948–956.

Habibi, A. (1972). Delta Modulation and DPCM Coding of Color Signals. *Proc. Int. Telemeter. Conf., Los Angeles, Calif., October* **8**, 333–343.

Habibi, A. (1974). Hybrid Coding of Pictorial Data. *IEEE Trans. Commun.* **22**, 614–621.

Habibi, A., and Hershel, R. S. (1974). A Unified Representation of Differential Pulse Code Modulation (DPCM) and Transform Coding Systems. *IEEE Trans. Commun.* **22**, 692–696.

Habibi, A., and Robinson, G. S. (1974). A Survey of Digital Picture Coding. *IEEE Comput.* **7**, No. 5, 21–34.

Habibi, A., and Wintz, P. A. (1971). Image Coding by Linear Transformation and Block Quantization. *IEEE Trans. Commun. Technol.* **19**, 50–62.

Hall, C. F. (1978). Digital Color Image Compression in a Perceptual Space. Ph.D. thesis, Univ. of Southern California, Los Angeles, February (unpublished).

Happ, W. W. (1969). Coding Schemes for Run-Length Information, Based on Poisson Distribution. *Natl. Telemeter. Conf. Rec., Washington, D.C., April*, pp. 257–261.

Harper, L. H. (1966). PCM Picture Transmissions. *IEEE Spectrum* **3** (June), 146.

Harris, J. L., Sr. (1966). Image Evaluation and Restoration. *J. Opt. Soc. Am.* **56**, No. 5, 569–574.

Harrison, C. W. (1952). Experiments with Linear Prediction in Television. *Bell Syst. Tech. J.* **31**, No. 4, 746–783.

Haskell, B. G., Mounts, F. W., and Candy, J. C. (1972). Interframe Coding of Videotelephone Pictures. *Proc. IEEE* **60**, No. 7, 792–800.

Hayes, J. F., Habibi, A., and Wintz, P. A. (1970). Rate Distortion Function for a Gaussian Source Model of Images. *IEEE Trans. Inf. Theory* **16** (July), 507–508.

Hochman, D., Katzman, H., and Weber, D. R. (1967). Application of Redundancy Reduction to Television Bandwidth Compression. *Proc. IEEE* **55**, No. 3, 263–267.

Huang, J. J. Y., and Schultheiss, P. M. (1963). Block Quantization of Correlated Gaussian Random Variables. *IEEE Trans. Commun. Syst.* **11**, No. 3, 289–296.

Huang, T. S. (1965). PCM Picture Transmission. *IEEE Spectrum* **2**, No. 12, 57–60.

Huang, T. S. (1966). Digital Picture Coding. *Proc. Natl. Electron. Conf., Chicago, October*, pp. 793–797.

Huang, T. S. (1972). Run-Length Coding and Its Extensions. *In* "Picture Bandwidth Compression" (T. S. Huang and O. J. Tretiak, eds.), pp. 221–266. Gordon & Breach, New York.

Huang, T. S., and Tretiak, O. J. (1968). A Pseudorandom Multiplex System for Facsimile Transmission. *IEEE Trans. Commun. Technol.* **16**, No. 3, 436–438.

Huang, T. S. and Tretiak, O. J., eds. (1972). "Picture Bandwidth Compression." Gordon & Breach, New York.

Huang, T. S., Tretiak, O. J. Prasada, B., and Yamaguchi, Y. (1967). Design Considerations in PCM Transmission of Low-Resolution Monochrome Still Pictures. *Proc. IEEE* **55**, No. 3, 331–335.

Huang, T. S., Schreiber, W. F., and Tretiak, O. J. (1971). Image Processing. *Proc. IEEE* **59**, No. 11, 1586–1609.

Huffman, D. A. (1952). A Method for the Construction of Minimum-Redundancy Codes. *Proc. IRE* **40**, No. 9, 1098–1101.

Huhns, M. N. (1975). Optimum Restoration of Quantized Correlated Signals. Ph. D. thesis, Univ. of Southern California, Los Angeles, August (unpublished).

Inose, H., and Yasuda, Y. (1963). A Unit Bit Coding Method by Negative Feedback. *Proc. IEEE* **51**, No. 11, 1524–1535.

Jain, A. K. (1972). Color Distance and Geodesics in Color 3 Space. *J. Opt. Soc. Am.* **62**, No. 11, 1287–1290.

Jain, A. K. (1974). A Fast Karhunen-Loeve Transform for Finite Discrete Images. *Proc. Nat. Electron. Conf., Chicago, Ill., October*, pp. 322–328.

Jain, A. K., and Pratt, W. K. (1972). Color Image Quantization. *Natl. Telecommun. Conf. Rec. Houston, Tex., December* IEEE Publ. No. 72CHO 601-5-NTC, pp. 34D1–34D6.

364 6 Digital Television, Encoding, and Data Compression

Jayant, N. S. (1970). Adaptive Delta Modulation with One-Bit Memory. *Bell Syst. Tech. J.* **49**, No. 3, 321–342.

Julesz, B. (1959). A Method of Coding TV Signals Based on Edge Detection. *Bell Syst. Tech. J.* **38**, No. 4, 1001–1020.

Kamunski, W., and Brown, E. F. (1971). An Edge-Adaptive Three-Bit Ten-Level Differential PCM Coder for Television. *IEEE Trans. Commun. Technol.* **19**, No. 6, 944–947.

Katz, R. L., and Sciulli, J. A. (1968). The Performance of an Adaptive Image Data Compression System in the Presence of Noise. *IEEE Trans. Inf. Theory* **14**, No. 2, 273–279.

Kaul, P., and Golding, L. (1972). A DPCM Code Using Edge Coding and Line Replacement. *Natl. Telecommun. Conf. Rec., Houston, Tex., December* IEEE Publ. No. 72CHO 601-5-NTC, pp. 34B-1–34B-6.

Kimme, E. G., and Kuo, F. F. (1963). Synthesis of Optimal Filters for a Feedback Quantization System. *IEEE Trans. Circuit Theory* **10**, No. 3, 405–413.

Knight, J. M. (1962). Maximum Acceptable Bit Error Rates for PCM Analog and Digital TV Systems. *Proc. Natl. Telemeter. Conf.*, pp. 1–9.

Kolmogorov, A. N. (1956). On the Shannon Theory of Information Transmission in the Case of Continuous Signals. *IRE Trans. Inf. Theory* **2**, 102–108.

Kortman, C. M. (1967a). Data Compression by Redundancy Reduction. *IEEE Spectrum* **4** (March), 133–139.

Kortman, C. M. (1967b). Redundancy Reduction: A Practical Method of Data Compression. *Proc. IEEE* **55**, No. 3, 253–263.

Kramer, H. P., and Mathews, M. V. (1956). A Linear Coding for Transmitting a Set of Correlated Signals. *IRE Trans. Inf. Theory* **2** (September), 41–46.

Kretz, F. (1975). Subjectively Optimal Quantization of Pictures. *IEEE Trans. Commun.* **23**, No. 11, 1288–1292.

Kretzmer, E. R. (1952). Statistics of Television Signals. *Bell Syst. Tech. J.* **31**, No. 4, 751–763.

Kutz, R. L., and Davisson, L. D. (1971). A Real-Time Programmable Data Compression System, *Proc. Int. Telemetering Conf., Washington, D. C., September*, pp. 20–25.

Laemmel, A. E. (1951). Coding Processes for Bandwidth Reduction in Picture Transmission. Rep. R 246-251, Microwave Inst., Polytechnic Inst. of Brooklyn, New York, August.

Landau, H. J., and Slepian, D. (1971). Some Computer Experiments in Picture Processing for Bandwidth Reduction. *Bell Syst. Tech. J.* **50**, 1525–1540.

Lewis, N. W. (1954). Waveform Responses of Television Links. *IEE Conf. Proc. (London)* **101**, Part 3, 258–270.

Limb, J. O. (1967). Source-Receiver Encoding of Television Signals. *Proc. IEEE* **55** No. 3, 364–379.

Limb, J. O. (1969) Design of Dither Waveforms for Quantized Visual Signals. *Bell Syst. Tech. J.* **48**, No. 7, 2555–2583.

Limb, J. O. (1970). Efficiency of Variable Length Binary Encoding. *Proc. UMR-Mervin J. Kelly Commun. Conf., Univ. Missouri, Rolla, October* pp. 13.3-1–13.3-9.

Limb, J. O. (1972). Adaptive Encoding of Picture Signals. *In* "Picture Bandwidth Compression" (T. S. Huang and O. J. Tretiak, eds.), pp. 341–382. Gordon & Breach, New York.

Limb, J. O., and Mounts, F. W. (1969). Digital Differential Quantizer for Television. *Bell Syst. Tech. J.* **48**, 2583–2599.

Limb, J. O., and Pease, R. F. W. (1971). A Simple Interframe Coder for Video Telephony. *Bell Syst. Tech. J.* **50**, No. 6, 1877–1888.

Limb, J. O., Rubinstein, C. B., and Walsh, K. A. (1971). Digital Coding of Color Picturephone Signals by Element-Differential Quantization. *IEEE Trans. Commun. Technol.* **19**, 992–1006.

Linfoot, E. H. (1958). Quality Evaluations of Optical Systems. *Opt. Acta* **5**, Nos. 1/2, 1–14.

Lippel, B., and Kurland, M. (1971). The Effect of Dither on Luminance Quantization of Pictures. *IEEE Trans. Commun. Technol.* **19**, No. 6, 879–889.

Lippel, B., Kurland, M., and Marsh, A. H. (1971). Ordered Dither Patterns for Coarse Quantization of Pictures. *Proc. IEEE* **59**, No. 3, 429–431.

McLean, F. C. (1966). Worldwide Color Television Standards. *IEEE Spectrum* **3**, No. 6, 59–60.

Maier, J. L., and Gardenhire, L. (1972). Redundant Area Coding System (REARCS). *Proc. Int. Telemeter. Conf., Los Angeles, Calif., October* **8**, 301–314.

Martin, R. S., and Wilkinson, J. W. (1965). Symmetric Decomposition of a Positive Definite Matrix. *Numer. Math.* **7**, 362–383.

Max, J. (1960). Quantizing for Minimum Distortion. *IRE Trans. Inf. Theory* **6**, 7–12.

May, C. L., and Spencer, D. J. (1972). Data Compression for Earth Resources Satellites. *Proc. Int. Telemeter. Conf., Los Angeles, Calif., October* **8**, 352–362.

Mertz, P., and Grey, F. (1934). A Theory of Scanning and Its Relation to the Characteristics of the Transmitted Signal in Telephotography and Television. *Bell Syst. Tech. J.* **13**, 464–515.

Mertz. P., Fowler, A. D., and Christoper, H. N., (1950). Quality Rating of Television Images. *Proc. IRE* **38**, No. 11, 1269–1283.

Millard, J. B., and Maunsell, H. I. (1971). Digital Encoding of Video Signals. *Bell Syst. Tech. J.* **50**, No. 2, 459–479.

Mounts, F. W. (1969). A Video Encoding System Using Conditional Picture-Element Replenishment. *Bell Syst. Tech. J.* **48**, No. 7, 2545–2555.

Mounts, F. W. (1972). Frame-to-Frame Digital Processing of TV Pictures to Remove Redundancy. *In* "Picture Bandwidth Compression" (T. S. Huang and O. J. Tretiak, eds.), pp. 653–673. Gordon & Breach, New York.

Nishikawa, A., Massa, R. J., and Mott-Smith, J. C. (1965). Area Properties of Television Pictures. *IEEE Trans. Inf. Theory* **11**, 348–352.

O'Neal, J. B. (1966). Predictive Quantizing Systems (Differential Pulse Code Modulation) for the Transmisson of Television Signals. *Bell Syst. Tech. J.* **45**, No. 5, 689–721.

O'Neal, J. B. (1967). A Bound on Signal-to-Quantizing Noise Ratios for Digital Encoding Systems. *Proc. IEEE* **55**, No. 3,287–292.

O'Neal, J. B. (1971). Entropy Coding in Speech and Television Differential PCM Systems. *IEEE Trans. Inf. Theory* **17**, No. 6, 758–761.

Pearl, J. (1971). Basis-Restricted Transformations and Performance Measures for Spectral Representations. *IEEE Trans. Inf. Theory* **17**, 751–752.

Pearson, D. E., (1972). Methods for Scaling Television Picture Quality. *In*, "Picture Bandwidth Compression" (T. S. Huang and O. J. Tretiak, eds.), pp. 47–96. Gordon & Breach, New York.

Pearson, D. E. (1975). "Transmission and Display of Pictorial Information." Wiley (Halsted Press), New York.

Pease, R. F. W., and Limb, J. O. (1971). Exchange of Spatial and Temporal Resolution in Television Coding. *Bell Syst. Tech. J.* **50**, No. 1, 191–200.

Peterson, D. P., and Middleton, D. (1963). Sampling and Reconstruction of Wave-Number-Limited Functions in *n*-dimensional Euclidean Spaces. *Inf. Control* **5**, 279–323.

Pratt, W. K. (1971). Spatial Transform Coding of Color Images. *IEEE Trans. Commun. Technol.* **19**, No. 6, 980–982.

Pratt, W. K. (1978). "Digital Image Processing." Wiley, New York.

Pratt, W. K., Chen, W. H., and Welch, L. R. (1974). Slant Transform Image Coding. *IEEE Trans. Commun.* **22**, No. 8, 1075–1093.

Preuss, D. (1975). Comparison of Two-Dimensional Facsimile Coding Schemes, *Int. Commun. Conf, San Francisco, Calif., June, Conf. Rec.* **1**, 7.12–7.16.

Prosser, R. D., Allnatt, J. W., and Lewis, N. W. (1964). Quality Grading of Impaired Television Pictures. *IEEE Conf. Proc. (London)* **111**, No. 3, 491–502.

Protonotarios, E. N. (1966). Slope Overload Noise in Differential Pulse Code Modulation Systems. *Bell Syst. Tech. J.* **46**, 689–721.

Rao, K. R., Narasimhan, M. A., and Revuluri, K. (1975). Image Data Processing by Hadamard–Haar Transforms. *IEEE Trans. Comput.* **23**, No. 9, 888–896.

Reis, J. J., Lynch, R. T., and Butman, J. (1976). Haar Transform Video Bandwidth Reduction System for RPVs. *Proc. SPIE Conf. Adv. Image Transmission Tech., San Diego, Calif., August*, Vol. 87.

Rice, R. F., and Plaunt J. R., (1971). Adaptive Variable-Length Coding for Efficient Compression of Spacecraft Television Data. *IEEE Trans. Commun. Technol.* **19**, 889–897.

Richards, G. P., and Bisignani, W. T. (1967). Redundancy Reduction Applied to Coarse–Fine Encoded Video. *Proc. IEEE* **55**, No. 10, 1707–1717.

Roberts, L. G. (1962). Picture Coding Using Pseudo-Random Noise. *IRE Trans. Inf. Theory* **8**, 145–154.

Robinson, A. H., and Cherry, C. (1967). Results of Prototype Television Bandwidth Compression Scheme. *Proc. IEEE* **55**, 356–563.

Robinson, G. S., and Roese, J. A. (1975). Interframe Image Coding. USC Tech. Rep., Univ. of Southern California, Los Angeles, March.

Roe, G. M. (1964). Quantizing for Minimum Distortion. *IEEE Trans. Inf. Theory* **10**, 384–385.

Roese, J. A., and Pratt, W. K. (1976). Theoretical Performance Models for Interframe Transform and Hybrid Transform/DPCM Coders. *Proc. SPIE Conf. Adv. Image Transmission Tech., San Diego, Calif., August* **87**, 172–179.

Roese, J. A., Pratt, W. K., and Robinson, G. S. (1977). Interframe Cosine Transform Image Coding. *IEEE·Trans. Commun.* **25**, No. 11.

Sakrison, D. J., and Algazi, V. R. (1971). Comparison of Line-by-Line and Two-Dimensional Encoding of Random Images. *IEEE Trans. Inf. Theory* **17** (July), 386–398.

Schade, O. H. (1964). Modern Image Evaluation and Television. *Appl. Opt.* **3**, No. 1, 17–23.

Schindler, H. R. (1970). Delta Modulation. *IEEE Spectrum* **7** (October), 69–78.

Schreiber, W. F., and Knapp, C. F. (1958). TV Bandwidth Reduction by Digital Coding. *IRE Natl. Conv. Rec.* **6**, Part 4, 88–89.

Schreiber, W. F., Huang, T. S., and Tretiak, O. J. (1972). Contour Coding of Images. *In* "Picture Bandwidth Compression" (T. S. Huang and O. J. Tretiak, eds.), pp. 443–448. Gordon & Breach, New York.

Schwartz, J. W., and Barker, R. C. (1966). Bit-Plane Encoding: A Technique for Source Encoding. *IEEE Trans. Aerosp. Electron. Syst.* **2**, No. 4, 385–392.

Scoville, F. W., and Huang, T. S. (1965). The Subjective Effect of Spatial and Brightness Quantization in PCM Picture Transmission. *NEREM Rec.* pp. 234–235.

Seitzer, D., Class, F., and Stucki, P. (1969). An Experimental Approach to Video Bandwidth Compression by Multiplexing. *IEEE Trans. Commun.* **17**, No. 5, 564–568.

Seyler, A. J. (1962). The Coding of Visual Signals to Reduce Channel-Capacity Requirements. *Proc. IEEE (London)* **109C** (September), 676–684.

Seyler, A. J. (1965). Statistics of Television Frame Differences. *Proc. IEEE* **53**, No. 42, 2127–2128.

Seyler, A. J., and Budrikis, Z. L. (1965). Detail Perception After Scene Changes in Television Image Presentations. *IEEE Trans. Inf. Theory* **11**, No. 1, 31–43.

Shannon, C. E., and Weaver, W. (1949). "The Mathematical Theory of Communication." Univ. of Illinois Press, Urbana. [Orig. publ. (1948)] *Bell Syst. Tech. J.* **27**, 379–423, 623, 656.

Snyder, H. L. (1973). Image Quality and Observer Performance. *In* "Perception of Displayed Information" (L. M. Biberman, ed.). Plenum, New York.

Spencer, D. R., and Huang, T. (1969). Bit Plane Encoding of Continuous-Tone Pictures. *Symp. Comput. Process. Commun., Polytechnic Inst. Brooklyn, New York, April*.

Stockham, T. G. (1972). Intra-Frame Encoding for Monochrome Images by Means of a Psychophysical Model Based on Nonlinear Filtering of Multiplied Signals. *In* "Picture Bandwidth Compression" (T. S. Huang and O. J. Tretiak, eds.), pp. 415–442. Gordon & Breach, New York.

Stucki, P. (1971). Limits of Instantaneous Priority Multiplexing Applied to Black-and-White Pictures. *IEEE Trans. Commun. Technol.* **19**, No. 2, 169–177.

Tasto, M., and Wintz, P. A. (1972). A Bound on the Rate-Distortion Function and Application to Images. *IEEE Trans. Inf. Theory* **18**, No. 1, 150–159.

Thompson, J. E. (1971). A 36-MBIT/S Television Codec Employing Pseudorandom Quantization. *IEEE Trans. Commun. Technol.* **19**, No. 6, 872–879.

Thompson, J. E., and Sparkes, J. J. (1967). A Pseudo-Random Quantizer for Television Signals. *Proc. IEEE* **55**, No. 3, 353–355

Weaver, L. E. (1968). The Quality Rating of Color Television Pictures. *J. Soc. Motion Pict. Telev. Eng.* **77**, No. 6, 610–612.

White, H. E., Lippman, M. D., and Powers, K. H. (1972). Dictionary Look-Up Encoding of Graphics Data. *In* "Picture Bandwidth Compression" (T. S. Huang and O. J. Tretiak, eds.), pp. 265–281. Gordon & Breach, New York.

Wilder, W. C. (1972). Subjectively Relevant Error Criteria for Pictorial Data Processing. Rep. TR-EE-72-34, Sch. Electr. Eng., Purdue Univ., Lafayette, Indiana, December.

Wilkins, L. C., and Wintz, P. A. (1970). Image Coding by Coding Contours. *Proc. Int. Conf. Commun., San Francisco, Calif., June,* Vol 1.

Winkler, M. R. (1963). High Information Delta Modulation. *IEEE Int. Conv. Rec.* Part 8, pp. 260–265.

Winkler, M. R. (1965). Pictorial Transmission with HIDM. *IEEE Int. Conv. Rec.* Part 1, pp. 285–291.

Wintz, P. A. (1972). Transform Picture Coding. *Proc. IEEE* **60**, 809–820.

Wintz, P. A., and Kurtenback, A. J. (1968). Waveform Error Control in PCM Telemetry. *IEEE Trans. Inf. Theory* **14**, No. 5, 650–661.

Wintz, P. A., and Wilkins, L. C. (1970). Studies of Data Compression. Part I: Picture Coding by Contours, Part II: Error Analysis of Run-Length Codes. Rep. TR-EE-70-17, Sch. Eng., Purdue Univ., Lafayette, Indiana, September.

Wood, R. C. (1969). On Optimum Quantization. *IEEE Trans. Inf. Theory* **15**, 248–252.

Woods, J. W., and Huang, T. S. (1972). Picture Bandwidth Compression by Linear Transformation and Block Quantization. *In* "Picture Bandwidth Compression" (T. S. Huang and O. J. Tretiak, eds.), pp. 555–573. Gordon & Breach, New York.

Young, I. T., and Mott-Smith, J. C. (1965). On Weighted PCM. *IEEE Trans. Inf. Theory* **11**, No. 4, 596–597.

Youngblood, W. A. (1968). Picture Processing. Q. Prog. Rep., MIT Res. Lab. Electron., January, pp. 95–100.

# 7 | Scene Understanding

## 7.1 Introduction

The purpose of this chapter is to describe an approach to a problem of great practical concern—describing physical objects in a scene from images of the scene. Since most physical objects do not produce images that possess a single measurable characteristic, the problem is very difficult. However, by noting that objects are usually composed of component parts that may possess a uniform characteristic *and* that are arranged in a particular way, a multistep subdivision of the problem may be devised. One possible set of subdivisions is shown in Fig. 7.1. The first step is simply observing the scene with imaging devices. The sensors could be sensitive to visible light, x rays, distance to objects, radar reflectance, or any other property of the physical world that may be imaged. No attempt will be made here to describe the sensors; rather, it will be assumed that at least one image function $f(x, y)$ of the scene may be recorded. The first processing step is to segment the image into regions of common characteristics. Only in rare cases will this step result in regions that correspond to objects. Therefore, one might next measure certain properties of the segmented regions, such as gray level, size, shape, etc. Also, the relations between regions must be considered. Thus, the next step consists of a tabulation of binary relationships between regions. At this point one has extracted a large amount of information about the scene. However, to "understand" the scene, that is, be able to answer reasonable questions about it, requires one further step. This step is called object formation in Fig. 7.1 and consists of assigning labels to particular structures in the scene corresponding to physical objects. A simple example might clarify the overall problem.

Suppose one wanted to "understand" a page of text from a book using a TV imaging system. The first step might be to image the page and store it in a memory for handy reference. Next we might segment the image into dark and light regions, and store the coordinates of each connected region. Note that

Physical                                                          Facts  About
World                                                             Physical World

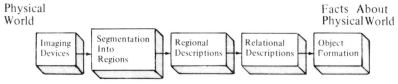

*Figure 7.1*   Scene understanding system.

with text, connectedness would be an important property since most characters except lower case "i" and "j" have only one connected region. At this
point we may have produced an edge image with ones at dark points and
zeros at light points. We could show this picture to a human who could
probably understand the information; however, the computer "knows" very
little at this point. For example, we could not easily locate the character "A."
The next step would attempt to get us at least to the single character
recognition stage. We might for regional descriptions locate the coordinates
of each connected region and store these in a table. By searching this table
with a matching set of character coordinates and proper normalization, we
should be able to locate most single characters. However we still are a long
way from understanding the page of text. We might next consider a binary
relationship such as "$x$ is adjacent to $y$ along a line." We may then form
another table for each segmented region that tabulates the adjacencies.
Another binary relationship such as "above" might also be tabulated to help
with the characters "i" and "j." At this point we could compress all the
measurements into one large relational table that contains everything the
computer "knows" about the scene. If "understanding" the scene is possible,
it must be done based upon the relational table. Could one locate a character? Yes, a search of the table for the regions forming a character should be
possible. How about a word? Again, given the characters that form the word
and, of course, the adjacencies of the characters, one should be able to search
for all occurrences of it. How about the meaning of a word? Sorry, this
requires information not available in the scene, unless the text is a page from
a dictionary. However, a dictionary table could be added to our storage. Note
that it would contain many more entries than the relational table that
contains one entry for each segmented region. How about a sentence? Given
the structure of a sentence, this should again be possible. How about what the
sentence means? Again, this may be possible—but not from the information
in the scene, but only from the information in the dictionary. Could we now
understand the scene? Not in the way a human does, since we have not
incorporated past experience, and many other factors. Have we built a useful
machine? Yes, because the machine could answer several useful questions.
For example, it could locate a character. This would permit scene matching, a
very important task that is described in Chapter 8. It could also locate words
or objects composed of several primitive elements. It could answer questions
about relationships between characters or words. It could even determine

spelling or syntax errors in sentences. Thus, even though this machine is limited in understanding, it is not limited in usefulness. The considerations for building such machines, not for understanding text but for understanding general physical scenes, are the main purpose of this chapter.

## 7.2  Segmentation

### 7.2.1  Overview

Segmentation consists of dividing an image into meaningful regions. Before considering specific techniques for segmentation, it is interesting to look at some broad notions in order to relate the problems to our visual experience. First, it is readily observed that a single scene may produce an indefinite variety of images. To simplify this variety, the concepts of subject and intention are useful. Subjects of a general image may be divided into a few types in photography such as the six categories listed below.

1. Objects.   The objects, persons or things that constitute the main center of interest in the image, for example, a picture of an object in a catalog.

2. Situations.   The general situation or location in which the recognizable objects in the picture will be seen, for example, a landscape whose primary subject is occupied space.

3. Events.   The event or events that are shown to be occurring in the image, for example, a photograph of a meeting.

4. Emotions.   The emotions or other interactions between the people or objects that are displayed because of the events taking place; for example, most newspaper photographs attempt to show emotion.

5. Comments.   The displayed opinion or attitude on any or all of the above subjects, for example, a simple comment—this is the way the object looks from this position and with this lighting—or an abstract comment—the object is beautiful.

6. Idea.   The idea which the originator is trying to express, for example, an image of a sunset that conveys an idea of beauty.

The intention or reason for considering, creating, or analyzing a scene should also be considered, since the global knowledge as well as a priori information for a scene description is usually determined by the intention or purpose of the image. The intention of photography may be derived from the Greek roots: *photo*—light, and *graph*—to describe. Most pictures are made to visually describe a scene. All scenes may be visually described by providing a set of images of the scene. However, some scenes can be described more efficiently or effectively by words. The combination of images and words is always more effective in communication than either alone; for example, one may compare silent movies or radio with television.

Many important decisions may be based upon the information contained in images of scenes. Since most objects encountered in the physical world are *three-dimensional*, a collection of these objects may be called a *scene* and a two-dimensional *picture* may be generated by viewing an *image* of the scene. The *intensity* at a given point in the image is affected by several variables including the viewing geometry, spectral composition of the illumination source, reflectance properties of the object surfaces, spectral characteristics, processing, and nonlinearities of the sensor systems, and the scene content.

Consider a scene that is a collection of three-dimensional objects. The content of the scene refers not only to the objects but also to their relationships in the three-dimensional environment. Since it is usually difficult or impossible to present the physical scene to describe its content, a useful approach is to construct a three-dimensional scale model that defines the physical objects and their relationships to some degree of precision. The three-dimensional model would thus define the scene content and permit descriptions to be extracted from images produced under different viewing conditions. The procedures for interpreting or describing the scene in terms of the model from measurements made from images of the scene is one of the main tasks in scene content analysis.

For some scenes, such as simple polyhedra, formal methods may be developed that permit concise descriptions of the scene. Two-dimensional images coupled with certain prior knowledge about the contents often permit a simple description of the scene. For a general physical scene complex models are required, and the description is difficult.

Given a three-dimensional scene, we would like to extract a complete description of the objects that make up the scene. Complete knowledge is likely to be unavailable or may require an unmanageably large number of tests to determine. Therefore, it may be satisfactory to select a simplified three-dimensional model, containing a finite number of objects that may be defined by a finite number of points in three-dimensional space. Then, from images of the scene, which are projections of the model onto an image plane, one may determine the model whose projection best matches the image or set of images. The problem of computing the best match can be divided into two subproblems: identifying corresponding regions between the model and the image, and computing the degree of match between these regions. The first subproblem of identifying corresponding regions may be approached by *segmenting*, or decomposing the images into separate components. The components may then be used to define the objects or match the regions. For example, information about only six components of a scene, the structural, height, spectral, statistical, shadow, and specular components, from images of the scene, should permit a high degree of inference about the scene model

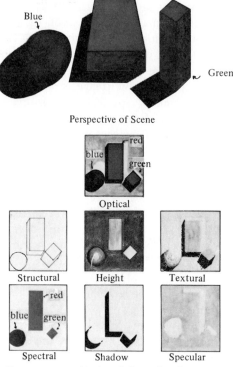

Red

Blue

Green

Perspective of Scene

red
blue
green

Optical

Structural    Height    Textural

red
blue    green

Spectral    Shadow    Specular

*Figure 7.2*    Scene segmentation into informational component images.

and a concise description of the scene content. An idealized example of scene segmentation into these image components is shown in Fig. 7.2.

The mathematical formulation for scene segmentation may be called clustering. Clustering is defined as finding "natural grouping" in a set of measurements $\{\mathbf{x}\}$, where the measurement vector $\mathbf{x} = (x_1, x_2, \ldots, x_n)'$ represents properties or attributes of some underlying set of patterns. In contrast to supervised pattern classification, the sample vectors are not labeled. Therefore, clustering is a form of unsupervised learning and consists of determining both the number of clusters and the group membership of the samples.

Segmentation may be considered as a special type of clustering in which some of the measurement components correspond to spatial locations. The remaining components may correspond to point properties such as gray level or spectral coordinates, or to regional properties such as edge measurements.

As an example, consider a sampled gray level image $f(x, y)$. One may consider the picture elements as a set of three-dimensional samples,

$$\mathbf{x} = \begin{bmatrix} x_1 \\ x_2 \\ x_3 \end{bmatrix} = \begin{bmatrix} x \\ y \\ f \end{bmatrix}.$$

A "natural grouping" in this data might be defined in several ways. Given two points, $x_1$ and $x_2$, we may consider the Euclidean distance between the points as a measure of dissimilarity, i.e., the smaller this distance, the greater the similarity in the three-dimensional space. The Euclidean distance between two points $x_1$ and $x_2$ is

$$d(x_1, x_2) = \left[ (x_1 - x_2)^2 + (y_1 - y_2)^2 + (f_1 - f_2)^2 \right]^{1/2}.$$

Note that a normalization is required in order to compare spatial distance to gray level distance. If this problem can be overcome, the Euclidean distance may provide a natural grouping into regions. Suppose we apply this distance measure to the image of Fig. 7.3. Note that the points in the dark region would form a cluster. However, the background points in the white region may not form a cluster even though they have the same gray level because they are separated by a large spatial distance. Suppose we now construct a histogram of the measurement vectors throughout the image. An idealized histogram for the $x$ and $f$ components is shown in Fig. 7.3b. Note that the

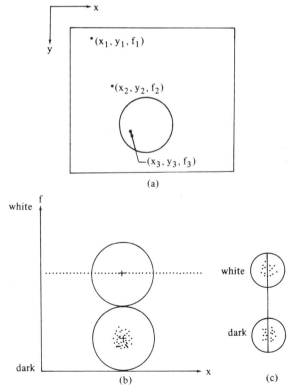

(a)

(b)

(c)

*Figure 7.3* Example of segmentation and clustering. (a) Original image, (b) two-dimensional space, and (c) one-dimensional measurement space.

object and background regions can be easily separated by a threshold or gray level and that the object values form a compact cluster in the two-dimensional space. However, the background values that do cluster along the gray level dimension are uniformly distributed over the $x$ coordinate values, which provides no discrimination value. The centroid of each distribution is also shown and a curve of constant Euclidean distance. This distance could easily describe the object distribution but is not appropriate for the background region in the given space. Finally, a histogram constructed only with the gray level values is shown in Fig. 7.3c. The Euclidean distance about the centroid is now seen to be an appropriate clustering measure.

This example illustrates that segmentation may be considered as a clustering problem whose solution depends highly upon the measurements and similarity criterion used. Although the purpose of segmentation is to locate meaningful regions, the spatial locations of points may not always provide discrimination for determining the regions. This also illustrates why segmentation is often approached as a two-pass or constrained clustering problem. During the first pass, clusters of similar gray level or colors are determined. A second operation is used to determine if these clusters form connected or contiguous regions.

The recognition of segmentation as a special clustering problem permits one to use, with some caution, the mathematical techniques developed for clustering. An excellent discussion of these techniques may be found in Tou and Gonzalez (1974). A few of the major concepts will now be reviewed.

In some cases, a single measurement such as gray level will permit adequate segmentation. However, in general, an $n$-dimensional measurement vector may be required. A notable example is multispectral images. In this case each pixel is characterized by an $n$ vector of spectral values and it becomes possible to construct an $n$-dimensional histogram. The concept of thresholding a single variable now becomes one of finding clusters of points in $n$-dimensional space. Suppose, for example, that we locate $K$ significant clusters of points in the histogram. The image can be segmented by assigning a unique number or gray level to pixels whose spectral components are nearest to one cluster center.

Clustering was defined as finding "natural groupings" in a set of data. To define natural grouping, we must determine a method to say that samples in one cluster are more like one another than samples in another cluster. The simplest measure of the similarity between two samples is the Euclidean distance between them.

If this distance is a good measure, then we may say that two samples belong to the same cluster if the distance between them is less than some threshold $d_0$. Obviously the choice of $d_0$ is important.

If $d_0$ is too large, then all samples will be placed in one cluster. If $d_0$ is too small, then each point will be placed in a singleton cluster. To obtain "natural" clusters, $d_0$ must be greater than the typical within-cluster distance and less than the typical between cluster distance.

The use of Euclidean distance as a measure of dissimilarity implies that the space is isotropic. Consequently, clusters defined by Euclidean distance will be invariant to translations or rotations—rigid body motions of the data points; however, they will not be invariant to linear transformations in general or to other transforms that distort the distance relationships. A simple scaling of the axes can result in different groupings and clusters.

If clusters are to mean something in general, they should be invariant to transformations natural to the problem. One way to achieve some invariance is to normalize the data prior to clustering, i.e., to translate to zero mean, unit variance to give invariance to translations or to rotate axes to correspond to eigenvectors of the sample covariance matrix. However this may not work well, especially for multicluster data. An alternative to normalizing the data and using Euclidean distance is to use some kind of normalized distance such as Mahalanobis distance. More generally, one may abandon the use of distance and introduce a nonmetric *similarity function* $S(\mathbf{X}, \mathbf{X}_1)$ to compare two vectors $\mathbf{X}$ and $\mathbf{X}_1$. Generally, we would want a symmetric function that is large when $\mathbf{X}$ and $\mathbf{X}_1$ are similar, e.g., the angle between the vectors may be meaningful. The normalized inner product

$$S(\mathbf{X}, \mathbf{X}_1) = \mathbf{X}'\mathbf{X}_1 / \|\mathbf{X}\| \, \|\mathbf{X}_1\| = \cos\theta$$

is a nonmetric similarity function.

### Criterion Functions for Clustering

Suppose we have a set $\chi$ of $n$ samples $\mathbf{X}_1, \ldots, \mathbf{X}_n$ and we want to partition $\chi$ into exactly $C$ disjoint subsets $\chi_1, \ldots, \chi_C$. Each subset is to represent a cluster, with samples in the same cluster being somehow more similar than samples in different clusters. To make this a well-defined problem, we must define a criterion function that measures the quality of any partition.

The problem is then to find a partition that extremizes the criterion function. An often used criterion is the sum of the squared error between the samples and cluster centers. Let $n_i$ be the number of samples in $\chi_i$ and

$$\boldsymbol{\mu}_i = \frac{1}{n_i} \sum_{\mathbf{X} \in \chi_i} \mathbf{X}.$$

Then the minimization of the sum of squared errors

$$J_e = \sum_{i=1}^{C} \sum_{\mathbf{X} \in \chi_i} \|\mathbf{X} - \boldsymbol{\mu}_i\|^2$$

results in a minimum variance partition.

### Hierarchical Clustering

Consider a sequence of partitions of $n$ samples into $C$ clusters. An exhaustive enumeration would show that there are

$$\frac{1}{C!} \sum_{i=1}^{C} \binom{C}{i}(-1)^{C-i} i^n \simeq \frac{C^n}{C!}$$

partitions of $n$ items into $C$ nonempty sets (Duda and Hart, 1973, p. 226). This number is usually much too large to permit an exhaustive search.

A special sequence of partitions of the possible subsets can be defined in the following way: The first of the sequence is a partition into $n$ clusters with each cluster containing exactly one sample. The next is a partition into $n-1$ clusters, the next a partition into $n-2$, and so on until the $n$th step, in which all samples form one cluster.

We shall say we are at level $k$ in the sequence when the number of clusters $C$ is

$$C = n - k + 1;$$

thus $k = 1$ or level 1 corresponds to $n$ clusters and level $n$ corresponds to 1 cluster.

Given any two samples $\mathbf{X}$ and $\mathbf{X}_1$, at some level they will be grouped together in the same cluster. If the sequence has the property that whenever two samples are in the same cluster at level $k$, they remain together at all higher levels, then the sequence is said to be *hierarchical* clustering.

For every hierarchical clustering there is a corresponding tree, called a *dendrogram*, that shows how the samples are grouped. In Fig. 7.4, a dendrogram for six samples is shown. The grouping at levels 2, 3, 4, 5 appears natural while that at level 6 appears forced because of the large difference in similarity values. At level 2, samples $\mathbf{X}_3$ and $\mathbf{X}_5$ merge to form a cluster and stay together for all higher levels. If it is possible to measure similarity between clusters, then the dendrogram is drawn to *scale* to show the similarity. The similarity values are used to tell whether the grouping is *forced*. Because of their conceptual simplicity, hierarchical clustering procedures are among the best known methods.

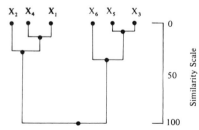

*Figure 7.4*   Dendrogram for six samples.

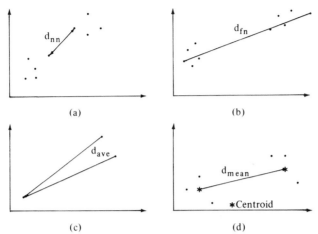

*Figure 7.5*  Measuring distance between clusters: (a) nearest neighbor, (b) farthest neighbor, (c) average distance, and (d) mean distance.

The approaches to clustering can be divided into two distinct classes: *agglomerative*, bottom-up or many-to-one; and *divisive*, top-down or one-to-many.

Agglomerative or bottom-up procedures start with $C$ singleton clusters and form the sequence by successively merging clusters. Divisive or top-down procedures start with all samples in one cluster and form the sequence by successively splitting clusters. The computations are usually simpler for bottom-up procedures; however, for segmentation the number of samples is usually very large, and the desired number of clusters is small so that top-down procedures may be more appropriate.

Let us now consider how to measure the distance between clusters. Any of several possible criteria may be used. Some of the simpler criteria are shown in Fig. 7.5 including the nearest neighbor distance, the farthest neighbor distance, the average distance, and the mean distance.

If the clusters are well separated, all of these yield the same results. However, if the clusters are close to one another, quite different results can be obtained.

### Nearest Neighbor Algorithm

Consider the following procedure: If we think of the data points as being nodes of a graph, with edges forming a path between nodes in the same subset $X_i$, then when $d_{nn}$ is used to measure distance between subsets, the nearest neighbors determine the nearest subsets. The merging of two clusters $X_i$ and $X_j$ corresponds to adding an edge between the nearest nodes in $X_i$ and $X_j$. Since edges linking clusters always go between distinct clusters the resulting graph never has any closed loops or circuits. In graph theory this graph is called a *tree*. If the procedure is allowed to continue until all of the subsets are linked, the result is called a *spanning tree*, i.e., a tree with a path from any node to any other node. It can be shown that the sum of the edge lengths of the spanning tree generated using $d_{nn}$ will not exceed the sum of the edge lengths for any other spanning tree for that set of samples. Thus, the bottom-up procedure using $d_{nn}$ becomes an algorithm for generating a *minimal spanning tree*. Given a minimal spanning tree, a natural division method would be to break at the *longest link*.

### Farthest Neighbor Algorithm

The use of $d_{fn}$ tends to reduce the growth of elongated clusters. Application of this procedure may be thought of as producing a graph in which edges connect all of the nodes in a cluster. In the terminology of graph theory, every cluster constitutes a *complete* subgraph. The distance between two clusters is determined by the two most distant nodes in the cluster.

If we define the *diameter* of a cluster as the largest distance between points in a cluster, then each iteration increases the diameter as little as possible. This is advantageous when the groups are compact and about equal in size, but if the groups are elongated it may result in meaningless groupings.

The distances $d_{nn}$ and $d_{fn}$ represent two extremes in measuring the distance between clusters; however, they tend to be oversensitive to "outliers."

The use of averaging is an obvious way to ameliorate these problems, and $d_{ave}$ and $d_{mean}$ are natural compromises.

### 7.2.2   Segmentation Using Point Properties

### Brightness Thresholding

The simplest measurement vector to use for segmentation is the picture gray level or brightness. We shall attempt to cluster picture points which have similar brightness values independent of their position. For example, consider the simple scene $f(x, y)$ shown in Fig. 7.6a. To segment the object from the background we may first form a histogram as shown in Fig. 7.6b. Note that the histogram has only two peaks and that the height of each peak is proportional to the area of the corresponding region in the scene. Also note that a threshold gray level value $T$ may easily be chosen that may be used to segment the picture into two regions. We may indicate the position of the regions by forming a new indicator scene $g(x, y)$, whose gray level corresponds to the region number. The decision rule

$$g(x, y) = \begin{cases} 1, & f(x, y) \ge T, \\ 0, & f(x, y) < T, \end{cases}$$

applied to each point of $f(x, y)$ would produce the indicator scene.

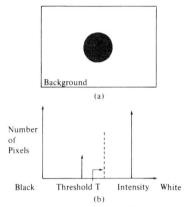

Background

(a)

Number
of
Pixels

Black          Threshold T          Intensity          White

(b)

*Figure 7.6*  Simple scene and histogram.

It is interesting to note the a priori information assumed in this solution. First, the number of gray level clusters was assumed known. Second, a single threshold decision function was used. Note that two threshold values could also be used to define a cluster of an object on a background. Finally, it was assumed that the gray level clusters would correspond to contiguous regions. A random dot image with the same relative areas could have the same histogram. If any of these factors are known a priori, the segmentation problem may usually be simplified. However, in the general scene nothing is known a priori, therefore, making these or other assumptions may restrict the generality of the solution.

One approach to the first part of the general problem is to consider the histogram as an estimate of a mixture distribution.

The problems associated with mixture probability distributions are well known in pattern recognition. To begin let us review some facts, assuming that we know the complete probability structure for the problem with the exception of a parameter vector.

We shall make the following assumptions:

1. The samples come from a known number of classes $N$.
2. The a priori probabilities for each class are known, $P(\omega_j), j = 1, 2, \ldots, N$.
3. The forms of the conditional probability densities $p(\mathbf{X}/\omega_j; \boldsymbol{\theta}_j)$ are known for $j = 1, 2, \ldots, N$.
4. All that is unknown are the values for the $N$ parameter vectors $\boldsymbol{\theta}_1, \boldsymbol{\theta}_2, \ldots, \boldsymbol{\theta}_N$.

Samples are assumed to be obtained by selecting a state of nature $\omega_j$ with probability $P(\omega_j)$ and then selecting an $\mathbf{X}$ according to the probability law $p(\mathbf{X}/\omega_j; \boldsymbol{\theta}_j)$. Thus, the probability density for the samples is given by

$$p(\mathbf{X}/\boldsymbol{\theta}) = \sum_{j=1}^{N} p(\mathbf{X}/\omega_j; \boldsymbol{\theta}_j) P(\omega_j), \quad \text{where} \quad \boldsymbol{\theta} = (\boldsymbol{\theta}_1, \boldsymbol{\theta}_2, \ldots, \boldsymbol{\theta}_j).$$

A density function of this form is called a *mixture density*. The conditional densities $p(X/\omega_j; \theta_j)$ are called the *component densities*, and the a priori probabilities $P(\omega_j)$ are called the *mixing parameters*. The mixing parameters can also be assumed to be among the unknown parameters, but for the moment we shall assume that only $\theta_j$ is unknown.

Our basic goal will be to use samples drawn from this mixture density to estimate the unknown parameter vector $\theta$. Once we know $\theta$, we can decompose the mixture into its components, and the problem is solved.

Before seeking explicit solutions to this problem, let us ask whether or not it is possible in principle to recover $\theta$ from the mixture. We could use some method to determine the value of $p(X/\theta)$ for every $X$. If there is only one value of $\theta$ that will produce the observed value of $p(X/\theta)$, then a solution is possible in principle. If there are several different values of $\theta$ that produce the same value for $p(X/\theta)$, then there is no hope of obtaining a unique solution.

A density $p(X/\theta)$ is said to be *identifiable* if $\theta \neq \theta^1$ implies that there exists an $X$ such that $p(X/\theta) \neq p(X/\theta^1)$.

Clearly the study of clustering is greatly simplified if we restrict ourselves to identifiable mixtures. Fortunately, most mixture distributions encountered in physical situations are identifiable. However, nonidentifiable examples are easily produced. For example, consider the discrete distribution, where $x$ is a binary random variable, and

$$P(x/\theta) = \tfrac{1}{2}\theta_1{}^x(1-\theta_1)^{1-x} + \tfrac{1}{2}\theta_2{}^x(1-\theta_2)^{1-x}$$

$$= \begin{cases} \tfrac{1}{2}(\theta_1+\theta_2) & \text{if } x=1 \\ 1-\tfrac{1}{2}(\theta_1+\theta_2) & \text{if } x=0. \end{cases}$$

If we know that $P(x=1/\theta)=0.8$ and hence $P(x=0/\theta)=0.2$, then we know $P(x/\theta)$; but we cannot determine $\theta$ since

$$\tfrac{1}{2}(\theta_1+\theta_2)=0.8, \qquad 1-\tfrac{1}{2}(\theta_1+\theta_2)=0.2$$

is one equation in two unknowns. This mixture distribution is not identifiable, and clustering is thus impossible in principle.

For the continuous case, the problems are less severe, although minor problems can arise. For example, the mixtures of two normal densities are usually identifiable; however, consider the following univariant distribution,

$$p(x/\theta) = \frac{P(\omega_1)}{\sqrt{2\pi}} \exp\left[ -\frac{1}{2}(x-\theta_1)^2 \right] + \frac{P(\omega_2)}{\sqrt{2\pi}} \exp\left[ -\frac{1}{2}(x-\theta_2)^2 \right].$$

If $P(\omega_1) = P(\omega_2)$, the mixture cannot be uniquely identified since

$$p\{x/(\theta_1,\theta_2)\} = p\{x/(\theta_2,\theta_1)\}.$$

Assuming that the mixture densities encountered during segmentation are identifiable, there are several approaches to determining the component distributions. Maximum likelihood estimation, Bayes estimation, and stochastic approximation theories may be used as guidelines for component estimation. However, heuristic techniques are often used in practice as illustrated in the following examples.

*Example.* In some cases it may be known or assumed a priori that a picture contains only two principal brightness values, light and dark. This would be true for an image of text or for some x-ray images that show only two density levels. The histogram of such a picture may be considered as an estimate of the mixture probability density function. If the component densities are assumed to be Gaussian or approximated by unimodal Gaussian densities, then the problem is to separate the mixture into the Gaussian component densities. An important observation is that for a fixed region size, the a priori probabilities or mixing parameters are proportional to the relative area of each brightness in the region. If these mixing parameters are not equal, the mixture distribution is identifiable and it should be possible to determine the components. If the components can be determined, it should be possible to determine a threshold for segmenting the image into light and dark points and perhaps regions.

If the picture brightness $x$ consists of two brightness levels, $\mu_1$ and $\mu_2$ combined with Gaussian noise with variances $\sigma_1^2$ and $\sigma_2^2$, respectively, the mixture probability density function is given by

$$p(x) = P_1 p_1(x) + P_2 p_2(x),$$

where

$$p_i(x) = \frac{1}{\sigma_i \sqrt{2\pi}} \exp\left[ -\frac{1}{2}\left(\frac{x - \mu_i}{\sigma_i}\right)^2 \right] \quad \text{and} \quad P_1 + P_2 = 1.$$

This mixture density has five unknown parameters, $\theta = (\mu_1, \mu_2, \sigma_1, \sigma_2, P_1)$. If these parameters could be determined, then a threshold $T$ that minimizes the error for segmenting the brightness into light and dark values could be determined. Suppose that all points with brightness less than the threshold are called dark and all points with brightness greater than the threshold light. The probability of error in classifying a light point as dark is

$$\varepsilon_1(T) = \int_{-\infty}^{T} p_2(x)\, dx.$$

Similarly, the probability of classifying a dark point as light is

$$\varepsilon_2(T) = \int_{T}^{\infty} p_1(x)\, dx.$$

The average probability of error is therefore

$$\varepsilon(T) = P_1 \varepsilon_1(T) + P_2 \varepsilon_2(T).$$

To determine the threshold value that minimizes the average error, we may simply differentiate $\varepsilon(T)$ with respect to $T$ (using Leibnitz's rule) and equate the result to zero. The well-known result is

$$P_1 p_1(T) = P_2 p_2(T).$$

Applying this condition to the Gaussian densities, taking logarithms, and simplifying gives the quadratic equation

$$AT^2 + BT + C = 0,$$

where

$$A = \sigma_1^2 - \sigma_2^2, \qquad B = 2(\mu_1 \sigma_2^2 - \mu_2 \sigma_1^2)$$

$$C = \sigma_1^2 \mu_2^2 - \sigma_2^2 \mu_1^2 + \sigma_1^2 \sigma_2^2 \ln(\sigma_1 P_1 / \sigma_2 P_2)$$

If the variances are equal, $\sigma_1^2 = \sigma_2^2 = \sigma^2$, then a single threshold value is indicated:

$$T = \frac{\mu_1 + \mu_2}{2} + \frac{\sigma}{\mu_1 - \mu_2} \ln(P_2 / P_1).$$

If the variances are not equal, then two threshold values are required to define the regions. A similar solution for the optimum threshold may be developed for other unimodal densities known parametrically.

The above solution depends upon knowing the parameters of the mixture density. To estimate these parameters from a histogram of the picture, one may use a minimum mean squared error approach. The mean squared error between the mixture density $p(x)$ and the experimental histogram $h(x)$ is given by

$$\overline{\epsilon^2} = \frac{1}{N} \sum_{i=1}^{N} \left[ p(x_i) - h(x_i) \right]^2,$$

where an $N$-point histogram is used. One may now substitute the expression for $p(x)$ and obtain the mean squared error in terms of the five unknown parameters. However, the straightforward computation of equating the partial derivatives with respect to the parameters to zero leads to a set of simultaneous transcendental equations that can be solved only by iterative numerical procedures. Since the gradient is easily computed, the conjugate gradient or Newton's method for simultaneous nonlinear equations may be used to minimize the mean squared error. With either of the methods starting values must be specified. Assuming the a priori probabilities to be equal may be sufficient. If two modes can be determined in the histogram, these would be excellent starting values for the means. Local variances about the means could be used to estimate the variances. If modes cannot be detected, the histogram range can be divided arbitrarily into two parts and the mean and variance of each part used for the starting value.

The previously described procedure was developed by Chow and Kaneko (1972) for outlining boundaries of the left ventricle in cardioangiograms (i.e., x-ray pictures of a heart that has been injected with a dye).

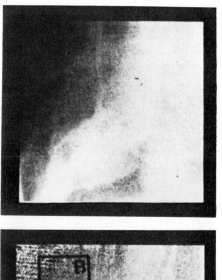

*Figure 7.7* A cineangiogram of the heart before and after processing. [From Chow and Kaneko (1972).]

Before thresholding, the images were first preprocessed by (1) taking the logarithm of every pixel to invert the exponential effects caused by radioactive absorption, (2) subtracting two images that were obtained before and after the dye agent was applied in order to remove the spinal column present in both images, and (3) averaging several angiograms to remove noise. Figure 7.7 shows a cardioangiogram before and after preprocessing (the regions marked A and B are explained below).

In order to compute the optimum thresholds, each preprocessed image was subdivided into $7 \times 7$ regions (the original images were of size $256 \times 256$) with 50% overlap. Each of the 49 resulting regions contained $64 \times 64$ pixels. Figures 7.8a and 7.8b are the histograms of the regions marked A and B in Fig. 7.7b. It is noted that the histogram for region A is very clearly bimodal, indicating the presence of a boundary. The histogram for region B, on the

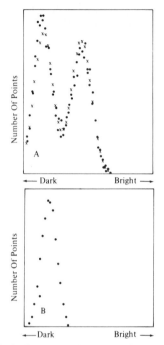

*Figure 7.8* Histograms (black dots) of regions A and B in Fig. 7.7(b). [From Chow and Kaneko (1972).]

other hand, is unimodal, indicating the absence of two markedly distinct regions.

After all 49 histograms were computed, a test of bimodality was performed to reject the unimodal histograms. The remaining histograms were then fitted by bimodal Gaussian density curves using a conjugate gradient hill-climbing method to minimize the mean squared error function. The $x$s in Fig. 7.8a represent a fit to the histogram shown in black dots. The optimum thresholds were then obtained.

At this stage of the process, only the regions with bimodal histograms were assigned thresholds. The thresholds for the remaining regions were obtained by interpolating the original thresholds. After this was done, a second interpolation was carried out in a point-by-point manner using neighboring threshold values so that, at the end of the procedure, every point in the image was assigned a threshold. Finally, a binary decision was carried out for each pixel using the rule

$$f(x,y) = \begin{cases} 1 & \text{if } f(x,y) \geq T_{xy} \\ 0 & \text{otherwise,} \end{cases}$$

where $T_{xy}$ was the threshold computed at location $(x,y)$ in the image. Boundaries were then obtained by taking the gradient of the binary picture. The results are shown in Fig. 7.9, in which the boundary was superimposed.

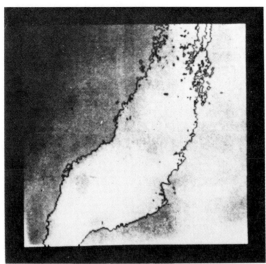

*Figure 7.9*  A cineangiogram showing superimposed boundaries. [From Chow and Kaneko (1972).]

### Multivariate Histogram Technique

In the previous example, the one-dimensional histogram of a region has been used to provide a method for segmentation into two regions. Although the method could be extended to search for several thresholds in a multimodal histogram, consider the picture shown in Fig. 7.10, which contains four objects on a background. The one-dimensional histogram might be as shown in Fig. 7.10b, in which only four peaks are indicated. Even if the four modes of this histogram could be detected, the red and blue objects could not be distinguished. A two-dimensional histogram of both intensity and hue is shown in Fig. 7.10c. Note that the blue and red objects could now be distinguished using the histogram modes. Riseman and Arbib (1975) developed such a two-dimensional histogram technique. The important point is that the two-dimensional joint probability density function as estimated from the histogram permits features to be distinguished that could not be distinguished from univariate densities alone.

The method could also be extended to three or more dimensions. For example, if a color picture of a scene were available, then a three-dimensional analysis using the red, green, and blue component images could be made. If both a monochrome image and an image depicting the distance or range to objects were available, then one might expect that a two-dimensional histogram analysis would permit superior segmentation. Even if only a monochrome image were available, a multidimensional technique could be used, perhaps with an edge gradient and local texture variance as additional images. Although the computational complexity of the multidimensional

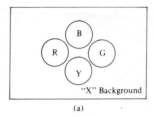

(a)

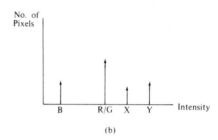

(b)

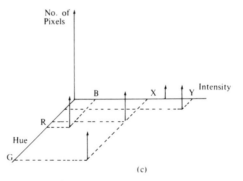

(c)

*Figure 7.10* Comparison of one- and two-dimensional histograms for segmenting a simple scene: (a) simple scene, (b) one-dimensional histogram, (c) two-dimensional histogram.

technique is greater than that of the one-dimensional technique, superior segmentation results often justify the extra cost. Some simple examples will now be presented to illustrate the method. In these examples the images are assumed to be perfectly registered; if not, then the geometric transformation techniques previously described should be used first.

Consider the three images of the same scene shown in Fig. 7.11, which consist of a visible band image, a radar image, and a range image. A one-dimensional segmentation technique was used in an attempt to isolate the east wall of the building from each original image. The results of this thresholding operation are shown in Fig. 7.12. Note that each obtains a rough outline of the wall with different noise regions in each attempt. A three-

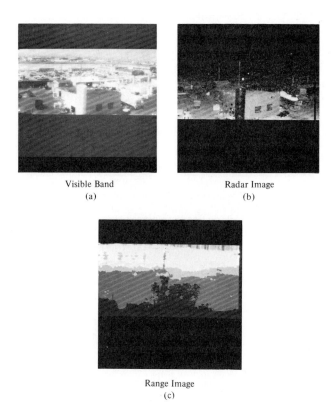

Visible Band
(a)

Radar Image
(b)

Range Image
(c)

*Figure 7.11* Original multispectral images.

dimensional histogram segmentation using 16-level visual, radar, and range components was used to locate objects in the Fig. 7.11 scene. Three slices from the $16 \times 16 \times 16$ histogram are shown in Fig. 7.13 with peak locations indicated. A list of the peak locations and the number of pixels in each is shown in Fig. 7.14. Each of these peaks corresponds to a major region. The results of this analysis for locating the east wall, the tower, and the airstrip are shown in Fig. 7.15. The automatic results shown on the left could be improved only slightly by a further interactive step. Note that these results are considerably better than those of the one-dimensional analysis.

As another example consider the segmentation of a color image using the registered red, green, and blue components shown in Fig. 7.16. The procedure involved first quantizing the intensity information of three red, green, and blue component pictures into 16 levels. A three-dimensional histogram was then computed. Thus, a cube of 4096 locations is formed. At each location, the number of pixels is indicated, thereby forming an estimate of the three-dimensional joint probability density function.

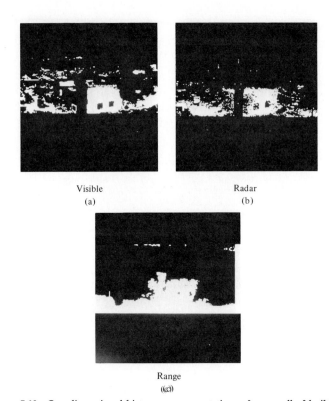

Visible                                    Radar
(a)                                       (b)

Range
((c))

*Figure 7.12*   One-dimensional histogram segmentations of east wall of building.

Once the histogram is computed, the detection of peaks in the histogram is the next step. One solution to the problem was implemented in the program and is explained as follows: Consider the detection of a peak along one dimension or axis of the histogram. To verify that this is a valid peak requires the observance of surrounding points in the remaining dimensions. For example, if red, green, and blue represent the sensors employed and a peak is detected along the red axis, then points along the blue axis and green axis on either side of the peak must be equal to or smaller than the peak on the red axis. Only the four surrounding points were examined.

Having located all the peaks, a classification procedure for each pixel must be introduced. At this point the results may be verified by displaying a picture of just the points corresponding to the location of peaks in the histogram. A threshold was arbitrarily set at 100 pixels (i.e., if at each peak in the histogram the pixel count was less than 100, this peak was ignored). This allowed the largest peaks to be viewed. Two segmented regions using this method are shown in Fig. 7.17. As a further refinement, the transformation from the original components to the $U^*, V^*, W^*$ space described in Chapter 2

*Figure 7.13* Three-dimensional histogram slices of building (visual, radar, and range).

| Visual | IR | Range | No. of Pixels |
|--------|----|-------|---------------|
| 1 | 1 | 1 | 4475 |
| 13 | 7 | 1 | 13 |
| 1 | 10 | 1 | 194 |
| 3 | 12 | 1 | 517 |
| 1 | 13 | 1 | 490 |
| 4 | 15 | 1 | 36 |
| 3 | 1 | 2 | 205 |
| 9 | 1 | 2 | 121 |
| 14 | 1 | 2 | 12 |
| 9 | 8 | 2 | 114 |
| 14 | 8 | 2 | 10 |
| 6 | 9 | 2 | 84 |
| 3 | 10 | 2 | 511 |
| 10 | 11 | 2 | 128 |
| 4 | 1 | 3 | 242 |
| 7 | 1 | 3 | 266 |
| 11 | 1 | 3 | 53 |
| 6 | 3 | 3 | 29 |
| 11 | 4 | 3 | 13 |
| 11 | 6 | 3 | 19 |
| 8 | 1 | 4 | 267 |
| 8 | 4 | 4 | 33 |
| 5 | 1 | 5 | 240 |
| 7 | 1 | 5 | 297 |
| 7 | 3 | 5 | 27 |
| 7 | 5 | 5 | 38 |
| 6 | 6 | 5 | 44 |
| 8 | 6 | 5 | 38 |
| 5 | 7 | 5 | 35 |
| 5 | 4 | 7 | 11 |
| 5 | 5 | 8 | 10 |
| 6 | 1 | 10 | 88 |
| 7 | 6 | 11 | 13 |
| 7 | 1 | 12 | 101 |
| 7 | 4 | 12 | 15 |
| 7 | 7 | 12 | 16 |
| 7 | 1 | 14 | 138 |
| 7 | 5 | 14 | 17 |
| 7 | 3 | 15 | 10 |

*Figure 7.14* Three-dimensional histogram peaks for building.

was implemented. This transformation should improve the comparison of detected to visually observed regions. The results of segmentation after the visual transformation are shown in Fig. 7.18 and are improved.

### 7.2.3 Region-Dependent Techniques

The previous segmentation methods have partitioned an image into subsets based upon point properties or predicates. In general, local regional properties may also be used for segmentation. The local region surrounding a point may be called the *context* of the point. Thus, the inclusion of regional properties is an attempt to account for the image context. Edge location is the most common example of a region dependent technique, since to define an edge transition at least two points must be used. Texture, which is often defined as a global repetition of a local pattern, is another regionally dependent concept. Other properties, which depend on a set of point offset conditions, such as a red point followed by a green point are also included.

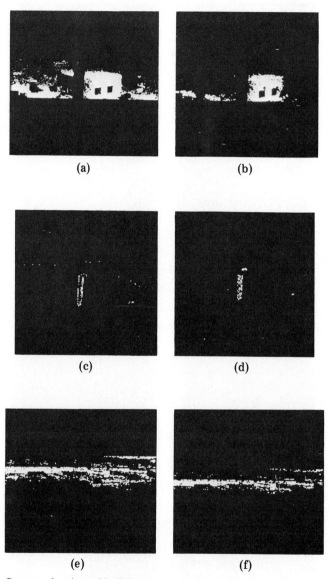

*Figure 7.15*   Segmented regions of building: (a) east wall, automatic, (b) east wall, interactive, (c) tower, automatic, (d) tower, interactive, (e) airstrip, automatic, (f) airstrip, interactive.

Edge information is important for both human and machine perception and a considerable amount of research has been done in attempts to extract edge information from pictures. Roberts (1965) in his pioneering work in machine perception extracted edges by computing a diagonal approximation to the local gradient and thresholding the result. Three different approaches to the location of edge information are shown in Fig. 7.19. The first method,

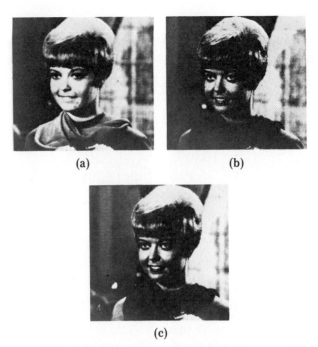

*Figure 7.16*  Color image components of girl. (a) Red, (b) green, and (c) blue.

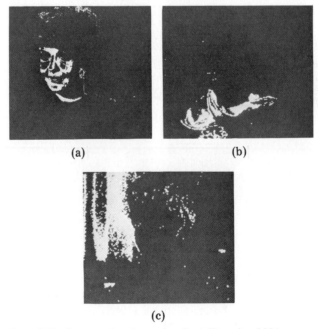

*Figure 7.17*  Segmented regions using three-dimensional histogram.

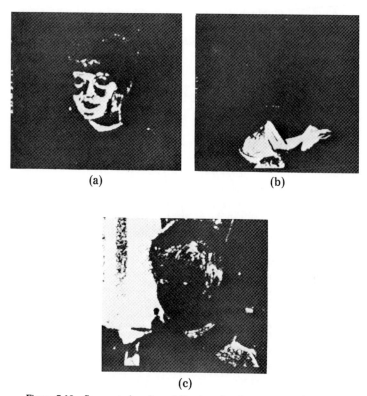

(a)

(b)

(c)

*Figure 7.18* Segmented regions following visual system transformation.

shown in Fig. 7.19a, involves edge enhancement followed by a threshold operation. It is characteristic of the gradient methods used by Roberts (1965), Kirsch (1971), and many others. The resulting edge images often produce excellent representations of the edge information; however, steps for thinning wide edges or connecting broken edges are often required. A similar method is shown in Fig. 7.19b and consists of edge detection followed by a threshold operation. This method may be called a correlation or template matching approach, since an edge template is matched to the image and the resulting correlation value is thresholded. A recent unification of this approach was developed by Frei and Chen (1977), which encompasses most of the previously developed edge templates for 3 by 3 regions. The results of the correlation approach may also require thinning and connection in order to produce reliable edge information. The third method, shown in Fig. 7.19c, may be called the region approach and is based on the fact that two adjacent regions of different properties define an edge at the adjacent boundary. The method involves two steps—region segmentation and boundary location. If properly implemented, the resulting edges will be thinned, connected, and closed.

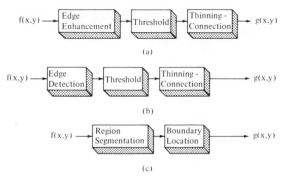

Figure 7.19   Edge location approaches: (a) gradient or filtering approach, (b) correlation or detection approach, and (c) region approach.

### Gradient or Filtering Approach

The creation of an edge picture can be considered as a process of transforming the original gray scale image into a binary-valued image. Let $\Phi$ be the operator of the transformation; then

$$g(x,y) = \Phi f(x,y),$$

where $f(x,y)$ is the intensity level of picture element $(x,y)$ of the original image and $g(x,y)$ is the intensity level of the picture element $(x,y)$ of the transformed image. An important class of operators has the general form

$$\Phi = \alpha + \beta^2 \nabla^2, \qquad \nabla^2 = \frac{\partial^2}{\partial x^2} + \frac{\partial^2}{\partial y^2},$$

where $\alpha$ and $\beta$ are constants and $\nabla^2$ is the Laplacian operator. Because of the physical results of its application, the operator is called the *contour enhancement operator*. Applying the above equation yields

$$g_1 = \alpha f + \beta^2 \left( \frac{\partial^2 f}{\partial x^2} + \frac{\partial^2 f}{\partial y^2} \right).$$

The simplest operator of the class takes the form

$$\Phi = \beta^2 |\nabla^2|.$$

The transformed image from the application of this operator is

$$g_2 = \beta^2 \left| \left( \frac{\partial f}{\partial x} \right)^2 + \left( \frac{\partial f}{\partial y} \right)^2 \right|.$$

Because $g_2$ is the square of the absolute magnitude of the gradient vector, its numerical value is invariant under translation and rotation. An edge image can be created by thresholding $g_2$ to create an image $g_3$,

$$g_3 = \begin{cases} 1 & \text{if } g_2 > T \\ 0 & \text{if } g_2 \leq T, \end{cases}$$

where $g_3$ is the resulting edge image and $T$ is a selected threshold.

The gradient vector may also be used for edge detection. Let $\nabla f$, with its directional cosines $m, n, l$, be expressed as

$$\nabla f = i\frac{\partial f}{\partial x} + j\frac{\partial f}{\partial y}, \qquad m = \left[1 + \left(\frac{\partial f}{\partial x}\right)^2 + \left(\frac{\partial f}{\partial y}\right)^2\right]^{-1/2}\frac{\partial f}{\partial x},$$

$$n = \left[1 + \left(\frac{\partial f}{\partial x}\right)^2 + \left(\frac{\partial f}{\partial y}\right)^2\right]^{-1/2}\frac{\partial f}{\partial y}, \qquad l = \left[1 + \left(\frac{\partial f}{\partial x}\right)^2 + \left(\frac{\partial f}{\partial y}\right)^2\right]^{-1/2}.$$

Since $\nabla f$ is a function of two variables, the occurrence of its maximum gradient can be calculated by requiring the following conditions to be satisfied:

$$\frac{\partial}{\partial x}|\nabla f|^2 = 0, \qquad \frac{\partial^2}{\partial x^2}|\nabla f|^2 < 0,$$

$$\frac{\partial}{\partial y}|\nabla f|^2 = 0, \qquad \frac{\partial^2}{\partial y^2}|\nabla f|^2 < 0,$$

$$\left[\frac{\partial^2}{\partial x^2}|\nabla f|^2\right]\left[\frac{\partial^2}{\partial y^2}|\nabla f|^2\right] - \left[\frac{\partial^2}{\partial x\,\partial y}|\nabla f|^2\right]^2 > 0.$$

The maximum gradient and magnitude $|\nabla f|_{\max}$ and $|\nabla f|$ may also be used for creating a binary-valued edge picture using the operations

$$g_3 = \begin{cases} 1 & \text{if} \quad |\nabla f| = \max \\ 0 & \text{if} \quad |\nabla f| \neq \max, \end{cases}$$

$$g_4 = \begin{cases} 1 & \text{if} \quad \nabla^2 f \leq 0 \\ 0 & \text{if} \quad \nabla^2 f > 0, \end{cases}$$

where $g_4$ is the transformed black and white picture, and $g_3$ is the transformed edge picture. The edges of $g_3$ so created will have varying thickness. However, the edges can be made constant in thickness through the use of edge thinning techniques as described in Shirai and Tsuji (1972) and Sakai *et al.* (1972).

Various digital filters have been proposed to approximate the previously described operators. The computation of the gradient involves a nonlinear combination of approximations of $\partial f(x,y)/\partial x$ and $\partial f(x,y)/\partial y$ to the partial derivatives in orthogonal directions.

The simplest filters of size $2\times 2$ become

$$\frac{\partial f(x,y)}{\partial x} = \begin{bmatrix} 1 & -1 \\ 0 & 0 \end{bmatrix}, \qquad \frac{\partial f(x,y)}{\partial y} = \begin{bmatrix} 1 & 0 \\ -1 & 0 \end{bmatrix}.$$

A better estimate can be realized by fitting a quadratic surface over a $3\times 3$ neighborhood by least squares and then computing the gradient for the fitted

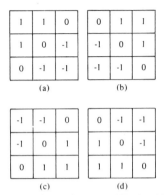

Figure 7.20 Directional gradient filters: (a) 45°, (b) 135°, (c) 225°, (d) 315°.

surface. These filters become

$$\frac{\partial f(x,y)}{\partial x} = \begin{bmatrix} 1 & 0 & -1 \\ 1 & 0 & -1 \\ 1 & 0 & -1 \end{bmatrix}, \qquad \frac{\partial f(x,y)}{\partial y} = \begin{bmatrix} 1 & 1 & 1 \\ 0 & 0 & 0 \\ -1 & -1 & -1 \end{bmatrix}.$$

The operations can be extended to extract the edge gradient along the diagonal directions. Filters that perform these operations are shown in Fig. 7.20.

## Edge Detection Approach

The concept of template matching has found wide acceptance in segmentation applications because of its simplicity. In terms of a digital image, a *template* (also called a *mask* or *window*) is an array designed to detect some regional property. Examples of some possible templates are shown in Fig. 7.21.

Let $w_1, w_2, \ldots, w_N$ represent the weights in an $N = n \times n$ mask, and let $x_1, x_2, \ldots, x_N$ be the gray levels of the pixels inside the mask. Template

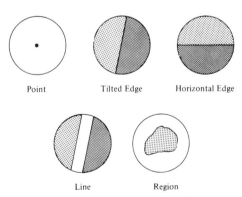

Point    Tilted Edge    Horizontal Edge

Line    Region

Figure 7.21 Examples of templates.

matching may be considered as taking the inner product of the vectors

$$\mathbf{w} = \begin{bmatrix} w_1 \\ w_2 \\ \cdot \\ \cdot \\ \cdot \\ w_N \end{bmatrix} \quad \text{and} \quad \mathbf{x} = \begin{bmatrix} x_1 \\ x_2 \\ \cdot \\ \cdot \\ \cdot \\ x_N \end{bmatrix},$$

where, for example, the first $n$ elements of $\mathbf{w}$ are the elements in the first row of the template, the next $n$ elements are from the second row, and so on. The inner product of $\mathbf{w}$ and $\mathbf{x}$, defined as

$$\mathbf{w}'\mathbf{x} = w_1 x_1 + w_2 x_2 + \cdots + w_N x_N,$$

is seen to be identical to the sum of products discussed above. We may say that an edge has been detected by template $\mathbf{w}$ if

$$\mathbf{w}'\mathbf{x} > T,$$

where $T$ is a specified threshold.

Point detection is a fairly straightforward procedure. The next level of complexity involves the detection of lines in an image. Directional masks may be established by noting that the preferred direction of each template is weighted with a larger coefficient than other possible directions.

Let $\mathbf{w}_1$, $\mathbf{w}_2$, $\mathbf{w}_3$, and $\mathbf{w}_4$ be $N$-dimensional vectors formed from the entries of the four directional templates. The individual responses of the line templates at any point in the image are given by $\mathbf{w}_i'\mathbf{x}$ for $i = 1, 2, 3, 4$. As before, $\mathbf{x}$ is the vector formed from the $N$ image pixels inside the template area. Given a particular $\mathbf{x}$, suppose that we wish to determine the closest match between the region in question and one of the four line templates. We say that $\mathbf{x}$ is closest to the $i$th template if the response of this template is the largest, in other words, if

$$\mathbf{w}_i'\mathbf{x} > \mathbf{w}_j'\mathbf{x}$$

for all values of $j$, excluding $j = i$. If, for example, $\mathbf{w}_1'\mathbf{x}$ were greater than $\mathbf{w}_j'\mathbf{x}$, $j = 2, 3, 4$, we would conclude that the region represented by $\mathbf{x}$ is characterized by a horizontal line since this is the feature to which the first template is most responsive.

The development of templates for edge detection follows essentially the same reasoning as above, with the exception that we are now interested in detecting transitions between regions. One approach often used for determining such transitions is to implement some form of two-dimensional derivative function. The following discussion considers the gradient concept for templates of size $3 \times 3$.

Let $G_x$ and $G_y$ be the directional discrete gradients in the $x$ and $y$ directions, respectively. The gradient at point $p$ is then defined as

$$G = \left[ G_x^2 + G_y^2 \right]^{1/2}.$$

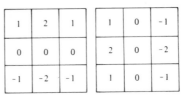

*Figure 7.22*   Gradient templates.

An alternative definition, using absolute values, is given by

$$G = |G_x| + |G_y|.$$

Implementation of the last equation simply requires a sum of the absolute values of the responses of the two templates. It is also important to note that edge detection can be expressed in vector form in exactly the same manner as discussed for line templates. Thus if $\mathbf{x}$ represents the image region in question, we have

$$G_x = \mathbf{w}_1' \mathbf{x} \quad\text{and}\quad G_y = \mathbf{w}_2' \mathbf{x},$$

where $\mathbf{w}_1$ and $\mathbf{w}_2$ are the vectors for the two masks shown in Fig. 7.22. The formulations given above for the gradient then become

$$G = \left[ (\mathbf{w}_1' \mathbf{x})^2 + (\mathbf{w}_2' \mathbf{x})^2 \right]^{1/2} \quad\text{and}\quad G = |\mathbf{w}_1' \mathbf{x}| + |\mathbf{w}_2' \mathbf{x}|.$$

The vector formulation for the detection of points, lines, and edges has the important advantage that it can be used to detect combinations of these features using a technique developed by Frei and Chen (1977). In order to see how this can be accomplished, let us consider two hypothetical templates with only three components. In this case, we would have two vectors $\mathbf{w}_1$ and $\mathbf{w}_2$, which are three dimensional. Assuming that $\mathbf{w}_1$ and $\mathbf{w}_2$ are orthogonal and normalized so that they have unit magnitude, we have that the terms $\mathbf{w}_1' \mathbf{x}$ and $\mathbf{w}_2' \mathbf{x}$ are equal to the projections of $\mathbf{x}$ onto the vectors $\mathbf{w}_1$ and $\mathbf{w}_2$, respectively. This follows from the fact that for $\mathbf{w}_1$

$$\mathbf{w}_1' \mathbf{x} = \|\mathbf{w}_1\| \, \|\mathbf{x}\| \cos\theta,$$

where $\theta$ is the angle between the two vectors. Since $\|\mathbf{w}_1\| = 1$,

$$\|\mathbf{x}\| \cos\theta = \mathbf{w}_1' \mathbf{x},$$

which is the projection of $\mathbf{x}$ onto $\mathbf{w}_1$. Similar comments hold for $\mathbf{w}_2$.

Now suppose that we have three orthogonal vectors of unit magnitudes $\mathbf{w}_1, \mathbf{w}_2, \mathbf{w}_3$, corresponding to three 3-point templates. The products $\mathbf{w}_1' \mathbf{x}$, $\mathbf{w}_2' \mathbf{x}$, and $\mathbf{w}_3' \mathbf{x}$ represent the projections of $\mathbf{x}$ onto the vectors $\mathbf{w}_1$, $\mathbf{w}_2$, and $\mathbf{w}_3$. According to our earlier discussion, these products also represent the *individual* responses of the three templates. Suppose that templates 1 and 2 are for lines and template 3 is for points. A reasonable question to ask is: Is the region represented by $\mathbf{x}$ more like a line or more like a point? Since there are two templates representing lines, and we are only interested in the line properties of $\mathbf{x}$, and not on what type of line is present, we could answer the question by projecting $x$ onto the subspace of $\mathbf{w}_1$ and $\mathbf{w}_2$ (which in this case is

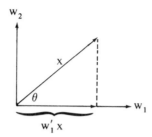

*Figure 7.23*  Projection of $x$ onto unit vector $\mathbf{w}_1$. (Reproduced from "Digital Image Processing and Recognition," 1977, by Rafael C. Gonzalez and Paul Wintz, with the permission of Addison-Wesley, Reading, Massachusetts.)

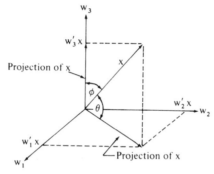

*Figure 7.24*  Projection of $x$ onto subspace determined by $\mathbf{w}_1$ and $\mathbf{w}_2$ and onto $\mathbf{w}_3$. (Reproduced from "Digital Image Processing and Recognition," 1977, by Rafael C. Gonzalez and Paul Wintz, with the permission of Addison-Wesley, Reading, Massachusetts.)

a plane) and also onto $\mathbf{w}_3$. The angle between $x$ and each of these two projections would tell us whether $x$ is closer to the line or the point subspace. This can be seen from the geometrical arrangement shown in Figs. 7.23 and 7.24. The magnitude of the projection of $x$ onto the plane determined by $\mathbf{w}_1$ and $\mathbf{w}_2$ is given by the quantity $\left[ (\mathbf{w}_1'x)^2 + (\mathbf{w}_2'x)^2 \right]^{1/2}$, while the magnitude (i.e., norm) of $x$ is

$$\|x\| = \left[ (\mathbf{w}_1'x)^2 + (\mathbf{w}_2'x)^2 + (\mathbf{w}_3'x)^2 \right]^{1/2}.$$

The angle between $x$ and its projection is then

$$\theta = \cos^{-1}\left( \left[ \sum_{i=1}^{2} |\mathbf{w}_i'x|^2 \right]^{1/2} / \|x\| \right).$$

A similar development would yield the angle of projection onto the $\mathbf{w}_3$ subspace,

$$\phi = \cos^{-1}(|\mathbf{w}_3'x| / \|x\|).$$

Thus if $\theta < \phi$, we say that the region represented by $x$ is closer to the characteristics of a line than of a point.

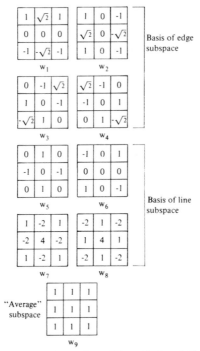

*Figure 7.25*  Orthogonal templates. [From Frei and Chen (1977).]

If we now consider $3 \times 3$ masks, the problem becomes nine dimensional, but the above concepts are still valid. We need, however, nine nine-dimensional orthogonal vectors to form a complete basis. The templates proposed by Frei and Chen (1977) shown in Fig. 7.25 satisfy this condition. The first four masks are suitable for detection of edges; the second set of four masks represents templates suitable for line detection; and the last template (added to complete the basis) is proportional to the average of the pixels in the region at which the mask is located in an image.

Given a $3 \times 3$ region represented by $\mathbf{x}$, and assuming that the vectors $\mathbf{w}_i$, $i = 1, 2, \ldots, 9$, have been normalized, we have from the above discussion that

$$P_e = \left[ \sum_{i=1}^{4} (\mathbf{w}_i' \mathbf{x})^2 \right]^{1/2}, \qquad P_l = \left[ \sum_{i=5}^{8} (\mathbf{w}_i' \mathbf{x})^2 \right]^{1/2}, \qquad \text{and} \qquad P_a = |\mathbf{w}_9' \mathbf{x}|,$$

where $P_e$, $P_l$, and $P_a$ are the magnitudes of the projection of $\mathbf{x}$ onto the edge, line, and average subspaces, respectively.

Similarly, we have that

$$\theta_e = \cos^{-1} \left[ \frac{1}{\|\mathbf{x}\|} \left( \sum_{i=1}^{4} (\mathbf{w}_i' \mathbf{x})^2 \right)^{1/2} \right], \qquad \theta_l = \cos^{-1} \left[ \frac{1}{\|\mathbf{x}\|} \left( \sum_{i=5}^{8} (\mathbf{w}_i' \mathbf{x})^2 \right)^{1/2} \right],$$

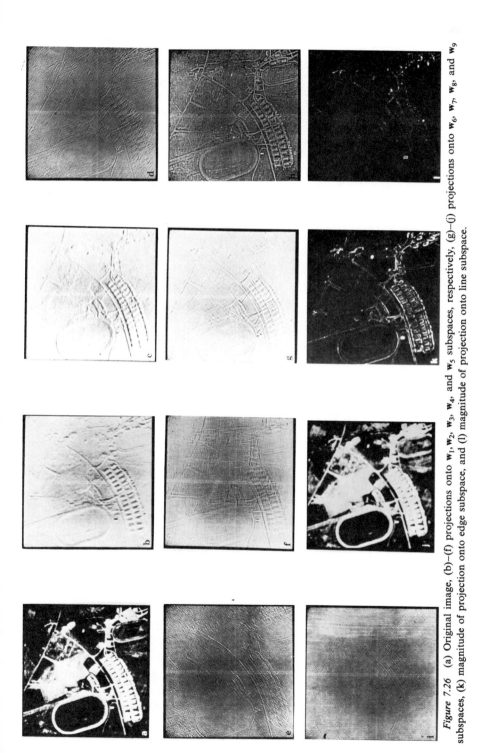

*Figure 7.26* (a) Original image, (b)–(f) projections onto $w_1$, $w_2$, $w_3$, $w_4$, and $w_5$ subspaces, respectively, (g)–(j) projections onto $w_6$, $w_7$, $w_8$, and $w_9$ subspaces, (k) magnitude of projection onto edge subspace, and (l) magnitude of projection onto line subspace.

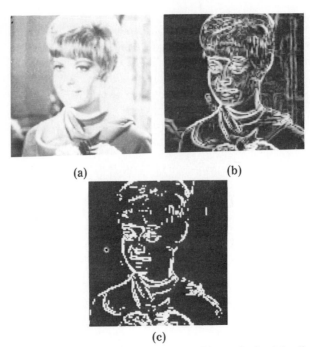

(a)                                                                    (b)

(c)

*Figure 7.27*   (a) Original $128 \times 128$ Kodak girl image, (b) magnitude of the discrete gray level gradient, and (c) thresholded edge image at 10% threshold. (Courtesy of David L. Davies, Dept. of Electrical Engineering, University of Tennessee.)

and

$$\theta_a = \cos^{-1}\left(\frac{1}{\|\mathbf{x}\|} |\mathbf{w}_9' \mathbf{x}|\right),$$

where $\theta_e$, $\theta_1$, and $\theta_a$ are the angles between $\mathbf{x}$ and its projections onto the edge, line, and average subspaces, respectively. These concepts are, of course, directly extendable to other bases and dimensions, as long as the basis vectors are orthogonal.

***Example.***   The image shown in Fig. 7.26a is a $256 \times 256$ aerial photograph of a football stadium site. Figure 7.26b–j gives the magnitudes of the projections along the individual basis vectors obtained by using each of the masks in Fig. 7.25 and, for each position of the $i$th mask, computing a pixel value equal to $|\mathbf{w}_i' \mathbf{x}|$. Figure 7.26k shows the magnitude of the projections onto the edge subspace, and Fig. 7.26l was formed from the magnitudes of projections onto the line subspace. In this example, the best results were obtained with the edge subspace projections, thus indicating a strong edge content in the original image.

### Comparison of Local Edge Operators

An illustrated comparison of ten local edge detection procedures will now be given. The first four methods operate on only a 2 by 2 region of pixels at each point. The next three operators are linear operators with a 3 by 3 convolution mask. The final three operators also use a 3 by 3 region but require a more complex calculation. These include the Frei–Chen system, the Sobel operator, and the Davies method. Both the edge enhanced and edge detected results will be presented. For comparison, a single threshold was set for each edge image that would produce 10% of the detected image points to be displayed as edge points. The picture selected for comparison was a 128 by 128 version of the Kodak girl shown in Fig. 7.27a.

Estimates of the magnitude of the picture gradient over a 2 by 2 region are perhaps the simplest edge operators. A direct estimation is the *discrete gray level gradient D* given by the square root of the squared average differences,

$$d_{1x} = f_{m+1,n} - f_{m,n}, \qquad d_{1y} = f_{m,n} - f_{m,n+1},$$
$$d_{2x} = f_{m+1,n+1} - f_{m,n+1}, \qquad d_{2y} = f_{m+1,n} - f_{m+1,n+1},$$
$$D = \tfrac{1}{2}(d_{1x} + d_{2x})^2 + (d_{1y} + d_{2y})^2 \big]^{1/2}.$$

The magnitude image is shown in Fig. 7.27b. The result of thresholding the edge enhanced image at the 10% threshold is shown in Fig. 7.27c.

Another well known method for estimating the gradient is the *Roberts's magnitude operator* $R_1$, which estimates the derivatives diagonally, i.e.,

$$d_1 = f_{m,n} - f_{m+1,n+1}, \qquad d_2 = f_{m+1,n} - f_{m,n+1}, \qquad R_1 = \big[d_1^2 + d_2^2\big]^{1/2}.$$

The results of the Roberts magnitude operator are shown in Fig. 7.28a, b.

A computationally simpler operator is the Roberts absolute value estimate $R_2$ of the gradient given by

$$R_2 = |d_1| + |d_2|,$$

which avoids the square root computation. The results of this operator are shown in Figs. 7.28c, d.

An alternative to the gradient estimates on a four-pixel group is the *maximum difference operator M*, which is defined as the difference between the maximum and minimum gray levels in a 2 by 2 region. Let

$$B = \max(f_{m,n}; f_{m+1,n}; f_{m,n+1}; f_{m+1,n+1}),$$
$$S = \min(f_{m,n}; f_{m+1,n}; f_{m,n+1}; f_{m+1,n+1}).$$

Then

$$M = B - S.$$

Although this procedure is slightly more complex than the gradient estimates,

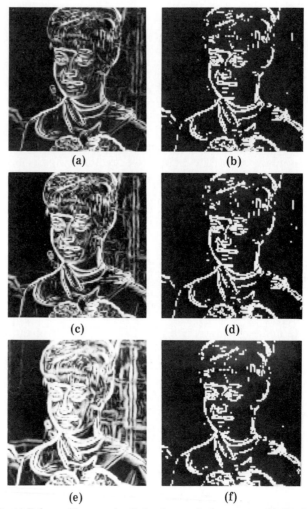

(a)                                          (b)

(c)                                          (d)

(e)                                          (f)

*Figure 7.28*   (a) Enhanced image using Robert's magnitude operator, (b) thresholded image of magnitude operator, (c) enhanced image using Robert's absolute value operator, (d) thresholded image of absolute value operator, (e) enhanced image using maximum difference operator, and (f) thresholded image of maximum difference method. (Courtesy of David L. Davies, Dept. of Electrical Engineering, University of Tennessee.)

it is also more sensitive. For example, if

$$f_{m,n} = f_{m+1,n+1} \neq f_{m+1,n} = f_{m,n+1},$$

then the Roberts operators will yield a result of zero while the maximum difference method will result in

$$M = |f_{m,n} - f_{m+1,n}|,$$

which is greater than zero. The results of the maximum difference operator are shown in Fig. 7.28e, f. Note that, in general, the maximum difference operator yields more continuous edges.

The next set of operators may be considered as Laplacian estimates. Since these operators use second derivative estimates, a 3 by 3 region is required. Several methods may be used to estimate the second derivatives. A second derivative may be estimated by the differences of the first differences, i.e.,

$$\frac{\partial^2 f}{\partial x^2} \simeq (f_{m,n} - f_{m+1,n}) - (f_{m+1,n} - f_{m+2,n}),$$

which simplifies to

$$f_{m,n} - 2f_{m+1,n} + f_{m+2,n}.$$

These estimates may be computed across the diagonals of a 3 by 3 region and summed to provide the Laplacian estimate. The mask shown in Fig. 7.29a is the equivalent convolution operator for this procedure.

If the derivative estimates were made both along the diagonals and the horizontal and vertical central elements and averaged, the negative of the mask shown in Fig. 7.29b would result. The negative mask is used to illustrate the similarity with the unsharp masking procedure. A final mask shown in Fig. 7.29c is derived by averaging the three horizontal and three vertical second derivative estimates.

The enhanced and 10% thresholded images resulting from the Laplacian masks are shown in Fig. 7.30. Note that a characteristic of the Laplacian is that it is zero at the locations at which the gradient is maximum or minimum. Therefore, points detected as gradient edges would generally not be detected as edge points with the Laplacian operator. Another characteristic of the Laplacian operators is that a single gray level transition may produce two distinct peaks, one positive and one negative in the Laplacian. These peaks may be offset from the gradient edge location. For example, this phenomenon is apparent in Fig. 7.30a, b, in which double edges are detected at the right shoulder of the girl.

The set of Frei–Chen basis vectors permit the representation of any 3 by 3 pattern. For edge detection, the four edge basis vectors are convolved with an image. The magnitude of the resulting vectors is estimated by the sum of the absolute values of the convolution results to produce an edge enhanced image. The magnitude and thresholded edge images are shown in Figs. 7.31a, b.

$$\begin{bmatrix} 1 & 0 & 1 \\ 0 & -4 & 0 \\ 1 & 0 & 1 \end{bmatrix} \quad \begin{bmatrix} -1 & -1 & -1 \\ -1 & 8 & -1 \\ -1 & -1 & -1 \end{bmatrix} \quad \begin{bmatrix} 1 & -2 & 1 \\ -2 & 4 & -2 \\ 1 & -2 & 1 \end{bmatrix}$$

(a)          (b)          (c)

*Figure 7.29* Laplacian masks.

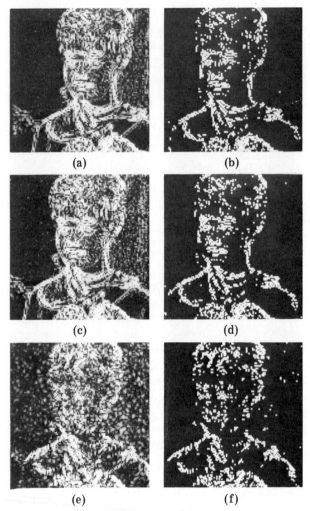

*Figure 7.30* Edge detection using Laplacian masks. (a) enhanced image using diagonal mask, (b) thresholded image using diagonal mask, (c) enhanced image using unsharp mask, (d) thresholded unsharp mask image, (e) enhanced image using averaged derivative mask, and (f) thresholded averaged mask image. (Courtesy of David L. Davies, Dept. of Electrical Engineering, University of Tennessee.)

The Sobel operator is a nonlinear computation of the edge magnitude at $(m,n)$ defined by

$$S(m,n) = \left(d_x^2 + d_y^2\right)^{1/2},$$

where

$$d_x = (f_{m-1,n-1} + 2f_{m,n-1} + f_{m+1,n-1}) - (f_{m-1,n+1} + 2f_{m,n+1} + f_{m+1,n+1}),$$
$$d_y = (f_{m+1,n-1} + 2f_{m+1,n} + f_{m+1,n+1}) - (f_{m-1,n-1} + 2f_{m-1,n} - f_{m-1,n+1}).$$

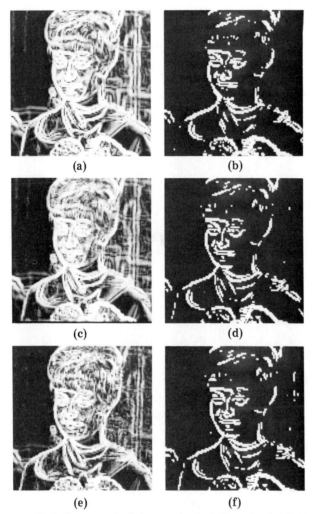

(a)                                          (b)

(c)                                          (d)

(e)                                          (f)

*Figure 7.31*  (a) Frei–Chen magnitude image, (b) Frei–Chen thresholded image, (c) Sobel magnitude image, (d) Sobel thresholded image, (e) Davies magnitude image, and (f) Davies thresholded image. (Courtesy of David L. Davies, Dept. of Electrical Engineering, University of Tennessee.)

Note that the Sobel operator does not make use of the value of $f_{m,n}$. Edge magnitude and thresholded results of the Sobel operator are shown in Fig. 7.31c, d.

The Davies operator was designed to permit edge detection between regions of constant gray level and irregular texture, a pattern that occurs quite often in natural scenes. For example, the boundary of a river and forest in an aerial photograph might produce this pattern. The operator explicitly considers the regularity of gray levels of each side of a possible edge—by performing the following calculations. First, the average gray level of three

|   | 1 | 2 | 3 | 4 | 5 |
|---|---|---|---|---|---|
| 1 | 0 | 0 | 4 | 6 | 7 |
| 2 | 1 | 1 | 5 | 8 | 7 |
| 3 | 0 | 1 | 6´ | 7 | 7 |
| 4 | 2 | 0 | 7 | 6 | 6 |
| 5 | 0 | 1 | 4 | 6 | 4 |

(a)

| | | | | |
|---|---|---|---|---|
| a | a | b | b | b |
| a | a | b | b | b |
| a | a | b | b | b |
| a | a | b | b | b |
| a | a | b | b | b |

(b)

| | | | | |
|---|---|---|---|---|
| a | a | a | a | a |
| a | a | a | a | a |
| a | a | a | a | a |
| a | a | a | a | a |
| a | a | a | a | a |

(c)

*Figure 7.32* Example of region growing using known starting point: (a) original array, (b) segmentation result using distance less than 3, and (c) segmentation result using distance less than 4.

contiguous pixels is calculated. Next, the sum of the deviations from the mean is determined for both the three pixels used to generate the mean and separately for the three opposite pixels. The absolute value of the difference between the deviation sums is used as the edge estimate at that particular orientation. The final edge estimate is the maximum of the absolute deviation differences for all eight orientations.

Consider the procedure for a single orientation. One may hypothesize that an edge exists between a region of constant gray level in the upper right and a texture region in the lower left of a 3 by 3 region. Therefore, the average gray level of the contiguous region is first determined as

$$\mu = \tfrac{1}{3}(f_{m,n-1} + f_{m+1,n-1} + f_{m+1,n}).$$

Next the sums of the deviation of the opposite regions are determined:

$$\sigma_1 = |f_{m,n-1} - \mu| + |f_{m+1,n-1} - \mu| + |f_{m+1,n} - \mu|,$$
$$\sigma_2 = |f_{m,n+1} - \mu| + |f_{m-1,n+1} - \mu| + |f_{m-1,n} - \mu|.$$

The edge estimate at this orientation is therefore $|\sigma_1 - \sigma_2|$. The results of this process, using all eight orientations, are shown in Fig. 7.31e, f.

### Region Growing

A direct approach to the problem of segmenting an image into regions that satisfy a similarity criterion is to start at a point within a region and "grow" a region by grouping all neighboring points that possess a similar property. If the number of regions, the location of a point within each region, and a similarity criterion are known a priori, then the region growing method results in a simple segmentation algorithm.

A numerical example of region growing is shown in Fig. 7.32. Based on gray level values and the gray level histogram, the original image shown in Fig. 7.32a may be assumed to consist of two regions. Starting points of $(3,2)$ and $(3,4)$, respectively, are assumed for the two regions. The segmentation results, using a Euclidean distance less than 3, are shown in Fig. 7.32b. The segmentation resulted in two regions, as expected, and the same segmentation would have resulted from any other starting point in the region. The segmentation results using the same starting points but grouping points with distance less than 4 is shown in Fig. 7.32c. Note the sensitivity to the similarity threshold.

### Region Clustering

The region clustering problem applies directly to image segmentation, and differs from general clustering only by the requirement that points within a cluster must be contiguous in the image plane as well as similar in properties. This requirement may also be used to define contiguous image points with similar properties, as distinguished from a cluster, which is simply a set of points with similar properties. The general procedure of region clustering is to first examine the set of image measurements to determine the number and location of clusters in measurement space, then apply these cluster definitions to the image to obtain region clusters. Criteria may be applied to both the clustering procedure and to the region clustering. Different criteria may be required at each step. For example, a color random dot image would provide clusters in the three-dimensional color space, but would not provide regional clusters.

The following procedure is illustrative of a region clustering technique. The reader should be aware, however, that cluster seeking is a major area of automatic information processing and, as such, can only be introduced in our present discussion. The interested reader can consult, for example, the book by Tou and Gonzalez (1974) for additional details and references on this topic.

A *partition* of a set $X$ is any collection of sets $\{R_1, R_2, \ldots, R_n\}$ such that the union of the sets $R_k$ is exactly $X$ and the pairwise intersection of the $R_k$ is null unless the two sets are identical. If we define some equivalence relation on the picture—a simple example is $P(i,j)$ is equivalent to $P(k,1)$ if their values are equal—then this relation induces a natural equivalence relation on the grid of points, given by $(i,j)$ is equivalent to $(k,1)$ if and only if $P(i,j)$ is equivalent to $P(k,1)$.

Any equivalence relation on the grid of points yields a partition of the grid into equivalence classes. For example, if the values of $P(i,j)$ range from 0 to 63, then 64 equivalence class masks could be produced whose values would be equal to 1 if the relation were satisfied and 0 otherwise. The mask images would be pairwise disjoint and the union of the 64 masks would fill the entire grid.

The equivalence classes can be further subdivided into maximally connected subsets called *connected components*. Connectedness may be defined in terms of the neighbors of a points $(i,j)$. The *4-connected neighbors of a point* are the four nondiagonally adjacent neighbors. The *8-connected neighbors* are the eight surrounding neighbor points. Using 4-connectedness, we say that two points $p_1$ and $p_2$ belonging to a subset $R$ of the grid are connected if there exists a sequence of points in $R$, the first of which is $p_1$ and the last of which is $p_2$, such that consecutive points are 4-connected neighbors. With this definition of connected points, we may define a region as a subset of $R$ in which any pair of points is connected with respect to $R$. The equivalence masks may be divided into maximally connected regions (sometimes called *atomic regions*). These can then be used as building blocks to form regions that respond to natural, meaningful segmentations.

A method for representing a region during segmentation, developed by Brice and Fennema (1970), is as follows. One may consider the picture grid $G$ to be a subgrid of a larger grid $S$. In particular, if $G$ is an $n \times m$ grid, then let $S$ be a $(2n+1) \times (2m+1)$ grid in which the points $(i,j)$ of $G$ are placed on the points $(2i+1, 2j+1)$ of $S$. The points in the picture correspond to points of $S$ in which each subscript is odd, and the remaining points may be used to represent the boundaries of regions.

Representing regions in this manner yields a simple algorithm for finding atomic regions of a picture. Each point of $G$, except for its edges, is compared with the one above it and the one to its right. If a difference in gray scale is encountered, the boundary segment is inserted between them. After each picture point is considered, the grid has been partitioned into regions. An example is shown in Fig. 7.33. Note that the equal gray levels along the diagonal produce separate regions because a connectedness of 4 was used for the equivalence relation.

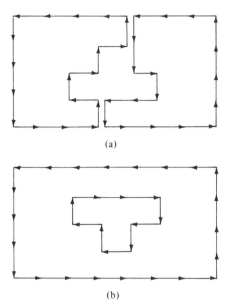

(a)

(b)

*Figure 7.33*   Example of Brice and Fennema segmentation method: (a) 3×3 array with gray level and grid values shown. (b) The 7×7 supergrid with gray levels placed on odd subscripted points and resulting segmentation.

Since the atomic regions produced in the previous step may not correspond to physical or visible boundaries, merging or splitting of regions may be necessary. If the region boundaries are assumed to be oriented so that the region always lies to the left of the boundary, then the operation of merging two regions into one may be accomplished by adding the boundaries. An example is shown in Fig. 7.34.

Segmentation may either be edge based or region based. The edge-based method works well but is sensitive to local noise. The global method is more stable since it considers the overall characteristics of the image, but often false regions may also be detected. For a perfect region segmentation, the global region boundaries should coincide with the local edges. A method called global–local edge coincidence (GLEC) segmentation has been developed by (Hwang *et al.*, 1979), which detects the coincidence of the region boundaries (or global edges) and local edges. Since the global edges are obtained from the global characteristics of the image, for instance, the histogram of the intensity of the image, the local noise edges will not be detected in the global edge map; however, perfect region segmentation is hardly obtained by using only intensity information. Basically, GLEC is a merge-oriented region segmentation method. Two regions are merged if the common boundary between these regions does not match the local edges. An example of this technique is shown in Fig. 7.35. Figure 7.35a shows a picture of an original scene containing several solid objects. Figure 7.35c shows the

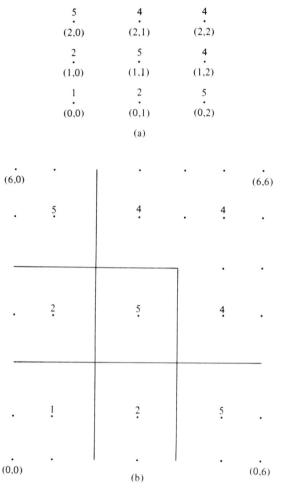

Figure 7.34 (a) Two regions with a common boundary and (b) result of merging by adding the directed boundary segments.

local edge-segmented scene using the Sobel operator. Figure 7.35b shows the global region boundaries obtained using the intensity histogram segmentation. Figure 7.35d shows the segmented picture using GLEC. Figure 7.35e shows the region boundaries of the GLEC segmented picture. Note that the resulting regions are much more meaningful than the original regions.

## 7.3 Description of Regions

The next step following the segmentation of a scene into regions, as shown in Fig. 7.1, is to develop regional descriptions. Since the results of segmentation may be either edge boundaries or regions, the types of descriptors are quite varied. It is usually desirable to choose descriptors that are invariant to

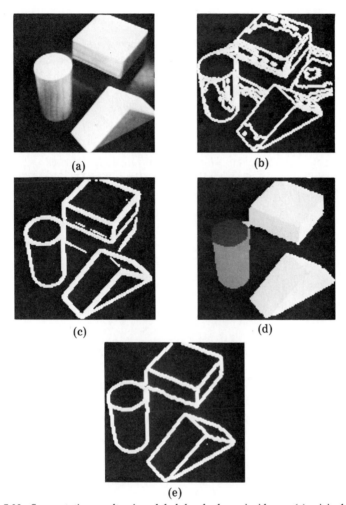

(a)                                   (b)

(c)                                   (d)

(e)

*Figure 7.35*  Segmentation result using global–local edge coincidence: (a) original scene, (b) region boundaries resulting from histogram segmentation, (c) edge boundaries resulting from Sobel operator, (d) segmentation after merging based on global–local edge coincidence, and (e) region boundaries of (d).

the normal variations in the scene. Invariance to translation, rotation, scale, and perspective might be desirable. Other desirable criteria might be the ability to discriminate between regions, computational simplicity, or data reduction.

### 7.3.1   Boundary Description

Given the boundary of a primitive or region, it is often desirable to extract measurements that characterize the length, direction, or shape of the region. Several methods, including linear and nonlinear curve fitting, chain coding, and Fourier boundary descriptors are available.

### Curve Fitting Using Linear and Nonlinear Least Squares

It is often possible to describe boundary properties by a curve such as a line, parabola, or ellipse. If the curve is specified parametrically, then linear or nonlinear least squares provides a method for determining the parameters.

Suppose that a set of edge points $\{(x_i, y_i), i = 1, 2, \ldots, n\}$ are given. These points represent a discrete boundary function $y_i = f(x_i)$ with domain

$$T = \{x_1, x_2, \ldots, x_n\}.$$

Let $g(x, p_1, p_2, \ldots, p_M)$ be an $M$-parameter family of curves of a particular form whose domain includes all $x$ in $T$. The discrete function $f$ may be considered as a point in Euclidean $n$-space,

$$\mathbf{y} = (y_1, y_2, \ldots, y_n)'.$$

The desired solution is also a point in $E^n$,

$$\mathbf{g} = [\, g(x_1), g(x_2), \ldots, g(x_k) \,]',$$

which minimizes the distance

$$D(\mathbf{p}) = \|\mathbf{y} - \mathbf{g}\| = \left\{ \sum_{k=1}^{n} [\, y_k - g(x_k) \,]^2 \right\}^{1/2}.$$

For each parameter vector $\mathbf{p}$, there is a corresponding point $\mathbf{g}$ in $E^n$ so that the set of all points $\{\mathbf{g} : \mathbf{p} \in E^M\}$ is an $M$ dimensional surface $S$ in $E^n$. If $\mathbf{y}$ lies in $S$, then for some $\mathbf{p}$, $D(\mathbf{p}) = 0$. If $\mathbf{y}$ does not lie in $S$, then there may be one or more places on the surface $S$ that are closest to $\mathbf{y}$.

Suppose that $g(x, \mathbf{p})$ is differentiable in all its variables. Then $D(\mathbf{p})$ is also differentiable with respect to $\mathbf{p}$. If $D(\mathbf{p}^*) \leq D(\mathbf{p})$, then

$$\frac{\partial D(\mathbf{p}^*)}{\partial p_1} = \frac{\partial D(\mathbf{p}^*)}{\partial p_2} = \cdots = \frac{\partial D(\mathbf{p}^*)}{\partial p_M} = 0.$$

This leads to the system of equations:

$$F_1(\mathbf{p}) = \sum_{k=1}^{n} [\, y_k - g(x_\mathbf{p}) \,] \frac{\partial g(x, \mathbf{p})}{\partial p_1} = 0,$$

$$F_2(\mathbf{p}) = \sum_{k=1}^{n} [\, y_k - g(x_\mathbf{p}) \,] \frac{\partial g(x, \mathbf{p})}{\partial p_2} = 0,$$

$$\vdots$$

$$F_M(\mathbf{p}) = \sum_{k=1}^{n} [\, y_k - g(x_\mathbf{p}) \,] \frac{\partial g(x, \mathbf{p})}{\partial p_M} = 0.$$

This is a system of equations of the form

$$\mathbf{F}(\mathbf{p}) = \begin{bmatrix} F_1(\mathbf{p}) \\ F_2(\mathbf{p}) \\ \vdots \\ F_M(\mathbf{p}) \end{bmatrix} = 0.$$

If $g$ is a linear function of $p_1, p_2, \ldots, p_M$, then the system of equations is linear and may be written in the form

$$g(x, \mathbf{p}) = p_1 * g_1(x) + p_2 * g_2(x) + \cdots + p_M * g_M(x),$$

and we define the vectors

$$\mathbf{g}_i(x) = \left[ g_1(x_1), g_1(x_2), \ldots, g_1(x_M) \right] \quad \text{and} \quad \mathbf{e} = \mathbf{y} - \sum_{j=1}^{M} \mathbf{g}_j p_j.$$

The solution may be written in the form

$$\mathbf{g}_i' \mathbf{y} - \sum_{j=1}^{M} \mathbf{g}_i' \mathbf{g}_j p_j = 0,$$

for $i = 1, 2, \ldots, M$ or $\mathbf{g}_i' \mathbf{e} = 0$, which illustrates that the error vector is orthogonal to each of the vectors $\mathbf{g}_i$.

If $g(x, p)$ is nonlinear in $\mathbf{p}$, then $\mathbf{F}(\mathbf{p})$ is a system of nonlinear equations and one may use Newton's method or steepest descent to find a $\mathbf{p}$ that minimizes $D(\mathbf{p})$. If $\mathbf{F}(\mathbf{p})$ is differentiable in $\mathbf{p}$ and has a nonsingular Jacobian matrix near the solution $\mathbf{p}^*$ and if a sufficiently good approximation $\mathbf{p}^0$ to $\mathbf{p}^*$ may be made, then Newton's method provides rapid convergence to a solution.

Newton's method for finding $\mathbf{p}^*$ is an iterative process defined by

$$\mathbf{F}'\left[\mathbf{p}^{(k)}\right]\left[\mathbf{p}^{(k+1)} - \mathbf{p}^{(k)}\right] = -\mathbf{F}\left[\mathbf{p}^{(k)}\right],$$

where $\mathbf{F}'$ is the $M \times M$ Jacobian matrix of partial derivatives

$$\mathbf{F}'(\mathbf{p}) = \begin{bmatrix} \dfrac{\partial F_1}{\partial p_1} & \dfrac{\partial F_1}{\partial p_2} & \cdots & \dfrac{\partial F_1}{\partial p_M} \\ & \vdots & & \\ \dfrac{\partial F_M}{\partial p_1} & \dfrac{\partial F_M}{\partial p_2} & \cdots & \dfrac{\partial F_M}{\partial p_M} \end{bmatrix}.$$

To use Newton's method, we begin with $k = 0$ and the starting point $\mathbf{p}^0$. At each iteration evaluate the vector function $\mathbf{F}(\mathbf{p}^{(k)})$ and the Jacobian matrix $\mathbf{F}'(\mathbf{p}^{(k)})$. Then solve the linear system

$$\mathbf{F}'(\mathbf{p}^{(k)}) \Delta^{(k)} = \mathbf{F}(\mathbf{p}^{(k)})$$

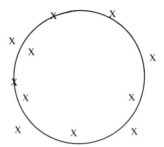

*Figure 7.36*   Example of nonlinear least squares fit of points to a circle.

for the increment vector $\mathbf{\Delta}^{(k)}$. Then set

$$\mathbf{p}^{(k+1)} = \mathbf{p}^{(k)} + \mathbf{\Delta}^{(k)}$$

and iterate again or stop with $\mathbf{p}^* \simeq \mathbf{p}^{(k+1)}$. A reasonable stopping criterion in many applications is to evaluate the distance function $D(\mathbf{p}^{(k+1)})$ and stop when the distance is no longer decreasing. For example, iterate until $D(\mathbf{p}^{(k+1)}) > 0.99 D(\mathbf{p}^{(k)})$.

An example of the use of the least squares for fitting points to a circle is shown in Fig. 7.36.

### Fourier Descriptors

It is shown in this section that the discrete Fourier transform (DFT) can be used for obtaining a set of regional descriptors.

Suppose that $M$ points on the boundary of a region are available. We may view the region as being in the complex plane, with the ordinate being the imaginary axis and the abscissa being the real axis, as shown in Fig. 7.37. The $x$–$y$ coordinates of each point in the contour to be analyzed become complex numbers $x + jy$. Starting at an arbitrary point on the contour, and tracing once around it, yields a sequence of complex numbers. The DFT of this sequence will be referred to in the following discussion as the Fourier descriptor (FD) of the contour.

Since the DFT is a reversible linear transformation, there is no information gained or lost by this process. However, certain simple manipulations of this

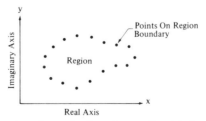

*Figure 7.37*   Representation of a region boundary in the complex plane. (Reproduced from "Digital Image Processing and Recognition," 1977, by Rafael C. Gonzalez and Paul Wintz, with permission of Addison-Wesley, Reading, Massachusetts.)

frequency domain representation of shape can eliminate dependence on position, size, and orientation. Given an arbitrary FD, several successive steps can normalize it so that it can be matched to a test set of FDs regardless of its original size, position, and orientation.

### Normalization

The frequency domain operations that affect the size, orientation, and starting point of the contour follow directly from properties of the DFT (see Sec. 3.3). To change the size of the contour, the components of the FD are simply multiplied by a constant. Due to linearity, the inverse transform will have its coordinates multiplied by the same constant.

To rotate the contour in the spatial domain simply requires multiplying each coordinate by $\exp(j\theta)$, where $\theta$ is the angle of rotation. Again, by linearity, the constant $\exp(j\theta)$ has the same effect when the frequency domain coefficients are multiplied by it.

To see how the contour starting point can be moved in the frequency domain, recall the periodicity property of the DFT. The finite sequence of numbers in the spatial domain actually represents one cycle of a periodic function. The DFT coefficients are actually coefficients of the Fourier series representation of this periodic function. Remembering these facts, it is easy to see that shifting the starting point of the contour in the spatial domain corresponds to multiplying the $k$th frequency coefficient in the frequency domain by $\exp(jkT)$, where $T$ is the fraction of a period through which the starting point is shifted. (As $T$ goes from 0 to $2\pi$, the starting point traverses the whole contour once.)

Given the FD of an arbitrary contour, the normalization procedure requires performing the normalization operations so that the contour has a standard size, orientation, and starting point. A standard size is easily defined by requiring the Fourier component $F(1)$ to have unit magnitude. If the contour is a simple closed figure, and it is traced in the counterclockwise direction, this coefficient will be the largest.

The orientation and starting point operations affect only the phases of the FD coefficients. Since there are two allowable operations, the definition of standard position and orientation must involve the phases of at least two coefficients. Let us denote the FD array of length M by $\{F(-\frac{1}{2}M+1),$ $\ldots, F(-1), F(0), F(1), \ldots, F(\frac{1}{2}M)\}$. One obvious coefficient to use is $F(1)$, already normalized to have unit magnitude. If we start by requiring the phase of $F(1)$ to be some value, say zero, it can be shown that if the $k$th coefficient is also required to have zero phase, there are $(k-1)$ possible starting point–orientation combinations that satisfy these restrictions.

The obvious procedure is to require $F(1)$ and $F(2)$ to have phases equal to some specified value, thereby achieving a unique standard normalization. This sounds like a solution to the problem, but while $F(1)$ is guaranteed to have unit magnitude after normalizing the FD for size, there is no such guarantee for $F(2)$. A consistent solution to this problem can be obtained by selecting a nonzero coefficient to use for normalization, and then using a third coefficient to resolve the ambiguity that may be caused by the multiple normalization effect.

### *Practical Considerations*

The practical implementation of this procedure requires paying attention to a few details not mentioned above. Theoretically, the procedure involves an exact representation of a contour that is sampled at uniform spacing. While nonuniform spacing can result in a frequency domain representation that converges faster, there are some serious difficulties involved in attempting to define a standard sampling strategy using nonuniform spacing.

Remembering that the FFT algorithm requires an input array whose length is an integer power of 2, it is clear that the length of an arbitrary chain representation must be adjusted before the FFT can be used. A procedure for doing this is to compute the perimeter of the contour, divide it by the desired length (desired power of 2), and starting at one point, trace around the contour saving the coordinates of appropriately spaced points. The desired power of 2 might be the smallest power of 2 larger than the length of the chain.

Practically, the input to the shape analysis algorithm will be a contour taken from a sampled picture. The perimeter of this contour will be an approximation to the actual perimeter of the contour. While it can be argued that, for high enough sampling density in the original picture, the chain is an arbitrarily good approximation to the contour, this argument breaks down if one considers the density of points around the approximate contour versus the exact contour.

Consider an equilateral right triangle oriented so that the legs line up with the $x$ and $y$ axes, with the hypotenuse at $45°$. The "length" of the contour, if an ordinary four neighbor chain is used, will be four times the length of one leg; the hypotenuse will be as long as both legs combined. Obviously, the density of points on the hypotenuse will depart from the proper value by a factor of $\sqrt{2}$. This error will cause the normalized Fourier descriptors (NFDs) of simple figures such as triangles to differ substantially, and render the algorithm virtually useless.

One solution to this problem is to use an 8-neighbor chain code, in which the four diagonal neighbors of a point can also be the next point in the chain. In the example just considered, this eliminates the point density error. Of course, for different orientations, there will still be a certain amount of error due to the chain code approximation, but this is reduced from a maximum of about 40% to a maximum of about 8%. Experimental results using the 8-neighbor chain code confirm that this error is tolerable. If a picture is contoured using a 4-neighbor chain code and it is desired to process the contours using the FD method, the four chain codes can be easily converted to approximate eight chain codes that are suitable for analysis.

Other practical considerations involve the normalization process. While, theoretically, any nonzero coefficient can be used with $F(1)$ to define standard orientation and starting point as outlined above, practical contours show the effects of noise and quantization error. This noise perturbs the phases of the FD coefficients so that the coefficients of lower amplitude can be substantially affected. It can be shown that the mean squared error in the frequency domain corresponds to the point-by-point mean squared error in the spatial domain. It follows from this result that slight shifts in orientation and/or starting point due to noise can have drastic effects on classification of shapes made using this criterion. One way to minimize this effect is to choose

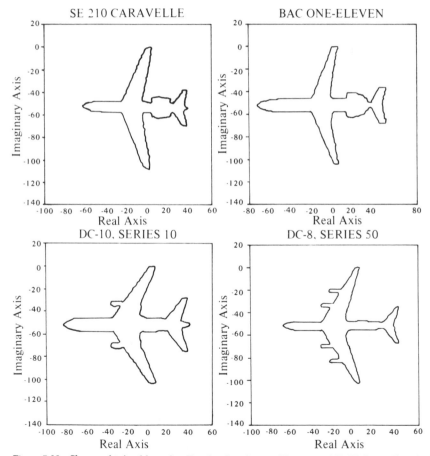

*Figure 7.38* Shapes obtained by using Fourier descriptors. (Courtesy of T. Wallace, Electrical Engineering Dept., Purdue University.)

the largest magnitude coefficients as normalization coefficients. $F(1)$ is already the largest, so the second largest is chosen to accompany $F(1)$. Generally a third coefficient will be required to decide which of the allowable normalizations is óptimum, as explained above. This coefficient can be chosen to be the largest remaining coefficient suitable for resolving the ambiguity.

The normalization procedure tends to reduce the proportion of the information contained in the phase, as compared to that contained in the magnitudes. Also, it can be shown that if the contour under analysis has bilateral symmetry, the resulting NFD will have phases equal to either the normalization phase (phase to which the normalization coefficients are constrained), or that value plus 180°. In view of these results classification using only the magnitudes of the NFD seems like a reasonable procedure. In this case the normalization procedure consists of simply dividing each coefficient by the magnitude of $F(1)$.

As an illustration of the use of Fourier descriptors for shape analysis, consider the region boundaries shown in Fig. 7.38a. Each region was first described by a shape profile or radial distance from the centroid to the boundary sampled in 512 equal intervals. The Fourier transform of each profile was then computed and only the 32 lowest frequency components were retained. The inverse Fourier transform was then computed and used to reconstruct the region boundaries, which are shown in Fig. 7.38b. Note that although the boundaries are somewhat distorted, the essential shapes have been retained, while the data has been compressed by a factor of 8.

### 7.3.2  Region Description Using Moments

The Fourier descriptions discussed in the previous section are based on the assumption that a set of boundary points is available. Sometimes a region may be given in the form of interior points and, as indicated earlier, one may be interested in finding descriptors that are invariant to variations in translation, rotation, and size. The moment approach discussed below is often used for this purpose.

Given a two-dimensional continuous function $f(x, y)$ we define the moment of order $p + q$ by the relation

$$m_{pq} = \int_{-\infty}^{\infty} \int_{-\infty}^{\infty} x^p y^q f(x, y) \, dx \, dy$$

for $p, q = 0, 1, 2, \ldots$ .

A uniqueness theorem (Papoulis, 1968) states that if $f(x, y)$ is piecewise continuous and has nonzero values only in a finite part of the $x$–$y$ plane, then moments of all orders exist and the moment sequence $(m_{pq})$ is uniquely determined by $f(x, y)$ and, conversely, $(m_{pq})$ uniquely determines $f(x, y)$. The *central moments* can be expressed as

$$\mu_{pq} = \int_{-\infty}^{\infty} \int_{-\infty}^{\infty} (x - \bar{x})^p (y - \bar{y})^q f(x,y) \, dx \, dy,$$

where $\bar{x} = m_{10}/m_{00}$, $\bar{y} = m_{01}/m_{00}$.

For a digital image the above equation becomes

$$\mu_{pq} = \sum_x \sum_y (x - \bar{x})^p (y - \bar{y})^q f(x,y).$$

The central moments of order 3 are as follows:

$$\mu_{10} = \sum_x \sum_y (x - \bar{x})^1 (y - \bar{y})^0 f(x,y) = m_{10} - \frac{m_{10}}{m_{00}}(m_{00}) = 0,$$

$$\mu_{11} = \sum_x \sum_y (x - \bar{x})^1 (y - \bar{y})^1 f(x,y) = m_{11} - \frac{m_{10}m_{01}}{m_{00}},$$

$$\mu_{20} = \sum_x \sum_y (x - \bar{x})^2 (y - \bar{y})^0 f(x,y) = m_{20} - \frac{2m_{10}^2}{m_{00}} + \frac{m_{10}^2}{m_{00}} = m_{20} - \frac{m_{10}^2}{m_{00}},$$

$$\mu_{02} = \sum_x \sum_y (x - \bar{x})^0 (y - \bar{y})^2 f(x,y) = m_{02} - \frac{m_{01}^2}{m_{00}},$$

$$\mu_{30} = \sum_x \sum_y (x - \bar{x})^3 (y - \bar{y})^0 f(x,y) = m_{30} - 3\bar{x}m_{20} + 2m_{10}\bar{x}^2,$$

$$\mu_{12} = \sum_x \sum_y (x - \bar{x})(y - \bar{y})^2 f(x,y) = m_{12} - 2\bar{y}m_{11} - \bar{x}m_{02} + 2\bar{y}^2 m_{10},$$

$$\mu_{21} = \sum_x \sum_y (x - \bar{x})^2 (y - \bar{y})^1 f(x,y) = m_{21} - 2\bar{x}m_{11} - \bar{y}m_{20} + 2\bar{x}^2 m_{01},$$

$$\mu_{03} = \sum_x \sum_y (x - \bar{x})(y - \bar{y})^3 f(x,y) = m_{03} - 3\bar{y}m_{02} + 2\bar{y}^2 m_{01}.$$

In summary

$$\mu_{00} = m_{00}, \qquad \mu_{11} = m_{11} - \bar{y}m_{10},$$

$$\mu_{10} = 0, \qquad \mu_{30} = m_{30} - 3\bar{x}m_{20} + 2m_{10}\bar{x}^2,$$

$$\mu_{01} = 0, \qquad \mu_{12} = m_{12} - 2\bar{y}m_{11} - \bar{x}m_{02} + 2\bar{y}^2 m_{10},$$

$$\mu_{20} = m_{20} - \bar{x}m_{10}, \qquad \mu_{21} = m_{21} - 2\bar{x}m_{11} - \bar{y}m_{20} + 2\bar{x}^2 m_{01},$$

$$\mu_{02} = m_{02} - \bar{y}m_{01}, \qquad \mu_{03} = m_{03} - 3\bar{y}m_{02} + 2\bar{y}^2 m_{01},$$

The *normalized central moments*, denoted by $\eta_{pq}$, are defined as

$$\eta_{pq} = \mu_{pq}/\mu_{00}^\gamma, \qquad \text{where} \quad \gamma = \tfrac{1}{2}(p+q) + 1 \quad \text{for} \quad p+q = 2,3,\dots.$$

From the second and third moments, a set of seven *invariant moments* can

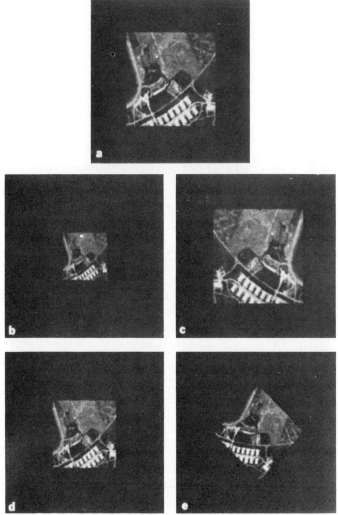

*Figure 7.39*  Images used to demonstrate invariant moment properties. [From Sadjadi and Hall (1978).]

be derived. They are given by

$$\phi_1 = \eta_{20} + \eta_{02},$$
$$\phi_2 = (\eta_{20} - \eta_{02})^2 + 4\eta_{11}^2,$$
$$\phi_3 = (\eta_{30} - 3\eta_{12})^2 + (3\eta_{21} - \eta_{03})^2,$$
$$\phi_4 = (\eta_{30} + \eta_{12})^2 + (\eta_{21} + \eta_{03})^2,$$

| Invariant (log) | Original | Half Size | Mirrored | Rotated($2°$) | Rotated($45°$) |
|---|---|---|---|---|---|
| $\varphi_1$ | 6.249 | 6.226 | 6.919 | 6.253 | 6.318 |
| $\varphi_2$ | 17.180 | 16.954 | 19.955 | 17.270 | 16.803 |
| $\varphi_3$ | 22.655 | 23.531 | 26.689 | 22.836 | 19.724 |
| $\varphi_4$ | 22.919 | 24.236 | 26.901 | 23.130 | 20.437 |
| $\varphi_5$ | 45.749 | 48.349 | 53.724 | 46.136 | 40.525 |
| $\varphi_6$ | 31.830 | 32.916 | 37.134 | 32.068 | 29.315 |
| $\varphi_7$ | 45.589 | 48.343 | 53.590 | 46.017 | 40.470 |

*Figure 7.40* Moment invariants for the images in Fig. 7.39a–e. [From Sadjadi and Hall (1978).]

$$\phi_5 = (\eta_{30} - 3\eta_{12})(\eta_{30} + \eta_{12})\left[(\eta_{30} + \eta_{12})^2 - 3(\eta_{21} + \eta_{03})^2\right],$$
$$+ (3\eta_{21} - \eta_{03})(\eta_{21} + \eta_{03})\left[3(\eta_{30} + \eta_{12})^2 - (\eta_{21} + \eta_{03})^2\right],$$
$$\phi_6 = (\eta_{20} - \eta_{02})\left[(\eta_{30} + \eta_{12})^2 - (\eta_{21} + \eta_{03})^2\right],$$
$$+ 4\eta_{11}(\eta_{30} + \eta_{12})(\eta_{21} + \eta_{03}),$$
$$\phi_7 = (3\eta_{21} - \eta_{03})(\eta_{30} + \eta_{12})\left[(\eta_{30} + \eta_{12})^2 - 3(\eta_{21} + \eta_{03})^2\right],$$
$$+ (3\eta_{12} - \eta_{30})(\eta_{21} + \eta_{03})\left[3(\eta_{30} + \eta_{12})^2 - (\eta_{21} + \eta_{03})^2\right].$$

This set of moments has been shown to be invariant to translation, rotation, and scale change by Hu (1962).

***Example.*** The image shown in Fig. 7.39a was reduced to half size (Fig. 7.39b), mirror imaged (Fig. 7.39c), and rotated by 2 and 45°, as shown in Fig. 7.39d, e. The seven-moment invariants given in the above equations were then computed for each of these images, and the logarithm of the results taken to reduce the dynamic range. As shown in Fig. 7.40, the results for Fig. 7.39b–e are in reasonable agreement with the invariants computed for the original image. The major cause of error can be attributed to the digital nature of the data.

### 7.3.3 Topological Description

Topological properties are useful for global descriptions of regions in the image plane. Simply defined, topology is the study of properties of a figure that are unaffected by any deformation without tearing or joining. Consider, for example, the region shown in Fig. 7.41. If we define as a topological descriptor the number of holes in the region, it is evident that this property

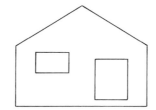

*Figure 7.41*   A region with two holes.

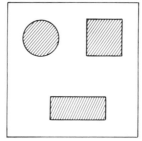

*Figure 7.42*   A region with three connected components.

will not be affected by a stretching or rotation transformation. However, the number of holes will, in general, change if we tear or fold the region. Note that since stretching affects distance, topological properties do not depend on any notion of distance or any properties that are implicitly based on the concept of a distance measure.

Another topological property useful for region description is the number of connected components. A connected component of a set is a subset of maximal size such that any two of its points can be joined by a connected curve lying entirely within the subset. Fig. 7.42 shows a region with three connected components.

The number of holes $H$ and connected components $C$ in a figure can be used to define the *Euler* number $E$ as

$$E = C - H.$$

The Euler number is also a topological property. The regions shown in Fig. 7.43, for example, have Euler numbers equal to 0 and $-1$, respectively, since Fig. 7.43a has one connected component and one hole and Fig. 7.43b has one connected component but two holes.

Regions represented by straight line segments (referred to as *polygonal networks*) have a particularly simple interpretation in terms of the Euler number. A polygonal network is shown in Fig. 7.44. It is often important to classify interior regions of such a network into faces and holes. If we denote the number of vertices by $V$, the number of edges by $Q$, and the number

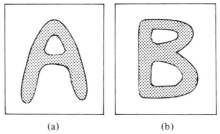

(a)                              (b)

*Figure 7.43*   Regions with Euler numbers equal to 0 and −1, respectively. (Reproduced from "Digital Image Processing and Recognition," 1977, by Rafael C. Gonzalez and Paul Wintz, with permission of Addison-Wesley, Reading, Massachusetts.)

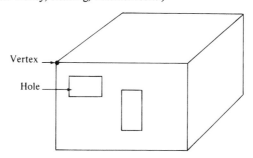

*Figure 7.44*   A region containing a polygonal network. Vertices, $V = 15$; edges, $Q = 17$; faces, $F = 3$; connected regions, $C = 3$; holes, $H = 2$. Euler number $E = V - Q + F = C - H = 1$.

of faces by $F$, we have the relationship, called the *Euler formula*,

$$V - Q + F = C - H,$$

which, in view of the previous equation, is related to the Euler number,

$$W - Q + F = C - H = E.$$

The network shown in Fig. 7.44 has 15 vertices, 17 edges, 3 faces, 3 connected regions, and 2 holes; thus,

$$E = 15 - 17 + 3 = 3 - 2 = 1.$$

Although topological concepts are rather general, they provide an additional feature that is often useful in characterizing regions in a scene.

## 7.4    Relationships between Regions

### 7.4.1    Relational Representations

If an image or scene has been segmented into regions or primitive components, the next step or level of complexity in scene understanding is to determine the important relations and structure in the scene and to organize these relations in a meaningful way. Although a unified theory for relating regions or components of a scene has yet to be developed, several important general concepts for using a relational representation will be introduced in this section.

The concept of the relational table was explicitly stated by Codd (1970, 1972) in order to establish a model that would allow a user to specify nontrivial operations on and with the data, yet not require the user to know all the details about how the data are actually organized in the computer. A relational model of a scene exists independently of the applications programs that operate on it.

In comparison with earlier models, Codd's concept overcomes problems like ordering dependency (many earlier models required data elements to be stored in a total ordering associated with the hardware-determined ordering of addresses) and access path dependency (many earlier systems employed tree-structured files or something similar, around which the applications programs had to be developed). Recent applications of the relational table approach for the representation of scenes may be found in Kunii *et al.* (1978) and Thomason *et al.* (1978).

Since we may consider a scene to be composed of a set of discrete objects and a picture to be composed of a set of discrete regions, we shall first review some elements of discrete mathematics (Liu, 1977).

An important concept is that of a *relation*. Although relations are often referred to by descriptive names such as "color," "above," "inside," etc., a formal definition is necessary. Let $A$ and $B$ be two sets. The *Cartesian product* of $A$ and $B$, which will be denoted $A \times B$ is the set of ordered pairs of the form $(a, b)$, where $a \in A$ and $b \in B$. A *binary relation* from $A$ to $B$ is a subset of $A \times B$. The binary relation is simply a formalism for the intuitive notion that some of the elements in $A$ are related to some of the elements in $B$. In fact if the ordered pair $(a, b)$ is in the subset that defines the relation $R$, then we may say that $a$ is related to $b$. A binary relation may be represented in several ways. The most direct method is to simply list the set of ordered pairs that are in $R$. This list is one type of relational table. As an example, consider the simple picture shown in Fig. 7.45a, which is segmented into three regions labeled $e$, $f$, and $g$. The set $A = \{e, f, g\} = B$. The relation "larger than" is a subset of $A \times A$ consisting of the ordered pairs listed in the relational table shown in Fig. 7.45b. Another representation of the binary relation is by a *relational matrix* corresponding to the Cartesian product space with indicators such as 1s for the ordered pairs that are in the relation. An example is shown in Fig. 7.45c. The binary relation may also be indicated graphically by a relational diagram that consists of two sets of nodes corresponding to the elements of the sets $A$ and $B$ and directed lines from the elements of $A$ to the elements of $B$ that form the subset. An example is shown in Fig. 7.45d.

Binary relations describe relationships between pairs of objects. We may also define ternary relations for ordered triples of objects, quaternary relations for ordered quadruples of objects, etc. A *ternary relation* may be defined

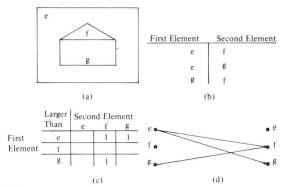

*Figure 7.45*   (a) Picture segmented into three regions, (b) relational table for relation "larger than," (c) relational matrix, and (d) relational diagram.

in terms of ordered pairs of the form $[(a,b),c]$ where the first component is an ordered pair. This could also be considered as an ordered triple; however, the previous definition permits us to consider only binary relations. Formally, a ternary relation among three sets $A$, $B$, and $C$ is defined as a subset of the Cartesian product $(A \times B) \times C$. Similarly, a quarternary relation among four sets $A$, $B$, $C$, and $D$ is a subset of $[(A \times B) \times C] \times D$. In general an ordered $n$-tuple may be considered as an ordered pair whose first component is an ordered $(n-1)$-tuple and an *n-ary relation* among the sets $A_1, A_2, \ldots, A_n$ is defined as a subset of $[(A_1 \times A_2) \times A_3] \times \cdots \times A_n]$. Thus, we may concentrate on the properties of binary relations.

A binary relation $R$ from $A$ to $B$ is said to be

(i)   *reflexive* if $(a,a)$ is in $R$ for every $a$ in $A$;
(ii)  *symmetric* if $(a,b)$ in $R$ implies that $(b,a)$ is in $R$;
(iii) *transitive* if whenever $(a,b)$ and $(b,c)$ are in $R$, then $(a,c)$ is also in $R$.

For ternary and higher-order relations, the simplest representation is a relational table that contains the list of ordered $n$-tuples. Let $A_1, A_2, \ldots, A_n$ be $n$ (not necessarily distinct) sets. The sets $A_1, A_2, \ldots, A_n$ are called the *domains* of the relational table and $n$ is called the *degree* of the table. Each domain may be given a name. These names are called *attributes*. A domain or attribute is called a *primary key* if its value in an ordered $n$-tuple uniquely identifies the ordered $n$-tuple in the table. A table may not contain a primary key. A *composite primary key* is the Cartesian product of the minimal number of domains or attributes such that its value uniquely identifies the ordered $n$-tuple in the table. Since, by definition of the Cartesian product, all $n$-tuples are distinct, at least one composite primary key exists for every table.

The relational table may be organized with $N$ rows with each row corresponding to an element, entry, or primitive and $n$ columns corresponding to the attributes. Attributes that are included in at least one composite primary key of $R$ are called *prime attributes*. Other attributes are called nonprime attributes. An attribute $A_j$ of a relation $R$ is said to be functionally dependent on attribute $A_i$ if and only if for each value of $A_i$ in $R$ there is a unique value of attribute $A_j$ also in $R$. A functional dependence may be defined between $n$-tuples in a similar manner.

Three important forms of a relational table were proposed by Codd (1971). The *first normal form* requires that no domain of the relational table have relations as elements. A relation that also satisfies the condition that the nonprime attributes are fully functionally dependent on each composite key is said to be in the *second normal form*. If in addition, a relation has all nonprime attributes independent of each other, it is said to be in the *third normal form*. A relational table in third normal form has a high degree of machine independence since nonprime attributes may be independently updated, and $n$-tuples may be added or deleted independently of one another.

Codd's original formulation also introduced the concept of a "world model," which is assumed to be commonly agreed upon among users of the relational data. The world model might describe the functional dependencies and other characteristics of the relational table. The world model may depend upon the scene analysis application; however, two generally agreed upon facts are that the physical world of structures in space is three dimensional and that images are two dimensional. An example will now be given that illustrates the importance of a world model and other important features of a relational table representation for scenes and images.

Consider the three-dimensional object illustrated in Fig. 7.46. The object may be considered as consisting of the connected surfaces of a unit cube floating in space. This object, which we call a cube, may be represented in several ways. The surfaces enclose a volume. Each surface is a portion of a plane. Each planar surface is enclosed by edges. Each edge is a line between two vertex points. Thus, the natural dimensionality hierarchy of volume, surface, line, and point provides a variety of representations. It is interesting to examine the relational table representations at the various levels such as surface, edge, and vertex to attempt to determine if one level is perhaps better than the others. We shall now consider some of the possible representations in a top-down manner.

Six surfaces define the cube. If the surfaces are considered continuous, then enumerating the points on the surface is not possible. If the surfaces are considered sampled, then a finite list of the coordinates $(x_i, y_i, z_i)$ is possible. A list of the unit cube surface points with a coarse sampling interval of $\frac{1}{2}$ in each direction is shown in Fig. 7.47. Note that even with coarse sampling the list is rather long. The list may be considered as a quarternary relation that might be called "surface and surface points of the unit cube." The number of elements or *size* of the relational table is important for storage considerations; however, for representing scenes several other features are also important. Suppose we wished to reconstruct the object or its projection from the relational table. Assuming a hidden surface algorithm to be used, the reconstruction from the table for this example would appear as shown in Fig. 7.48.

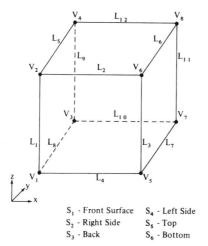

S₁ - Front Surface    S₄ - Left Side
S₂ - Right Side       S₅ - Top
S₃ - Back             S₆ - Bottom

*Figure 7.46*  A three-dimensional object—a unit cube—with vertices, edges, and surfaces labeled for reference.

| Surface $S_1$ : Front | | | Surface $S_4$ : Left Side | | |
|---|---|---|---|---|---|
| x | y | z | x | y | z |
| 0 | 0 | 0 | 0 | 0 | 0 |
| ½ | 0 | 0 | 0 | ½ | 0 |
| 1 | 0 | 0 | 0 | 1 | 0 |
| 0 | 0 | ½ | 0 | 0 | ½ |
| ½ | 0 | ½ | 0 | ½ | ½ |
| 1 | 0 | ½ | 0 | 1 | ½ |
| 0 | 0 | 1 | 0 | 0 | 1 |
| ½ | 0 | 1 | 0 | ½ | 1 |
| 1 | 0 | 1 | 0 | 1 | 1 |

| Surface $S_2$ : Right Side | | | Surface $S_5$ : Top | | |
|---|---|---|---|---|---|
| x | y | z | x | y | z |
| 1 | 0 | 0 | 0 | 0 | 1 |
| 1 | ½ | 0 | ½ | 0 | 1 |
| 1 | 1 | 0 | 1 | 0 | 1 |
| 1 | 0 | ½ | 0 | ½ | 1 |
| 1 | ½ | ½ | ½ | ½ | 1 |
| 1 | 1 | ½ | 1 | ½ | 1 |
| 1 | 0 | 1 | 0 | 1 | 1 |
| 1 | ½ | 1 | ½ | 1 | 1 |
| 1 | 1 | 1 | 1 | 1 | 1 |

| Surface $S_3$ : Back | | | Surface $S_6$ : Bottom | | |
|---|---|---|---|---|---|
| x | y | z | x | y | z |
| 0 | 1 | 0 | 0 | 0 | 0 |
| ½ | 1 | 0 | ½ | 0 | 0 |
| 1 | 1 | 0 | 1 | 0 | 0 |
| 0 | 1 | ½ | 0 | ½ | 0 |
| ½ | 1 | ½ | ½ | ½ | 0 |
| 1 | 1 | ½ | 1 | ½ | 0 |
| 0 | 1 | 1 | 0 | 1 | 0 |
| ½ | 1 | 1 | ½ | 1 | 0 |
| 1 | 1 | 1 | 1 | 1 | 0 |

*Figure 7.47*  Enumeration of sampled points on the surface of the unit cube considered as the relation "surface point."

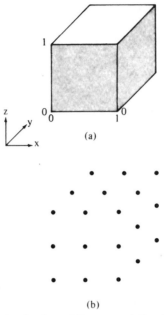

(a)

(b)

*Figure 7.48*   A three-dimensional cube and its representation by surface points. (a) Original cube and (b) sampled points.

Note the considerable difference from the cube as reconstructed from this relational table and the original object. Thus *accuracy* of reconstruction is another important consideration. Accuracy is difficult to define but might be measured as the mean squared error between the original and reconstructed objects. An accurate representation could be obtained by interpolating a continuous surface between the sample points on each surface of the cube. This would require extra computation and brings up the question of how much information must be stored in the table versus how much information can be derived from the table. The ability and ease of accurate reconstruction would therefore be another consideration in the design of a relational table.

The object surfaces could also be described by six intersecting planes. Each plane may be described by a polynomial of the form

$$S_i = \mathbf{w}'\mathbf{x} = w_{i1}x + w_{i2}y + w_{i3}z + w_{i4} = 0,$$

where $i = 1, 2, \ldots, 6$ and $\mathbf{w} = (w_{i1}, w_{i2}, w_{i3}, w_{i4})'$, $\mathbf{x} = (x, y, z, 1)'$. A relational table containing the polynomial coefficients and information about the intersections is shown in Fig. 7.49. In terms of size this relational table is about the same size as for the coarsely sampled grid. However, the accuracy and efficiency of this representation are clearly superior to those of the enumerated table. Note that in addition to the coefficients of the plane, a relational matrix for the intersecting planes has also been included. This information may be considered redundant since one may determine which

|        | $W_{i1}$ | $W_{i2}$ | $W_{i3}$ | $W_{i4}$ |
|--------|------|------|------|------|
| $S_1$ : | 0 | 1 | 0 | 0 |
| $S_2$ : | 1 | 0 | 0 | -1 |
| $S_3$ : | 0 | 1 | 0 | -1 |
| $S_4$ : | 1 | 0 | 0 | 0 |
| $S_5$ : | 0 | 0 | 1 | -1 |
| $S_6$ : | 0 | 0 | 1 | 0 |

(a)

|        | $S_1$ | $S_2$ | $S_3$ | $S_4$ | $S_5$ | $S_6$ |
|--------|----|----|----|----|----|----|
| $S_1$ |   | 1 |   | 1 | 1 | 1 |
| $S_2$ | 1 |   | 1 |   | 1 | 1 |
| $S_3$ |   | 1 |   | 1 | 1 | 1 |
| $S_4$ | 1 |   | 1 |   | 1 | 1 |
| $S_5$ | 1 | 1 | 1 | 1 |   |   |
| $S_6$ | 1 | 1 | 1 | 1 |   |   |

(b)

*Figure 7.49*   (a) Surface-based relational table and (b) intersection matrix.

planar surfaces are parallel and therefore would not intersect from the coefficients. However, by storing the matrix, this processing would not be required.

A lower level of representation is in terms of lines. Since the edges of the cube are straight lines, the problem of curves can be avoided temporarily. A set of 12 lines defines the stick figure of the cube. Each line may be represented by an equation of the form

$$L_i : \alpha \mathbf{V}_j + (1 - \alpha)\mathbf{V}_k,$$

where $\mathbf{V}_j$ and $\mathbf{V}_k$ are two points on the line and $\alpha$ is a parameter proportional to the distance along the line. A relational table based upon the edges is shown in Fig. 7.50. The vertices have been used as the starting and ending points of the lines. A relational matrix that indicates the connection of lines is also included. This matrix may also be considered as redundant. A relational matrix indicating the surfaces is also given. This matrix is not redundant but necessary for obtaining the correct set of surfaces for the unit cube. For example, diagonal planes are not included.

If the edges are represented by discrete sample points, a relational table containing the sequence and coordinates of the edge points could be constructed. The connection information for the edges can also be stored as a pointer to the common vertex. The enumeration of edge points could be interpolated to obtain lines that must also be processed to obtain the surfaces of the object.

The final level to be considered is the vertex or point level. A relational table containing the eight vertices and the connection matrix is shown in Fig. 7.51. The conciseness of this table is readily apparent. Since lines may be drawn between the connected vertices, it is also possible to obtain the line representation from the table. To define planes additional information must be added, as shown in the higher level table of Fig. 7.51. Also, the planes are

| Line | Starting Point | Ending Point |
|------|----------------|--------------|
| $L_1$ | $V_1$ | $V_2$ |
| $L_2$ | $V_2$ | $V_6$ |
| $L_3$ | $V_5$ | $V_6$ |
| $L_4$ | $V_1$ | $V_5$ |
| $L_5$ | $V_2$ | $V_4$ |
| $L_6$ | $V_6$ | $V_8$ |
| $L_7$ | $V_5$ | $V_7$ |
| $L_8$ | $V_1$ | $V_3$ |
| $L_9$ | $V_3$ | $V_4$ |
| $L_{10}$ | $V_3$ | $V_7$ |
| $L_{11}$ | $V_7$ | $V_8$ |
| $L_{12}$ | $V_4$ | $V_8$ |

(a)

|  | $L_1$ | $L_2$ | $L_3$ | $L_4$ | $L_5$ | $L_6$ | $L_7$ | $L_8$ | $L_9$ | $L_{10}$ | $L_{11}$ | $L_{12}$ |
|------|---|---|---|---|---|---|---|---|---|---|---|---|
| $L_1$ |  | 1 |  | 1 | 1 |  |  | 1 |  |  |  |  |
| $L_2$ | 1 |  | 1 |  | 1 | 1 |  |  |  |  |  |  |
| $L_3$ |  | 1 |  | 1 |  | 1 | 1 |  |  |  |  |  |
| $L_4$ | 1 |  | 1 |  |  |  | 1 | 1 |  |  |  |  |
| $L_5$ | 1 | 1 |  |  |  |  |  |  | 1 |  |  | 1 |
| $L_6$ |  | 1 | 1 |  |  |  |  |  |  |  | 1 | 1 |
| $L_7$ |  |  | 1 | 1 |  |  |  |  |  | 1 | 1 |  |
| $L_8$ | 1 |  |  | 1 |  |  |  |  | 1 | 1 |  |  |
| $L_9$ |  |  |  |  | 1 |  |  | 1 |  | 1 |  | 1 |
| $L_{10}$ |  |  |  |  |  |  | 1 | 1 | 1 |  | 1 |  |
| $L_{11}$ |  |  |  |  |  | 1 | 1 |  |  | 1 |  | 1 |
| $L_{12}$ |  |  |  |  | 1 | 1 |  |  | 1 |  | 1 |  |

(b)

|  | $S_1$ | $S_2$ | $S_3$ | $S_4$ | $S_5$ | $S_6$ |
|------|---|---|---|---|---|---|
| $L_1$ | 1 |  |  | 1 |  |  |
| $L_2$ | 1 |  | 1 |  |  |  |
| $L_3$ | 1 | 1 |  |  |  |  |
| $L_4$ | 1 |  |  |  | 1 |  |
| $L_5$ |  |  | 1 | 1 |  |  |
| $L_6$ |  | 1 |  | 1 |  |  |
| $L_7$ |  | 1 |  |  | 1 |  |
| $L_8$ |  |  | 1 |  | 1 |  |
| $L_9$ |  | 1 | 1 |  |  |  |
| $L_{10}$ |  | 1 |  |  | 1 |  |
| $L_{11}$ | 1 | 1 |  |  |  |  |
| $L_{12}$ |  | 1 |  | 1 |  |  |

(c)

*Figure 7.50* (a) Line-based relational table, (b) line connections, and (c) line to surface connections.

related and form a solid as shown in the still higher level representation. Note that the vertex-based table starts at the lowest level and includes relations up to the highest level.

## Projections

Now consider two-dimensional projections of the object as shown in Fig. 7.52. It is interesting to consider which of the relational tables is affected the least by projection of the object onto an image plane. For the isometric projection, three of the six surfaces are visible, nine of the twelve lines, and seven of the eight vertices. Thus, in this isometric projection, for a surface-based relational table only 50% of the prime attributes are visible. For the line-based table 75% of the prime attributes are visible. However, for the

**Vertices** / **Connections**

| | x | y | z | $V_1$ | $V_2$ | $V_3$ | $V_4$ | $V_5$ | $V_6$ | $V_7$ | $V_8$ |
|---|---|---|---|---|---|---|---|---|---|---|---|
| $V_1$ | 0 | 0 | 0 | 1 | 1 | 1 | | 1 | | | |
| $V_2$ | 0 | 0 | 1 | 1 | 1 | | 1 | | 1 | | |
| $V_3$ | 0 | 1 | 0 | 1 | | 1 | 1 | | | 1 | |
| $V_4$ | 0 | 1 | 1 | | 1 | 1 | 1 | | | | 1 |
| $V_5$ | 1 | 0 | 0 | 1 | | | | 1 | 1 | 1 | |
| $V_6$ | 1 | 0 | 1 | | 1 | | | 1 | 1 | | 1 |
| $V_7$ | 1 | 1 | 0 | | | 1 | | 1 | | 1 | 1 |
| $V_8$ | 1 | 1 | 1 | | | | 1 | | 1 | 1 | 1 |

(a)

**Lines**

| | $L_1$ | $L_2$ | $L_3$ | $L_4$ | $L_5$ | $L_6$ | $L_7$ | $L_8$ | $L_9$ | $L_{10}$ | $L_{11}$ | $L_{12}$ |
|---|---|---|---|---|---|---|---|---|---|---|---|---|
| $V_1$ | 1 | | | 1 | | | | 1 | | | | |
| $V_2$ | 1 | 1 | | | 1 | | | | | | | |
| $V_3$ | | | | | | | | 1 | 1 | 1 | | |
| $V_4$ | | | | | 1 | | | | 1 | | | 1 |
| $V_5$ | | | 1 | 1 | | | 1 | | | | | |
| $V_6$ | | 1 | 1 | | | 1 | | | | | | |
| $V_7$ | | | | | | | 1 | | | 1 | 1 | |
| $V_8$ | | | | | | 1 | | | | | 1 | 1 |

(b)

**Surfaces**

| | $S_1$ | $S_2$ | $S_3$ | $S_4$ | $S_5$ | $S_6$ |
|---|---|---|---|---|---|---|
| $V_1$ | 1 | | | 1 | | 1 |
| $V_2$ | 1 | | | 1 | 1 | |
| $V_3$ | | | 1 | 1 | | 1 |
| $V_4$ | | | 1 | 1 | 1 | |
| $V_5$ | 1 | 1 | | | | 1 |
| $V_6$ | 1 | 1 | | | 1 | |
| $V_7$ | | 1 | 1 | | | 1 |
| $V_8$ | | 1 | 1 | | 1 | |

(c)

*Figure 7.51* (a) Vertex-based relational table and vertex connections, (b) vertex to line connections, and (c) vertex to surface connections.

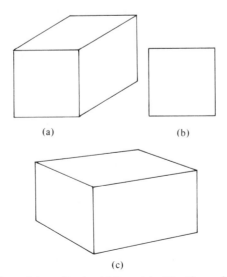

(a)  (b)

(c)

*Figure 7.52* Projections of the unit cube. (a) Isometric, (b) orthographic, and (c) perspective.

vertex-based table 87.5% of the prime attributes are visible. Perhaps the worst case projection would be an orthographic projection parallel to a surface, which would show one surface, four lines, and four vertices. The percentages of visible prime attributes would then be 16.7, 33.3, and 50% for the surface-, line-, and vertex-based tables. Note that the same ordering of percentages would occur for perspective projections as well. Therefore, we may conclude that, at least for this example, the vertex-based relational table would provide the greatest percentage of visible prime attributes when the scene is projected onto an image plane. This would be an important feature for any object recognition, scene matching, or scene understanding system.

### Extension to Multiple Objects

For consideration of physical scenes, the relational table representation must be extended to multiple objects of various shapes, colors, textures, etc. A distinct table could be prepared for each three-dimensional object using prismatic solids or other approximations.

The projection of multiple objects also introduces the problem of superimposition or occlusion, in which portions of an object are hidden by other objects. Again the relational table representation that is least affected by projection is a vertex-based table.

Another interesting feature introduced by projection of multiple objects is that if a camera or observer position is given, a range or distance from the observer may be computed and used to order the objects.

The extension of this example to a more-complex object representation including relations of color, material, or other properties of the surfaces could be made by adding these relations to the previous tables. Further extensions and research are required in order to fully develop this method of scene representation. The relational representation appears to be a natural step between complex physical scenes and images and the formal syntactic description methods.

A second example of scene representation will now be considered in which the only information given is a single image of the scene (Hwang and Hall, 1978). The representation will be based upon the relation of adjacency of regions, as represented by a relational matrix called the adjacency matrix, and shape descriptors. Each region is described by the shape of its boundary. To determine the accuracy of the representation, the relational representation is used to reconstruct the major attributes of the scene. Data compression ratios of 10:1 were achieved, comparing the relational representation to the original boundary points. This ratio is much higher if the number of points in the representation is compared to the original image.

The technique presented illustrates that simple methods may be used to concisely represent a complex image of a scene. Certain portions of the representation such as the adjacency matrix may be used as invariant features of the scene or as a basis for further analysis.

Some basic definitions from graph theory that will be used in this section will be briefly reviewed. More complete descriptions may be found in Harary (1969) and Pavlidis (1977).

A *graph* $G = (V, B)$ consists of a set of nodes $V$ together with a set $B$, whose members are unordered pairs of nodes, the branches of $G$. Two nodes of a graph are said to be *adjacent* if there is a branch between them. A graph $G$ can be represented by an *adjacency matrix* whose $ij$th element is set to one if there is a branch from node $i$ to node $j$.

A *weighted graph* is a graph with a real number or weight $W(i,j)$ assigned to the branch $(V_i, V_j)$. A *chain* in $G$ is a sequence $V_1, b_1, V_2, b_2, \ldots, V_t, b_t, V_{t+1}$, where $t \geq 0$, each $V_i$ is a node, and each $b_i$ is the branch $(V_i, V_{i+1})$. A chain is called *simple* if all the $V_i$ are distinct and *closed* if $V_{t+1} = V_1$. The *length* of a chain is the number of branches in it. A simple closed chain is a *circuit*. A graph $G$ is *complete* if for every $V_i \neq V_j$, $(V_i, V_j) \in B$. That is, $G$ is complete if all possible branches of $G$ are in $G$.

Consider a scene represented by a structural model in which the regions of the scene are the elements. A scene descriptor thus consists of a set of relational graphs and a set of property descriptors of the regions. A relational graph can be constructed by a specified binary relation. One of the important properties used to characterize a region is a shape description. There are various methods of shape description: the polygon approximation used by Pavlidis and Ali (1975), chain encoding and centroidal profiles used by Freeman (1961, 1977), and the Fourier descriptor used by Persoon and Fu (1977). All are successful in many aspects. In order to reduce the computation time and simplify the representation of a shape a relaxation description technique is used. The shape of the region is first approximated by a sequence of sampled points that are obtained by sampling the boundary of the region in some angular interval. The vector connecting the centroid with the boundary point of the region with maximum distance has been chosen as the reference axis. The sampled boundary points are then ordered clockwise following the same sequence as the original boundary points. Then the shapes of the regions can be described by a set of shape variation profiles and angular variation profiles. They can also be represented by graphical circuits.

*Adjacency* is an important binary relationship in scene description. All the eight neighboring points of a point $f(x, y)$ may be considered adjacent to $f(x, y)$, as shown in Fig. 7.53. Two regions in a scene are considered adjacent if at least one pair of boundary points $p_i, p_j$ is adjacent, $p_i \in p_{bi}, p_j \in p_{bj}$, where $p_{bi}$ and $p_{bj}$ are the boundary point sets of regions $i$ and $j$. Using the adjacency relation a region adjacency graph may be obtained from the boundary point sets. For example, a simple scene consisting of a block, a pyramid, and a cylinder is shown in Fig. 7.54. The scene is segmented into regions using only

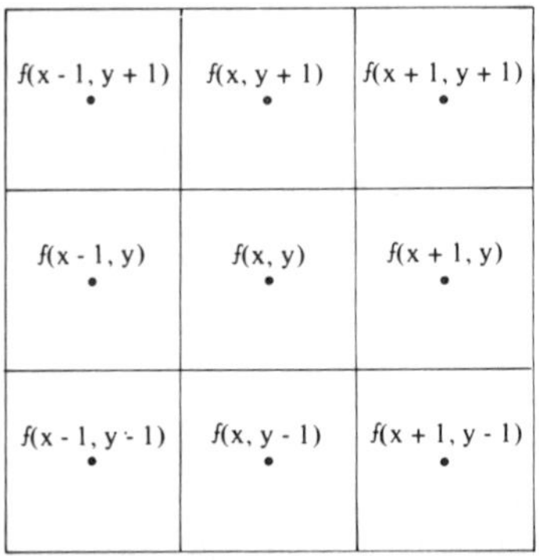

| $f(x-1, y+1)$ • | $f(x, y+1)$ • | $f(x+1, y+1)$ • |
|---|---|---|
| $f(x-1, y)$ • | $f(x, y)$ • | $f(x+1, y)$ • |
| $f(x-1, y-1)$ • | $f(x, y-1)$ • | $f(x+1, y-1)$ • |

*Figure 7.53* Adjacent neighbors of point $f(x, y)$. [From Hwang and Hall (1978).]

*Figure 7.54* Scene consisting of block, pyramid, and cylinder. [From Hwang and Hall (1978).]

*Figure 7.55* Scene segmented into regions. [From Hwang and Hall (1978).]

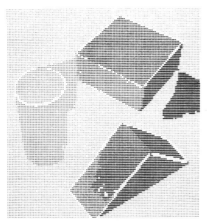

*Figure 7.56*   Thresholded image. [From Hwang and Hall (1978).]

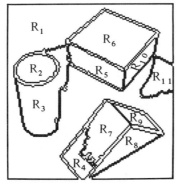

*Figure 7.57*   Resulting 11 regions and boundaries. [From Hwang and Hall (1978).]

histogram segmentation, as shown in Fig. 7.55. By thresholding the smaller regions that are less than 30 pixels, the boundary points of each region are then traced and stored. Figure 7.56 shows the thresholded image. Figure 7.57 shows the resulting 11 regions and their boundaries. $R_1$ denotes the background region. $R_2$ and $R_3$ are two adjacent regions of the cylinder. $R_5$, $R_6$, and $R_{10}$ are three adjacent regions corresponding to the block. $R_5$, $R_7$, $R_8$, and $R_9$ are the adjacent regions corresponding to the pyramid. $R_{11}$ is an isolated region. The adjacency graph is constructed by connecting the centroids of the adjacent regions. The region adjacency graph is shown superimposed on Fig. 7.58. The adjacency matrix is shown in Fig. 7.59. Notice that the background region $R_1$ is adjacent to all the regions; therefore all the entries in the first row of the matrix are 1s. $R_1$ is the only region to which $R_{11}$ is adjacent; therefore $R_{11}$ is an isolated region. The geometric information can also be represented in the relational graph by using a weighted relational graph. The degree of adjacency between two regions may be defined as the reciprocal of

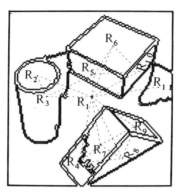

*Figure 7.58*  Connections of centroids of regions. [From Hwang and Hall (1978).]

the Euclidean distance between the centroids: the shorter the distance, the more adjacent the regions. Define the weight of each branch in the weighted adjacency graph by $W(R_i, R_j) = d_{max}/d(R_i, R_j)$, where $d(R_i, R_j)$ is the Euclidean distance between two adjacent regions $R_i$ and $R_j$, and $d_{max} = \max d(R_i, R_j)$ for all adjacent regions. The weighted adjacency matrix of the scene is shown in Fig. 7.60. The weighted relational graph is shown in Fig. 7.61. Since the weighted relational graph is normalized with respect to $d_{max}$, it is invariant under rotation and translation of the scene.

To describe the shape, consider a simply connected boundary of a region as shown in Fig. 7.62. Let $(B_{mx}, B_{my})$ be a boundary point with maximum distance to the centroid $(C_x, C_y)$. The vector connecting $(C_x, C_y)$ and $(B_{mx}, B_{my})$ is chosen as the reference axis for angular boundary sampling. If $n$ samples are taken in the $2\pi$ radius, the angular sampling interval is $\theta = 2\pi/n$. The slope of the reference axis can be calculated by

$$M_r = (B_{my} - C_y)/(B_{mx} - C_x).$$

The slope of any straight line connecting the boundary point $(B_x, B_y)$ and the

|     | R1 | R2 | R3 | R4 | R5 | R6 | R7 | R8 | R9 | R10 | R11 |
|-----|----|----|----|----|----|----|----|----|----|-----|-----|
| R1  | 0  | 1  | 1  | 1  | 1  | 1  | 1  | 1  | 1  | 1   | 1   |
| R2  | 1  | 0  | 1  | 0  | 0  | 0  | 0  | 0  | 0  | 0   | 0   |
| R3  | 1  | 1  | 0  | 0  | 1  | 0  | 0  | 0  | 0  | 0   | 0   |
| R4  | 1  | 0  | 0  | 0  | 0  | 0  | 1  | 1  | 0  | 0   | 0   |
| R5  | 1  | 0  | 1  | 0  | 0  | 1  | 0  | 0  | 0  | 1   | 0   |
| R6  | 1  | 0  | 0  | 0  | 1  | 0  | 0  | 0  | 0  | 1   | 0   |
| R7  | 1  | 0  | 0  | 1  | 0  | 0  | 0  | 1  | 1  | 0   | 0   |
| R8  | 1  | 0  | 0  | 1  | 0  | 0  | 1  | 0  | 1  | 0   | 0   |
| R9  | 1  | 0  | 0  | 0  | 0  | 0  | 1  | 1  | 0  | 0   | 0   |
| R10 | 1  | 0  | 0  | 0  | 1  | 1  | 0  | 0  | 0  | 0   | 0   |
| R11 | 1  | 0  | 0  | 0  | 0  | 0  | 0  | 0  | 0  | 0   | 0   |

*Figure 7.59*  The adjacent matrix of the scene. [From Hwang and Hall (1978).]

|      | R1   | R2   | R3   | R4   | R5   | R6   | R7   | R8   | R9   | R10  | R11  |
|------|------|------|------|------|------|------|------|------|------|------|------|
| R1   | 0.0  | 1.15 | 1.47 | 1.08 | 3.23 | 1.30 | 1.49 | 1.03 | 1.32 | 1.28 | 1    |
| R2   | 1.15 | 0.0  | 3.13 | 0.0  | 0.0  | 0.0  | 0.0  | 0.0  | 0.0  | 0.0  | 0.0  |
| R3   | 1.47 | 3.13 | 0.0  | 0.0  | 1.28 | 0.0  | 0.0  | 0.0  | 0.0  | 0.0  | 0.0  |
| R4   | 1.08 | 0.0  | 0.0  | 0.0  | 0.0  | 0.0  | 2.50 | 1.56 | 0.0  | 0.0  | 0.0  |
| R5   | 3.23 | 0.0  | 1.28 | 0.0  | 0.0  | 2.13 | 0.0  | 0.0  | 0.0  | 1.52 | 0.0  |
| R6   | 1.30 | 0.0  | 0.0  | 0.0  | 2.13 | 0.0  | 0.0  | 0.0  | 0.0  | 1.92 | 0.0  |
| R7   | 1.49 | 0.0  | 0.0  | 2.50 | 0.0  | 0.0  | 0.0  | 2.63 | 1.79 | 0.0  | 0.0  |
| R8   | 1.03 | 0.0  | 0.0  | 1.56 | 0.0  | 0.0  | 2.63 | 0.0  | 2.08 | 0.0  | 0.0  |
| R9   | 1.32 | 0.0  | 0.0  | 0.0  | 0.0  | 0.0  | 1.79 | 2.08 | 0.0  | 0.0  | 0.0  |
| R10  | 1.28 | 0.0  | 0.0  | 0.0  | 1.52 | 1.92 | 0.0  | 0.0  | 0.0  | 0.0  | 0.0  |
| R11  | 1.   | 0.0  | 0.0  | 0.0  | 0.0  | 0.0  | 0.0  | 0.0  | 0.0  | 0.0  | 0.0  |

*Figure 7.60*　The weighted adjacency matrix of the scene. [From Hwang and Hall (1978).]

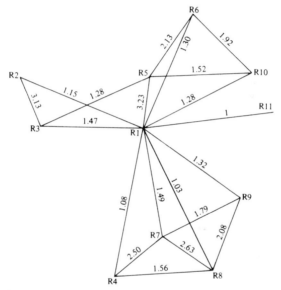

*Figure 7.61*　The weighted adjacency graph of the scene. [From Hwang and Hall (1978).]

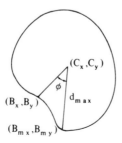

*Figure 7.62*　Sampling a boundary at angular increments. [From Hwang and Hall (1978).]

centroid is given by

$$M_b = (B_y - C_y)/(B_x - C_x).$$

Then the sampled boundary point at $\theta$ must satisfy the sampling line equation

$$\left[\left(\frac{B_{my} - C_y}{B_{mx} - C_x} - \frac{B_y - C_y}{B_x - C_x}\right) \bigg/ \left(1 + \frac{B_{my} - C_y}{B_{mx} - C_x}\frac{B_y - C_y}{B_x - C_x}\right)\right] - \tan\theta = 0$$

and

$$(B_x - C_x)(B_{my} - C_y) - (B_y - C_y)(B_{mx} - C_x) > 0, \qquad \text{for} \quad \theta < \pi,$$

$$(B_x - C_x)(B_{my} - C_y) - (B_y - C_y)(B_{mx} - C_x) < 0, \qquad \text{for} \quad \theta > \pi.$$

Since the boundaries of the regions are not continuous curves but are quantized, the point that satisfies the equation may not be a boundary point. Therefore, a point in the boundary point set with minimum distance to the sampling line equation is chosen as the sampled boundary point at $\theta$. For a region with a spiral boundary, as shown in Fig. 7.63, the nonadjacent boundary points are chosen as the sampled boundary points if the distance from these points to the sampling line is less than $1/\sqrt{2}$ times the unit pixel scale. If two adjacent boundary points satisfy the above criterion, the one with the smallest distance is chosen.

Since all the boundary points are ordered clockwise sequentially, the sampled boundary points can also be ordered by following the sequential order of the original boundary point set. For example, assume that the boundary point sequence in Fig. 7.63 is *abcdefghijklmna*, and $C_t a$ is the reference axis. If samples are taken for 0, $\frac{2}{3}\pi$, and $\frac{4}{3}\pi$ rad, the sampled boundary points will be *abigfkhdm*. If one then compares the sampled boundary points with the original boundary point set, the ordered boundary point sequence *abdfghikma* is obtained. A piecewise linear approximation

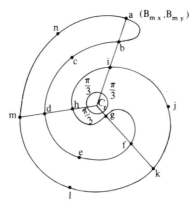

*Figure 7.63* Sampling a region with spiral boundary. [From Hwang and Hall (1978).]

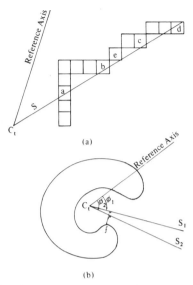

*Figure 7.64* (a) Sampling discrete boundary requires selection of nearest neighbor and (b) sampling at discrete angles can introduce more information at some angles than others. [From Hwang and Hall (1978).]

shape may then be obtained by connecting the consecutive sampled boundary points.

Notice that the boundary points are quantized since for sampling interval $\theta$, the number of sampled boundary points $n$ might be larger than $2\pi/\theta$ but smaller than the total number of the boundary points $N$, that is, $2\pi/\theta \leq n \leq N$. This is illustrated by Fig. 7.64a. For a sampling line $S$ at $\theta$, four sampled boundary points abcd may be obtained by this sampling. The point e is not a sampled boundary point, since e is adjacent to b but b is closer to the sampling line. For $n = 2\pi/\theta$ the sampling points are adequate only when the shape of the region is convex and the angular sampling interval is reasonably large. Figure 7.64b illustrates another characteristic of the sampled boundary point. For the sampling lines $S_1$ at $\theta_1$ and $S_2$ at $\theta_2$ the sampled boundary point is the same point f, since f is the minimum distance point to both $S_1$ and $S_2$. One may also argue that more information has been lost in some sampling intervals than in others. The basic problem involves techniques for adaptive sampling of the region boundary.

Figure 7.65 shows the reference axis of sampling for the scene of Fig. 7.54. Figure 7.66 shows the boundary of the background region. Figure 7.67 shows the piecewise linear approximation of the background region if the sampling interval is 5°. The total number of sampling points required to draw Fig. 7.67 was 293. The number of boundary points for the background region of Fig. 7.66 was 1011; therefore, the data compression ratio is approximately $\frac{3}{10}$. Figure 7.68 shows the piecewise linear approximated region boundary of the

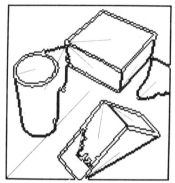

*Figure 7.65*  Reference axes selected for segmented regions. [From Hwang and Hall (1978).]

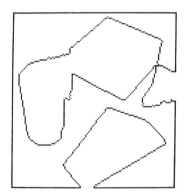

*Figure 7.66*  Boundary of background region. [From Hwang and Hall (1978).]

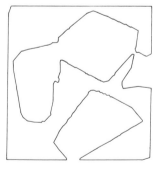

*Figure 7.67*  Sampled boundary of background region sampled at 5° intervals. [From Hwang and Hall (1978).]

442

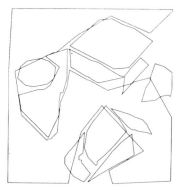

*Figure 7.68* Piecewise-linear approximated region boundaries sampled at 30° intervals. [From Hwang and Hall (1978).]

*Figure 7.69* Sampled boundaries for 15° intervals. [From Hwang and Hall (1978).]

*Figure 7.70* Sampled boundaries for 5° intervals. [From Hwang and Hall (1978).]

scene for the sampling interval of 30°. Figure 7.69 is the sampled boundary scene for 15° and Fig. 7.70 for 5° sampling interval, which shows that the sampled boundaries do give a good approximation. A comparison of the sampled boundary points used is shown in Fig. 7.71. The average data compression ratio for 5° sampling was 0.47, for 15° sampling 0.19, and for 30° sampling 0.10.

As discussed previously, the boundaries of the regions in a scene can be piecewise linearly approximated by a sampled shape. The information of the shape of a region of a scene is stored in a sequence of sampled boundary points. Consider the sampled boundary points as the primitive elements of a shape descriptor. If the sampled boundary points are consecutively labeled, the graphic representation can be used to describe the sampled shape of a region. For example, given the region boundary shown in Fig. 7.72a, if the boundary is sampled every 30°, 12 sampled boundary points are obtained. Since

$$\overline{C_t V_1} = \max_i \left( \overline{C_t V_i} \right),$$

$\overline{C_t V_1}$ is chosen as the reference axis. The 12 sampled boundary points are then labeled $V_1, V_2, \ldots, V_{12}$ clockwise. Since all the line segments between two nodes carry the geometric information of the shape of a region, one may assume that all the nodes are related to one another. Therefore, a complete graph corresponding to the twelve nodes $V_1, V_2, \ldots, V_{12}$ can be drawn as shown in Fig. 7.72b. Assume the weight of the relationship to be $d_{max}/d(V_i, V_j)$, where

$$d_{max} = \max_{i,j} d(V_i, V_j);$$

then a weighted adjacency matrix can be constructed. The sampled region boundary is then a circuit of $V_1, B_1(V_1, V_2), V_2, B_2(V_2, V_3), \ldots, V_{12}, B_{12}(V_{12}, V_1), V_1$.

For example, the weighted adjacency matrix of region 2 of Fig. 7.60 for 30° sampling is shown in Fig. 7.72c. The circuit of $V_1, 3.70, V_2, 4.76, V_3, 5.88, V_4, 5, V_5, 4.76. V_6, 4.76, V_7, 3.7, V_8, 4.76, V_9, 5.88, V_{10}, 5, V_{11}, 5, V_{12}, 3.85$, represents the sampled boundary of region 2.

The sampled shapes of the regions can also be described by sampled shape variation profiles and angular variation profiles, which originated from Freeman's centroidal profile (Freeman, 1961). The shape variation profile is simply a normalized plot of the distance from the sampled boundary points to the centroid of the region versus angle. The angular variation profile is a plot of the angle between the lines connecting the sampled boundary points to the centroid and the reference axis. For example, the sampled shape variation

| Regions | No. of Boundary Points | No. of Sampled Boundary Points for 5° Sampling | Data Compression ratio for 5° | No. of sampled Boundary Points for 15° sampling | Data Compression ratio for 15° | No. of Sampled Boundary Points for 30° Sampling | Data Compression ratio for 30° |
|---|---|---|---|---|---|---|---|
| R1 | 1011 | 293 | 0.29 | 107 | 0.11 | 60 | 0.06 |
| R2 | 75 | 66 | 0.88 | 24 | 0.32 | 12 | 0.16 |
| R3 | 192 | 104 | 0.541 | 46 | 0.24 | 16 | 0.08 |
| R4 | 88 | 65 | 0.74 | 31 | 0.35 | 15 | 0.18 |
| R5 | 111 | 67 | 0.60 | 28 | 0.25 | 16 | 0.14 |
| R6 | 132 | 74 | 0.56 | 26 | 0.20 | 12 | 0.09 |
| R7 | 113 | 72 | 0.64 | 29 | 0.26 | 14 | 0.12 |
| R8 | 114 | 65 | 0.57 | 32 | 0.28 | 15 | 0.13 |
| R9 | 71 | 52 | 0.73 | 30 | 0.42 | 15 | 0.21 |
| R10 | 74 | 47 | 0.64 | 24 | 0.32 | 14 | 0.19 |
| R11 | 84 | 65 | 0.77 | 25 | 0.30 | 13 | 0.15 |
| TOTAL | 2065 | 970 | 0.47 | 402 | 0.19 | 203 | 0.10 |

*Figure 7.71* Comparison of number of boundary points used for region reconstruction. [From Hwang and Hall (1978).]

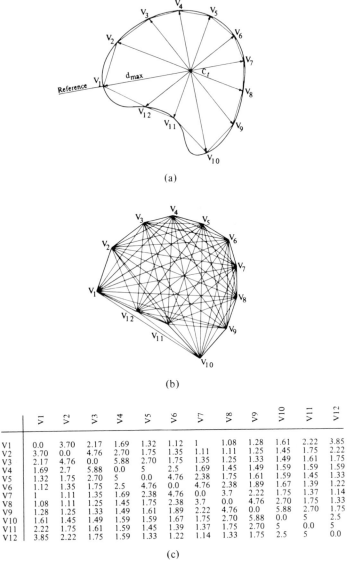

(a)

(b)

| | V1 | V2 | V3 | V4 | V5 | V6 | V7 | V8 | V9 | V10 | V11 | V12 |
|------|------|------|------|------|------|------|------|------|------|------|------|------|
| V1 | 0.0 | 3.70 | 2.17 | 1.69 | 1.32 | 1.12 | 1 | 1.08 | 1.28 | 1.61 | 2.22 | 3.85 |
| V2 | 3.70 | 0.0 | 4.76 | 2.70 | 1.75 | 1.35 | 1.11 | 1.25 | 1.33 | 1.45 | 1.75 | 2.22 |
| V3 | 2.17 | 4.76 | 0.0 | 5.88 | 2.70 | 1.75 | 1.35 | 1.25 | 1.33 | 1.49 | 1.61 | 1.75 |
| V4 | 1.69 | 2.7 | 5.88 | 0.0 | 5 | 2.5 | 1.69 | 1.45 | 1.49 | 1.59 | 1.59 | 1.59 |
| V5 | 1.32 | 1.75 | 2.70 | 5 | 0.0 | 4.76 | 2.38 | 1.75 | 1.61 | 1.59 | 1.45 | 1.33 |
| V6 | 1.12 | 1.35 | 1.75 | 2.5 | 4.76 | 0.0 | 4.76 | 2.38 | 1.89 | 1.67 | 1.39 | 1.22 |
| V7 | 1 | 1.11 | 1.35 | 1.69 | 2.38 | 4.76 | 0.0 | 3.7 | 2.22 | 1.75 | 1.37 | 1.14 |
| V8 | 1.08 | 1.11 | 1.25 | 1.45 | 1.75 | 2.38 | 3.7 | 0.0 | 4.76 | 2.70 | 1.75 | 1.33 |
| V9 | 1.28 | 1.25 | 1.33 | 1.49 | 1.61 | 1.89 | 2.22 | 4.76 | 0.0 | 5.88 | 2.70 | 1.75 |
| V10 | 1.61 | 1.45 | 1.49 | 1.59 | 1.59 | 1.67 | 1.75 | 2.70 | 5.88 | 0.0 | 5 | 2.5 |
| V11 | 2.22 | 1.75 | 1.61 | 1.59 | 1.45 | 1.39 | 1.37 | 1.75 | 2.70 | 5 | 0.0 | 5 |
| V12 | 3.85 | 2.22 | 1.75 | 1.59 | 1.33 | 1.22 | 1.14 | 1.33 | 1.75 | 2.5 | 5 | 0.0 |

(c)

*Figure 7.72* (a) Region boundary sampled every 30°, (b) complete graph for the 12 sampled points, and (c) weighted adjacency matrix of $R_2$ for 30° sampling. [From Hwang and Hall (1978).]

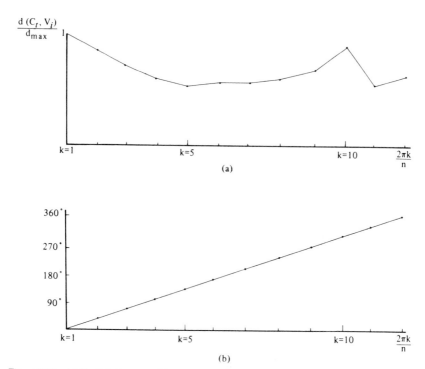

*Figure 7.73*   (a) Radial shape profile and (b) angular variation profile for region shown in Fig. 7.72a. [From Hwang and Hall (1978).]

profile and angular variation profile of the shapes in Fig. 7.72a are shown in Fig. 7.73a, b, where $k$ is the sampling sequence number and $n$ is the total number of sampled boundary points. Notice that the angular variation profile in Fig. 7.73 is linear. However, in the practical case, a linear profile would rarely happen since the boundary points are quantized. Exceptions might be a circular shape or a sampling interval so large that only one sampled boundary point is obtained at each sampling. The sampled shape variation profile and angular variation profile of the scene are shown in Fig. 7.74a–d for 30 and 5° sampling.

It is easy to find that $R_2$ is the region whose angular plot is linear, since its shape is regular. The variation of the sampled shape profile and angular variation profile can also indicate the irregularity of the shapes of a region. One may also see that the angular variation profile of $R_1$ crosses 180° four times, whereas the profile of $R_2$ crosses only once. The sampled shape profile of $R_1$ also has more peaks than that of $R_2$. Hence, the shape of $R_1$ is more irregular than that of $R_2$. In addition, the sampled shapes' variation profiles and angular variation profiles are normalized plots and are therefore invariant under translation and rotation.

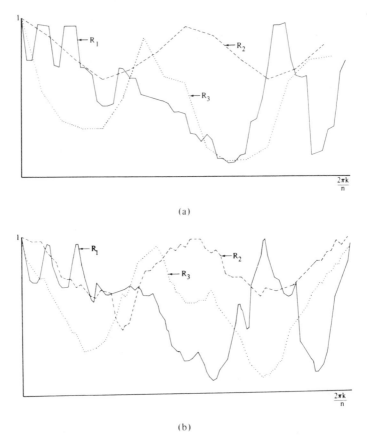

(a)

(b)

*Figure 7.74*   (a) The shape variation profiles of regions $R_1$, $R_2$, and $R_3$ for 30° sampling, and (b) the shape variation profiles of regions $R_1$, $R_2$, and $R_3$ for 5° sampling.

### 7.4.2  Syntactic Description

As previously indicated, natural scenes may be described in terms of hierarchical structures such as the scene–object–surface–boundary-point hierarchy, in which each pattern is described in terms of simpler patterns until a terminal level is reached. The formal approach to these methods of pattern composition is often called syntactic or structural description and recognition. The techniques are analogous to those used in the study of language, such as parsing a sentence into phrases that are built up of words that are composed of characters. The typical parsing result is a tree structure that may be used to describe the sentence or verify its correctness with respect to a set of grammatical rules or grammar.

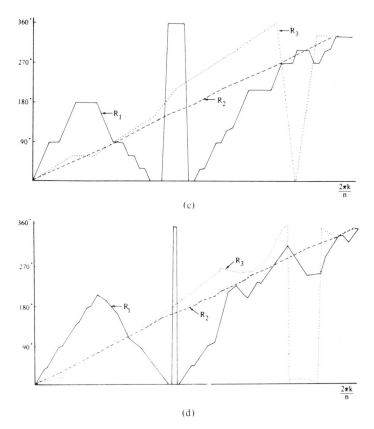

*Figure 7.74*   (c) the angular variation profiles of regions $R_1$, $R_2$, and $R_3$ for 30° sampling, and (d) the angular variation profiles of regions $R_1$, $R_2$, and $R_3$ for 5° sampling. [From Hwang and Hall (1978).]

The syntactic approach to scene analysis offers a possibility for describing a large set of complex patterns using small sets of simple pattern primitives and structural rules. An introduction to this approach will be described in this section. Excellent starting points for a more complete study may be found in the books by Fu (1974) and Gonzalez and Thomason (1978). Since many of the concepts have been adapted from mathematical language theory some important definitions will be briefly reviewed.

As an example of a large class of patterns that are commonly recognized syntactically consider FORTRAN programs. Each statement in a program is a string of characters; only a small number of the possible strings are valid and these must be distinguished from invalid ones. The FORTRAN compiler may therefore be called a structural pattern recognizer. Most images do not consist of character strings; however, certain pictorial patterns such as chromosome images, bubble chamber images, and many computer graphics images can be made to fit the string model. Generalizations of the relationships from "concatenating characters" to more general types have resulted in more powerful syntactic methods.

Syntactic description is most easily applied to scenes or images that have the following properties:

(a)   They can be decomposed into a relatively small number of unique elements called primitives, such as vertices or line segments.

(b)   The primitives must be recognizable by an automatic procedure.

(c)   The scenes must have an underlying structure.

An *alphabet* $V$, is any finite set of symbols. A *sentence*, *string*, or *word* over an alphabet $V$ is any string of finite length composed of symbols from the alphabet. For example, given the alphabet $V = \{0, 1\}$, the following are valid sentences: $0, 1, 00, 01, 10, 11, 000, 011, \ldots$ .

The sentence with no symbols is called the *emptysentence*, which we shall denote by $\lambda$. For an alphabet $V$, we shall use $V^*$ to denote the set of all sentences composed of symbols from $V$, including the empty sentence. The symbol $V^+$ will denote the set of sentences $V^* - \lambda$. For example, given the alphabet $V = \{a, b\}$, we have $V^* = \{\lambda, a, b, aa, ab, ba, \ldots\}$ and $V^+ = \{a, b, aa, ab, ba, \ldots\}$. A *language* is any set (not necessarily finite) of sentences over an alphabet. For example, the FORTRAN language consists of a set of sentences, each constructed according to a set of rules. As is true in natural languages, a serious study of formal language theory must be focused on grammars and their properties.

We define a *formal string grammar* (or simply, *grammar*) as the 4-tuple $G = (N, \Sigma, P, S)$, where $N$ is a set of *nonterminals* (variables), $\Sigma$ a set of terminals (constants), $P$ a set of productions or rewriting rules, and $S$ is the *start* or *root* symbol. It is assumed that $S$ belongs to the set $N$, that $N$ and $\Sigma$ are disjoint sets, and that $N$, $\Sigma$, and $P$ are finite sets. The alphabet $V$ is the union of $N$ and $\Sigma$.

The *language* generated by $G$, denoted by $L(G)$, is the set of strings that satisfy two conditions: (1) each string is composed only of terminals (i.e., each string is a *terminal* sentence), and (2) each string can be derived from $S$ by suitable applications of productions from the set $P$.

The following notation will be used throughout this section: Nonterminals will be denoted by capital letters: $S, A, B, C, \ldots$ . Lower-case letters at the beginning of the alphabet will be used for terminals: $a, b, c, \ldots$ . Strings of terminals will be denoted by lower-case letters toward the end of the alphabet: $v, w, x, \ldots$ . Strings of mixed terminals and nonterminals will be represented by lowercase Greek letters: $\alpha, \beta, \gamma, \delta, \ldots$ .

The set $P$ of productions consists of expressions of the form $\alpha \to \beta$, where $\alpha$ is a string in $V^+$ and $\beta$ is a string in $V^*$. In other words, the symbol $\to$

indicates replacement of the string $\alpha$ by the string $\beta$. The symbol $\Rightarrow$ will be used to indicate operations of the form $\gamma\alpha\delta \underset{G}{\Rightarrow} \gamma\beta\delta$ in the grammar $G$; that is, $\underset{G}{\Rightarrow}$ indicates the replacement of $\alpha$ by $\beta$ by means of the production $\alpha \rightarrow \beta$, $\gamma$ and $\delta$ being left unchanged. It is customary to drop the $G$ and simply use the symbol $\Rightarrow$ when it is clear which grammar is being considered. Also, if $\alpha$ is a string, $\alpha^n$ is $\alpha$ written $n$ times. The length of a string is denoted $|\alpha|$.

The production rules are an integral part of the definition of the grammar. Also, note that a production rule may have either a terminal or nonterminal symbol or string on the right, but it never has an isolated terminal symbol on the left. To *generate* a valid sentence according to the rules, we begin with a string consisting of a single symbol and then use the rules to make replacements. Any symbol occurring in the string and on the left side of a production rule may be replaced within the string by the corresponding right side of the rule. Conversely, a set of rules may be used to recognize a string as a valid sentence if it is possible to make replacements of the right sides of production rules with their corresponding left sides until the resulting string is a single-symbol sentence.

Grammars may be categorized into four types depending on the type of restriction placed on the production rules. A grammar in which the productions have the general form $\alpha \rightarrow \beta$ is called a *phrase structure grammar*.

A grammar in which the general form of productions $\alpha \rightarrow \beta$ ($\alpha$ in $V^+$ and $\beta$ in $V^*$) is allowed is called an *unrestricted grammar*.

A *context-sensitive grammar* has productions of the form $\alpha_1 A \alpha_2 \rightarrow \alpha_1 \beta \alpha_2$, where $\alpha_1$ and $\alpha_2$ are in $V^*$, $\beta$ is in $V^+$, and $A$ is in $N$. This grammar allows replacement of the nonterminal $A$ by the string $\beta$ only when $A$ appears in the context $\alpha_1 A \alpha_2$ of strings $\alpha_1$ and $\alpha_2$.

A *context-free grammar* has productions of the form $A \rightarrow \beta$, where $A$ is in $N$ and $\beta$ is in $V^+$. The name "context free" arises from the fact that the variable $A$ may be replaced by a string $\beta$ regardless of the context in which $A$ appears.

Finally, a *regular* (or *finite-state*) *grammar* is one with productions of the form $A \rightarrow aB$ or $A \rightarrow a$, where $A$ and $B$ are variables in $N$ and $a$ is a terminal in $\Sigma$. Alternative valid productions are $A \rightarrow Ba$ and $A \rightarrow a$. However, once one of the two types has been chosen, the other set must be excluded.

These grammars are sometimes called *type 0, 1, 2, and 3 grammars*, respectively. It is interesting to note that all regular grammars are context free, all context-free grammars are context sensitive, and all context-sensitive grammars are unrestricted.

Although unrestricted grammars are considerably more powerful than the other three types, their generality presents some serious difficulties from both a theoretical and a practical point of view. To a large extent, this is also true of context-sensitive grammars. For these reasons, most of the work dealing with the use of grammatical concepts for image description and pattern recognition has been limited to context-free and regular grammars.

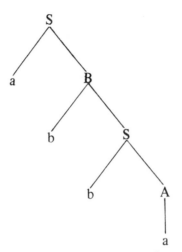

*Figure 7.75*   Derivation tree corresponding to the production *abba*.

Any derivation in a context-free grammar may be described by the use of a *derivation tree*. A derivation or parse tree may be constructed by the following procedure:

(1)   The root of the tree has the label $S$.

(2)   Every node of the tree has a label that is a symbol in $N$ or $\Sigma$.

(3)   If a node labeled $A$ has at least one descendant other than itself, then $A$ is an element of $N$.

(4)   If nodes $n_1, n_2, \ldots, n_k$ are the direct descendants of node $n$ with label $A$ ordered, from left to right, with labels $A_1, A_2, \ldots, A_k$, respectively, then $A \rightarrow A_1 A_2 \cdots A_k$ must be a production in $P$.

As an example consider the context-free grammar that generates strings consisting of an equal number of $a$s and $b$s (Fu, 1977):

$$G = (N, \Sigma, P, S), \quad \text{where} \quad N = \{S, A, B\}, \quad \Sigma = \{a, b\}$$

and productions $P$:

| (1) $S \rightarrow aB$, | (3) $A \rightarrow aS$, | (5) $A \rightarrow a$, | (7) $B \rightarrow aBB$, |
|---|---|---|---|
| (2) $S \rightarrow bA$, | (4) $A \rightarrow bAA$, | (6) $B \rightarrow bS$, | (8) $B \rightarrow b$. |

Typical generations or derivations of sentences may be generated by starting with $S$ and applying the production rules. For examples, the application of productions (1), (6), (2), (5) gives

$$S \Rightarrow aB \Rightarrow abS \Rightarrow abbA \Rightarrow abba.$$

The derivation tree corresponding to this production is shown in Fig. 7.75. Note that an infinite number of productions with corresponding derivation trees could be generated by the simple grammar.

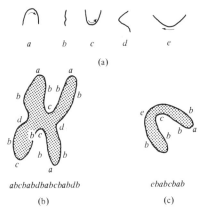

abcbabdbabcbabdb                          cbabcbab

(b)                                           (c)

Figure 7.76   (a) Primitives of a chromosome grammar, (b) submedian chromosome, and (c) telocentric chromosome. [From Ledley *et al.* (1965). Reprinted from "Optical and Electro-Optical Information Processing," edited by J. T. Tippett *et al.*, by permission of The MIT Press, Cambridge, Massachusetts. Copyright © 1965 by the MIT Press.]

The use of a string grammar restricts the relations between subpatterns or primitives to one-dimensional concatenation, that is, each subpattern can be connected to another only at the left or right. Since images and scenes are inherently two- and three-dimensional, one must either reduce the scene patterns to strings or generalize the types of relations permitted, to apply syntactic methods to scene analysis. Both approaches will now be illustrated.

An interesting illustration of image description by boundary tracking is the grammar proposed by Ledley (1964; Ledley *et al.*, 1965) to characterize submedian and telocentric chromosomes. Typical submedian and telocentric chromosome shapes are shown in Fig. 7.76b, c, along with the string representation obtained by tracking the boundary of each chromosome. This grammar utilizes the primitive elements shown in Fig. 7.76a, which are detected as a chromosome boundary is tracked in a clockwise direction. The complete grammar is given by $G = (N, \Sigma, P, \hat{S})$, where $\Sigma = \{a, b, c, d, e\}, N = \{S, T, A, B, C, D, E, F\}$, where $\hat{S}$ equals $S$ or $T$ and

| | | |
|---|---|---|
| $S$ | submedian chromosome, | $C$ side, |
| $T$ | telocentric chromosome, | $R$ right part, |
| $A$ | arm, | $L$ left part, |
| $B$ | bottom, | $M$ arm pair. |

The productions are $P$:

(1) $S \rightarrow M \cdot M$,   (6) $M \rightarrow L \cdot A$,   (10) $B \rightarrow B \cdot b$,   (15) $C \rightarrow d$,
(2) $T \rightarrow B \cdot M$,   (7) $L \rightarrow A \cdot c$,   (11) $B \rightarrow e$,   (16) $A \rightarrow b \cdot A$,
(3) $M \rightarrow C \cdot M$,   (8) $R \rightarrow c \cdot A$,   (12) $C \rightarrow b \cdot C$,   (17) $A \rightarrow A \cdot b$,
(4) $M \rightarrow M \cdot C$,   (9) $B \rightarrow b \cdot B$,   (13) $C \rightarrow C \cdot b$,   (18) $A \rightarrow a$.
(5) $M \rightarrow A \cdot R$,   (14) $C \rightarrow b$,

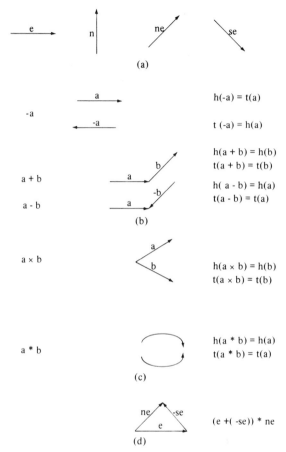

*Figure 7.77* Steps in the construction of a picture description language. (Note the connections of heads and tails in the composite structures.) (a) primitives, (b) negative, operators *h(a)* and *t(a)* indicate head and tail of directed primitives, (c) binary operators, (d) a PDL expression.

The operator "·" is used to describe simple connectivity of the terms in a production as the boundary is tracked in a clockwise direction.

The above grammar is in reality a combination of two grammars with starting symbols $S$ and $T$, respectively. Thus, starting with $S$ allows generation of structures that correspond to submedian chromosomes, and $T$ produces structures that correspond to telocentric chromosomes.

Another interesting example of a grammar that uses a different form of concatenation is the picture description language (PDL) developed by Shaw (1970) for describing objects using a directed line or curve primitive and four binary operators for defining concatenation relations between the primitives. The primitives of a PDL is the directed line segments shown in Fig. 7.77a. For disconnected subpatterns, the "blank" primitive is also used. Five basic operations, as shown in Fig. 7.77b, c, are also used. Each line-type pictorial pattern is then represented by a labeled branch-oriented graph in which the branches represent primitives. A sample construction is shown in Fig. 7.77d.

The grammar that generates the PDL expressions in the PDL is a context-free grammar

$$G = (N, \Sigma, P, S),$$

where the nonterminal set is

$$N = \{ S, L \},$$

and the set of terminals is

$$\Sigma = \{ b \} \cup \{ +, \times, -, /, (,) \} \cup \{ l \},$$

where $b$ may be any primitive, including a "blank" primitive, that is useful in describing disconnected subpatterns. The blank primitive is considered to have an identical tail and head.

The productions of the PDL are

$$
\begin{array}{lll}
S \to b & L \to Sl & \phi \to + \\
S \to (S \phi S) & L \to (L \phi L) & \phi \to \times \\
S \to (\sim S) & L \to (\sim L) & \phi \to - \\
S \to L & L \to (/L) & \phi \to *, \\
S \to /L & &
\end{array}
$$

where $l$ is a label designator that is used to permit cross reference to the expressions within a description, and $/$ is an operator that is used to enable the head and tail of an expression to be arbitrarily located.

### Higher-Dimensional Grammars

The types of grammars previously discussed have related primitives only by a type of concatenation, that is, the collections of symbols have been inherently ordered by their position in a string. Since there is no natural ordering for two- or higher-dimensional space, more complex relations are required for image and scene analysis. For example, a line in a picture has not only a "value" such as "straight line", but also a set of coordinates that describe its location and extent in two or three dimensions. Tree, web, and graph grammars will now be described briefly to illustrate the generalizations that have been developed. More complete discussions may be found in Gonzalez and Thomason (1978) and Fu (1977).

A *tree T* is a finite set of one or more nodes such that

(1)   there is a unique node designated the root;

(2)   the remaining nodes are partitioned into $m$ disjoint sets, $T_1, \ldots, T_m$, each of which in turn is a tree called a *subtree* of $T$.

The *tree frontier* is the set of nodes at the bottom of the tree (the *leaves*), taken in order from left to right. For example, the tree shown in Fig. 7.78 has root $ and frontier $xy$.

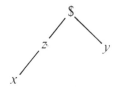

*Figure 7.78* Tree with root $ and frontier *xy*.

Generally, two types of information in a tree are important, the information about a node stored as a set of words describing the node, and the information relating the node to its neighbors stored as a set of pointers to those neighbors. As used in image descriptions, the first type of information identifies a pattern primitive, while the second type defines the physical relationship of the primitive to other substructures.

The structure shown in Fig. 7.79a can be represented by a tree by using the relationship "inside of." Thus, denoting the root of the tree by the symbol $, we see from Fig. 7.79a that the first level of complexity involves *a* and *c* inside $. This produces two branches emanating from the root, as shown in Fig. 7.79b. The next level involves *b* inside *a* and *d* and *e* inside *c*. Finally, we complete the tree by noting that *f* is inside *e*.

A *tree grammar* is defined as a 5-tuple $G = (N, \Sigma, P, r, S)$, where $N$ and $\Sigma$ are, as before, sets of nonterminals and terminals, respectively; $S$ is the start symbol, which can, in general, be a tree; $P$ is a set of productions of the form $\Omega \rightarrow \Psi$, where $\Omega$ and $\Psi$ are trees; and $r$ is a *ranking function*, which denotes the number of direct descendants of a node whose label is a terminal in the grammar.

The form of production $\Omega \rightarrow \Psi$ is analogous to that given for unrestricted string grammars and, as such, is usually too general to be of much practical use. A type of production that has found wide acceptance in the study of tree systems is an *expansive* production, which is shown in Fig. 7.80, where $A, A_1, A_2, \ldots, A_n$ are nonterminals and $a$ is a terminal. A tree grammar that has only productions of this form is called an *expansive tree grammar*.

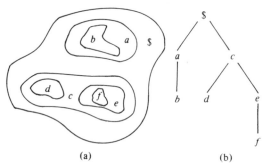

(a)                                          (b)

*Figure 7.79* (a) A simple composite region and (b) tree representation obtained by using the relation "inside of." (Reproduced from "Digital Image Processing and Recognition," 1977, by Rafael C. Gonzalez and Paul Wintz, with permission of Addison-Wesley, Reading, Massachusetts.)

*Figure 7.80*   Expansive production. (Reproduced from "Syntactic Pattern Recognition," 1978, by Rafael C. Gonzalez and Michael G. Thomason, with permission of Addison-Wesley, Reading, Massachusetts.)

## Web Grammars

As illustrated in Fig. 7.81, webs are undirected graph structures whose nodes are labeled. When used for image description, webs allow representations at a level considerably more abstract than that afforded by string or tree formalisms.

In a conventional phrase-structure string grammar rewriting rules of the form $\alpha \rightarrow \beta$ are used to replace one string by another. Such a rule is completely specified by specifying strings $\alpha$ and $\beta$; any string $\gamma\alpha\delta$ that contains $\alpha$ as a substring can be immediately rewritten as $\gamma\beta\delta$. Similarly, the productions of expansive tree grammars are interpreted without difficulty. The definition of rewriting rules involving webs, however, is much more complicated. Thus, if we want to replace a subweb $\alpha$ of the web $\omega$ by another subweb $\beta$, it is necessary to specify how to *embed* $\beta$ in $\omega$ in place of $\alpha$. As will be seen below, this can be done by using embedding rules. An important point, however, is that the definition of an embedding rule must not depend on the "host web" $\omega$ because we want to be able to replace $\alpha$ by $\beta$ in any web containing $\alpha$ as a subweb.

Let $V$ be a set of labels and $N_\alpha$ and $N_\beta$ the set of nodes of webs $\alpha$ and $\beta$, respectively. Based on the above concepts we define a *web rewriting rule* as a triple $(\alpha, \beta, \phi)$, where $\phi$ is a function from $N_\beta \times N_\alpha$ into $2^V$ (the set of subsets of labels). This function specifies the embedding of $\beta$ in place of $\alpha$; that is, it specifies how to join to nodes of $\beta$ to the neighbors of each node of the *removed* subweb $\alpha$. Since $\phi$ is a function from the set of ordered pairs $N_\beta \times N_\alpha$, its argument is of the form $(n, m)$, for $n$ in $N_\beta$, and $m$ in $N_\alpha$. The values of $\phi(n, m)$ specify the allowed connections of $n$ to the neighbors of $m$. For example, $\phi(B, A) = \{C, D\}$ means "join node $B$ (in $\beta$) to the neighbors of node $A$ (in $\alpha$) whose labels are either $C$ or $D$." We shall omit the embedding specification and instead use the term "normal" to denote situations in which there is no ambiguity in a rewriting rule.

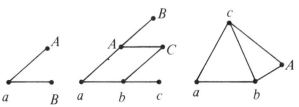

*Figure 7.81*   Some simple webs. (Reproduced from "Syntactic Pattern Recognition," 1978, by Rafael C. Gonzalez and Michael G. Thomason, with permission of Addison-Wesley, Reading, Massachusetts.)

A *web grammar* is defined as a 4-tuple $G = (N, \Sigma, P, S)$, where $N$ is the nonterminal vocabulary, $\Sigma$ the terminal vocabulary, $P$ a set of web productions, and $S$ the set of starting webs. As usual, $S$ is in $N$, and the vocabulary $V$ is the union of $N$ and $\Sigma$.

The above definition of a web grammar is analogous to that of an unrestricted string grammar, and as such is too broad to be of much practical use. As in the case of string grammars, however, it is possible to define restricted types of web grammars by limiting the generality of the productions.

We shall call a web rewriting rule $(\alpha, \beta, \phi)$ *context sensitive* if there exists a nonterminal point $A$ of $\alpha$ such that $(\alpha - A)$ is a subweb of $\beta$. In this case the rule rewrites only a single point of $\alpha$, regardless of how complex $\alpha$ is. If $\alpha$ contains a single point, we shall call $(\alpha, \beta, \phi)$ a *context-free* rule. It is noted that this is a special case of a context-sensitive rule since $(\alpha - A)$ is empty when $\alpha$ contains a single point. By representing strings as webs (e.g., the string $aAbc\ldots$ may be expressed as the web $a \Rightarrow A \Rightarrow b \Rightarrow c \cdots$) it may be shown that the above rewriting rules are analogous to context-sensitive and context-free string productions.

If the terminal vocabulary of a web grammar consists of a single symbol, every point of every web generated by the grammar will have the same label. In this case we can ignore the labels and identify the webs by their underlying graphs. This special type of web grammar is sometimes referred to as a *graph grammar*.

## Bibliographical Guide

Segmentation is an important and difficult step in the analysis of scenes. A variety of techniques and approaches are described by Prewitt and Mendelsohn (1966), Bartz (1968), Wolfe (1969), Prewitt (1970), Zahn (1971), Chow and Kaneko (1972), Hall (1972), Pavlidis (1972a,b, 1977), Strong and Rosenfeld (1973), Tomita *et al.* (1973), Yakimovsky (1973), Yakimovsky and Feldman (1973), Weszka *et al.* (1974), Pavlidis and Horowitz (1974), Tenenbaum and Weyl (1975), Ohlander (1975), Price (1976), Riseman and Arbib (1977), and Duda and Hart (1972).

Segmentation techniques for detecting edges, curves, texture boundaries and general region dependent techniques are described by Brodatz (1956), Hough (1962), Blum (1962, 1967), Butler *et al.* (1963), Kubba (1963), Roberts (1965), Glucksman (1965), Golay (1969), Belson (1969), Levialdi (1970), Macleod (1970), Brice and Fennema (1970), Billingsley *et al.* (1970), Zusne (1970), Levi and Montanari (1970), Pickett (1970), Rosenfeld (1969, 1970 ), Gray (1971), Stefanelli and Rosenfeld (1971), Kelly (1971), Hueckel (1971, 1973), Argyle (1971), Clowes (1971), Barrow *et al.* (1972), Martelli (1972),

Rosenfeld *et al.* (1972), Sutton and Hall (1972), Griffith (1973), Shirai (1973), Fram and Deutsch (1975), Zobrist and Thompson (1975), Zucker *et al.* (1975), Nevatia (1976), Hall and Frei (1976), O'Gorman and Clowes (1976), Robinson (1976), and Frei and Chen (1977).

Descriptors of regions are described by Attneave and Arnoult (1956), Attneave (1957), Preston (1961), Hu (1961, 1962), Freeman (1961), Novikoff (1962), Holmes *et al.* (1962), Alt (1962), Rosenfeld (1962, 1970a, b, 1973, 1974a, b, c), Greanis *et al.* (1963), Bradshaw (1963), Tenery (1963), Becker *et al.* (1964), Kirsch (1964, 1971), Rosenfeld *et al.* (1965), Rosenfeld and Pfaltz (1966), Pfaltz and Rosenfeld (1967), Grenander (1967, 1970), Guzman (1968), Munson (1968), Montanari (1968, 1969, 1971), Moore (1968, 1972), Brill (1968), Calabi and Hartnett (1968), Rosenfeld *et al.* (1969), Beckenbach and Desilets (1969), Haralick and Kelley (1969), Cook and Rosenfeld (1970), Hawkins (1970), Hodes (1970a, b), Kolers (1970), Lendaris and Stanley (1970), Sklansky (1970), Rosenfeld and Lipkin (1970), Paton (1970), Nagy and Tuong (1970), Muerle (1970), Mott-Smith (1970), Arcelli and Levialdi (1971), Hall *et al.* (1971), Preston (1971), Falk (1972), Kruger *et al.* (1972, 1974), Levialdi (1972), Moore (1972), Ramer (1972), Zahn and Roskies (1972), Binford and Agin (1973), Haralick and Shanmugan (1973), Haralick *et al.* (1973), Julesz *et al.* (1973), Widrow (1973), Tsuji and Tomita (1973), Nevatia and Binford (1973), Hayes *et al.* (1974), Hall *et al.* (1975), Bennett and MacDonald (1975), Feng and Pavlidis (1975), Galloway (1975), Weszka *et al.* (1976), Sklansky *et al.* (1976), and Nevatia (1977).

Relational descriptions are described by Ledley (1964), Narasimhan (1964, 1966, 1969), Ledley *et al.* (1965), Bazin and Benoit (1965), Anderson (1968), Miller and Shaw (1968), Guzman (1968), Rutovitz (1968), Rosenfeld and Pfaltz (1968), Cofer and Tou (1969), Kasvand (1969), Shaw (1969, 1970), Kaneff (1970), Barrow and Popplestone (1971), Feder (1971), Fu and Swain (1971), Huffman (1971), Williams (1971), Gonzalez (1972), Nake and Rosenfeld (1972), Winograd (1972), Duda and Hart (1973), Duff *et al.* (1973), Kruse (1973), Fu (1974), Thomason and Gonzalez (1975), Winston (1975, 1977), Gonzalez *et al.* (1976), Fu and Rosenfeld (1976), and Gonzalez and Thomason (1978).

## PROBLEMS

**7.1** Compute an edge enhanced image for Fig. 3.21a using
  (a) the Roberts magnitude operator,
  (b) The Frei–Chen magnitude operator, and
  (c) the Sobel operator.
Select a threshold that produces 10% of the picture points as edge points and produce edge images for each method. Which method produces the most continuous boundary?

**7.2**  Use the Brice–Fennema threshold method to segment the image of Fig. 3.21a. Merging and splitting of regions is not required. Perform a least squares fit to a circle for the boundary point between gray levels 4 and 5. **7.3**  Use the picture description language (PDL) shown in Fig. 7.77 to construct an expression for the image shown in Fig. 7.44. Use a "blank" primitive for the disconnected patterns.

## References

Alt, F. L. (1962). Digital Pattern Recognition by Moments. *J. Assoc. Comput. Mach.* 9, No. 2, 240–258.

Anderson, R. H. (1968). Syntax-Directed Recognition of Handprinted Two-Dimensional Mathematics. *In* "Interactive Systems for Experimental Applied Mathematics" (M. Klerer and J. Reinfelds, eds.), pp. 436–459. Academic Press, New York.

Arcelli, C., and Levialdi, S. (1971). Concavity Extraction by Parallel Processing. *IEEE Trans. Syst., Man Cybern.* 1, 349–396.

Argyle, E. (1971). Techniques for Edge Detection. *Proc. IEEE* 59, No. 2, 285–287.

Attneave, F. (1957). Physical Determinants of the Judged Complexity of Shape. *J. Exp. Psychol.* 53, 221–227.

Attneave, F., and Arnoult, M. (1956). The Quantitative Study of Shape and Pattern Perception. *Psychol. Bull.* 53, 452–471.

Barrow, H. G., and Popplestone, R. J. (1971). Relational Descriptions in Picture Processing. *In* "Machine Intelligence" (B. Meltzer and D. Michie, eds.), Vol. 6, pp. 377–396. Edinburgh Univ. Press, Edinburgh.

Barrow, H. G., Ambler, A. P., and Burstall, R. M. (1972). Some Techniques for Recognizing Structure in Pictures. *In* "Frontiers of Pattern Recognition" (S. Watanabe, ed.), pp. 1–29. Academic Press, New York.

Bartz, M. R. (1968). The IBM 1975 Optical Page Reader, Part II: Video Thresholding System. *IBM J. Res. Dev.* 12 (September), 354–363.

Bazin, M. J., and Benoit, J. W. (1965). Off-Line Global Approach to Pattern Recognition for Bubble Chamber Pictures. *IEEE Trans. Nucl. Sci.* 12 (August), 291–295.

Beckenbach, E. S., and Desilets, D. T. (1969). The Computerization of High Speed Cineangio-cardiographic Left Ventricular Volume Determination. *In* "Pattern Recognition Studies" (H. L. Kasnitz and G. C. Cheng, eds.), pp. 173–192. SPIE, Redondo Beach, California.

Becker, H. C., Nettleton, W. I., and Meyers P. H. (1964). Digital Computer Determination of a Medical Diagnostic Index Directly from Chest X-Ray Images. *IEEE Trans. Biomed. Eng.* 11, No. 3, 67–72.

Belson, M. (1969). A New Boundary Method for Pictorial Pattern Recognition. *EASCON Rec., October IEEE Publ. 69-C-31-AES*, pp. 274–279.

Bennett, J. R., and MacDonald, J. S. (1975). On the Measurement of Curvature in a Quantized Environment. *IEEE Trans. Comput.* 25, No. 8, 803–820.

Billingsley, F. C., Goetz, A. F. H., and Lindsley, J. N. (1970). Color Differentiation by Computer Image Processing. *Photo. Sci. Eng.* 14, No. 1, 28–35.

Binford, T. O., and Agin, G. J. (1973). Computer Description of Curved Objects. *Proc. Int. J. Conf. Artif. Intell., 3rd, Menlo Park, Calif.*, pp. 629–640.

Blum, H. (1962). An Associative Machine for Dealing with the Visual Field and Some of its Biological Implications. *In* "Biological Prototypes and Synthetic Systems" (E. E. Bernard and M. R. Kare, eds.), pp. 244–260. Plenum, New York.

Blum, H. (1967). A Transformation for Extracting New Descriptors of Shape. *In* "Models for the Perception of Speech and Visual Form" (W. Wathen-Dunn, ed.), pp. 362–380. MIT Press, Cambridge, Massachusetts.

Bradshaw, J. A. (1963). Letter Recognition Using a Captive Scan. *IEEE Trans. Electron. Comput.* **12**, 26.

Brice, C. R., and Fennema, C. L. (1970). Scene Analysis Using Regions. *Artif. Intell.* **1**, 205–226.

Brill, E. L. (1968). Character Recognition Via Fourier Descriptors. *WESCON Conv. Rec., Los Angeles, Calif.* Pap. 25/3.

Brodatz, P. (1956). "Texture: A Photograph Album for Artists and Designers." Dover, New York.

Butler, J. W., Butler, M. K., and Stroud, A. (1963). Automatic Analysis of Chromosomes. "Data Acquisition and Processing in Biology and Medicine," Vol. 3, pp. 261–275. Pergamon, Oxford.

Calabi, L., and Hartnett, W. E. (1968). Shape Recognition, Prairie Fires, Convex Deficiencies and Skeletons. *Am. Math. Mon.* **75**, No. 4, 335–342.

Chow, C. K., and Kaneko, T. (1972). Automatic Boundary Detection of the Left Ventricle from Cineangiograms. *Comput. Biomed. Res.* **5**, 388–410.

Clowes, M. (1971). On Seeing Things. *Artif. Intell.* **4**, No. 1, 79–116.

Codd, E. F. (1970). A Relational Model for Large Shared Data Banks, *Commun. ACM* **13**, No. 6, 377–387.

Codd, E. F. (1972). Further Normalization of the Data Base Relational Model, *Courant Comput. Sci. Symp. Ser.* **6**, *Data Base Systems*, pp. 33.

Cofer, R. H., Jr., and Tou, J. T. (1969). Some Approaches Toward Scene Extraction. *In* "Pattern Recognition Studies" (H. L. Kasnitz and G. C. Cheng, eds.), pp. 63–79. SPIE, Redondo Beach, California.

Cook, C. M., and Rosenfeld, A. (1970). Size Detectors. *Proc. IEEE* **58**, No. 12, 1956–1957.

Davies, D. L. (1978). Edge Detection in Digital Images Using Small Mask and Vector Operators and the Method of Polar Histograms. M. S. Thesis, Univ. of Tennessee, Knoxville, December (unpublished).

Duda, R. O., and Hart, P. E. (1972). Use of the Hough Transformation to Detect Lines and Curves in Pictures. *Commun. ACM* **15**, No. 1, 11–15.

Duda, R. O., and Hart P. E. (1973). "Pattern Classification and Scene Analysis." Wiley, New York.

Duff, M. J. B., Watson, D. M., Fountain, T. J., and Shaw, A. K. (1973). A Cellular Logic Array for Image Processing. *Pattern Recognition* **5**, 229–247.

Falk, G. (1972). Interpretation of Imperfect Line Data as a Three-Dimensional Scene. *Artif. Intell.* **3**, No. 2, 101–144.

Feder, J. (1971). Plex Languages. *Inf. Sci.* **3**, 225–241.

Feng, H.-Y. F., and Pavlidis, T. (1975). Decomposition of Polygons into Simpler Components: Feature Generation for Syntactic Pattern Recognition. *IEEE Trans. Comput.* **24**, No. 6, 636–650.

Fram, J. R., and Deutsch, E. S. (1975). On the Quantitative Evaluation of Edge Detection Schemes and Their Comparison with Human Performance. *IEEE Trans. Comput.* **24**, No. 6, 616–628.

Freeman, H. (1961). On the Encoding of Arbitrary Geometric Configurations. *IEEE Trans. Electron. Comput.* **10**, 260–268.

Freeman, H. (1977). Computer Processing of Line-Drawing Images, *ACM Comput. Surv.* **6**, 57–97.

Frei, W., and Chen, C. C. (1977). Fast Boundary Detection: A Generalization and a New Algorithm. *IEEE Trans. Comput.* **26**, 988–998.

Fu, K. S. (1974). "Syntactic Methods in Pattern Recognition." Academic Press, New York.

Fu, K. S. (ed.) (1977). "Syntactic Pattern Recognition, Applications." Springer-Verlag, New York.

Fu, K. S., and Rosenfeld, A. (1976). Pattern Recognition and Image Processing. *IEEE Trans. Comput.* **25**, No. 12, 1336–1346.

Fu, K. S., and Swain, P. H. (1971). On Syntactic Pattern Recognition. *In* "Software Engineering" (J. T. Tou, ed.), Vol. 2, pp. 155–182. Academic Press, New York.

Galloway, M. M. (1975). Texture Analysis Using Gray Level Run Lengths. *Comput. Graphics Image Process.* **4**, No. 2, 172–179.

Glucksman, H. A. (1965). A Parapropagation Pattern Classifier. *IEEE Trans. Electron. Comput.* **14**, 434–443.

Golay, M. J. E. (1969). Hexagonal Pattern Transformation. *IEEE Trans. Comput.* **18**, No. 8, 733–740.

Gonzalez, R. C. (1972). Syntactic Pattern Recognition—Introduction and Survey. *Proc. Natl. Electron. Conf.* **27**, 27–31.

Gonzalez, R. C., and Thomason, M. G. (1978). "Syntactic Pattern Recognition: An Introduction." Addison-Wesley, Reading, Massachusetts.

Gonzalez, R. C., and Wintz, P. (1977). "Digital Image Processing." Addison-Wesley, Reading, Massachusetts.

Gonzalez, R. C., Edwards, J. J., and Thomason, M. G. (1976). An Algorithm for the Inference of Tree Grammars. *Int. J. Comput. Inf. Sci.* **5**, No. 2, 145–163.

Gray, S. B. (1971). Local Properties of Binary Images in Two Dimensions. *IEEE Trans. Comput.* **20**, No. 5, 551–561.

Greanis, E. C., Meagher, P. F., Norman, R. J., and Essinger, P. (1963). The Recognition of Handwritten Numerals by Contour Analysis. *IBM J. Res. Dev.* **7**, No. 1, 14–21.

Grenander, U. (1967). Toward a Theory of Patterns. *Proc. Symp. Probab. Methods Anal.*, New York, pp. 79–111.

Grenander, U. (1970). A Unified Approach to Pattern Analysis. *Adv. Comput.* **10**, 175–216.

Griffith, A. K. (1973). Mathematical Models for Automatic Line Detection. *J. Assoc. Comput. Mach.* **20**, 62–80.

Guzman, A. (1968). Decomposition of a Visual Scene into Three-Dimensional Bodies. *Proc. Fall J. Comput. Conf., AFIPS*, **33**, 291–304.

Hall, E. L. (1972). Automated Computer Diagnosis Applied to Lung Cancer. *Proc. Int. Conf. Cybern. Soc., New Orleans, La.*

Hall, E. L., and Frei, W. (1976). Invariant Features for Quantitative Scene Analysis. Final Rep., Contract F 08606-72-C-0008, Image Process. Inst., Univ. of Southern California, Los Angeles, 1976.

Hall, E. L., Kruger, R. P., Dwyer, S. J., McLaren, R. W., and Lodwick, G. S. (1971). A Survey of Preprocessing and Feature Extraction Techniques for Radiographic Images. *IEEE Trans. Comput.* **20**, No. 9, 1032–1044.

Hall, E. L., Roberts, F. E., and Crawford, W. O. (1975). Computer Classification of Coal Worker's Pneumoconiosis, *IEEE Trans. Biomed. Eng.* **BME22**, No. 6, 518–527.

Haralick, R. M., and Kelley, G. L. (1969). Pattern Recognition with Measurement Space and Spatial Clustering for Multiple Images. *Proc. IEEE* **57**, No. 4, 654–665.

Haralick, R. M., and Shanmugan, K. (1973). Computer Classification of Reservoir Sandstones. *IEEE Trans. Geosci. Electron.* **11** (October), 171–177.

Haralick, R. M., Shanmugan, K., and Dinstein, I. (1973). Texture Features for Image Classification. *IEEE Trans. Syst. Man Cybern.* **3** (November), 610–621.

Harary, F. (1969). "Graph Theory." Addison-Wesley, Reading, Massachusetts.

Hawkins, J. K. (1970). Textural Properties for Pattern Recognition. *In* "Picture Processing and Psychopictorics" (B. S. Lipkin and A. Rosenfeld, eds.), pp. 347–370. Academic Press, New York.

Hayes, K. C., Jr., Shah, A. N. and Rosenfeld, A. (1974). Texture Coarseness: Further Experiments. *IEEE Trans. Syst., Man Cybern.* **4**, No. 5, 467–472.

Hodes, L. (1970a). Discrete Approximation of Continuous Convex Blobs. *SIAM (Soc. Ind. Appl. Math.) J. Appl. Math.* **19**, 477–485.

Hodes, L. (1970b). The Logical Complexity of Geometric Properties in the Plane. *J. Assoc. Comput. Mach.* **17**, 339–347.

Holmes, W. S., Leland, H. R., and Richmond, G. E. (1962). Design of a Photo Interpretation Automaton. *Proc. East. J. Comput. Conf.*, 27–35.

Hough, P. V. C. (1962). Method and Means for Recognizing Complex Patterns. U.S. Patent No. 3,069,654, December 18.

Hu, M. K. (1961). Pattern Recognition by Moment Invariants. *Proc. IRE* **49**, 1428.

Hu, M. K. (1962). Visual Pattern Recognition by Moment Invariants. *IRE Trans. Inf. Theory* **8**, 179–187.

Hueckel, M. H. (1971). An Operator which Locates Edges in Digital Pictures. *J. Assoc. Comput. Mach.* **18**, No. 1, 113–125.

Hueckel, M. H. (1973). A Local Visual Operator which Recognizes Edges and Lines. *J. Assoc. Comput. Mach.* **20**, 634–647.

Huffman, D. A. (1971). Impossible Objects as Nonsense Sentences. *In* "Machine Intelligence" (B. Meltzer and D. Mitchie, eds.), Vol. 6, pp. 295–323. Edinburgh Univ. Press, Edinburgh.

Hwang, J. J., Lee, C. C., and Hall, E. L. (1979). Segmentation of Solid Objects Using Global and Local Edge Coincidence, *Proc. IEEE Conf. Pattern Recognition Image Process., Chicago, August.*

Hwang, J. J., and Hall, E. L. (1978). Scene Representation Using Adjacency Matrices and Sampled Shapes of Regions. *Proc. IEEE Conf. Pattern Recognition Image Process., Chicago, May,* pp. 250–261.

Kaneff, S. (1970). "Picture Language Machines." Academic Press, New York.

Julesz, B., Gilbert, E. N., Shepp, L. A., and Frisch, H. L. (1973). Inability of Humans to Discriminate Between Visual Textures that Agree in Second-Order Statistics-Revisited. *Perception* **2**, 391–405.

Kasvand, T. (1969). Some Thoughts and Experiments on Pattern Recognition. *In* "Automatic Interpretation and Classification of Images" (A. Grasselli, ed.), pp. 391–398. Academic Press, New York.

Kelly, M. D. (1971). Edge Detection in Pictures by Computer Using Planning. *In* "Machine Intelligence" (B. Meltzer and D. Michie, eds.), Vol. 6, pp. 379–409. Edinburgh Univ. Press, Edinburgh.

Kirsch, R. A. (1964). Computer Interpretation of English Text and Picture Patterns. *IEEE Trans. Electron. Comput.* **13**, No. 4, 363–376.

Kirsch, R. (1971). Computer Determination of the Constituent Structure of Biological Images. *Comput. Biomed. Res.* **4**, 315–328.

Kolers, P. A. (1970). The Role of Shape and Geometry in Picture Recognition. *In* "Picture Processing and Psycopictorics" (B. S. Lipkin and A. Rosenfeld, eds.), pp. 181–202. Academic Press, New York.

Kruger, R. P., Townes, J. R., Hall, D. L., Dwyer, S. J., and Lodwick, G. S. (1972). Radiographic Diagnosis via Feature Extraction and Classification of Cardiac Size and Shape Descriptors. *IEEE Trans. Biomed. Eng.* **19**, No. 3, 174–186.

Kruger, R. P., Thompson, W. B., and Turner, A. F. (1974). Computer Diagnosis of Pneumoconiosis. *IEEE Trans. Syst. Man Cybern.* **4**, No. 1, 40–49.

Kruse, B. (1973). A Parallel Picture Processing Machine. *IEEE Trans. Comput.* **22**, No. 12, 1075–1087.

Kubba, M. H. (1963). Automatic Picture Detail Detection in the Presence of Random Noise. *Proc. IEEE* **51**, No. 11, 1518–1523.

Kunii, T. L., Weyl, S., and Tenenbaum, J. M. (1978). A Relational Database Schema for Describing Complex Pictures with Color and Texture, *Policy Anal. Info.* **1**, No. 2, 127–142.

Ledley, R. S. (1964). High-Speed Automatic Analysis of Biomedical Pictures. *Science* **146**, 216–223.

Ledley, R. S., Rotolo, L. S., Golab, T. J., Jacobsen, J. D., Ginsberg, M. D., and Wilson, J. B. (1965). FIDAC: Film Input to Digital Automatic Computer and Associated Syntax-Directed Pattern-Recognition Programming System. *In* "Optical and Electro-Optical Information Processing" (J. T. Tippett *et al*., eds.), pp. 591–614. MIT Press, Cambridge, Massachusetts.

Lendaris, G. G., and Stanley, G. L. (1970). Diffraction Pattern Sampling for Automatic Pattern Recognition. *Proc. IEEE* **58**, No. 2, 198–216.

Levi, G., and Montanari, U. (1970). A Grey Weighted Skeleton. *Inf. Control* **17**, 62–91.

Levialdi, S. (1970). Parallel Counting of Binary Patterns. *Electron. Lett.* **6**, 798–800.

Levialdi, S. (1972). On Shrinking Binary Picture Patterns. *Commun. ACM* **15**, 7–10.

Liu, C. L. (1977). "Elements of Discrete Mathematics." McGraw-Hill, New York.

Macleod, I. D. G. (1970). On Finding Structure in Pictures. *In* "Picture Language Machines" (S. Kaneff, ed.), p. 231. Academic Press, New York.

Martelli, A. (1972). Edge Detection Using Heuristic Search Methods. *Comput. Graphics Image Process.* **1**, 169–182.

Miller, W. F., and Shaw, A. C. (1968). Linguistic Methods in Picture Processing—A Survey. *Proc. Fall J. Comput. Conf. AFIPS, December*, pp. 279–290.

Montanari, U. (1968). A Method for Obtaining Skeletons Using a Quasi-Euclidean Distance. *J. Assoc. Comput. Mach.* **15** (October), 600–624.

Montanari, U. (1969). Continuous Skeletons from Digitized Pictures. *J. Assoc. Comput. Mach.* **16**, No. 4, 534–549.

Montanari, U. (1971). On the Optimum Detection of Curves in Noisy Pictures. *Commun. ACM* **14**, 335–345.

Moore, D. J. H. (1972). An Approach to the Analysis and Extraction of Pattern Features Using Integral Geometry. *IEEE Trans. Syst. Man Cybern.* **2**, 97–102.

Moore, G. A. (1968). Automatic Scanning and Computer Processes for the Quantitative Analysis of Micrographs and Equivalent Subjects. *In* "Pictorial Pattern Recognition" (G. C. Cheng *et al.*, eds.), pp. 275–326. Thompson Book Co., Washington, D. C.

Mott-Smith, J. C. (1970). Medial Axis Transforms. *In* "Picture Processing and Psychopictorics" (B. S. Lipkin and A. Rosenfeld, eds.), pp. 267–288. Academic Press, New York.

Muerle, J. L. (1970). Some Thoughts on Texture Discrimination by Computer. *In* "Picture Processing and Psychopictorics" (B. S. Lipkin and A. Rosenfeld, eds.), pp. 371–379. Academic Press, New York.

Munson, J. H. (1968). Experiments in the Recognition of Handprinted Text: Part I—Character Recognition. *Proc. Fall J. Comput. Conf. AFIPS, December* pp. 1125–1138.

Nagy, G., and Tuong, N. (1970). Normalization Techniques for Handprinted Numerals. *Commun. ACM* **13**, 475–481.

Nake, F., and Rosenfeld, A., eds. (1972). "Graphic Languages." North-Holland Publ., Amsterdam.

Narasimhan, R. (1964). Labelling Schemata and Syntactic Description of Pictures. *Inf. Control* **7** (June), 151–179.

Narasimhan, R. (1966). Syntax-Directed Interpretation of Classes of Pictures. *Commun. ACM* **9**, 166–173.

Narasimhan, R. (1969). On the Description, Generation, and Recognition of Classes in Pictures. *In* "Automatic Interpretation and Classification of Images" (A. Graselli, ed.), pp. 1–42. Academic Press, New York.

Nevatia, R. (1976). Locating Object Boundaries in Textured Environments. *IEEE Trans. Comput.* **25**, No. 11, 1170–1175.

Nevatia, R. (1977). "Structured Descriptions of Complex Curved Objects for Recognition and Visual Memory." Springer-Verlag, Berlin and New York.

Nevatia, R., and Binford, T. O. (1973). Structural Descriptions of Complex Objects. *Proc. Int. J. Conf. Artif. Intell., 3rd, Menlo Park, Calif.*, pp. 641–657.

Novikoff, A. B. J. (1962). Integral Geometry as a Tool in Pattern Perception. *In* "Principles of Self-Organization" (H. von Foerster and G. W. Zopf, eds.), pp. 347–368. Pergamon, New York.

O'Gorman, F., and Clowes, M. B. (1976). Finding Picture Edges Through Collinearity of Feature Points. *IEEE Trans. Comput.* **25**, No. 4, 449–456.

Ohlander, R. B. (1975). Analysis of Natural Scenes. Ph. D. thesis, Carnegie-Mellon Univ., Pittsburgh, Pennsylvania, April (unpublished).

Papoulis, A. (1969). "Systems and Transforms with Applications to Optics." McGraw-Hill, New York.

Paton, K. (1970). Conic Sections in Chromosome Analysis. *Pattern Recognition* 2, 39–51.

Pavlidis, T. (1972a). Piecewise Approximation of Functions of Two Variables through Regions with Variable Boundaries. *Proc. ACM Natl. Conf. Boston,* pp. 652–662.

Pavlidis, T. (1972b). Segmentation of Pictures and Maps Through Functional Approximation. *Comput. Graphics Image Process.* 1, 360–372.

Pavlidis, T. (1977). "Structural Pattern Recognition." Springer-Verlag, New York.

Pavlidis, T., and Ali, F. (1975). Computer Recognition of Handwritten Numerals by Polygonal Approximation, *IEEE Trans. Syst. Man Cybern.* SMC6, 610–614.

Pavlidis, T., and Horowitz, S. L. (1974). Segmentation of Plane Curves. *IEEE Trans. Comput.* 23, No. 8, 860–870.

Persoon, E., and Fu, K. S. (1977). Shape Discrimination Using Fourier Descriptors, *IEEE Trans. Syst. Man Cybern.* SMC7, 170–179.

Pfaltz, J. L., and Rosenfeld, A. (1967). Computer Representation of Planar Regions by Their Skeletons. *Commun. ACM* 10, 119–122, 125.

Pickett, R. M. (1970). Visual Analysis of Texture in the Detection and Recognition of Objects. *In* "Picture Processing and Psychopictorics" (B. C. Lipkin and A. Rosenfeld, eds.), pp. 289–308. Academic Press, New York.

Preston, K., Jr. (1961). The CELLSCAN System, a Leucocyte Pattern Analyzer. *Proc. West. J. Comput. Conf., May,* pp. 173–183.

Preston, K., Jr. (1971). Feature Extraction by Golay Hexagonal Pattern Transforms. *IEEE Trans. Comput.* 20, No. 9, 1007–1014.

Prewitt, J. M. S. (1970). Object Enhancement and Extraction. *In* "Picture Processing and Psychopictorics" (B. S. Lipkin and A. Rosenfeld, eds.), pp. 75–149. Academic Press, New York.

Prewitt, J. M. S., and Mendelsohn, M. L. (1966). The Analysis of Cell Images. *Ann. N.Y. Acad. Sci.* 128, 1035–1053.

Price, K. E. (1976). Change Detection and Analysis of Multispectral Images. Dep. Comput. Sci., Carnegie-Mellon Univ., Pittsburgh, Pennsylvania.

Ramer, U. (1972). An Iterative Procedure for the Polygonal Approximation of Plane Curves. *Comput. Graphics Image Process.* 1, No. 3, 244–256.

Riseman, E. A., and Arbib, M. A. (1977). Computational Techniques in Visual Systems, Part II: Segmenting Static Scenes. *IEEE Comput. Soc. Repository* R77–87.

Roberts, L. G. (1965). Machine Perception of Three-Dimensional Solids. *In* "Optical and Electro-Optical Information Processing" (J. T. Tippett *et al.,* eds.), pp. 159–197. MIT Press, Cambridge, Massachusetts.

Robinson, G. S. (1976a). Color Edge Detection. *Proc. SPIE Symp. Adv. Image Transmission Techn., San Diego, Calif., August* 87, 126–133.

Robinson, G. S. (1976b). Detection and Coding of Edges Using Directional Masks. *Proc. SPIE Conf. Adv. Image Transmission Techn., San Diego, Calif., August* 87, 117–125.

Rosenfeld, A. (1962). Automatic Recognition of Basic Terrain Types from Aerial Photographs. *Photogramm. Eng.* 28, No. 1, 115–132.

Rosenfeld, A. (1969). "Picture Processing by Computer." Academic Press, New York.

Rosenfeld, A. (1970a). A Nonlinear Edge Detection Technique. *Proc. IEEE* 58, No. 5, 814–816.

Rosenfeld, A. (1970b). Connectivity in Digital Pictures. *J. Assoc. Comput. Mach.* 17, No. 1, 146–160.

Rosenfeld, A. (1973). Arcs and Curves in Digital Pictures. *J. Assoc. Comput. Mach.* 20, 81–87.

Rosenfeld, A. (1974a). Compact Figures in Digital Pictures. *IEEE Trans. Syst., Man Cybern.* 4, 211–223.

Rosenfeld, A. (1974b). Digital Straight Line Segments. *IEEE Trans. Comput.* 23, 1264–1269.

Rosenfeld, A. (1974c). Adjacency in Digital Pictures. *Inf. Control* 26, 24–33.

Rosenfeld, A., and Lipkin, B. S. (1970). Texture Synthesis. *In* "Picture Processing and Psychopictorics" (A Rosenfeld and B. S. Lipkin, eds.), pp. 309–345. Academic Press, New York.

Rosenfeld, A., and Pfaltz, J. L. (1966). Sequential Operations in Digital Picture Processing. *J. Assoc. Comput. Mach.* **13**, 471–494.

Rosenfeld, A., and Pfaltz, J. L. (1968). Distance Functions on Digital Pictures. *Pattern Recognition* **1** (July), 33–62.

Rosenfeld, A., Fried, C., and Orton, J. N. (1965). Automatic Cloud Interpretation. *Photogramm. Eng.* **31**, 991–1002.

Rosenfeld, A., Huang, H. K., and Schneider, V. B. (1969). An Application of Cluster Detection to Text and Picture Processing. *IEEE Trans. Inf. Theory* **15**, 672–681.

Rosenfeld, A., Thurston, M., and Lee, Y. (1972). Edge and Curve Detection: Further Experiments. *IEEE Trans. Comput.* **21**, No. 7, 677–715.

Rutovitz, D. (1968). Data Structures for Operations on Digital Images. *In* "Pictorial Pattern Recognition" (G. C. Cheng *et al.*, eds.), pp. 105–133. Thompson Book Co., Washington, D.C.

Sadjadi, F. A. and Hall, E. L. (1978), Invariant Moments for Scene Analysis, *Proc. IEEE Cont. Pattern Recognition Image Process, Chicago*, pp. 181–187.

Sakai, T., Nagao, M., and Matsushima, H. (1972). Extraction of Invariant Picture Sub-Structure by Computer, *Comput. Graphics Image Process.* **4**, 81–96.

Shaw, A. C. (1969). A Formal Picture Description Scheme as a Basis for Picture Processing Systems. *Inf. Control* **14**, 9–52.

Shaw, A. C. (1970). Parsing of Graph-Representable Pictures. *J. Assoc. Comput. Mach.* **17**, No. 3, 453–481.

Shirai, Y. (1973). A Context Sensitive Line Finder for Recognition of Polyhedra. *Artif. Intell.* **4**, 95–119.

Shirai, Y., and Tsuji, S. (1972). Extraction of the Line Drawing of Three-Dimensional Objects by Sequential Illumination from Several Directions, *Pattern Recognition* **4**, 343–351.

Sklansky, J. (1970). Recognition of Convex Blobs. *Pattern Recognition* **2**, 3–10.

Sklansky, J., Cordella, L. P., and Levidaldi, S. (1976). Parallel Detection of Concavities in Cellular Blobs. *IEEE Trans. Comput.* **25**, No. 2, 187–196.

Stefanelli, R., and Rosenfeld, A. (1971). Some Parallel Thinning Algorithms for Digital Pictures. *J. Assoc. Comput. Mach.* **18**, 255–264.

Strong, J. P., III, and Rosenfeld, A. (1973). A Region Coloring Technique for Scene Analysis. *Commun. ACM* **16**, 237–246.

Sutton, R. N., and Hall, E. L. (1972). Texture Measures for Automatic Classification of Pulmonary Disease. *IEEE Trans. Comput.* **21** (July), 667–676.

Tenenbaum, J. M., and Weyl, S. (1975). A Region Analysis Subsystem for Interactive Scene Analysis. *Proc. Int. J. Conf. Artif. Intell., 4th, Cambridge, Mass., September.*

Tenery, G. R. (1963). A Pattern Recognition Function of Integral Geometry. *IEEE Trans. Mil. Electron.* **7**, 196–199.

Thomason, M. G., and Gonzalez, R. C. (1975). Syntactic Recognition of Imperfectly Specified Patterns. *IEEE Trans. Comput.* **24**, No. 1, 93–96.

Thomason, M. G., Barrero, A., and Gonzalez, R. C. (1978). Relational Database Table Representation of Scenes, *Proc. IEEE SOUTHEASCON, Atlanta.*, pp. 32–37.

Tomita, F., Yachida, M., and Tsuji, S. (1973). Detection of Homogeneous Regions by Structural Analysis. *Proc. Int. J. Conf. Artif. Intell., 3rd Stanford, Calif., August* pp. 564–571.

Tou, J. T., and Gonzalez, R. C., (1974). "Pattern Recognition Principles." Addison-Wesley, Reading, Massachusetts.

Tsuji, S., and Tomita, F. (1973). A Structural Analyzer for a Class of Textures. *Comput. Graphics Image Process.* **2**, No. 3/4, 216–231.

Wallace, T. (1978). Private correspondence. Purdue Univ., Lafayette, Ind.

Weszka, J. S., Nagel, R. N., and Rosenfeld, A. (1974). A Threshold Selection Technique. *IEEE Trans. Comput.* **23**, No. 12, 1322–1326.

Weszka, J. S., Dyer, C. R., and Rosenfeld, A. (1976). A Comparative Study of Texture Measures for Terrain Classification. *IEEE Trans. Syst. Man Cybern.* **6**, No. 4, 269–285.

Widrow, B. (1973). The "Rubber-Mask" Technique. *Pattern Recognition* **5**, 175–211.

Williams, R. (1971). A Survey of Data Structures for Computer Graphics Systems. *Comput. Surv.* **3**, 1–21.

Winograd, T. (1972). "Understanding Natural Language." Academic Press, New York.

Winston, P. H. (1975). "The Psychology of Computer Vision." McGraw-Hill, New York.

Winston, P. H. (1977). "Artificial Intelligence." Addison-Wesley, New York.

Wolfe, R. N. (1969). A Dynamic Thresholding Technique for Quantization of Scanned Images. *Proc. Symp. Autom. Pattern Recognition, Washington, D. C., May*, pp. 142–162.

Yakimovsky, Y. (1973). Scene Analysis Using a Semantic Base for Region Growing. Rep. AIM-209, Stanford Univ., Stanford, California.

Yakimovsky, Y., and Feldman, J. A. (1973). A Semantics-Based Decision Theory Region Analyzer. *Proc. Int. J. Conf. Artif. Intell., 3rd, Stanford, Calif., August*, pp. 580–588.

Zahn, C. T. (1971). Graph-Theoretic Methods for Detecting and Describing Gestalt Clusters. *IEEE Trans. Comput.* **20**, 68–86.

Zahn, C. T., and Roskies, R. Z. (1972). Fourier Descriptors for Plane Closed Curves. *IEEE Trans. Comput.* **21**, 269–281.

Zobrist, A. L., and Thompson, W. B. (1975). Building a Distance Function for Gestalt Grouping. *IEEE Trans. Comput.* **4**, No. 7, 718–728.

Zucker, S. W., Rosenfeld, A., and Davis, L. S. (1975). Picture Segmentation by Texture Discrimination. *IEEE Trans. Comput.* **24**, No. 12, 1228–1233.

Zusne, L. (1970). "Visual Perception of Form." Academic Press, New York.

# 8 | Scene Matching and Recognition

## 8.1 Introduction

Many important *decisions* may be made based upon the information contained in images of scenes. Most objects encountered in the physical world are *three-dimensional* structures. A collection of these objects may be called a *scene* and the two-dimensional picture generated by viewing the scene, an *image*. The *intensity* at a given point in the image is affected by several variables, including

(a) viewing geometry: distance and angle of object surfaces with respect to the illumination source,
(b) spectral composition of the illumination source,
(c) reflectance properties of the object surfaces,
(d) spectral characteristics, processing, and nonlinearities of the sensor systems,
(e) scene content.

The desire to make a decision based upon information contained in a scene leads us directly into the study of pattern recognition. Pattern recognition is a well-developed mathematical discipline concerned with models and methods for decision processes. These methods will be applied to the problem of scene matching and provide guidelines not only for technique development but also for performance evaluation.

In the previous chapter, we considered segmentation, regional descriptions, and relational descriptions that apply directly to scenes containing objects.

Techniques that also apply to scenes in which objects may be indistinct or irrelevant to the pattern of the scene will now be discussed. For example, consider the scene shown in Fig. 8.1, which is a derived population distribution map. This scene shows a situation rather than distinct objects. Even in scenes of this complexity, some very interesting problems may be solved. One general class of problems is called *scene matching*: Given a pictorial descrip-

*Figure 8.1*   Typical image encountered in scene matching, population distribution in eastern United States. (From the US Bureau of the Census, US Maps, GE-70, No. 1. Courtesy of J. R. Carter.)

tion of a region of a scene, determine which region in another image is similar. The method used to solve this problem is, in its simplest form, called template matching. An image of a template is given, and it is desired to determine all locations of the template in another image. The general problem may involve geometrical and sensor variations. For example, given a region from the image shown in Fig. 8.1, determine the corresponding region in the image shown in Fig. 8.2. Note that if corresponding regions could be determined, then the images could be accurately registered, corresponding objects located, and perhaps the relations between the objects in the scene could be described.

The basic approaches to scene matching will now be considered. The correlation and template matching techniques will first be reviewed. Then methods of sequential similarity will be described. Finally, a new approach

*Figure 8.2* Typical image corresponding to Fig. 8.1, satellite image taken at night of the eastern US and Canada showing lights of the urban areas and moonlight reflected from clouds and snow. (From the USAF Air Weather Service photograph for March 1972. Courtesy of J. R. Carter.)

involving hierarchical search techniques will be described and several examples given. Methods for comparing the performance of scene matching techniques will be described and illustrated.

### 8.1.1 Recognition Models Used in Scene Matching

Before considering the specific problems of scene recognition, it is important to review several pattern recognition models. Each of these models will later be applied to scene recognition; however, the models are of a very general nature and apply to nonpictorial as well as pictorial patterns.

The first model to be considered is the decision process called the *basic recognition model*: Given a measurement vector **x**, determine to which of several classes $C_i$ the measurement belongs. In the statistical recognition formulation, an optimum decision may be made if sufficient information about the joint probability function is known. Often only a set of training samples is given. If these samples are labeled according to membership, this problem may be called *supervised* classification. If the samples are not labeled, the problem may be called unsupervised classification or *clustering*.

The supervised pattern recognition problem may be formulated as follows. Consider a group of objects $G = \{O_i\}$, each belonging to one of $K$ classes

$$C_1, C_2, \ldots, C_K.$$

It is desired to be able to decide which object belongs to which class.

The information available for solving this problem may consist of a set of measurements $\{x\}$ made from some objects from $G$ that are labeled with their true classification. This set of measurements is called the training set. The training sample matrix is shown in Table 8.1.

*Table 8.1*

*Training Sample*

| | |
|---|---|
| Class $c_1$: | $x_1, x_2, \ldots, x_{L_1}$ |
| Class $c_2$: | $x_{L_1+1}, x_{L_1+2}, \ldots, x_{L_1+L_2}$ |
| | $\vdots$ |
| Class $c_K$: | $x_{L_{K-1}+1}, x_{L_{K-2}}+2, \ldots, x_{L_{K-1}+L_K}$ |

Ordinarily, the training set is an extremely small subset of $G$. Typically, the training set is finite, whereas $G$ is infinite. By examining a few examples provided by the training set, the pattern recognizer must formulate a rule for classifying all the objects in $G$. It is often assumed that all the objects in $G$ are distinguishable with respect to their observable characteristics. It may also be assumed that all the training samples are correctly labeled. A decision rule must be formulated in terms of these observable characteristics.

The pattern recognition solution consists of a rule for determining the class membership for each object in $G$. Performance may be measured by the percentage of correct classification for each group, an average classification, or by a cost analysis.

Pattern recognition is a fundamental component of intelligent behavior. Pattern recognition in general may be considered as the classification of objects based upon characterization knowledge and may be roughly separated into four phases:

(1)   observing the attributes or characteristics of the objects,
(2)   selecting useful features from the set of characteristics,
(3)   making the classification decision, and
(4)   evaluating the classification performance.

A pattern classifier must accordingly be equipped with receptors or sensors with which the pertinent information may be observed. The characteristics observed by the receptors may be expressed as numbers, either naturally or artificially. For example, the presence or absence of some attribute may be coded as 0 or 1, respectively. In general, the numbers obtained from the receptors are called *measurements*.

Although it is possible to make the classification decision on the basis of the measurements provided by the receptors, frequently these measurements are operated upon to produce another set of numbers, called *features*.

Whereas measurements and features are both numbers used to characterize the objects to be classified, features are meant to be more directly relevant to the intended classification. Each object is characterized by its measurement values and thus by an ordered set of numbers. This ordered set of numbers is called the *measurement vector*. Similarly, the feature values constitute the *feature vector*. The *classification rule* is a mapping of the pattern space into the set of classification decisions.

At present, there is no standard terminology in the field of pattern recognition, and variations may be encountered from publication to publication. The following definitions from Webster's New World Dictionary [1960] are, however, relevant.

> Recognize:   To perceive or identify from past experience or knowledge, or by some detail.
>
> Pattern:   A representative sample of a class or type.

Three distinct, but not necessarily independent, aspects of a pattern recognition problem may be identified.

(1) *Characterization* is that aspect of the problem that involves the specification of the measurement vector. Insufficient characterization will preclude a good recognition solution while superfluous characterization will overburden limited recognition capabilities. The characterization problem is also called *measurement selection* or *feature extraction*.

(2) *Abstraction* is that aspect that involves the formulation of a classification rule once the feature extraction has been completed. From the total information provided by the pattern vector, the classification information must be abstracted. The majority of the pattern recognition literature has dealt with this aspect.

(3) *Generalization* is that aspect that deals with evaluating a proposed solution. The capability of a classification rule to generalize beyond the training set to the entire pattern population must be assessed. Often, this involves obtaining another set of patterns called the *testing sample*, with known classification, and applying a proposed classification rule to this testing set.

An underlying tendency in pattern recognition is the *law of parsimony*. Traditionally, scientists have had a strong faith in the ultimate simplicity of the basic laws. Hawkins, the editor of Bongard's book "Pattern Recognition" stressed the idea that the simplest explanation of the known facts may greatly reduce the effort that others must expend to understand these facts. "Society rightly recognizes the importance of simplicity, and applauds labor saving insights and devices" (Bongard, 1970, p. 53).

Einstein chose the simplest set of tensor equations for his theory of gravitation with complete confidence and remarked to John Kemeny that

> God would not have passed up an opportunity to make nature that simple.

Newton also believed in simplicity.

> Nature is pleased with simplicity and affects not the pomp of superfluous causes.

The block diagram of the basic pattern recognition system is shown in Fig. 8.3. The set of objects may be considered as the pattern space. Measurements sensed from some property of the objects, after suitable preprocessing, form the measurement space. By selecting the relevant information and excluding irrelevant data, the feature vector is formed. The set of feature vectors forms the feature space. The decision maker, or classifier, uses the feature vector to make a decision about the object. The classification space is simply a set of discrete points corresponding to the set of possible decisions.

All realistic pattern recognition problems may be considered as statistical in nature. The principal statistic upon which a decision may be based is the *likelihood ratio*. For any two classes $\omega_i$ and $\omega_j$, the likelihood ratio $\lambda_{ij}$ is defined by

$$\lambda_{ij} = p(\mathbf{x}/\omega_i)/p(\mathbf{x}/\omega_j),$$

where $p(\mathbf{x}/\omega_i)$ is the conditional probability of the feature vector $\mathbf{x}$ given that it came from class $\omega_i$. If $\lambda_{ij} > 1$, then the conditional probability or likelihood of the measurement vector belonging to class $\omega_i$ is greater than the likelihood that it belongs to class $\omega_j$. If the classes are not equally probable, the likelihood ratio may be weighted by the a priori class probabilities to form a

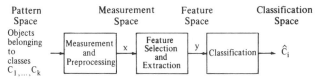

*Figure 8.3*   Pattern recognition system.

total probability ratio. Since

$$P(\mathbf{x}, \omega_i) = p(\mathbf{x}/\omega_i) P(\omega_i),$$

the weighted ratio may be used to determine which class is most probable. If

$$\lambda_{ij} > P(\omega_j)/P(\omega_i),$$

then class $\omega_i$ is the most probable class. A further generalization may be useful if the costs of correct or incorrect decisions are not equal. Let $G_i$ and $G_j$ represent the gains obtained from correct decisions about class $\omega_i$ and $\omega_j$, respectively. Then the likelihood ratio may be compared to the threshold

$$T = P(\omega_i) G_j / P(\omega_i) G_i$$

to determine which decision will produce the greatest gain or least cost. If $\lambda_{ij} > T$, then choosing class $\omega_i$ should produce the greatest gain.

In the previous model we have supposed that all $N$ features are observed by the classifier at one stage. However, the cost of feature measurement may preclude this formulation. An insufficient number of features will not give satisfactory correct classification results. An arbitrarily large number of features is impractical in data collection cost and in analysis. If the cost of taking features is high, or if the problem is sequential in nature, the appropriate formulation is in terms of *sequential decision* procedures.

Examples of high cost for feature measurement are found in industrial processes when elaborate equipment must be shut down to make a measurement, or in risky diagnostic medical procedures.

There is a balance that must be established between the information provided by the measurements and the cost of taking it.

Furthermore, a trade-off between the recognition error and the measurement cost or number of features must be made. The sequential procedure takes features sequentially, terminating the process and making a decision when an acceptable accuracy of classification can be achieved. If the feature measurements are to be taken sequentially, the order of the features to be measured is important. The features should be ordered so that the measurements taken will produce the earliest terminal decision.

If there are two pattern classes, then Wald's sequential probability ratio test (SPRT) can be applied directly (Fu, 1968). At the $n$th stage of the sequential process, i.e., when the $n$th measurement has been taken, the classifier computes the sequential probability ratio

$$\lambda_n = p_n(\mathbf{x}/\omega_1) / P_n(\mathbf{x}/\omega_2),$$

where $p_n(\mathbf{x}/\omega_i)$, $i = 1, 2$, is the multivariate conditional probability density of the $n$-dimensional pattern vector $\mathbf{x}$ for class $\omega_i$. The likelihood ratio $\lambda_n$ is then compared with two stopping boundaries $A$ and $B$.

If   $\lambda_n \geq A$, a decision $\mathbf{x} \in \omega_1$ is made;
$\qquad \lambda_n \leq B$, a decision $\mathbf{x} \in \omega_2$ is made.
If   $B < \lambda_n < A$, then an additional feature is taken and the process proceeds to the $(n+1)$th stage.

The two stopping boundaries are related to the recognition errors $e_{21}$ and $e_{12}$ by

$$A = (1 - e_{21})/e_{12}, \qquad B = e_{21}/(1 - e_{12}),$$

where $e_{ij}$ is the probability of deciding $\mathbf{x} \in \omega_i$ when actually $\mathbf{x} \in \omega_j$ is true.

Following Wald's sequential analysis, it has been shown that a classifier using the SPRT has an optimal property for the two-class problem. That is, for a given $e_{12}$ and $e_{21}$ there is no other procedure with lower error probability or expected risk with a smaller average number of features.

For more than two classes, the generalized sequential probability ratio test may be used. In the generalized test the pattern classes are rejected sequentially until only one is left that is accepted as the recognized class. This rejection criterion is somewhat conservative but will usually lead to a high percentage of correct recognition because only the classes that are the most unlikely to be true are rejected.

The sequential test partitions the feature space into three regions. The decision boundaries in a sequential process vary with the number of features $n$ in both dimensionality and surface characteristics.

For example, suppose that $x_1, x_2, \ldots$ are independent feature measurements with densities $p(x_j/\omega_i)$; $j = 1, 2, \ldots$; $i = 1, 2$ which form a univariate Gaussian density with mean $m_i$ and common variance $\sigma^2$. For computational simplicity, $\log \lambda_n$ is computed rather than $\lambda_n$. After the first feature $x_1$ is measured,

$$\log \lambda_1 = \left[ (m_1 - m_2)x_1 - \tfrac{1}{2}(m_1^2 - m_2^2) \right]/\sigma^2.$$

Compare $\log \lambda_1$ with $\log A$ and $\log B$, i.e., let

$$T_1 = \frac{\sigma^2}{m_1 - m_2} \log A + \frac{1}{2}(m_1 + m_2)_1$$

and

$$T_2 = \frac{\sigma^2}{m_1 - m_2} \log B + \frac{1}{2}(m_1 + m_2)_2;$$

then if

$$x_1 \geq T_1, \qquad \text{then} \quad x_1 \in \omega_1,$$
$$x_1 \leq T_2, \qquad \text{then} \quad x_1 \in \omega_2;$$

and if

$$T_2 < x_1 < T_1,$$

then another measurement $x_2$ will be taken and the process proceeds to the second stage.

After the second feature is measured,

$$\mathbf{x} = \begin{bmatrix} x_1 \\ x_2 \end{bmatrix} \quad \text{and} \quad \log \lambda_2 = \log \frac{p(\mathbf{x}/\omega_1)}{p(\mathbf{x}/\omega_2)}.$$

However,

$$p(\mathbf{x}/\omega_i) = p(x_1/\omega_i)p(x_2/\omega_i);$$

therefore

$$\log \lambda_2 = [(m_1 - m_2)/\sigma^2][x_1 + x_2 - (m_1 + m_2)];$$

thus, as before, let

$$T_1 = [\sigma^2/(m_1 - m_2)] \log A + (m_1 + m_2)$$

and

$$T_2 = [\sigma^2/(m_1 - m_2)] \log B + (m_1 + m_2),$$

and if

$$x_1 + x_2 \geq T_1, \qquad \text{then} \quad \mathbf{x} \in \omega_1,$$
$$x_1 + x_2 \leq T_2, \qquad \text{then} \quad \mathbf{x} \in \omega_2;$$

and if

$$T_2 < x_1 + x_2 < T_1,$$

then measurement $x_3$ will be made.

At the $n$th stage of the process, the log likelihood ratio is

$$\log \lambda_n = \sum_{i=1}^{n} \log \frac{p(x_i/\omega_1)}{p(x_i/\omega_2)} = \frac{m_1 - m_2}{G^2} \sum_{i=1}^{n} \left[ x_i - \frac{1}{2}(m_1 + m_2) \right],$$

and if the thresholds are

$$T_1 = [\sigma^2/(m_1 - m_2)] \log A + \tfrac{1}{2} n(m_1 + m_2),$$
$$T_2 = [\sigma^2/(m_1 - m_2)] \log B + \tfrac{1}{2} n(m_1 + m_2),$$

then $\mathbf{x} \in \omega_2$.

The classification procedure is if

$$\sum_{i=1}^{n} x_i \geq T_1, \qquad \text{then} \quad \mathbf{x} \in \omega_1,$$

$$\sum_{i=1}^{n} x_i \leq T_2, \qquad \text{then} \quad \mathbf{x} \in \omega_2,$$

and if

$$T_2 < \sum_{i=1}^{n} x_i < T_1,$$

then take measurement $x_{n+1}$.

A pattern classifier using a standard sequential decision process may be unsatisfactory because

(1)   an individual classification may require more feature measurements than can be obtained; and

(2)   the average number of feature measurements may become extremely large if the $e_{ij}$s are chosen to be very small.

In practical situations, it may become necessary to interrupt the standard procedure at a finite stage and resolve the various courses of action. This can be achieved by truncating the sequential process at $n = N$. For example, one may carry out the regular SPRT until either a terminal decision is made or stage $n = N$ is reached. At $n = N$, the input pattern is classified as belonging to the class with the largest generalized sequential probability ratio. Under the truncated procedure, the process must terminate with no more than $N$ stages. Truncation is a compromise between an entirely sequential method and a classical fixed-sample-size decision procedure.

### Structural and Syntactic Models

The models previously considered apply directly to object location and pattern classification, which are among the easiest problems encountered in visual recognition. In fact, locating objects is so elementary for a human that the difficulty of machine object detection is almost always underestimated. Also, the classification problem actually models only a few elementary situations encountered in visual experiences. What does a human do with the wealth of visual information impinging upon his retinas? He not only locates objects and classifies patterns, but also notes the relationships between them. The properties and relationships of visual elements provide the basis for describing pictures, scenes, or situations. The image description problem models a large portion of visual experiences. Again, some of the problems considered as examples of machine description may appear tediously elementary, simply because the task is so readily solved by a human. Nonetheless, the general image description problem is important, simply because it serves as a model for many visual situations.

The most common approach to the description problem may be called *syntactic recognition*. Given a set of symbols or primitive elements and relationships defining acceptable combinations of the primitives, one can generate an infinite set of configurations. Furthermore, one may test a given configuration to determine the specific relationships involved, and from this knowledge, describe or answer questions about the configuration. We shall now consider the syntactic problem and its application to scene matching.

The syntactic approach to scene matching may be contrasted to an estimation approach in which a parameter vector is the desired result, or to a recognition approach in which a set of classifications is the desired result.

*Figure 8.4* Illustration that a syntactic model may be required for scene recognition. (Courtesy of R. C. Gonzalez.). (Careful study is required to recognize this picture of a cow.)

An example may clarify the relationship of the syntactic approach to the previous models. Consider the scene recorded by the image shown in Fig. 8.4. Suppose that it is desired to first characterize the content of the scene so that a set of similar scenes could be generated or a new scene tested for similarity. A structural description of the objects and the relationships between the major objects is shown in Fig. 8.5. The overall scene as inferred from the image consists of a cow in front of a fence in a grassy field. Thus, the scene can be divided into two major regions, each of which contains a physical object and system noise. Note also that since the cow is in front of the background, a range image would aid greatly in characterizing the scene. Each "in front of" relation indicates the usefulness of range information in segmenting the scene. Each node in the graph represents a physical object that could be further characterized by measurements. A structure outline in three dimensions could characterize each object. Measurements of spectral reflectance characteristics, such as in a color image, would easily distinguish between the grass, sky, and cow. Textural measurements could be used to

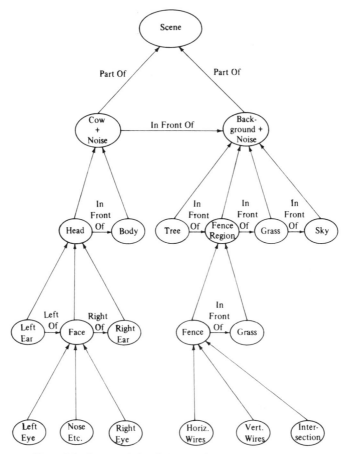

*Figure 8.5*   Structural description of objects and relationships.

distinguish certain regions such as the edge boundary at the left side of the face. Shadows in the upper right portion of the image tend to obscure the right ear boundary from the background, and if detected, could be used to define this boundary. Finally, specular reflectances occur at the intersection of the fence wires and give rise to the highest intensities in the scene. These specular reflectances could perhaps be predicted and thus eliminated if the viewing geometry parameters were determined from the image.

This structural model along with appropriate measurements of the objects could easily be used to generate a set of similar scenes. For example, one could vary the object characteristics, viewing parameters, or noise level. Also, a new scene could be examined to determine a similarity to the entire graph or a subgraph.

It is also interesting to consider certain problems encountered in scene matching in the context of this description. Suppose that we would like to determine a unique region to use as a reference region for matching a similar image. In the image shown many of the prominent features are caused by noise and not by a major physical object. Obviously, these would not be good reference regions, as indicated by the model—not the image. Principal nodes in the structural description should be used as reference regions. Note that the number of nodes to be considered is much smaller than the number of possible subregions in a digital picture. This example presents only a brief structural description, but hopefully illustrates that the syntactic model is complementary rather than competitive to the other recognition models.

## 8.1.2 Template Matching and Correlation Techniques

The simplest approach to scene matching is called *template matching*. Given a template of a scene, determine the location of this template in another scene. The elements of this method will be illustrated by the example shown in Fig. 8.6. Suppose the letter "T" shown in Fig. 8.6a is the template. It is desired to determine the location of this template in the larger image shown in Fig. 8.6b. The upper left-hand corner of the template and large image will be taken as the two reference coordinate systems. The location of the template may then be specified by two translation parameters with respect to the large image coordinate system. For example, it may be observed that if the template is first placed at the $(1,1)$ location in the larger image then translated 2 units in the $x$ and 6 units in the $y$ directions, then the two letters will superimpose exactly. Also note that the template cannot be superimposed at each location in the large image. The dotted line in Fig. 8.6b defines the limits of possible overlap comparison.

In order to develop a program that will determine the translation offset parameters, we must first decide upon a measure of similarity at each test location. One could insist that each template location exactly equal the corresponding location in the large image. Since the template is a 7 by 5 array, 35 decision conditions could be used. A simpler approach is to count the matching locations. If both "1" and "0" matches are counted then the position of exact match would result in a count of 35 and all mismatch locations would produce a count less than 35. The count values for all possible match locations may be ordered in another image corresponding to the $x$ and $y$ offset values used to determine the count value. This "count" image is shown in Fig. 8.6c. Note that the count value of 35 does indeed occur at offset locations $(2,6)$ and indicates the correct match position. If only the matching 1s had been counted, the count image shown in Fig. 8.6d would have resulted. Note that although the peak value of 11 would still have indicated the correct location, the magnitude of this value relative to the mismatching count values would have been less; therefore, the first procedure would indicate the correct offset with a greater confidence.

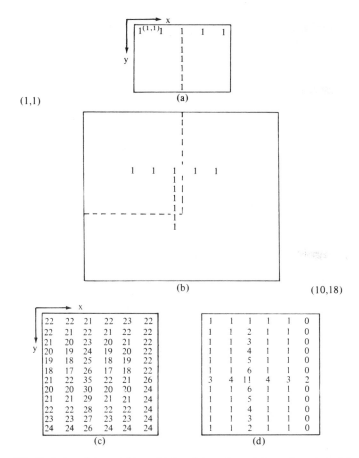

*Figure 8.6* Basic template matching: (a) template pattern, (b) search picture, (c) results of counting "1" and "0" matches at each offset location, and (d) result of counting only "1" matches at each offset location.

A pictorial example of this procedure is shown in Fig. 8.7. Note that the offset position is indicated by the peak in the count image. It will be shown that several generalizations of this procedure are required in order to improve the detection performance, reduce sensitivity to distortions and noise, and reduce the number of computations.

In order to generalize the method to gray level rather than binary images, it is important to introduce the concept of cross-correlation and the Cauchy–Schwarz inequality.

Given a template image vector $\mathbf{W}$, we wish to determine by the use of a digital computer whether the same image appears in an image $S$.

Some of the characteristic methods that have been developed for detecting similar regions and scene matching are the basic correlator, the statistical

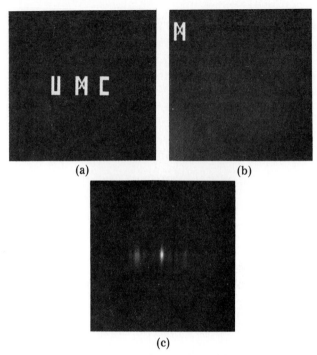

Figure 8.7   Correlation of letters UMC.

correlator, the sequential similarity comparison, and hierarchical matching. In the following sections, each of these methods will be described.

The basic correlator method consists of forming a correlation measure between two picture functions and determining the location of correspondence by finding the location of the maximum correlation (Wong, 1976). In applying this technique, the correlation measure $R(u,v)$ at the reference location $(u,v)$ is defined as

$$R(u,v) = \frac{\sum_{j=1}^{K}\sum_{i=1}^{J} S(i,j)W(i-u,j-v)}{\left[\sum_{j=1}^{K}\sum_{i=1}^{J} S^2(i,j)\right]^{1/2}\left[\sum_{j=1}^{K}\sum_{i=1}^{J} W^2(i-u,j-v)\right]^{1/2}}.$$

The Cauchy–Schwarz inequality states that $-1 \leq R(u,v) \leq 1$. For a given window $W$, the term $\sum\sum W^2(i-u,j-v)$ is a constant. Therefore the computation can be reduced to

$$R(u,v) = \frac{\sum_{j=1}^{K}\sum_{i=1}^{J} S(i,j)W(i-u,j-v)}{\left[\sum_{j=1}^{K}\sum_{i=1}^{J} S^2(i,j)\right]^{1/2}}.$$

To determine the location of maximum correlation, $R(u,v)$ must be computed at each location $(u,v)$, $1 \leq U \leq (M-J+1)$, $1 \leq v \leq (N-K+1)$. No decision can be made until the correlation array $R(u,v)$ is computed for all values of $u$ and $v$. The performance of this correlator can be described as follows:

(1)   This method is relatively sensitive to image noise (Wong *et al.*, 1977). In the presence of image noise, the correlation function produces a relatively broad peak, thus making a selection of a correlation peak difficult.

(2)   A great amount of computation must be performed since the window and search areas are usually large in an actual photograph. With this technique, no decision can be made until the correlation array $R(u,v)$ is computed for all $u,v$.

To overcome the first difficulty mentioned in the previous method, the statistical knowledge of the spatial relationships of picture elements within each image may be used in this statistical correlation (Pratt, 1974). The statistical correlation measure $R_s(u,v)$ is:

$$R_s(u,v) = \frac{\sum_{j=1}^{K} \sum_{i=1}^{J} F_s(i,j) F_w(i-u, j-v)}{\left[ \sum_{j=1}^{K} \sum_{i=1}^{J} F_s^2(i,j) \right]^{1/2} \left[ \sum_{j=1}^{K} \sum_{i=1}^{J} F_w^2(i,j) \right]^{1/2}},$$

where $F_s(i,j)$ and $F_w(i,j)$ are obtained by spatially convolving the images $S(i,j)$ and $W(i,j)$ with spatial filter functions $D_s(i,j)$ and $D_w(i,j)$,

$$F_s(i,j) = S(i,j) \circledast D_s(i,j), \qquad F_w(i,j) = W(i,j) \circledast D_w(i,j).$$

The spatial filter functions $D_s(i,j)$ and $D_w(i,j)$ are chosen to maximize the correlation peak. The first step in the spatial filter design is to decorrelate or whiten the images as follows:

$$\mathbf{A} = \mathbf{H}_s^{-1}\mathbf{S}, \qquad \mathbf{B} = \mathbf{H}_w^{-1}\mathbf{W},$$

where $\mathbf{S}$ and $\mathbf{W}$ are column vector representations of the images $S(i,j)$ and $W(i,j)$ obtained by column scanning the images. $\mathbf{H}_s$ and $\mathbf{H}_w$ are obtained by a factorization of the image covariance matrices $\mathbf{K}_s$ and $\mathbf{K}_w$:

$$\mathbf{K}_s = \mathbf{H}_s\mathbf{H}_s^{\mathrm{T}}, \qquad \mathbf{K}_w = \mathbf{H}_w\mathbf{H}_w^{\mathrm{T}}.$$

$\mathbf{H}_s$ and $\mathbf{H}_w$ may be formulated in terms of the eigenvectors and eigenvalues of $\mathbf{K}_s$ and $\mathbf{K}_w$ as follows:

$$\mathbf{K}_s = \mathbf{E}_s\Lambda_s\mathbf{E}_s^{\mathrm{T}} = (\mathbf{E}_s\Lambda_s)^{1/2}(\mathbf{E}_s\Lambda_s)^{1/2} = \mathbf{H}_s\mathbf{H}_s^{\mathrm{T}},$$

$$\mathbf{K}_w = \mathbf{E}_w\Lambda_w\mathbf{E}_w^{\mathrm{T}} = (\mathbf{E}_w\Lambda_w)^{1/2}(\mathbf{E}_w\Lambda_w)^{1/2} = \mathbf{H}_w\mathbf{H}_w^{\mathrm{T}}.$$

The correlation operation is performed on the whitened vectors $\mathbf{A}$ and $\mathbf{B}$ to yield the statistical correlation measure

$$R_s(u,v) = \mathbf{A}^{\mathrm{T}}\mathbf{B} / (\mathbf{A}^{\mathrm{T}}\mathbf{A})^{1/2}(\mathbf{B}^{\mathrm{T}}\mathbf{B})^{1/2}$$

or

$$R_s(u,v) = \frac{(\mathbf{K}^T)^{-1}\mathbf{S}^T\mathbf{W}}{\left\{\left[(\mathbf{K}^T)^{-1}\mathbf{S}\right]^T\left[(\mathbf{K}^T)^{-1}\mathbf{S}\right]\mathbf{W}^T\mathbf{W}\right\}^{1/2}},$$

where

$$(\mathbf{K}^T)^{-1} = (\mathbf{H}_s\mathbf{H}_w{}^T)^{-1}.$$

If the image elements are assumed to be samples of a Markov process, then

$$(\mathbf{K}^T)^{-1} = \mathbf{K}^{-1}$$

$$= \frac{1}{(1-\rho^2)}\begin{bmatrix} -1 & -\rho\Sigma^{-1} & 0 & 0 & \cdots & 0 \\ -\rho\Sigma^{-1} & (1+\rho^2)\Sigma^{-1} & -\rho\Sigma^{-1} & 0 & \cdots & 0 \\ 0 & -\rho\Sigma^{-1} & (1+\rho^2)\Sigma^{-1} & -\rho\Sigma^{-1} & \cdots & 0 \\ 0 & & \cdots & & & 1 \end{bmatrix},$$

where $\rho =$ (correlation between adjacent image elements), and

$$\Sigma^{-1} = \frac{1}{(1-\rho^2)}\begin{bmatrix} 1 & -\rho & 0 & 0 & \cdots & 0 \\ -\rho & 1+\rho^2 & -\rho & 0 & \cdots & 0 \\ 0 & -\rho & 0 & 0 & \cdots & 0 \\ \vdots & & & & & \vdots \\ 0 & \cdots & & & & 1 \end{bmatrix}.$$

Multiplying $\mathbf{S}$ by the $(\mathbf{K}^T)^{-1}$ is equivalent to convolving the image $S(i,j)$ with the spatial filter function $D(i,j)$,

$$D(i,j) = \begin{bmatrix} 0 & 0 & 0 & 0 & 0 \\ 0 & \rho^2 & -\rho(1+\rho^2) & \rho^2 & 0 \\ 0 & -\rho(1+\rho^2) & (1+\rho^2)^2 & -\rho(1+\rho^2) & 0 \\ 0 & \rho^2 & -\rho(1+\rho^2) & \rho^2 & 0 \\ 0 & 0 & 0 & 0 & 0 \end{bmatrix}.$$

Therefore, the filtered image functions are

$$F_s(i,j) = S(i,j) * D(i,j), \qquad F_w(i,j) = W(i,j) * D(i,j).$$

The performance of this correlator on selected images indicated that the statistical method does provide better performance in terms of providing a sharper peak at the location of image matching. In order for the method to be used, some knowledge of the picture statistics is required.

It is anticipated that performance of this correlator would be worse than that of the basic correlator if the statistics of the input data differ from statistics used in the design of the correlator.

### 8.1.3   Sequential Decision Techniques

A common criticism of both the basic and statistical correlators is the great amount of computation that must be performed. A method of sequential correlation has been proposed (Barnea and Silverman, 1972) to reduce the computation time. The form of this algorithm is similar to the sequential pattern recognition formulation. The feature vector is compared to the reference feature vector, one component at a time. An error function for the comparison may be defined as follows:

$$\varepsilon(u,v) = \sum_{j=1}^{K} \sum_{i=1}^{J} |S(i,j) - W(i-u,j-v)|.$$

Instead of testing all of the elements in the window area, elements of the region are selected in some order. An easy to implement order is a random arrangement. The error is accumulated as each of the elements are compared. If the error exceeds a predetermined threshold value before all the elements in the window area are tested, the window location $(u,v)$ is rejected and a new window is tested. The test procedure is depicted in Fig. 8.8. Curves A, B, and C depict the cumulative errors for three different reference points. A and B accumulate errors rapidly and the tests terminate early. Curve C, however, accumulates error more slowly. It is, therefore, much more likely to be a candidate for a matching point. Theoretical analyses and simulation tests (Barnea and Silverman, 1972) indicated that a saving of computation time is possible.

Extending the basic concept described above, a more complicated sequential detecting method was developed by Ramapriyan (1976). A set of templates was matched to a given image. Instead of matching each template of the set to an image at every location, the templates are partitioned and a representative template is defined for each of the partitions. Several levels of

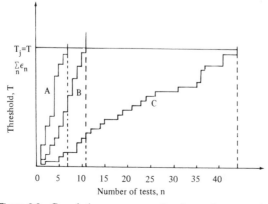

*Figure 8.8*   Cumulative error curves for three reference regions.

partitions are defined. Elimination of mismatching locations and termination of computation can take place at each level of detection. Each level of testing is over a more restricted subset of template class than the previous level. The matching process terminates when the cumulative template matching error exceeds a threshold level. A location that has gone through successive levels of matching without rejection is declared a likely candidate. The performance of the sequential template matching method indicates that computation time is reduced compared to basic correlation. However, several levels of templates may be needed; thus the task of template selection is difficult.

### 8.1.4  Hierarchical Search Techniques

The previous approaches to scene matching required the comparison levels of derived features of the reference region with the template at each possible shift position of the sensed image as the possible match location. Since a template of size $M \times M$ can be shifted into $(N - M + 1)^2$ possible positions in an $N \times N$ image, as shown in Fig. 8.9, the number of correlation computations can be extremely large. Fast Fourier transform techniques for correlation and edge correlation techniques decrease the correlation time at each shift position; however, these methods still require a computation at each of the $(N - M + 1)^2$ positions. In this section, hierarchical search techniques will be described that are logarithmically efficient, and reduce the number of search positions to $K \log(N - M + 1)$.

Several advantages of a structured approach are apparent. First, it is not necessary to examine each pixel in a high-resolution image to locate a region

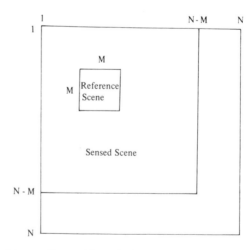

*Figure 8.9*  Template matching positions of reference region size $M \times M$ in several regions of $N \times N$.

at high resolution. The selectivity of the hierarchical techniques, especially coarse–fine search methods, is similar to the perceptual operation of the efficient human visual system. Also, the method in which a match region is obtained at different levels is extendable to other problems such as edge or object location. The method also provides a high degree of precision in locating a region, which is limited only by the highest resolution size and the uniqueness of the match region. Finally, the method permits an efficient decomposition of the sensed scene into "informative" and "irrelevant" regions.

## 8.2   Sequential Decision Rules for Hierarchical Search

### 8.2.1   Coarse–Fine Search

In the hierarchical technique a structured set of pictures at different resolutions is used, as illustrated in Fig. 8.10. The high-resolution sensed scene is denoted $f_L(i,j)$. The index $L$ is called the *level of the search*. The reference region at level $L$, $g_L(i,j)$ is assumed to be smaller than the sensed scene. The agglomerative rule by which the level-$K$ scene is reduced to a

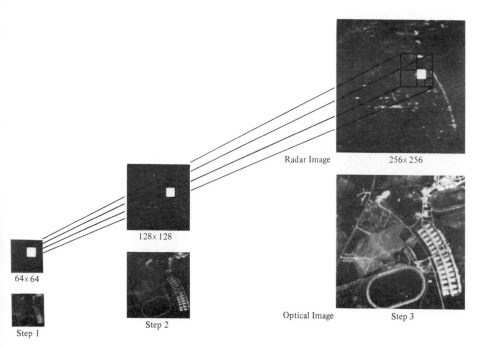

Radar Image          256×256

128×128

64×64

Optical Image        Step 3

Step 2

Step 1

*Figure 8.10*   Hierarchical search technique.

level $K-1$ scene is simple four-point averaging, i.e.,

$$f_{K-1}(i,j) = \tfrac{1}{4}\{ f_K(2i,2j) + f_K(2i,2j+1) + f_K(2i+1,2i) + f_K(2i+1,2j+1)\}$$

for $i,j = 0,1,2,\ldots,2^{K-1}-1$. Other methods could also be used. The reference image region size varies with the level and is an important consideration. Obviously, objects present at one level may not be recognizable at a lower level. However, allowing the reference image size to change with the level alleviates this problem. When looking for a forest one need not look at the leaves. A reference match region must be selected for each level and the performance of the algorithm depends upon the uniqueness of these reference regions.

A matching rule must also be specified to guide the search from level $K-1$ to level $K$. Several possible match functions, such as correlation may be used. The match function $x{\sim}y$, for vectors $x$ and $y$ should have certain properties, such as

   (1)   identity:   $x{\sim}x$;
   (2)   symmetry:   if $x{\sim}y$, then $y{\sim}x$;
   (3)   shift invariance:   if $x{\sim}y$, then $x+a{\sim}y+a$, where $a$ is a constant vector;
   (4)   scale invariance:   if $x{\sim}y$, then $\lambda x{\sim}\lambda y$, where $\lambda$ is a scalar;
   (5)   rotation invariance:   if $x{\sim}y$, then $Tx{\sim}y$, where $T$ is a rotational matrix;
   (6)   sensor invariance:   if $x{\sim}y$, then $\tau(x){\sim}y$, where $\tau$ is the sensor transformation.

The vectors $x$ and $y$ may be the gray level values or derived measurements from corresponding sensed and reference regions. Note that normalized correlation satisfies properties 1–4 but not 5 and 6. Invariant measurements such as moment invariants used with normalized correlation satisfy properties 1–5. Certain derived functions such as edges or sensor corrected images used with invariant measurements and normalized correlation may satisfy all the properties. Also, note that other properties such as invariance to natural or even man-made changes may be desirable.

A process may be developed to create a set of images that are decreasingly lower in resolution and smaller in size. This procedure results in a quadrant subdivision as shown in Fig. 8.11. Two sets of these images are created, one for the window and the other for the search region. The low spatial frequencies preserved in the low-resolution pictures are utilized to find the scene of interest at considerably lower cost. At the lowest level of resolution the number of possible test locations is reduced to $[(N-M+1)/2^L+1]^2$, where $L$ is the search level. This may be compared to the number of possible test locations of $(N-M+1)^2$ at the highest resolution, when $L=0$. There is a reduction of almost $2^{2L}$ in possible locations where computations must be made.

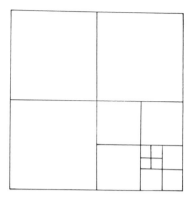

*Figure 8.11*   Quadrant subdivision for coarse–fine search.

Matching rules are used to guide the search from search level $L = K$ to search level $L = K - 1$. The rules are used to select the most promising test locations at level $K$. At level $K - 1$, only the locations selected in level $K$ needed to be tested. Therefore, this search technique is logarithmically efficient, i.e., the number of search positions is reduced to $K \log(N - M + 1)^2$, where $K$ is a constant.

At each test location, an ordering algorithm is used to order the $(M/2^L)^2$ window point pairs. An algorithm similar to the one previously described (Barnea and Silverman, 1972) can be used to select the window point pairs in a random, nonrepeating sequence, or the ordering may be adopted in a data-dependent fashion. For example, for the matching of edge pictures, ordering may be done along the edge gradient. An error measure is established and the errors of the window pairs are accumulated at each test location. At the $K$th window pair computation, the accumulated error is compared to a predetermined threshold $T_k$. If the accumulated error is equal to or greater than $T_k$, the computation terminates and the test location is eliminated. If all window pairs are exhausted and the accumulated error is less than $T_K$, the test location is recorded for further consideration at the next higher resolution level. It can be seen that for a properly chosen $T_k$, many fewer than $(M/2^L)^2$ window pairs are needed to eliminate the test locations that do not match.

At each search level $K$, except for the highest level at the lowest resolution, a search is made only at test locations recorded in the previous $(K + 1)$th level. As the search level decreases or the resolution of the image increases, increasingly fewer test locations need be considered. This is due to the fact that as the search advances, the distances separating the true location from the other test locations increase, and the threshold sequence $\{T_K\}$ is designed to eliminate those test locations that are furthest away from the true matching point. By this technique, only the most promising or probable test locations are examined at the higher resolutions.

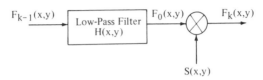

*Figure 8.12* Model of resolution reduction. $S(x,y)=\Sigma_{j_2=-\infty}^{\infty}\Sigma_{j_1=-\infty}^{\infty}\delta(x-j_12\Delta x,y-j_22\Delta y)$, $S(x,y)$=sample function at $\frac{1}{2}$ of sample rate of level $k-1$.

In the hierarchical search technique, a structured set of pictures at different resolutions is created. The process of creating a lower-resolution picture at the $k$th search level may be accomplished by two-dimensional low-pass filtering of the picture data of the $(k-1)$th level, then sampling of the filtered data at one-half the sampling rate of that of the $(k-1)$th level. Unless the low-pass filter is ideal, spurious low-spatial-frequency components are introduced due to the effect of aliasing. In this section, the hierarchical structure is modeled mathematically and analyzed to select the proper filtering characteristics to minimize aliasing errors.

Figure 8.12 shows the mathematical model. An image at the $(k-1)$th level, $f_{k-1}(x,y)$ with Fourier spectrum $F_{k-1}(u,v)$ is linearly filtered by a low-pass filter $h(x,y)$ with the transfer function $H(u,v)$. The filtered image is then spatially sampled at the spatial distances twice those of the image at the $k-1$ level. The resulting data $f_k(x,y)$ represent the picture function at the search level $k$,

$$f_0(x,y)=f_{k-1}(x,y)*h(x,y),$$

where $*$ is the convolution operation. The frequency domain relation is

$$F_0(u,v)=F_{k-1}(u,v)H(u,v).$$

The sampling signal $s(x,y)$ is

$$s(x,y)=\sum_{j_1=-\infty}^{\infty}\sum_{j_2=-\infty}^{\infty}\delta(x-j_12\Delta x,y-j_22\Delta y),$$

where $\Delta x$ and $\Delta y$ are the sampling intervals at the $(k-1)$th level. The Fourier spectrum of $s(x,y)$ is

$$S(u,v)=\frac{\pi^2}{\Delta x\,\Delta y}\sum_{j_1=-\infty}^{\infty}\sum_{j_2=-\infty}^{\infty}\delta\left(u-\frac{2\pi j_1}{2\Delta x},v-\frac{2\pi j_2}{2\Delta y}\right).$$

The Fourier spectrum of the sampled image field is

$$F_k(u,v)=F_0(u,v)*S(u,v)$$

$$=\frac{\pi^2}{\Delta x\,\Delta y}\sum_{j_2=-\infty}^{\infty}\sum_{j_1=-\infty}^{\infty}F_{k-1}\left(u-\frac{\pi j_1}{\Delta x},v-\frac{\pi j_2}{\Delta y}\right)H\left(u-\frac{\pi j_1}{\Delta x},v-\frac{\pi j_2}{\Delta y}\right).$$

The transfer function of one form of the low-pass filter can be expressed as

$$H(u,v)=[\cos(u/2)\cos(v/2)]^n.$$

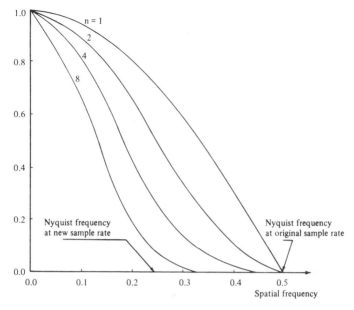

*Figure 8.13*   Low-pass filter characteristics.

The numerical value of $n$ is determined by the higher frequency attenuation rate of the filter. Figure 8.13 shows the spectral characteristics $H(u,v)$ for $n = 1$, 2, 4, and 8. For $n = 1$, a considerable amount of high-frequency energy will pass through the filter, resulting in large aliasing error. This effect is shown in Fig. 8.14, in which the spectral distribution of $F_k(u,v)$ is sketched in one dimension. This figure shows a large amount of spectral overlap resulting in a large aliasing error. For $n = 8$, the high-frequency energy is greatly attenuated. As shown in Fig. 8.15, the amount of spectral overlap is small,

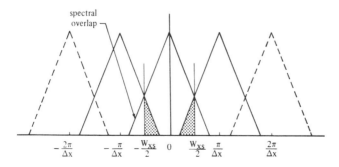

*Figure 8.14*   Spectral distribution for small value of $n$.

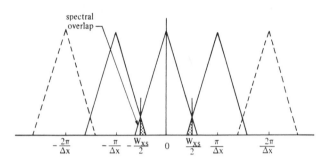

*Figure 8.15* Spectral distribution for large value of *n*.

resulting in a smaller aliasing error. The use of a filter with large *n* also has the effect of further blurring the image, thus further reducing the resolution of the picture.

Noise in an image generally has a higher spatial frequency content than the normal image components. Therefore, low-pass filtering has the effect of noise smoothing. Experiments performed on edge extraction indicated that a considerably better edge picture is obtained from the lower-resolution picture than is obtained from the high-resolution picture. Figure 8.16 shows the spectrum of $F_k(u,v)$ as a two-dimensional function. Let the spectral energy passed through a low-pass filter be $E_R$

$$E_R = \int_{-\infty}^{\infty} \int_{-\infty}^{\infty} F_{k-1}(u,v) H(u,v) \, du \, dv.$$

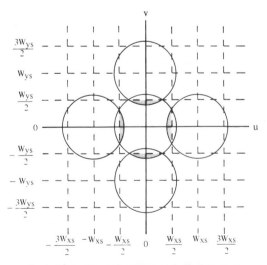

*Figure 8.16* Spectral overlap and aliasing error.

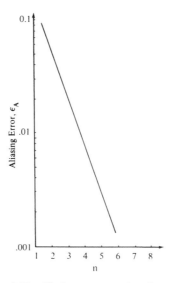

*Figure 8.17* Aliasing error as a function of $n$.

For an ideal filter with an infinitely sharp cutoff and bandwidth $U$ and $V$, the output of the filter becomes

$$E_0 = \int_{-U/2}^{U/2} \int_{-V/2}^{V/2} F_{k-1}(u,v) H(u,v)\, du\, dv.$$

The aliasing error energy, as shown by the shaded regions of Fig. 8.16, is

$$E_A = E_R - E_0.$$

An aliasing error can be defined as

$$\varepsilon_A = (E_R - E_0)/E_R.$$

Figure 8.17 shows a plot of the aliasing error $\varepsilon_A$ for a low-pass filter having various values of $n$. In this calculation, $F_{k-1}(u,v)$ was assumed to have an amplitude of one in the passband of the filter.

$$\varepsilon_A = \int_{0.25}^{0.5} \int_{0.25}^{0.5} (\cos \pi u \cos \pi v)^n\, du\, dv \Big/ \int_{0}^{0.5} \int_{0}^{0.5} (\cos \pi u \cos \pi v)^n\, du\, dv.$$

Figure 8.18 shows various implementations of digital low-pass filters. The coefficients of the filters are included to normalize the filters to unit weighting so that the low-pass filtering process does not introduce a brightness bias in the processed image. Figure 8.18a–c represents filters having an approximate value of $n = 4$. Figure 8.18d has an approximate value of $n = 8$.

A digital version of the sampled image can be written by letting

$$f_{k-1}(n_1, n_2) = \text{input image array},$$

$$f_k(m_1, m_2) = \text{output image array},$$

$$h(l_1, l_2) = \text{low-pass filter}.$$

$$\frac{1}{4}\begin{bmatrix} 1 & & 1 \\ & & \\ 1 & & 1 \end{bmatrix} \qquad \frac{1}{9}\begin{bmatrix} 1 & 1 & 1 \\ 1 & 1 & 1 \\ 1 & 1 & 1 \end{bmatrix}$$

(a)                          (b)

$$\frac{1}{16}\begin{bmatrix} 1 & 2 & 1 \\ 2 & 4 & 2 \\ 1 & 2 & 1 \end{bmatrix} \qquad \frac{1}{10}\begin{bmatrix} 1 & 1 & 1 \\ 1 & 2 & 1 \\ 1 & 1 & 1 \end{bmatrix}$$

(c)                          (d)

*Figure 8.18*  Digital low-pass filter.

Then

$$f_k(m_1,m_2) = \sum_{n_1=m_1}^{m_1-l+1} \sum_{n_2=m_2}^{m_2-l+1} f_{k-1}(n_1,n_2)h(n_1-m_1+1,n_2-m_2+1)$$

for $h$ corresponding to a four-point average, and the above equation becomes

$$f_k(m_1,m_2) = \sum_{n_1=m_1}^{m_1+1} \sum_{n_1=m_2}^{m_2+1} f_{k-1}(n_1,n_2)h(n_1-m_1+1,n_2-m_2+1).$$

## 8.2.2  Location Matrix

At each resolution level, the most promising test locations are selected for further testing at the next higher resolution level. The selection of the most promising test locations can be considered as a clustering problem. This problem is equivalent to maximizing the distance measures between the set of most promising test locations and the set of all other test locations. At the final level of testing, the problem is to maximize the distance measures between the true match location and all other test locations.

In the derivation of the decision rules the cumulative error at each test location is used as the distance measure. The maximization of this distance is performed in terms of deriving a threshold sequence such that

(a)  it maximizes the probability of a match, i.e., it maximizes the probability that the true matched location remains below the threshold;

(b)  it minimizes the false alarm probability, i.e., it minimizes the probability that a test location other than the true matched location remains under the threshold as long as the true matched location.

Let the window and search region be scanned row-by-row in a manner such that $S$ corresponds to the search region of size $N \times N$,

$$S = (s_1, s_2, \ldots, s_{N \times N})',$$

and $W$ corresponds to a reference window of size $M \times M$,

$$W = (w_1, w_2, \ldots, w_{M \times M})'.$$

Then the test location map $Q_k$ is of size $(N - M + 1)^2$, and test location value $Q_k(u, v)$ corresponds to the value when the top left corner of the window is located at $(u, v)$ in the search region.

Let $e_{u,v}^k(s_i, w_i)$ be the error measure of the $i$th window pair of test location $(u, v)$ at the resolution level $k$. The cumulated error after $j$ window pair computations becomes

$$E_j^k(u, v) = \sum_{i=1}^{j} e_{u,v}^k(s_i, w_i).$$

Let $N_k$ be a set of $(u, v)$ locations such that

$$N_k = \left\{ (u, v) \mid E_j^k < T_k, j = 1, \ldots, M^2 \right\}.$$

Now generate a $(2N - 2M + 1)^2$ matrix $G_{k-1}$ such that

$$G_{k-1}(2i, 2j) = \begin{cases} 1 & \text{if } (i, j) \in N_k \\ 0 & \text{if } (i, j) \notin N_k. \end{cases}$$

All other entries to $G_{k-1}$ are set to zero. Matrix $G_{k-1}$ is used as the test location map at the search level $k-1$ and is used as a guide in the search for the matched location at level $k-1$. Tests are to be performed only at the locations $(u, v)$ for which $G_{k-1}(u, v) = 1$. The search continues until one of two cases is encountered:

1. At the search level $n$, $G_n(u, v) = 1$ for only one value of $(u, v)$. Location $(u, v)$ is declared the matched location. If $n$ is not the highest resolution level, a local search can be made to locate a point of minimal cumulated error. This is done by searching the four adjacent locations of the declared location at the next search level and selecting a point with minimal accumulated error as the most likely candidate. This process is repeated until the highest resolution level is reached.

2. At the search level $l = 0$, there exist several locations $(u, v)$ such that $G_0(u, v) = 1$. Select the location with the smallest accumulated error as the most likely match location.

### 8.2.3  Error Measures

The error measure $e(s_j, w_j)$ at the $j$th window pair computation can be implemented in various ways (Rosenfeld and Pfaltz, 1968; Lissack and Fu, 1976). Among these are

$$e_1(s_i, w_i) = |s_i - w_i|,$$

$$e_2(s_i, w_i) = |(s_i - \bar{s}) - (w_i - \bar{w})|,$$

$$e_3(s_i, w_i) = |(s_i - \bar{s})/\sigma_s - (w_i - \bar{w})/\sigma_w|,$$

where $\bar{s}$ is the mean intensity of the image elements of the subimage $(u, v)$ in the search region,

$$\bar{s} = \frac{1}{M^2} \sum_{i=1}^{M^2} s_i,$$

and $\bar{w}$ is the mean intensity of the window,

$$\bar{w} = \frac{1}{M^2} \sum_{i=1}^{M^2} w_i,$$

and $\sigma_s$ and $\sigma_w$ are the standard deviations of the search and window

$$\sigma_s = \left[ \frac{1}{M^2} \sum_{i=1}^{M^2} (s_i - \bar{s})^2 \right]^{1/2}, \qquad \sigma_w = \left[ \frac{1}{M^2} \sum_{i=1}^{M^2} (w_i - \bar{w})^2 \right]^{1/2}.$$

The $L_1$ norm error measure $|s_i - w_i|$ was selected for the implementations because of its computational simplicity. Let the location of the true matched location be $(u^*, v^*)$. In the ideal case,

$$E(u^*, v^*) = \sum_{i=1}^{M^2} e(s_i, w_i) = 0 \qquad \text{and} \qquad 0 = E(u^*, v^*) \leq E(u, v).$$

Therefore for this error measure, no normalization is needed.

### 8.2.4  Derivation of Threshold Sequence

Let the sequence of error measurements be $e_1^k, e_2^k, \ldots, e_j^k$. Without loss of generality, the superscript $k$ will be deleted with the understanding that the derived threshold sequence is to be used at the search level $l = k$. Let

$$E_j = e_1 + \cdots + e_j.$$

The determination of the optimal threshold sequence $\{T_j\}$ for independent errors involves the evaluation of the integrals

$$\int_0^{T_1} \int_0^{T_2 - E_1} \cdots \int_0^{T_j - E_{j-1}} p_{u,v}(e_1) \cdots p_{u,v}(e_j) d(e_1) \cdots d(e_j),$$

where $p_{u,v}(e_j)$ is the probability density of the $j$th error term evaluated at the test location $(u,v)$. $T_j$ is the threshold used in the $j$th error computation. The form and the parameters of $p_{u,v}(e_j)$ depend upon the method of error measurement, the scene content, and the sensor transformations.

Let the error measure be based on the $L_1$ norm, or

$$e_j = |s_j - w_j|.$$

At the location of true match $(u^*,v^*)$, the sum of $e_j$ is a cumulative measure of the difference between the two images. In order to make a correct decision, one assumes that the total cumulative error is minimal at $(u^*,v^*)$. Therefore the knowledge of the shape of the error growth curve and the density distribution of $e_j$ at $(u^*,v^*)$ is important in determining the threshold sequence. Theoretical analyses and experimental verifications in the field of communication (Balakrishnan, 1968; Raemer, 1969), and in image processing (Barnea and Silverman, 1972) have shown that density distribution of an error measure of this type can be approximated by an exponential function. Figures 8.19 and 8.20 show the error density distributions at the matched locations $(u^*,v^*)$ for the matching of a $64 \times 64$ window and a $256 \times 256$ search region of optical to optical and radar to optical matching. It is seen that the density distributions can be represented by an exponential function with mean $\bar{e}$ as follows:

$$p_{u^*v^*}(e_j) = \frac{1}{r}\exp\left(-\frac{e_j}{r}\right), \qquad \bar{e} = \int_0^\infty \frac{e_j}{r}\exp\left(-\frac{e_j}{r}\right)de_j = r.$$

The variance of the error function is

$$\sigma^2 = \frac{1}{r}\int_0^\infty (e_j - r)^2 \exp\left(-\frac{e_j}{r}\right)de_j = r^2.$$

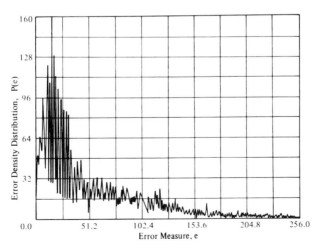

*Figure 8.19* Error density distribution at matched location of optical to optical scene matching.

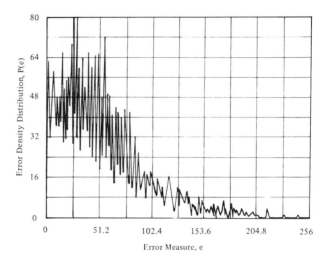

*Figure 8.20*   Error density distribution at matched location of radar to optical scene matching.

After $n$ measurements, let the cumulative error be $E_n$, where $E_n$ can be expressed as

$$E_n = e_1 + \cdots + e_n.$$

Although the error density function of the individual measurements is exponentially distributed, that of $E_n$ can be approximated by a Gaussian function by invoking the central limit theorem for large values of $n$ and assuming that the individual error measurements are statistically independent. $E_n$ then has a mean of $n\bar{e}_j$ and variance of $n\sigma_j^2$, where

$$\bar{E}_n = n\bar{e}_j = nr, \qquad \sigma_n = \sqrt{n}\,\sigma_j = \sqrt{n}\,r.$$

Let the threshold $T_n$ have the value of the expected error $\bar{E}_n$:

$$\bar{E}_n = nr = T_n \qquad \text{or} \qquad T_n/r = n.$$

The threshold $T_n/r$ can be modified to account for the possible variations in $\bar{E}_n$. For a variation of $g$ deviations from $\bar{E}_n$ the threshold becomes

$$T_n/r = n + g\sqrt{n}\,.$$

Figure 8.21 shows the curves computed using the above equation for $g = 0, 1, 2, 4$.

The process of creating a low-resolution image at the $k$th search level involved the low-pass filtering of the higher resolution image of the $(k-1)$th level and then sampling the filtered data at one-half of the sampling rate of that of the $(k-1)$th level. For a low-pass filter with a high attenuation rate, this has the effect of reducing the noise bandwidth to one-half. As the search resolution increases, the threshold sequence previously derived must be mod-

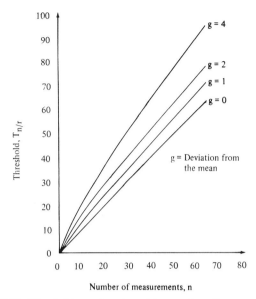

*Figure 8.21*    Threshold as a function of the number of measurements.

ified to account for the increase in noise. Assuming that these error measurements $e_j$ are statistically independent, the error measurement will have an amplitude of $\sqrt{2}\ e_j$ at the next higher resolution. The previous equation can now be modified to account for this effect. At the $k$th search level, the threshold sequence $(T_n^{\ k})$ becomes

$$T_n^{\ k}/r_m(\sqrt{2}\ )^{m-k} = n + g_k\sqrt{n} \qquad \text{or} \qquad T_n^{\ k}/r_m = (\sqrt{2}\ )^{m-k}(n + g_k\sqrt{n}\ ),$$

where $r_m$ is the amplitude of the average error measurement at the matched location of the lowest search level $m$. The value $g_k$ is the number of possible deviations from $r_m$ at search level $k$.

The value of $g_k$ is a function of the probability of a match $p_k$ at search level $k$. If the number of error measurements is large, which is usually true in the case of scene matching, the density distribution of the cumulative error $E_n^{\ k}$ can be approximated by a normal curve. Therefore $p_k$ can be expressed by

$$p_k = \frac{1}{\sqrt{2\pi}\ (\sqrt{n}\ r_m)} \int_{-\infty}^{T_n^{\ k}} \exp\left[ \frac{-(E_n^k - nr_m)^2}{2(\sqrt{n}\ r_m)^2} \right] dE_n^{\ k}$$

$$= 1 - \frac{1}{\sqrt{2\pi}} \int_y^{\infty} \exp\left[ \frac{-(E_n^{\ k})^2}{2} \right] dE_n^{\ k},$$

where $y = (T_n^{\ k} - nr_m)/(\sqrt{n}\ r_m)$.

The value of $p_k$ as a function of $g_k$ was computed using the previous two equations, and the result is shown in Fig. 8.22.

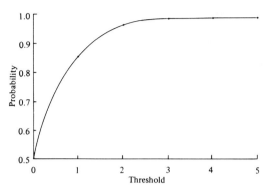

*Figure 8.22*   Probability of a match $p_k$ as a function of the number of deviations of threshold $y$.

It is seen that probability of a match increases with $g_k$. However, the computational efficiency decreases as $g_k$ increases. This is due to the fact that threshold $T_n^{\ k}$ increases with $g_k$, and more test locations need to be evaluated with a higher threshold.

The overall probability of a match $p_d$ is the product of the individual $p_k$ computed at each of the search levels

$$p_d = \prod_{k=m}^{n} p_k,$$

where $m$ is the lowest resolution level and $n \geq 0$ is the level at which a match decision is declared and independence of levels has been assumed.

## 8.3   Hierarchical Search Techniques for Scene Matching

A sequence of experiments were made to demonstrate the use of the hierarchical search for image matching. To illustrate the application to widely different images, radar and optical images of size $256 \times 256$ were used. Since the two images were taken at different view angles and altitudes, it is required that the optical image be geometrically corrected to be in registration with the radar image. The radar image was intensity corrected to match as closely as possible to that of the optical image. The methods used for these corrections are described by Wong (1977). The corrected images are as shown in Figs. 8.23 and 8.24.

For the first sequence of experiments, Gaussian noise with zero mean and $\sigma = 80$ was added to the optical picture. Then a sequence of structured pictures of sizes $128 \times 128$, $64 \times 64$, and $32 \times 32$ was created through the repeated applications of low-pass filtering and sampling. These are as shown in Fig. 8.25. Three regions of interest were selected for matching. Three $8 \times 8$ subimages were extracted from the $32 \times 32$ noise-corrupted optical picture.

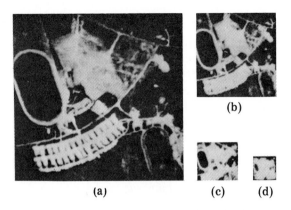

*Figure 8.23* Structured optical images: (a) 256×256, (b) 128×128, (c) 64×64, (d) 32×32. [From Wong and Hall (1978).]

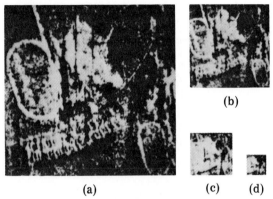

*Figure 8.24* Structured radar image: (a) 256×256, (b) 128×128, (c) 64×64, (d) 32×32. [From Wong and Hall (1978).]

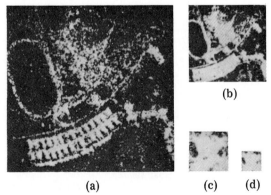

*Figure 8.25* Structured noise-corrupted images: (a) 256×256, (b) 128×128, (c) 64×64, (d) 32×32. [From Wong and Hall (1978).]

These subimages correspond to (i) the upper part of the large stadium at the left upper corner, (ii) the lower part of the large stadium at the middle left, and (iii) the small stadium at the right lower corner.

The parameters used in these experiments are summarized in Table 8.2. The test locations that satisfied the decision rule are recorded for further tests at the next higher resolution level. The testing continues until the registration location is found. The performance has been summarized as shown in Table 8.3a.

Similarly, an $8 \times 8$ subimage of the small stadium was also extracted from the $32 \times 32$ radar image. A matching was performed on the $32 \times 32$ optical

*Table 8.2*

*Experimental Threshold Values for Optical to Noisy Optical and Radar to Optical Search*

| Search level, $k$ | $y$ | $r_m$ |
|---|---|---|
| Optical to noisy optical | | |
| 3 | 2.5 | 15 |
| 2 | 2.0 | 15 |
| 1 | 2.0 | 15 |
| 0 | 2.0 | 15 |
| Radar to optical | | |
| 3 | 3.0 | 14 |
| 2 | 2.0 | 14 |
| 1 | 2.0 | 14 |
| 0 | 1.0 | 14 |

*Table 8.3a*

*Experimental Results for Optical to Noisy Optical Match*

| Search level, $k$ | No. of successful test location | No. of window pairs evaluated |
|---|---|---|
| Upper portion of large stadium | | |
| 3 | 30 | 26,871 |
| 2 | 10 | 7,138 |
| 1 | 3 | 9,847 |
| 0 | 1 | 7,771 |
| Lower portion of large stadium | | |
| 3 | 26 | 27,214 |
| 2 | 2 | 6,071 |
| 1 | 1 | 1,957 |
| Small stadium | | |
| 3 | 9 | 26,048 |
| 2 | 2 | 2,112 |
| 1 | 1 | 1,933 |

*Table 8.3b*

*Experimental Results for Optical to Radar Match*

| Search level, $k$ | No. of successful test location | No. of window pairs evaluated |
|---|---|---|
| | Small stadium | |
| 3 | 28 | 27,760 |
| 2 | 3 | 6,529 |
| 1 | 2 | 3,072 |
| 0 | 1 | 8,115 |

image. The parameters used in these experiments are also shown in Table 8.2 and the performance of this experiment is also summarized in Table 8.2. The number of locations examined and window pairs evaluated is shown in Table 8.3b.

### 8.3.1  Scene Matching Using Invariant Moments

The ability to extract invariant features from an image is important in the field of pattern recognition. An object in the image may then be identified independent of its position, size, and orientation. The mathematical foundations of invariant features is based on the theory of algebraic invariants [Gureuich, 1964]. This theory deals with algebraic functions of a certain class that remain unchanged under certain coordinate transformations. Hu (1962) has shown results relating to two-dimensional moment invariants to algebraic invariants. Sadjadi and Hall (1978) developed techniques by which seven separate moments were computed from digital images.

In Chapter 7, an analysis of the techniques for using invariant moments is described. The performances of these measurements for scene matching will now be considered.

A sequence of experiments was performed in matching radar to optical scenes using hierarchical search techniques and invariant moments as features. The two sets of structured optical and radar images shown in Figs. 8.23 and 8.24 were also used in these experiments. The three subimages, extracted from the radar image, were identical to those used in the previous experiment.

In these experiments, a standard product correlator was used to correlate the seven invariant moments of the window with those computed at each of the test locations in the optical search region. A threshold sequence $\{R_T{}^k\}$ was selected, and a sequential decision rule was established. Let $N_k$ be a set of test locations $(u,v)$ at search level $k$ such that

$$N_k = \left\{ (u,v) \mid R_k(u,v) \geq R_T{}^k,\ 1 \leq u,v \leq M^2 \right\},$$

where $R_T{}^k$ is the threshold selected to be used at search level $k$ and $R_k(u,v)$

is the moment correlation at test location $(u,v)$. $M^2$ is the number of picture elements in the window. For a search region of size $N \times N$, an $(2N - 2M + 1)^2$ matrix $\mathbf{G}_{k-1}$ was generated by

$$\mathbf{G}_{k-1}(2i, 2j) = \begin{cases} 1 & \text{if} \quad (i,j) \in N_k \\ 0 & \text{if} \quad (i,j) \notin N_k. \end{cases}$$

All other entries of $\mathbf{G}_{k-1}$ are set to zero. Matrix $\mathbf{G}_{k-1}$ is used as the test location map at the next search level $k-1$. Tests are to be performed at the test locations for $\mathbf{G}_{k-1}(u,v) = 1$. The search continues until one of two conditions is encountered:

(1)   at search level $n$, $G_n(u,v) = 1$ for one value of $(u,v)$. Location $(u,v)$ is declared the matched location.

(2)   at the search level $L = 0$, there exist several locations $(u,v)$ such that $\mathbf{G}_0(u,v) = 1$. Select the location with the highest correlation as the most likely matched location.

To select the threshold $R_T{}^k$, the invariant moments of the optical image were correlated with the moments of the image reduced by a factor of 2 and rotated by 2° and 45°. The averaged correlation of the three cases was then used as a bound to estimate a threshold sequence. The selected thresholds are shown in Table 8.4. It is seen that these thresholds take on a very narrow range of values. This is due to the fact that the correlations were made on the logarithms of amplitude of the moments rather than the amplitude itself. The

Table 8.4

Scene Matching with Invariant Moments

| Search level, $k$ | Threshold level, $R_T{}^k$ | No. of successful test locations |
|---|---|---|
| Upper part of large stadium | | |
| 3 | 0.990 | 12 |
| 2 | 0.993 | 4 |
| 1 | 0.996 | 2 |
| 0 | 0.998 | 1 |
| Lower part of large stadium | | |
| 3 | 0.990 | 52 |
| 2 | 0.993 | 34 |
| 1 | 0.996 | 21 |
| 0 | 0.998 | 6 |
| Small stadium | | |
| 3 | 0.990 | 63 |
| 2 | 0.993 | 62 |
| 1 | 0.996 | 49 |
| 0 | 0.998 | 25 |

third significant figure of these correlations is important in determining the threshold values.

The results of the experiments are also summarized in Table 8.4. For each of the three test windows, a search at the highest resolution was required and resulted in sets that contained more than one candidate. However, the correct test location is contained in each of the sets. Selecting the highest correlation as the most likely candidate resulted in the correct decision for the upper portion of the large stadium. For the other two test windows, there were several test locations of equally high correlations. This made a selection of the correct location difficult. The results of these experiments can be summarized as follows: Scene matching with invariant moments is costly in computation due to the calculations needed to obtain the seven invariant features. This is particularly true in scene matching at high resolution. However, at low resolution the computation is reasonable. Scene matching with invariant moments can be used to great advantage at the low resolution level at which other methods such as scene matching with edge images are not possible. The utilization of both feature types will be described in the next section.

### 8.3.2   Scene Matching with Edge Features

In the hierarchical search techniques two sets of structured images at different resolutions are created. As the resolution is reduced at each search level, the details of the images are diminished and only the salient outlines of the objects in the image remain. A natural extension of this concept is to form a binary-valued edge picture at the highest resolution and then to create a set of structured edge pictures decreasing in resolution and in size. Scene matching can then be performed on two such structured sets using the hierarchical scene matching techniques.

Edge pictures transformed from the geometrically corrected optical and intensity corrected radar images are shown in Fig. 8.26. Since the optical and radar images have different but distinguishable characteristics, many of the edges that appear in one image may not appear in the other. This is particularly true for edges extracted from background and shadows around the objects of interest. In order to match two-edge images, near-perfect edge registration is required. This is due to the fact that edges are usually thin, and any uncorrected geometric distortion on the order of one or two picture elements will produce a considerably poorer match than those that are perfectly registered. Therefore, the edge images created for scene matching must be capable of meeting the following requirements:

(1)   retain salient edges of the objects to be matched,
(2)   be relatively free of spurious edges extracted from the background, and
(3)   tolerate minor geometric misregistration.

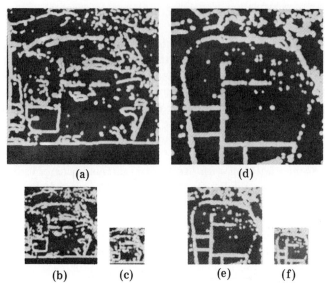

*Figure 8.26*   Structured optical and radar edge images: optical, (a) $256 \times 256$, (b) $128 \times 128$, (c) $64 \times 64$; radar, (d) $256 \times 256$, (e) $128 \times 128$, (f) $64 \times 64$. [From Wong and Hall (1978).]

A process was developed to provide edge images capable of meeting the three requirements. This process involves the following three major steps:

(1) eliminate background edges by making the bright background uniformly bright and dark background uniformly dark;

(2) retain only salient edges by adjusting the edge. threshold to minimize the weak edges;

(3) thicken the remaining edges before scene matching to minimize the effect of minor misregistration.

To remove the variations in background intensity, the following operations are performed on the gray-scale image $f(x, y)$:

$$f^*(x, y) = \begin{cases} f(x, y) & \text{if} \quad T_1 \le f(x, y) \le T_2 \\ I_{\max} & \text{if} \quad f(x, y) > T_2 \\ I_{\min} & \text{if} \quad f(x, y) < T_1. \end{cases}$$

Picture elements with intensity higher than the threshold $T_2$ are forced to have an intensity of $I_{\max}$. Picture elements with intensity lower than the threshold $T_1$ are forced to have an intensity of $I_{\min}$. In general, $I_{\max}$ is selected to have the intensity of the highest quantization level and $I_{\min}$ to have the intensity of the lowest quantization level. These operations are designed to make the bright and dark regions more homogeneous. In this method the crucial problem is in selecting the optimal thresholds $T_1$ and $T_2$. In many cases it is not possible to do this in advance since different images may

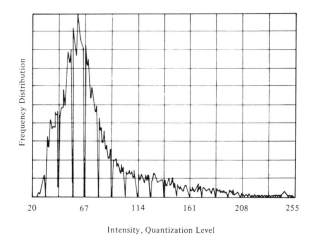

Intensity, Quantization Level

*Figure 8.27*   Frequency distribution of intensity of optical image in Fig. 8.23c.

require different values of thresholds. However, it is often possible to select good thresholds by examining the frequency distribution of gray levels. Figure 8.27 shows the frequency distribution of the gray levels of the optical image. For this image $T_1 = 25$ and $T_2 = 205$ were selected. Comparisons of edge picture extracted from the original and the modified images indicate that a considerable number of weak edges extracted from the background have been removed by this process.

Further reduction of weak edges can then be made by thresholding the gray scale edge image. The threshold for this operation was selected to minimize the noisy edges without significantly reducing the salient edges. All picture elements with amplitude greater than the threshold are set to the value of $I_{max}$ and all picture elements less than the threshold to $I_{min}$. An edge thickening process was then performed by setting the four neighbors of an edge point equal to $I_{max}$. The resulting edge image is shown in Fig. 8.26a.

Following the same procedures, the radar image was converted into an edge picture as shown in Fig. 8.26d. For this image, the thresholds were selected at $T_1 = T_2 = 51$. The frequency distribution of the gray levels of this image is shown in Fig. 8.28.

Two sets of structured edge images of size $128 \times 128$ and $64 \times 64$ were then created using methods described in Section 8.1. These are also shown in Fig. 8.26.

In the matching of radar to optical images, it is the degree of similarity between the two images that is important. This is due to the fact that two images of the same object taken by the two sensors are far from identical even under the most ideal conditions. Therefore it is appropriate to derive a measure of similarity that provides a large separation between the matched point and all other test locations. The statistical correlator previously described is capable of providing the desired performance in terms of giving a

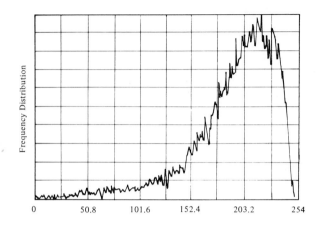

*Figure 8.28*   Frequency distribution of intensity of radar image in Fig. 8.23b.

sharp peak at the matched location. However, in order for the method to work well, prior knowledge of the picture statistics is required. This information is usually not available or extensive computations are required so that the correlator can be designed to tailor to the input data.

A simple but effective similarity correlation method (Hall *et al.*, 1976) has been developed. In this method, the concept of pairing functions is introduced. At each test location $(u,v)$ let the window and matching area in the search region be scanned row by row in such a manner that $S$ corresponds to the radar image region of size $N \times N$,

$$S = (s_1, \ldots, s_{N \times N}),$$

and $W$ corresponds to the optical matching area in the search region of size $N \times N$,

$$W = w_1, \ldots, w_{N \times N}$$

Each of the window pairs $(s_k, w_k)$, $1 \leq k \leq N^2$, is examined. If $s_k$ has a quantization level $i$ and $w_k$ has a quantization level $j$, the pairing function $F_{ij}(u,v)$, $0 \leq i, j \leq 2^n - 1$, is incremented by one. Here $n$ is the number of bits used in the quantization. Therefore $F_{ij}(u,v)$ for $i=j$ is the number of window pairs that are matched in intensity and $F_{ij}(u,v)$ for $i \neq j$ is the number of window pairs that are mismatched in intensity. If $S$ and $W$ were identical, then

$$\sum_{j=0}^{2^n-1} \sum_{j=0}^{2^n-1} F_{ij}(u,v) = \begin{cases} N^2 & \text{for} \quad i = j \\ 0 & \text{for} \quad i \neq j. \end{cases}$$

A similarity correlation measure $R(u,v)$ can be constructed as follows:

$$R(u,v) = \prod_{i=0}^{2^n-1} \left[ F_{ii}(u,v) \Big/ \sum_{j=0}^{2^n-1} F_{ij}(u,v) \right].$$

$R(u,v)$ is the product of the ratios of the number of actual matched window pairs to the number of possible matches of each type. Note that the number of pairing functions increases as the square of the number of quantization levels. Therefore it is desirable to keep the number of levels to a minimum. For binary-valued scene matching, only four pairing functions are required and the above equation becomes

$$R(u,v) = \left( \frac{F_{00}}{F_{00} + F_{01}} \right) \left( \frac{F_{11}}{F_{11} + F_{10}} \right) = \left( \frac{1}{1 + F_{01}/F_{00}} \right) \left( \frac{1}{1 + F_{10}/F_{11}} \right).$$

The simplicity of this equation is obvious. To implement this method, only four counters to accumulate comparisons are needed. When the counting operations are completed, the pairing functions represented by the contents of the counters can be used to compute $R(u,v)$ according to the above equation.

To demonstrate the performance of the similarity correlator, an autocorrelation was performed on an optical edge image with a search region size of $32 \times 32$ and a window size $8 \times 8$. The value of $R(u,v)$ at the test locations adjacent to the matched point is plotted in Fig. 8.29. To form a basis of comparison, autocorrelations using the standard product correlator were also made and results are as shown in the same figure. It is seen that the similarity correlation provides a larger measure of distance separating the matched point from the other test locations. Since the background level is low, a larger scale of correlation values up to unity is available for determining the degrees of similarity between two images.

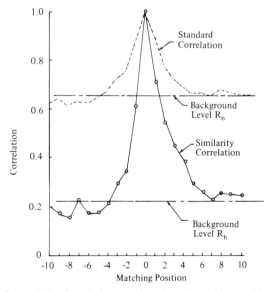

*Figure 8.29*    Correlation as a function of matching position.

A threshold sequence for the edge correlator will now be derived. Let $A$ be the event that there are $F_{ij}$ window pairs of picture elements of intensity $i$ in the window region $W$ paired with picture elements of intensity $j$ in the matching area $S$ of the search region. Let the probability of occurrence of $A$ be

$$P(A) = p_{ij}.$$

Then the probability that $F_{ij}$ takes on the exact value $n_{ij}$ is

$$P(N_{ij} = n_{ij}) = f(n_{ij}) = \binom{N_i}{n_{ij}} (p_{ij})^{n_{ij}} (1 - p_{ij})^{N_i - n_{ij}},$$

where $N_i$ is the number of picture elements in the window region that have intensity $i$. The above equation is formulated by considering that each of the matching areas $S$ is drawn at random from a set of all possible arrays that can be formed with the same dimensions and quantization levels as the window region $W$. This drawing process is the same as that of generating a random array, element by element, from a sequence of uniformly distributed quantization levels. Under these conditions, $n_{ij}$ is binomially distributed with mean and variance equal to

$$\bar{n}_{ij} = N_i p_{ij}, \qquad \sigma_{ij}^2 = N_i p_{ij}(1 - p_{ij}).$$

Correlation between a given window region $W$ and an array with $F_{ij} = n_{ij}$ can be considered as correlation between $S$ and a background scene. Correlation with the background level $R_b$ can be expressed as

$$R_b = \prod_{i=0}^{2^n - 1} \left( \bar{n}_{ii} \Big/ \sum_{j=0}^{2^n - 1} \bar{n}_{ij} \right).$$

Combining these results gives

$$R_b = \prod_{i=0}^{2-1} \left( N_i p_{ii} \Big/ \sum_{j=0}^{2^n - 1} N_i p_{ii} \right).$$

For a binary-valued image,

$$R_b = \left( \frac{1}{1 + p_{01} / p_{00}} \right) \left( \frac{1}{1 + p_{10}/p_{11}} \right).$$

For a given window region, $N_i$ can be obtained by counting the number of picture elements with intensity level $i$. $p_{ij}$ is a function of the intensity distribution of the search region. As an estimate $p_{ij}$ can be computed by the following equation:

$$p_{ij} = M_j \Big/ \sum_{k=0}^{2^n - 1} M_k,$$

where $M_k$ is the number of picture elements in the search region with intensity level $k$.

Let the similarity correlation at the match location $(u^*, v^*)$ be $R(u^*, v^*)$. Then the probability density function of $R(u,v)$ can be considered as Gaussian distributed with a mean of $R(u^*, v^*)$ and a variance of $R_b^2$. For a given threshold $R_T$, let $P_k$ be the probability that $R(u^*, v^*)$ exceeds $R_T$. $P_k$ can be expressed by

$$P_k = \frac{1}{\sqrt{2\pi}\, R_b} \int_{R_T}^{\infty} \exp\left\{ \frac{-[R(u,v) - R(u^*, v^*)]^2}{2R_b^2} \right\} dR(u,v)$$

$$= \frac{1}{\sqrt{2\pi}} \int_{y}^{\infty} \exp\left[ \frac{-R(u,v)}{2} \right] dR,$$

where $y = [R_T - R(u^*, v^*)]/R_b$.

Solving the above equation for $R_T$ yields

$$R_T = yR_b + R(u^*, v^*).$$

The numerical value of $y$ can be computed to achieve a given probability of a match. The threshold can then be related to $R(u^*, v^*)$, $y$, and $R_b$. $R_b$ can

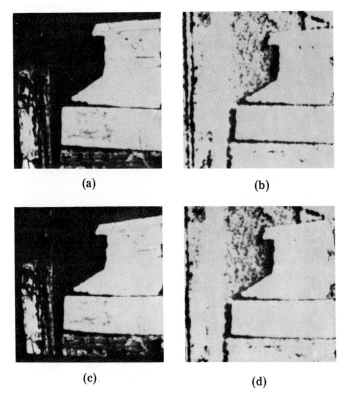

(a)                                        (b)

(c)                                        (d)

*Figure 8.30*   Optical and radar scene 2: (a) optical image, (b) radar image, (c) geometrically corrected optical image, and (d) superimposed (b) and (c). [From Wong and Hall (1978).]

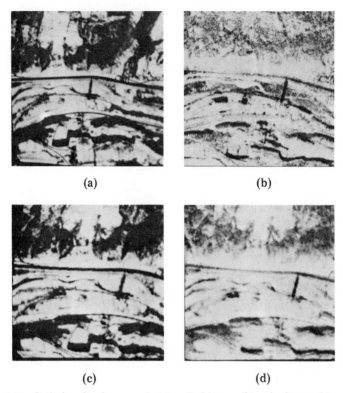

(a)

(b)

(c)

(d)

*Figure 8.31* Optical and radar scene 3: (a) optical image, (b) radar image, (c) geometrically corrected optical image, (d) superimposed (b) and (c). [From Wong and Hall (1978).]

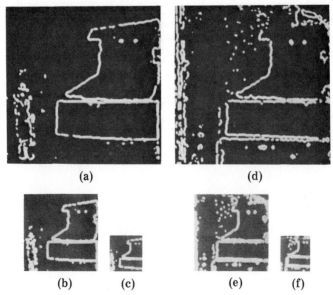

(a)

(d)

(b)

(c)

(e)

(f)

*Figure 8.32* Structured optical and radar images for scene 2: optical, (a) 256×256, (b) 128×128, (c) 64×64; radar, (d) 256×256, (e) 128×128, and (f) 64×64. [From Wong and Hall (1978).]

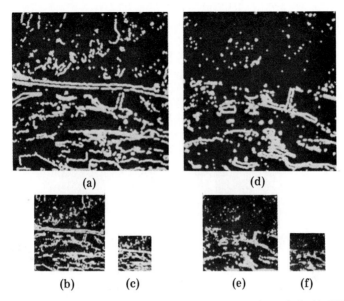

(a)                                          (d)

(b)              (c)                    (e)              (f)

*Figure 8.33* Structured optical and radar images of scene 3: optical, (a) $256 \times 256$, (b) $128 \times 128$, (c) $64 \times 64$; radar, (d) $256 \times 256$, (e) $128 \times 128$, (f) $64 \times 64$. [From Wong and Hall (1978).]

then be computed. In general $R(u^*, v^*)$ can be considered as the highest correlation value obtained in the scene matching. Therefore in hierarchical scene matching, a new threshold can be computed for each search level based on the highest correlation value obtained on that level.

Three sets of optical and radar images were used in an experiment of edge matching. These are as shown in Figs. 8.23, 8.24, 8.30 and 8.31. Edge extraction techniques described were applied to these images. The results are shown in Figs. 8.26, 8.32 and 8.33. Structured edge images of sizes $128 \times 128$ and $64 \times 64$ were created and are also shown in Figs. 8.26, 8.32 and 8.33.

A set of $32 \times 32$ gray level images was also created from the geometrically corrected optical images and the intensity corrected images through the repeated application of filtering. These images will be used at the lowest level of the hierarchical search with invariant moments.

### Scene Matching with Image 1

For a binary-valued image, the background correlation level at search level $k$ can be evaluated using

$$R_b{}^k = \left( \frac{1}{1 + p_{01}^k / p_{00}^k} \right) \left( \frac{1}{1 + p_{10}^k / p_{11}^k} \right) = \left( \frac{p_{00}^k}{p_{01}^k + p_{00}^k} \right) \left( \frac{p_{11}^k}{p_{10}^k + p_{11}^k} \right).$$

*Figure 8.34* Windows selected for scene matching for scene 1. [From Wong and Hall (1978).]

Since $p_{01}^k + p_{00}^k = p_{10}^k + p_{11}^k = 1$, using the above equation gives

$$R_b{}^k = \left(p_{00}^k\right)\left(p_{11}^k\right).$$

$p_{00}^k$ and $p_{11}^k$ can be computed using

$$p_{00}^k = M_0{}^k / \left(M_0{}^k + M_1{}^k\right), \qquad p_{11}^k = M_1{}^k / \left(M_0{}^k + M_1{}^k\right),$$

where $M_0{}^k$ and $M_1{}^k$ are the number of picture elements in the search region with intensities of $I_{max}$ and $I_{min}$, respectively.

Two windows extracted from the radar image were selected for the match. These windows, labeled as locations 1 and 2, are shown in Fig. 8.34. The results of these matches are summarized in Table 8.5. The match locations were correctly detected at each of the two matches. In the matching of location 1, two levels of search were required. However for the matching of location 2 three levels of search were needed.

*Table 8.5*
*Scene Matching with Image No. 1*

| Location | Search level, $k$ | Threshold | No. of successful test locations | Feature used |
|---|---|---|---|---|
| 1 | 3 | 0.984 | 56 | Moment |
|   | 2 | 0.285 | 1 | Edge |
| 2 | 3 | 0.984 | 62 | Moment |
|   | 2 | 0.396 | 5 | Edge |
|   | 1 | 0.382 | 1 | Edge |

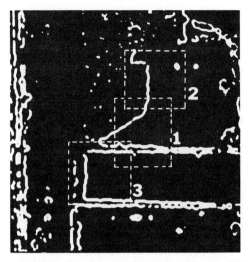

*Figure 8.35*   Window selected for scene 2. [From Wong and Hall (1978).]

### Scene Matching with Image 2

Three windows extracted from the radar image were selected for the match. These windows, labeled locations 1, 2, and 3, are shown in Fig. 8.35. The results of these matches are summarized in Table 8.6. In the match of locations 1 and 3, two levels of search were required. In the matching of location 2, three levels of search were required.

*Table 8.6*
*Scene Matching with Image No. 2*

| Location | Search level, $k$ | Threshold | No. of successful test locations | Feature used |
|----------|-------------------|-----------|----------------------------------|--------------|
| 1        | 3                 | 0.990     | 74                               | Moment       |
|          | 2                 | 0.699     | 1                                | Edge         |
| 2        | 3                 | 0.990     | 94                               | Moment       |
|          | 2                 | 0.635     | 2                                | Edge         |
|          | 1                 | 0.661     | 1                                | Edge         |
| 3        | 3                 | 0.990     | 44                               | Moment       |
|          | 2                 | 0.595     | 1                                | Edge         |

### Scene Matching with Image 3

For the matching of image 3, two subimages were extracted from the optical image and the search was performed on the radar image. The two subimages, labeled as locations 1 and 2, are shown in Fig. 8.36. The results of

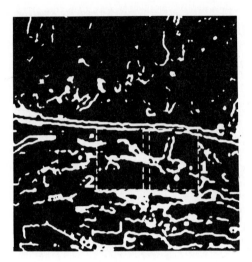

*Figure 8.36*  Window selected for scene matching image number 3. [From Wong and Hall (1978).]

these matches are summarized in Table 8.7. In the matching of location 2, only two levels of search were needed, whereas, in the matching of location 1, three levels of search were required.

*Table 8.7*
*Scene Matching with Image No. 3*

| Location | Search level, $k$ | Threshold | No. of successful test locations | Feature used |
|----------|-------------------|-----------|----------------------------------|--------------|
| 1 | 3 | 0.996 | 77 | Moment |
|   | 2 | 0.370 | 3 | Edge |
|   | 1 | 0.322 | 1 | Edge |
| 2 | 3 | 0.985 | 21 | Moment |
|   | 2 | 0.434 | 1 | Edge |

## 8.4 Performance Evaluation of Scene Matching Methods

Any given matching algorithm will be based upon certain measurements taken from imagery at given resolutions containing a certain class of features, and using a similarity measure. Since any scene matching algorithm is a decision process, the performance may be characterized by means of a *receiver operating characteristic* (ROC) curve that may be derived from statistical evaluations. Suppose, for example, that we are matching a road intersection area, using moment measurements. Four situations can occur at

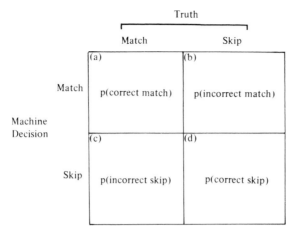

*Figure 8.37*  Matrix of conditional probabilities of possible decisions. (Note that the sum of the columns is unity.)

each comparison location as shown in Fig. 8.37 and listed below:

(a)  The true matching position is recognized.
(b)  A wrong matching position is classified as a match.
(c)  The true matching position is skipped.
(d)  A mismatching position is skipped.

An ideal algorithm would *maximize* the probabilities of (a) and (d) above, and consequently *minimize* the probabilities of (b) and (c). For any algorithms and parameters considered, these probabilities may be estimated and tabulated. Note that the sum of the column terms of Fig. 8.37 must add to unity. Variations of the decision rule, however, will affect the probability of a correct skip versus incorrect skip or, conversely, the probability of a correct match versus incorrect match. This trade-off may be shown in the ROC curve of Fig. 8.38. In Fig. 8.38a, it is easy to see that algorithm A is absolutely superior to algorithm B under the given set of parameters because, given a certain desired probability of match, the probability of false match of algorithm A is smaller than that of algorithm B. A comparison of performance of different measurements for the same scene features and resolution is shown in Fig. 8.38b. Plotting such curves for different parameter combinations allows one to evaluate the performance of any given algorithm under each given set of conditions. The effects of resolution, type of measurements, or type of scene features can be individually assessed by plotting parametric ROC curves. To illustrate the use of the ROC curve for performance evaluation, the previous scene matching approaches will be compared. The assumptions required to evaluate the statistics are illustrative of the methods available for obtaining statistical descriptions.

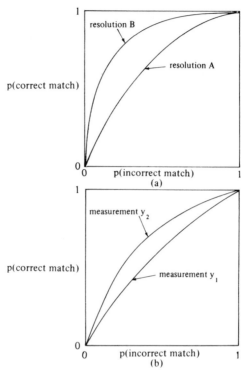

*Figure 8.38*  Receiver operating characteristic curves: (a) Resolution *B* is superior to resolution *A* for feature type *x* and measurement type *y* and (b) measurement type $y_2$ is superior to measurement $y_1$ for feature type *x* and resolution *A*.

Several approaches and techniques for scene matching have now been considered, including the basic template matching method, sequential similarity comparisons, matching with invariant moments, and matching with edge features.

To further improve searching efficiency and increase map matching performance, sequential hierarchical search may be incorporated as an integral part of the matching methods. To evaluate the performance of any of these techniques one may consider the probability distributions, as shown in Figs. 8.39 and 8.40, that could be at the *k*th level in the hierarchical search or the first level for template matching. The distribution $p_k(R)$ is the probability that the true match location takes on a specific similarity value *R*. $R_k(u^*,v^*)$ is the similarity value at the true match location for a particular match under consideration. At the search levels where the image resolution is low, the number of test locations is large, and the $p_k(R)$ will be assumed to have a Gaussian distribution with a variance of $(\sigma_R{}^k)^2$. Let $P_k$ be the probability of detection of the *k*th search level (i.e., the probability that the similarity measure at the true match location exceeds the threshold $R_T{}^k$). Then $P_k$ can

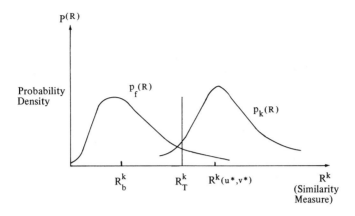

$R^k(u*,v*)$ = Similarity measure at match location

$R_b^k$ = Backround similarity measure

$R_T^k$ = Threshold

*Figure 8.39* Probability density function at match location $p_k(R)$ and background $p_f(R)$. $R^k(u*,v*)$ is the similarity measure at the match location, $R_b{}^k$ is the background similarity measure, and $R_T{}^k$ is the threshold.

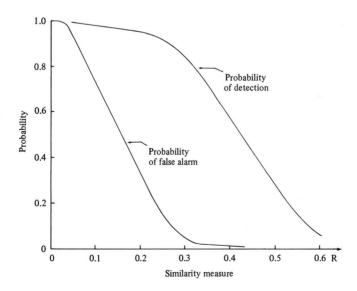

*Figure 8.40* $p_f$ and $p_d$ as functions of $R$ for a typical match.

**519**

be expressed as

$$P_k = \int_{R_T^k}^{\infty} p_k(R) \, dR,$$

$$P_k = \frac{1}{\sqrt{2\pi}\,\sigma_R^{\ k}} \int_{R_T^k}^{\infty} \exp\left\{ -\frac{1}{2} \left[ \frac{R - R_k(u^*, v^*)}{\sigma_R^{\ k}} \right]^2 \right\} dR = \frac{1}{\sqrt{2\pi}} \int_y^{\infty} \exp\left( \frac{-R}{2} \right)^2 dR,$$

where $y = [R_T^k - R_k(u^*, v^*)]/\sigma_R^{\ k}$.

Similarly, the probability of false fix at the $k$th search level can be computed by

$$P_f = \int_{R_T^k}^{\infty} p_f R \, dR = \frac{1}{\sqrt{2\pi}\,\sigma_k^{\ f}} \int_{R_T^k}^{\infty} \exp\left[ -\frac{1}{2} \left( \frac{R - R_b^{\ k}}{\sigma_h^{\ f}} \right)^2 \right] dR$$

$$= 1 - \frac{1}{\sqrt{2\pi}} \int_{-\infty}^{y} \exp\left( \frac{-R}{2} \right)^2 dR = 1 - \phi(y),$$

where $y = (R_T^k - R_b^{\ k})/\sigma_k^{\ f}$, $p_f(R)$ is the probability density distribution of the similarity measures of all test locations except the true match location, $R_b^{\ k}$ the similarity measure averaged over all test locations (similarity measure of the background), and $(\sigma_k^{\ f})^2$ the variance of $p_k(R)$.

For small values of $p_k(R)$, the error introduced by the Gaussian assumption for a non-Gaussian distribution may be large. Therefore either the actual distribution must be used in the computation or the Gaussian-assumed computation must be modified through the use of the Edgeworth expansion (Stuart and Kendall, 1969). Here the mean, variance, and the higher-order moments of the distribution are used in the computation of $P_f$. Let

$$\mu_n = E\left\{ [R - E(R)]^n \right\} = \text{the } n\text{th moment},$$

$$P_f = 1 - \phi(y) + \sum_{n=3}^{l} P_n(y),$$

where $P_n(x) = f(\mu_1, \ldots, \mu_n, x)$ and $l$ is the number of terms in the expansion;

$$P_k = 1 - \frac{1}{\sqrt{2\pi}} \int_{-\infty}^{y} \exp\left[ \frac{-(E_n^{\ k})^2}{2} \right] dE_n^{\ k},$$

where the normalized threshold $y$ is given by

$$y = (T_n^{\ k} - n r_m)/\sqrt{n}\, r_m,$$

and $r_m$ is the amplitude of the average error measurement at the match location of the lowest search level $m$. For these computations $n$ is set equal to the number of picture elements in the window. Table 8.8 shows the numerical value of $P_k$ as a function $T_k^{\ n}$ for the matching of the optical to noisy optical images and radar to optical images at the lowest resolution level. The images are as shown in Figs. 8.16–8.18. Subimages selected for the match are shown in Fig. 8.41.

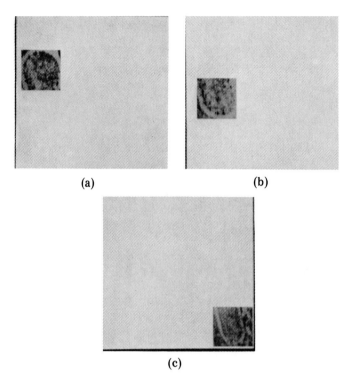

(a)                              (b)

(c)

*Figure 8.41* Subimages selected for map matching: (a) upper portion of large stadium, (b) lower portion of large stadium, and (c) small stadium. [From Wong and Hall (1978).]

*Table 8.8*
*Performance of Optical to Noisy Optical Scene Matching[a]*

|  |  | $T_n^k$ | |
|---|---|---|---|
| $y$ | $P_k$ | Optical to noisy optical match | Radar to optical match |
| 3.0 | 0.500 | 960 | 896 |
| 2.0 | 0.841 | 1080 | 1008 |
| 1.5 | 0.933 | 1140 | 1064 |
| 1.0 | 0.977 | 1200 | 1120 |
| 0.0 | 0.999 | 1320 | 1232 |

[a] $r_m$ is the amplitude of the average error measurement at the match location, and $n$ is the number of picture elements in the window. $r_m = 15$ for noisy optical and $r_m = 14$ for radar match. $n = 64$.

To compute the probability of false fix $P_f$ the actual probability density distribution $p_f$ is estimated from the experimental data. This is accomplished by computing the frequency of occurrence of test locations as a function of the accumulated error. Numerical integration is then performed to obtain $P_f$ as a function of $T_n^k$.

For $n = 5$, $P_f$ (Smith and Rockmore, 1976) becomes

$$P_f = 1 - \phi(x) + \left\{ \left[ \tfrac{1}{6} P_3(x^2 + 1) \right] + \left[ \tfrac{1}{24} P_4 x(x^2 + 3) \right] \right.$$
$$+ \left[ \tfrac{1}{72} P_3^2 x(x^4 - 10x^2 + 15) \right] + \left[ \tfrac{1}{120} P_5(x^4 - 6x + 3) \right]$$
$$\left. + \left[ \tfrac{1}{144} P_3 P_4(x^6 - 15x^4 + 45x - 15) \right] \right\},$$

where

$$P_3 = \frac{\mu_3}{(\mu_2)^{3/2}}, \qquad P_4 = \frac{\mu_4 - 3\mu_2^2}{\mu_2^2}, \qquad P_5 = \frac{\mu_5 - 10\mu_2\mu_3}{(\mu_2)^{5/2}}.$$

Figure 8.40 shows a graph of $P_f$ and $P_d$ as a function of the normalized similarity measure $R$ for a typical match. It is seen from these plots that either of the probabilities, $P_k$ or $P_f$ can be made to take any desired value by an appropriate choice of threshold $R_T^k$, and the other will then be determined. However, the lower $P_f$ is made, the lower $P_k$ will be. A map matching performance curve can be generated by plotting $P_k$ as a function of $P_f$. Two of these curves are shown in Fig. 8.42; note that the curve with a sharper rise produces a higher probability of a match for a given probability of false fix

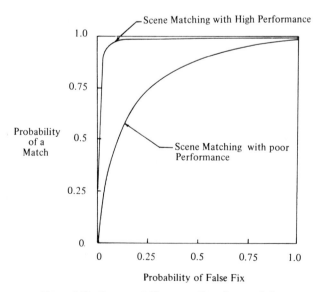

*Figure 8.42*   Scene matching operating characteristics.

and, therefore, represents a map matching method of higher performance. A set of curves can be generated for the matches by each of map matching methods. The relative performance can then be compared in the matching of images that have a variety of scene features.

### 8.4.1   Performance of Basic Sequential Map Matching

The basic sequential map matching with hierarchical search has been described in Section 8.1.3. In this method, the similarity between two images is determined by the accumulated sum of the absolute difference in gray level of the window pairs. The smaller the accumulated error, the more likely the match is achieved. The probability of a match $P_k$ at the search level $k$ as a function of the accumulated error or, equivalently, the threshold level $T_n^{\ k}$ can be computed.

The performance characteristics of the optical to noisy optical map matching of the upper portion of the large stadium, lower portion of the large stadium, and the small stadium are shown in Figs. 8.43, 8.44, and 8.45, respectively. Figure 8.46 shows the performance characteristics of the radar to optical matching of the small stadium. $P_k$ and $P_f$ as a function of $T_n^{\ k}$ for the four matches are tabulated in Table 8.9. The operating characteristics presented as the probability of a match as a function of the probability of false fix are shown in Fig. 8.47. The sharp rise of these curves indicated a relatively high performance by this method in matching the type of image under considerations.

Following the same procedures, the operating characteristics of map matching by the basic sequential method at the higher search level $k=2$ are shown in Figs. 8.48–8.51. $P_k$ and $P_f$ as a function of $T_n^{\ k}$ for these matches are

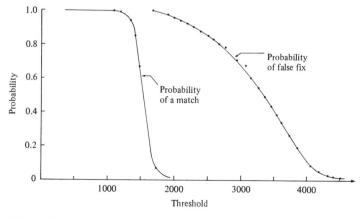

*Figure 8.43*   Optical to noisy optical scene matching upper portion of large stadium; search level $k=3$.

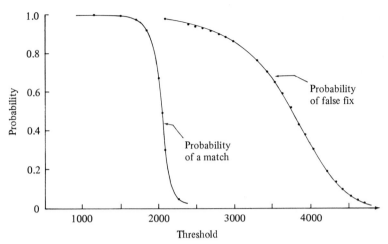

*Figure 8.44* Optical to noisy optical scene matching lower portion of large stadium; search level $k = 3$.

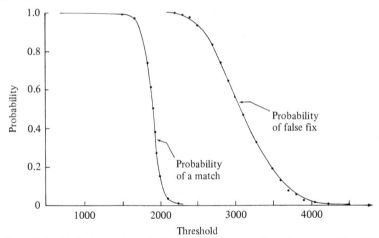

*Figure 8.45* Optical to noisy optical scene matching small stadium; search level $k = 3$.

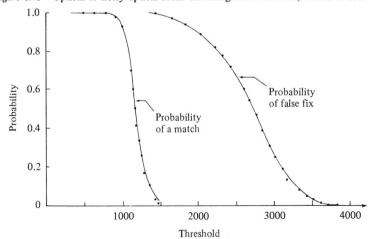

*Figure 8.46* Optical to radar scene matching small stadium; search level $k = 3$.

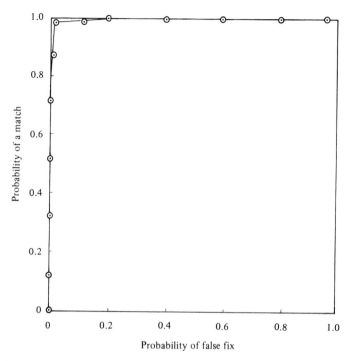

Figure 8.47 Optical to noisy optical scene matching for the upper portion of the large stadium.

Table 8.9

Performance Characteristics of Basic Sequential Map Matching, Search Level $k = 3$

| Threshold | Upper portion[a] of large stadium | | Lower portion[a] of large stadium | | Small stadium[a] | | Small stadium[b] | |
|---|---|---|---|---|---|---|---|---|
| | $P_k$ | $P_f$ | $P_k$ | $P_f$ | $P_k$ | $P_f$ | $P_k$ | $P_f$ |
| 1200 | 0.461 | 0.0002 | 0.461 | 0.0001 | 0.461 | 0.0001 | 0.670 | 0.0008 |
| 1150 | 0.634 | 0.0007 | 0.634 | 0.0005 | 0.634 | 0.0003 | 0.818 | 0.001 |
| 1100 | 0.763 | 0.001 | 0.763 | 0.001 | 0.763 | 0.0006 | 0.917 | 0.003 |
| 1050 | 0.875 | 0.004 | 0.875 | 0.003 | 0.875 | 0.003 | 0.976 | 0.006 |
| 1000 | 0.949 | 0.007 | 0.949 | 0.005 | 0.949 | 0.006 | 0.998 | 0.010 |
| 950 | 0.998 | 0.011 | 0.998 | 0.008 | 0.998 | 0.010 | 0.999 | 0.015 |

[a]Optical to noisy optical match.
[b]Radar to optical match.

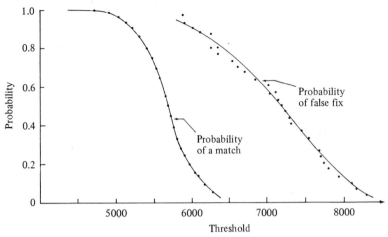

*Figure 8.48* Optical to noisy optical scene matching upper portion of large stadium; search level $k=2$.

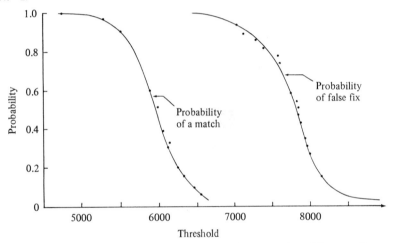

*Figure 8.49* Optical to noisy optical scene matching lower portion of large stadium; search level $k=2$.

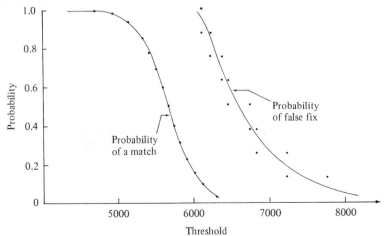

*Figure 8.50* Optical to noisy optical scene matching small stadium; search level $k=2$.

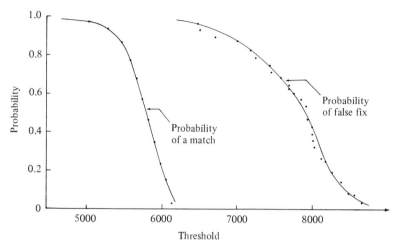

*Figure 8.51*  Optical to radar scene matching small stadium; search level $k=2$.

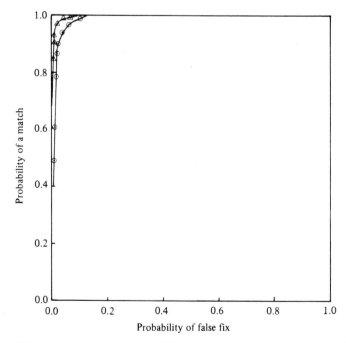

*Figure 8.52*  Optical to noisy optical and optical to radar scene matching; search level $k=2$. ⊙ optical to noisy optical match using lower portion of large stadium, △ optical to radar match using small stadium region.

*Table 8.10*

*Performance Characteristics of Basic Sequential Map Matching, Search Level k = 2*

| Threshold | Upper portion[a] of large stadium | | Lower portion[a] of large stadium | | Small stadium[a] | | Small stadium[b] | |
|---|---|---|---|---|---|---|---|---|
| | $P_k$ | $P_f$ | $P_k$ | $P_f$ | $P_k$ | $P_f$ | $P_k$ | $P_f$ |
| 6300 | 0.490 | 0.002 | 0.490 | 0.003 | 0.490 | 0.001 | 0.850 | 0.001 |
| 6200 | 0.601 | 0.003 | 0.601 | 0.005 | 0.601 | 0.001 | 0.902 | 0.001 |
| 6100 | 0.699 | 0.005 | 0.699 | 0.008 | 0.699 | 0.002 | 0.931 | 0.002 |
| 6000 | 0.792 | 0.008 | 0.792 | 0.012 | 0.792 | 0.004 | 0.952 | 0.005 |
| 5900 | 0.862 | 0.010 | 0.862 | 0.016 | 0.862 | 0.007 | 0.963 | 0.008 |
| 5800 | 0.890 | 0.020 | 0.890 | 0.018 | 0.890 | 0.01 | 0.984 | 0.012 |
| 5700 | 0.942 | 0.039 | 0.942 | 0.02 | 0.942 | 0.02 | 0.995 | 0.023 |
| 5600 | 0.970 | 0.06 | 0.970 | 0.021 | 0.970 | 0.03 | 0.998 | 0.040 |
| 5500 | 0.982 | 0.090 | 0.982 | 0.028 | 0.982 | 0.06 | 0.999 | 0.062 |
| 5400 | 0.996 | 0.120 | 0.996 | 0.03 | 0.996 | 0.09 | 0.999 | 0.088 |

[a] Optical to noisy optical match.
[b] Optical to radar match.

tabulated in Table 8.10. This data indicated that scene matching at this search level produces a slightly higher probability of a false fix for a given probability of a match than that obtained in the previous search level. This is primarily due to the fact that at the high resolution level fewer test locations are needed to be evaluated. The probability of a false fix $P_f$ is a function of the ratio of the number of test locations that passed the threshold test and the number of locations tested. Each false location would contribute to a higher

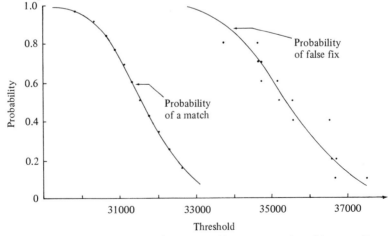

*Figure 8.53* Optical to noisy optical scene matching upper portion of large stadium; search level $k = 1$.

$P_f$ at the higher resolution level. Figure 8.52 shows the operating characteristics at search level $k=2$. The rapid rise of these curves indicated a high performance by this method in matching the type of images under consideration.

A typical performance characteristic curve of map matching at the search level $k=1$ is shown in Fig. 8.53. Here, the probability of a false fix becomes statistically less accurate due to the fact that very few test locations are used at the high resolution level. The numerical value of $P_f$, however, can still be obtained by theoretical computations or through the extrapolation of matching data at the lower resolution levels.

### 8.4.2   Performance of Map Matching with Invariant Moments

The mathematical foundations of invariant moment techniques and the derivation of invariant moment equations have been described in Chapter 7. These equations were applied to the matching of three test windows. The results of the experiments indicated that scene matching is successful in some cases, while in other cases it produces several equally likely candidates at the highest resolution used, thus making the selection of the correct location difficult. However, scene matching with invariant moments can be used to great advantage at the low resolution level at which other methods, such as scene matching with edge features, are not possible. The map matching method with invariant moments was applied to the matching of three sets of images at the lowest resolution level of $k=3$. The analyses presented in this section are based on the results from these matches.

In the map matching with invariant moments, the similarity between two images is determined by the correlation of the seven invariant moments of the radar subimages with those computed at each of the test locations in the optical search scene,

$$R(u,v) = \sum_{i=1}^{7} M_i N_i(u,v) \Bigg/ \left[ \sum_{i=1}^{7} M_i^2 \sum_{i=1}^{7} N_i^2(u,v) \right]^{1/2}.$$

$R(u,v)$ is the moment correlation at the test location $(u,v)$, $M_i$ the $i$th invariant moment of the radar subimage, and $N(u,v)$ the $i$th invariant moment of the optical subimage at test location $(u,v)$. Therefore the larger the correlation $R(u,v)$, the more likely that a match will be achieved.

The probability of a match $P_k$ as a function of the correlation $R$ can be computed using the previous equations. The standard derivation $\sigma_R^k$ is computed based on the variations of the correlation values around $R(u^*,v^*)$, where $R(u^*,v^*)$ is the correlation at the true match location $(u^*,v^*)$. To compute the probability of a false fix $P_f$ the actual probability density distribution derived from the experimental data of each match is used. This is

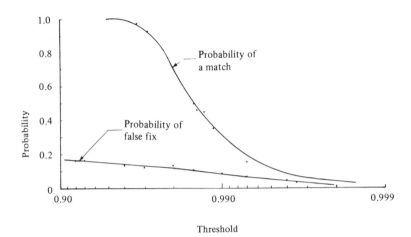

*Figure 8.54*  Scene matching with invariant moments, radar to optical matching image No. 1, location 1.

done by computing the frequency of occurrence of the test locations as a function of the correlation $R$.

Figures 8.54 and 8.55 show the performance characteristics of the radar to optical map matching at locations 1 and 2 of image 1 shown in Fig. 8.56. In these curves, the correlation threshold is plotted on a log scale. The log scale is used to compensate for the fact that the correlation was computed on the logarithm of amplitudes of the seven moments rather than on the amplitudes themselves. $P_k$ and $P_f$ as a function of the threshold for the two matches are tabulated in Table 8.11. The operating characteristics are shown in Fig. 8.57.

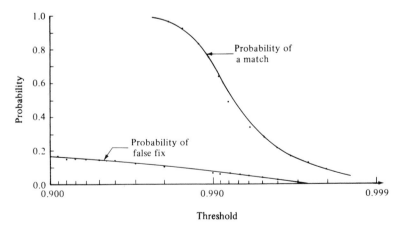

*Figure 8.55*  Scene matching with invariant moments, radar to optical matching image No. 1, location 2.

*Figure 8.56* Windows selected for scene matching image No. 1. [From Wong and Hall (1978).]

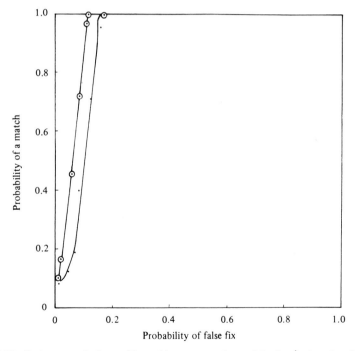

*Figure 8.57* Radar to optical matching with moments, image No. 1: ⋯, location 1; —⊙—, location 2.

*Table 8.11*

*Radar to Optical Scene Matching with Invariant Moments for Image 1 [a]*

| Threshold | Location 1 | | Location 2 | |
|-----------|---------|---------|---------|---------|
|           | $P_k$   | $P_f$   | $P_k$   | $P_f$   |
| 0.900 | 1.000 | 0.165 | 1.000 | 0.172 |
| 0.950 | 1.000 | 0.156 | 1.000 | 0.121 |
| 0.980 | 0.705 | 0.112 | 0.978 | 0.108 |
| 0.990 | 0.400 | 0.082 | 0.720 | 0.072 |
| 0.993 | 0.195 | 0.064 | 0.458 | 0.061 |
| 0.995 | 0.120 | 0.052 | 0.280 | 0.002 |
| 0.998 | 0.080 | 0.001 | 0.100 | 0.001 |
| 0.999 | 0.020 | 0.000 | 0.020 | 0.000 |

[a] $P_k$ is the probability of detection; $P_f$ is the probability of false fix.

Figures 8.58–8.60 show the performance characteristics of the radar to optical map matching at locations 1, 2, and 3 of image 2 shown in Fig. 8.61. $P_k$ and $P_f$ as a function of the threshold for the three matches are tabulated in Table 8.12. The operating characteristics are shown in Fig. 8.62.

Similarly, Figs. 8.63 and 8.64 show the performance characteristics of the radar to optical map matching at locations 1 and 2 of image 3 shown in Fig. 8.65. $P_k$ and $P_f$ as a function of the threshold for the two matches are tabulated in Table 8.13. The operating characteristics are shown in Fig. 8.66.

It is seen from the operating characteristics in Figs. 8.63, 8.64, and 8.66 that map matching at different locations within the same image produces a set of curves that are consistent in character. The matching of images 2 and 3 produces similar characteristics. For a given probability of false fix, the matching of image 1 produced a higher probability of a match than those that resulted from the matching of images 2 and 3. This is true in the operating range where a probability of a match is significant (i.e., $P_k \geq 0.6$).

### 8.4.3 Performance Comparison of Scene Matching with the Two Types of Intensity-Transformed Radar Images

In this section, a comparison of the radar images produced by the two methods of intensity transformation is made. The two transformation methods are the Karhunen–Loeve transform and the method with sensor-dependent parameters. Radar to optical map matching is performed by matching regions extracted from the two types of radar images to the optical images. Performance of map matching with invariant moments, described previously, resulted in the generations of sets of map matching performance characteristic curves. Radar regions used in these matches were extracted from the Karhunen–Loeve transformed images shown in Fig. 8.67d. Figures 8.68 and 8.69 show the characteristic curves of the matching of locations 1

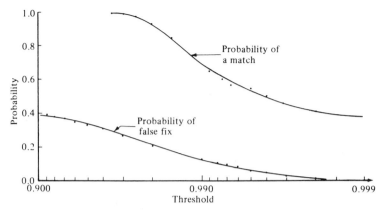

**Figure 8.58** Scene matching with invariant moments, radar to optical matching image No. 2, location 1.

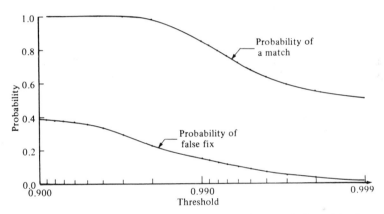

**Figure 8.59** Scene matching with invariant moments, radar to optical matching image No. 2, location 2.

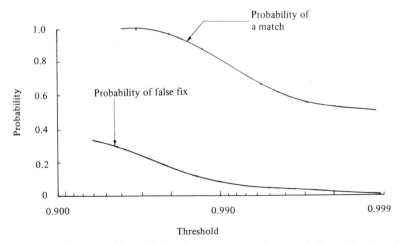

**Figure 8.60** Scene matching with invariant moments, radar to optical matching image No. 2, location 3.

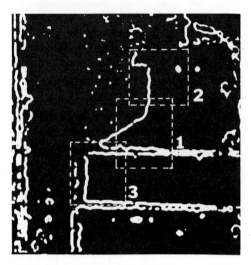

*Figure 8.61*  Window selected for scene matching image No. 2. [From Wong and Hall (1978).]

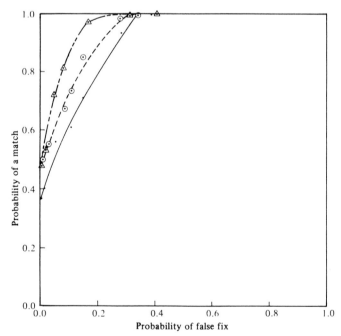

*Figure 8.62*  Radar to optical matching with moment image No. 2: —, location 1; --⊙--, location 2; —△—△—, location 3.

Table 8.12

### Table 8.12

#### Radar to Optical Scene Matching with Invariant Moments for Image 2 [a]

| | Location 1 | | Location 2 | | Location 3 | |
|---|---|---|---|---|---|---|
| Threshold | $P_k$ | $P_f$ | $P_k$ | $P_f$ | $P_k$ | $P_f$ |
| 0.900 | 1.000 | 0.390 | 1.000 | 0.385 | 1.000 | 0.410 |
| 0.950 | 1.000 | 0.338 | 1.000 | 0.350 | 1.000 | 0.320 |
| 0.980 | 0.933 | 0.208 | 0.978 | 0.228 | 0.968 | 0.170 |
| 0.990 | 0.702 | 0.123 | 0.845 | 0.150 | 0.808 | 0.070 |
| 0.993 | 0.610 | 0.092 | 0.733 | 0.118 | 0.720 | 0.050 |
| 0.995 | 0.560 | 0.058 | 0.675 | 0.090 | 0.650 | 0.040 |
| 0.998 | 0.400 | 0.018 | 0.553 | 0.030 | 0.530 | 0.020 |
| 0.999 | 0.370 | 0.002 | 0.500 | 0.008 | 0.520 | 0.008 |

[a] $P_k$ is the probability of detection; $P_f$ is the probability of a false fix.

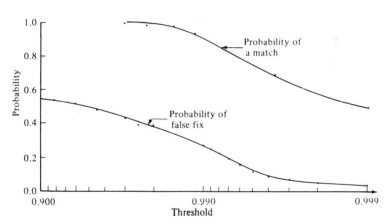

*Figure 8.63*  Radar to optical scene matching with invariant moments, image No. 3, location 1.

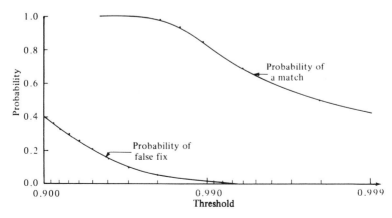

*Figure 8.64*  Radar to optical scene matching with invariant moments, image No. 3, location 2.

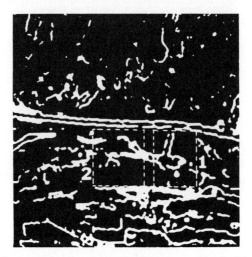

*Figure 8.65* Windows selected for scene matching image No. 3. [From Wong and Hall (1978).]

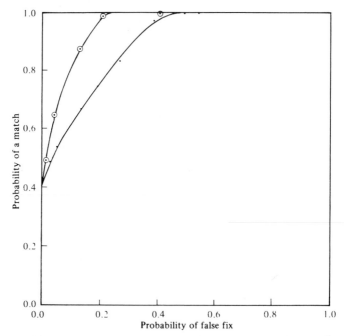

*Figure 8.66* Radar to optical scene matching with moments, image No. 3: —, location 1; —⊙—, location 2.

Table 8.13

Radar to Optical Scene Matching with Invariant Moments for Image 3 [a]

| Threshold | Location 1 | | Location 2 | |
|---|---|---|---|---|
| | $P_k$ | $P_f$ | $P_k$ | $P_f$ |
| 0.900 | 1.000 | 0.541 | 1.000 | 0.401 |
| 0.950 | 1.000 | 0.499 | 1.000 | 0.215 |
| 0.980 | 0.995 | 0.385 | 0.979 | 0.059 |
| 0.990 | 0.902 | 0.275 | 0.830 | 0.021 |
| 0.993 | 0.856 | 0.195 | 0.718 | 0.012 |
| 0.995 | 0.740 | 0.124 | 0.645 | 0.005 |
| 0.998 | 0.580 | 0.051 | 0.502 | 0.002 |
| 0.999 | 0.502 | 0.032 | 0.420 | 0.001 |

[a] $P_k$ is the probability of detection; $P_f$ is the probability of a false fix.

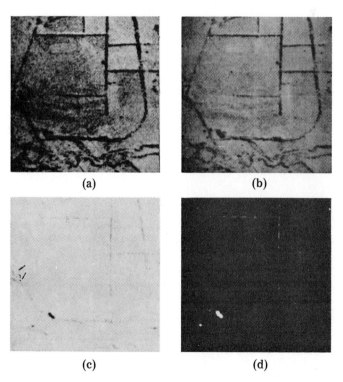

(a)　　　　　　(b)

(c)　　　　　　(d)

Figure 8.67  Intensity transformation of image No. 1: (a) original radar, (b) intensity reversed, (c) intensity transformed with sensor-dependent parameters, (d) intensity transformed with Karhunen–Loeve transform. [From Wong and Hall (1978).]

537

and 2 of Fig. 8.70. To form a basis for comparison, the regions of locations 1 and 2 are also extracted from the radar image in Fig. 8.67c. The results of these matches are presented as characteristic curves in Figs. 8.71 and 8.72. Probability of a match $P_k$ and probability of a false fix $P_f$ as a function of the threshold for these matches are shown in Table 8.14. The operating characteristics, presented as $P_d$ as a function of $P_f$, are shown in Fig. 8.73. Comparison of map matching performance for the two types of image is shown in Table 8.15. It is seen that the two types of image provided similar performance in terms of the probability of a match and probability of a false fix. Map matching by the Karhunen–Loeve transformed images provided a higher probability of a match at a lower probability of a false fix. This transformation, however, requires the knowledge of the statistics of both the optical and radar image. When these statistics are available, intensity correction with the

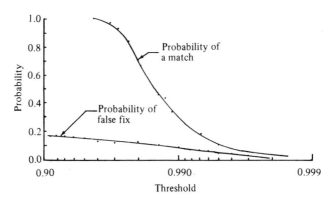

*Figure 8.68*  Scene matching with invariant moments, Karhunen–Loeve intensity-transformed radar image to optical matching image No. 1, location 1.

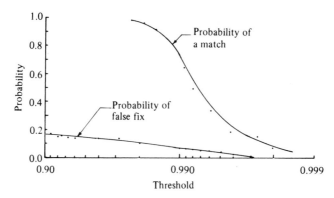

*Figure 8.69*  Scene matching with invariant moments, Karhunen–Loeve intensity-transformed radar image to optical matching image No. 1, location 2.

*Figure 8.70* Windows selected for scene matching image No. 1. [From Wong and Hall (1978).]

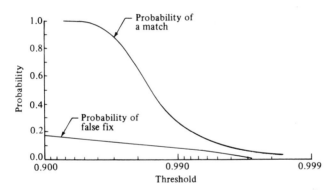

*Figure 8.71* Scene matching with invariant moments, radar image transformed with sensor-dependent parameters to optical matching image No. 1, location 1.

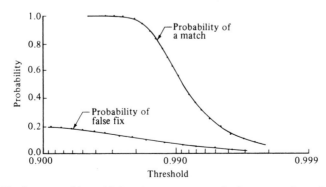

*Figure 8.72* Scene matching with invariant moments, radar image transformed with sensor-dependent parameters to optical matching image No. 1, location 2.

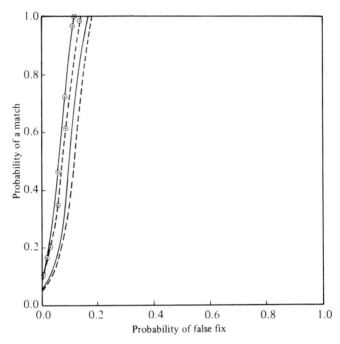

*Figure 8.73* Radar to optical map matching with invariant moments, image No. 1: Karhunen–Loeve transformed image, —, location 1; --⊙--, location 2; image transformed with sensor-dependent parameters, ---, location 1; --⊙--, location 2.

*Table 8.14*

*Radar to Optical Scene Matching with Invariant Moments*

| Threshold | Location 1[a] | | Location 1[b] | | Location 2[a] | | Location 2[b] | |
|---|---|---|---|---|---|---|---|---|
| | $P_k$ | $P_f$ | $P_k$ | $P_f$ | $P_k$ | $P_f$ | $P_k$ | $P_f$ |
| 0.900 | 1.000 | 0.165 | 1.000 | 0.173 | 1.000 | 0.172 | 1.000 | 0.193 |
| 0.950 | 1.000 | 0.156 | 0.990 | 0.168 | 1.000 | 0.121 | 0.995 | 0.136 |
| 0.980 | 0.705 | 0.112 | 0.656 | 0.121 | 0.978 | 0.108 | 0.970 | 0.122 |
| 0.990 | 0.400 | 0.082 | 0.282 | 0.085 | 0.720 | 0.072 | 0.612 | 0.086 |
| 0.993 | 0.195 | 0.064 | 0.172 | 0.066 | 0.458 | 0.061 | 0.350 | 0.065 |
| 0.995 | 0.120 | 0.052 | 0.115 | 0.053 | 0.280 | 0.002 | 0.212 | 0.002 |
| 0.998 | 0.080 | 0.001 | 0.072 | 0.001 | 0.100 | 0.001 | 0.092 | 0.001 |
| 0.999 | 0.020 | 0.000 | 0.018 | 0.000 | 0.020 | 0.000 | 0.018 | 0.000 |

[a] Radar image intensity-transformed with Karhunen–Loeve transform.
[b] Radar image intensity-transformed with sensor-dependent parameters.

*Table 8.15*

*Comparison of Map Matching Results of the Two Intensity-Transform Methods*

| | Location 1 | | Location 2 | |
|---|---|---|---|---|
| Threshold | $\Delta P_k$ | $\Delta P_f$ | $\Delta P_k$ | $\Delta P_f$ |
| 0.900 | 0.000 | 0.049 | 0.000 | 0.122 |
| 0.950 | 0.010 | 0.077 | 0.005 | 0.124 |
| 0.980 | 0.069 | 0.080 | 0.008 | 0.129 |
| 0.990 | 0.295 | 0.036 | 0.150 | 0.194 |
| 0.993 | 0.118 | 0.019 | 0.236 | 0.065 |
| 0.995 | 0.042 | 0.000 | 0.243 | 0.000 |
| 0.998 | 0.100 | 0.000 | 0.210 | 0.000 |
| 0.999 | 0.100 | 0.000 | 0.100 | 0.000 |
| Averaged | $0.092^a$ | $0.033^b$ | $0.119^a$ | $0.079^b$ |

[a] Average increase in probability of a match by the Karhunen–Loeve transform method over that of the method with sensor-dependent parameters.
[b] Average decrease in probability of false fix of the Karhunen–Loeve transform method over that of the method with sensor-dependent parameters.

Karhunen–Loeve transform can be used to great advantage. Intensity transformation using the method with sensor-dependent parameters provided suboptical intensity corrections. Radar images transformed by this method are shown to have sufficiently high quality to have resulted in map matching with a relatively high probability of a match.

### 8.4.4   Relative Performance of the Different Map Matching Methods

Although experimental work was performed on the images, there were too few images to be able to reach any specific final conclusions. Many remarks and tentative conclusions can be made based on the results obtained in the previous sections. Results from Section 8.8 indicated that the basic sequential matching appeared to be an excellent method in the matching of the images used in the experiments. Scene contents in these images consist of relatively well defined man-made objects located in areas of varying background. This method is useful in matching images taken by the same type of sensors under different operating conditions. These sensors may include x ray, radar, and radiometers. The true significance of this method, however, is that scene matching can be accomplished even in cases that are difficult for humans or standard correlation techniques and can be accomplished with greatly reduced computation. This was explicitly demonstrated in the matching of the noisy small stadium.

A comparison of the operating characteristics of map matching with invariant moments and those of the basic sequential method indicated that map matching with the basic sequential method generally produces higher performance than map matching with invariant moments for scenes of many types.

Of the three methods analyzed, map matching with edge features appeared to be the best candidate for the matching of radar to optical images.

When a radar and an optical picture are examined in detail, it is often found that the most distinguishable common features of the two types of images are the outlines of the various natural as well as man-made objects. Therefore, edge features can be used to great advantage in radar to optical scene matching. However, features extracted by the existing edge operators contain many spurious background edges. This difficulty has been largely overcome through the development of the edge processing technique. As indicated by the results, excellent performance was obtained in the matching of three image sets with scenes that have a variety of content. Since identical scenes from the same image sets were used in the matching with invariant moments and with edge features, a direct comparison of the matching performance by the two methods is possible. The results indicated a better performance by the matches with edge features in all scenes.

The overall probability of a match $P_D$ is the product of the individual $P_k$ computed at each of the search levels:

$$P_D = \prod_{k=m}^{n} P_k,$$

where $m$ is the lowest resolution level and $n \geq 0$ the level at which a match decision is declared.

The overall probability of false fix $P_F$ is the sum of the individual $P_f$ computed at each search level:

$$P_F = \sum_{f=m}^{n} P_f.$$

The overall performance of map matching by the three methods is tabulated as

| Method | Performance |
|---|---|
| Basic sequential | Table 8.16 |
| Invariant moment | Table 8.17 |
| Edge features | Table 8.18 |

In summary, data from these tables indicate map matching with the three methods produces overall averaged $P_D$ and $P_F$ as

| Method | $P_D$ | $P_F$ |
|---|---|---|
| Basic sequential | 0.915 | 0.079 |
| Invariant moment | 0.807 | 0.095 |
| Edge features | 0.954 | 0.059 |

Table 8.16

*Overall Performance of Map Matching by the Basic Sequential Method*

| Location | Search level $k=3$ | | Search level $k=2$ | | Search level $k=1$ | | Search level $k=0$ | | | |
|---|---|---|---|---|---|---|---|---|---|---|
| | $P_k$ | $P_f$ | $P_k$ | $P_f$ | $P_k$ | $P_f$ | $P_k$ | $P_f$ | $P_D$ | $P_F$ |
| Upper portion[a] of large stadium | 0.998 | 0.018 | 0.977 | 0.052 | 0.977 | 0.018 | 0.977 | 0.007 | 0.931 | 0.095 |
| Lower portion[b] of large stadium | 0.998 | 0.010 | 0.977 | 0.028 | 0.977 | 0.013 | — | — | 0.953 | 0.051 |
| Small stadium[a] | 0.998 | 0.014 | 0.977 | 0.043 | 0.977 | 0.015 | — | — | 0.953 | 0.072 |
| Small stadium[b] | 0.999 | 0.010 | 0.977 | 0.019 | 0.977 | 0.013 | 0.841 | 0.006 | 0.818 | 0.096 |
| Averaged | | | | | | | | | 0.914 | 0.079 |

[a] Optical to noisy optical match.
[b] Radar to optical match.

Table 8.17

*Performance of Map Matching with Invariant Moments*

| Image No. | Location | $P_k$ | $P_f$ |
|---|---|---|---|
| 1 | 1 | 0.810 | 0.123 |
| | 2 | 0.920 | 0.081 |
| 2 | 1 | 0.702 | 0.123 |
| | 2 | 0.720 | 0.098 |
| | 3 | 0.808 | 0.070 |
| 3 | 1 | 0.740 | 0.124 |
| | 2 | 0.952 | 0.043 |
| Averaged | | 0.807 | 0.095 |

Table 8.18

*Performance of Map Matching with Edge Features*

| Image No. | Location | Search level $k=2$ | | Search level $k=1$ | | | |
|---|---|---|---|---|---|---|---|
| | | $P_k$ | $P_f$ | $P_k$ | $P_f$ | $P_D$ | $P_F$ |
| 1 | 1 | 0.977 | 0.067 | | | 0.977 | 0.067 |
| | 2 | 0.977 | 0.039 | 0.977 | 0.043 | 0.955 | 0.082 |
| 2 | 1 | 0.998 | 0.020 | 0.977 | 0.038 | 0.975 | 0.058 |
| | 2 | 0.998 | 0.031 | 0.977 | 0.054 | 0.975 | 0.085 |
| | 3 | 0.988 | 0.030 | | | 0.988 | 0.030 |
| 3 | 1 | 0.933 | 0.029 | 0.933 | 0.042 | 0.870 | 0.071 |
| | 2 | 0.933 | 0.022 | | | 0.933 | 0.022 |
| Averaged | | | | | | 0.954 | 0.0593 |

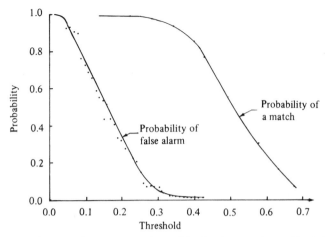

*Figure 8.74* Radar to optical scene matching with edge features, image No. 1, location 1.

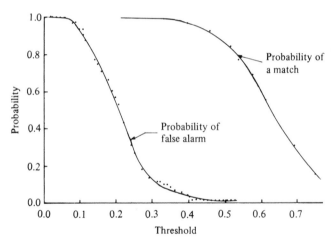

*Figure 8.75* Radar to optical scene matching with edge features, image No. 1, location 2: $\dot{-}$, location 1; $-\odot-$, location 2.

*Table 8.19*

*Radar to Optical Scene Matching with Edge Features for Image 1*

| Threshold | Location 1 | | | Location 2 | |
|---|---|---|---|---|---|
| | $P_k$ | $P_f$ | | $P_k$ | $P_f$ |
| 0.10 | 1.000 | 0.74 | | 1.000 | 0.920 |
| 0.15 | 1.000 | 0.54 | | 1.000 | 0.740 |
| 0.20 | 0.992 | 0.34 | | 1.000 | 0.540 |
| 0.25 | 0.990 | 0.15 | | 1.000 | 0.250 |
| 0.30 | 0.969 | 0.05 | | 0.998 | 0.110 |
| 0.40 | 0.842 | 0.015 | | 0.962 | 0.037 |
| 0.50 | 0.490 | 0.005 | | 0.860 | 0.010 |
| 0.60 | 0.22 | 0.001 | | 0.602 | 0.008 |

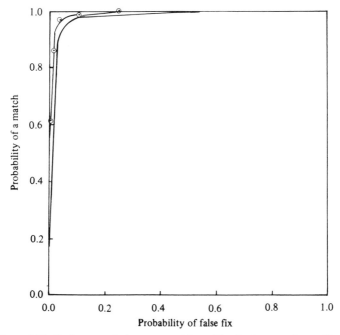

*Figure 8.76*   Radar to optical scene matching with edge features, image No. 1.

Figures 8.74 and 8.75 show the performance characteristics of the radar to optical map matching at locations 1 and 2 of image 1 as shown in Fig. 8.56. $P_k$ and $P_f$ as a function of the threshold or similarity correlation for the two matches are tabulated in Table 8.19. The operating characteristics are shown in Fig. 8.76.

Figures 8.77–8.79 show the performance characteristics of the radar to optical matches at locations 1, 2, and 3 of image 2 as shown in Fig. 8.61. $P_k$ and $P_f$ as a function of the threshold for the three matches are tabulated in Table 8.20. The operating characteristics are shown in Fig. 8.80.

Similarly, Figs. 8.81 and 8.82 show the performance characteristics of the matches at locations 1 and 2 of image 3 as shown in Fig. 8.65. $P_k$ and $P_f$ as a function of the threshold are tabulated in Table 8.21. The operating characteristics are shown in Fig. 8.83.

### 8.4.5   Performance of Map Matching with Edge Features

Radar to optical map matching with edge features has been applied to the three sets of images shown in Figs. 8.27, 8.31, and 8.33. The matchings started at search level $k=2$. This is due to the fact that as the resolution of the edge image is reduced to level $k=1$ the effect of the blurring on the image simplifies the image to such an extent that a matching decision cannot be made with sufficiently high confidence. The analyses presented in the section are based on results from the matches at the search level $k=2$.

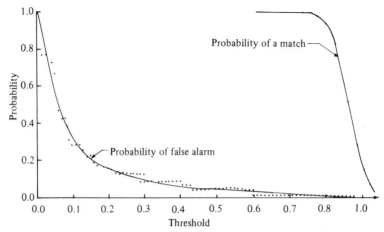

*Figure 8.77* Radar to optical scene matching with edge features, image No. 2, location 1.

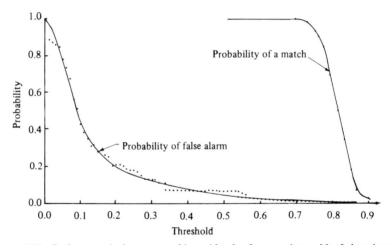

*Figure 8.78* Radar to optical scene matching with edge features, image No. 2, location 2.

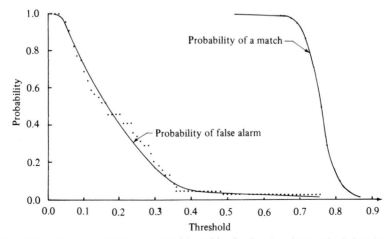

*Figure 8.79* Radar to optical scene matching with edge features, image No. 2, location 3.

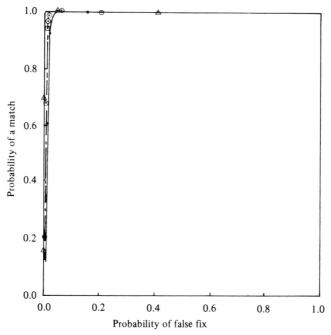

*Figure 8.80* Radar to optical scene matching with edge features, image No. 2: —, location 1; —⊙—, location 2; —△— , location 3.

*Table 8.20*

*Radar to Optical Scene Matching with Edge Features for Image 2[a]*

| Threshold | Location 1 | | Location 2 | | Location 3 | |
|---|---|---|---|---|---|---|
| | $P_k$ | $P_f$ | $P_k$ | $P_f$ | $P_k$ | $P_f$ |
| 0.20 | 1.000 | 0.160 | 1.000 | 0.208 | 1.000 | 0.410 |
| 0.40 | 1.000 | 0.070 | 1.000 | 0.075 | 1.000 | 0.050 |
| 0.60 | 1.000 | 0.030 | 0.998 | 0.018 | 1.000 | 0.030 |
| 0.70 | 0.992 | 0.021 | 0.977 | 0.015 | 0.940 | 0.020 |
| 0.75 | 0.981 | 0.007 | 0.940 | 0.010 | 0.640 | 0.015 |
| 0.80 | 0.933 | 0.001 | 0.680 | 0.008 | 0.170 | 0.005 |
| 0.85 | 0.610 | 0.005 | 0.200 | 0.006 | 0.040 | 0.001 |

[a] $P_d$ is the probability of detection; $P_f$ is the probability of a false fix.

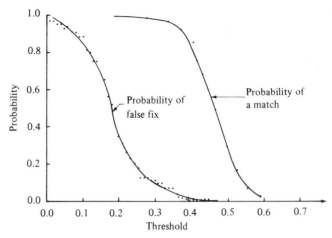

*Figure 8.81* Radar to optical scene matching with edge features, image No. 3, location 1.

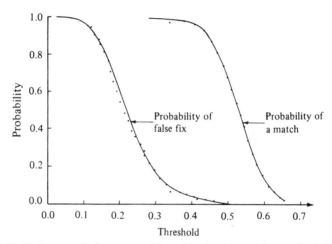

*Figure 8.82* Radar to optical scene matching with edge features, image No. 3, location 2.

*Table 8.21*

*Radar to Optical Scene Matching with Edge Features for Image 3[a]*

| Threshold | Location 1 | | Location 2 | |
|---|---|---|---|---|
| | $P_k$ | $P_f$ | $P_k$ | $P_f$ |
| 0.10 | 1.000 | 0.825 | 1.000 | 0.972 |
| 0.15 | 1.000 | 0.689 | 1.000 | 0.836 |
| 0.20 | 1.000 | 0.361 | 1.000 | 0.603 |
| 0.25 | 1.000 | 0.156 | 1.000 | 0.348 |
| 0.30 | 0.982 | 0.083 | 1.000 | 0.180 |
| 0.40 | 0.842 | 0.018 | 0.980 | 0.042 |
| 0.45 | 0.663 | 0.009 | 0.905 | 0.021 |
| 0.50 | 0.267 | 0.003 | 0.850 | 0.007 |
| 0.60 | 0.018 | 0.001 | 0.123 | 0.001 |

[a] $P_d$ is the probability of detection; $P_f$ is the probability of a false fix.

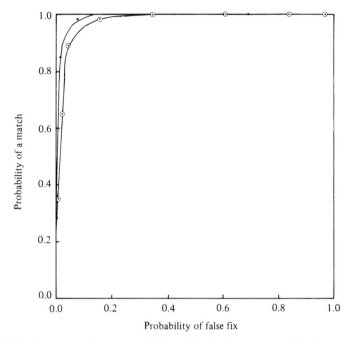

*Figure 8.83*   Radar to optical scene matching with edge features, image No. 3: —, location 1; —⊙— , location 2.

In the map matching with edge features, the similarity between two images is determined by a similarity correlation

$$R(u,v) = \left[ F_{00}/(F_{00} + F_{01}) \right]\left[ F_{11}/(F_{11} + F_{01}) \right].$$

$F_{ij}$ is the number of window pairs with intensity $i$ in the radar image and intensity $j$ in the optical image. Therefore, the larger the correlation, the more likely that the match will be achieved. To compute the probability of a false fix $P_f$ the actual probability density distribution derived from the experimental data of each match is used. This is done by computing the frequency of occurrence of the test locations as a function of the similarity correlation $R(u,v)$.

## 8.5   Conclusions and Discussion

The problem of matching two images of some scene taken by different sensors under different operating conditions is a challenging problem in image processing and pattern recognition. Geometric and intensity corrections are often necessary in preparing the images for scene matching. Of the two methods of intensity correction, transformation using the Karhunen–Loeve transform provided the best intensity match. This transformation, however, required the knowledge of the statistics of both the optical and radar images. When the statistics of both images are available, intensity correction with the Karhunen–Loeve transform can be used to great advantage. Intensity transformation, using methods with sensor-dependent parameters, may also be used.

Experimental work was performed on too few images to be able to reach any final conclusions; however, tentative conclusions can be made based on the results of the analyses and experiments. The techniques of scene matching with hierarchical search have proved to be effective in matching scenes of many types. The development of these techniques has provided an important method toward solving the difficult problem of scene matching in an efficient manner. Scene matching with hierarchical search has resulted in an averaged computational saving of three orders of magnitude over that of the standard correlation for the matching of $256 \times 256$ images. Of the three matching methods using hierarchical search, map matching with edge features appeared to be the best candidate for radar to optical matching. As indicated by the results from Sec. 5.3 excellent performances were obtained in the matching of all three image sets for regions that have a variety of contents. Scene matching with the edge features has provided a probability of a match of 0.954 and a probability of a false fix of 0.059 when averaged over the match regions.

Scene matching with the basic sequential method provided good performance in the matching of scenes that contain relatively well-defined manmade objects of varying background. This method is particularly useful in matching images taken by the same type of sensors under different operating conditions. These may include

radiometric images taken at different view angles and altitude,
medical x rays taken at different viewing geometries,
radar images taken at different viewing geometries, and
optical images taken at different time and viewing geometries through the earth's time-varying atmosphere and cloud formations.

However, the true significance of this method is that scene matching can be accomplished even in cases that are difficult for humans and for standard correlation techniques and can be accomplished with greatly reduced computations. This was demonstrated in the matching of the noisy small stadium. Scene matching with the basic sequence method has resulted in an averaged probability of a match of 0.915 and an averaged probability of false fix of 0.079.

Depending on the scene content, scene matching with invariant moments was successful in some cases. In others, it produces several equally likely candidates, thus making the selection of the correct location more difficult.

However, scene matching with invariant moments can be used to great advantage at the low resolution level at which other methods, such as scene matching with edge features, are not possible. Scene matching with invariant moments has resulted in an averaged probability of a match of 0.81 and an averaged probability of false fix of about 0.1. This suggests a possible need for improvement in this method. Two possible improvements by which a better and more efficient match may be accomplished are the following:

(1) Weight each of the seven moments with an appropriate weighting factor before correlation. The numerical values of these weighting factors may be related to the information content or the frequency distribution of the grey levels of the images.

(2) Generate higher-order moments. Select a set of moments for correlation computation with the selection based on the information contents of the images.

Scene matching using structured features has resulted in superior performance in terms of providing a sharper correlation peak at the location of true registration; at the same time, this was accomplished at greatly reduced computation and memory storage requirements as compared to those required by the standard correlator. Several algorithms based on this matching technique are developed. Experimental results indicated that this method can successfully be applied to the matching of images in the presence of a high noise level.

Although the experiments were performed primarily on optical and radar images, the methods described in this report should be equally applicable to images taken by other types of sensors, such as radiometers, x ray, or infrared imaging systems.

## Bibliographical Guide

The mathematical model for scene matching is pattern recognition, therefore, the books on pattern recognition listed in the bibliography of Chapter 1 are excellent background. The general techniques of sequential recognition are described by Fu (1968). A recent description of the basic correlation as well as modern methods is given by Wong (1976). Filtering to whiten the spectra is described by Pratt (1974).

Sequential consideration of measurements was used successfully by Barnea and Silverman (1972). Another form of sequential template matching is described by Ramapriyan (1976).

The coarse–fine sequential technique was introduced by Wong *et al.* (1977) as a modern approach to scene matching that uses both the variable resolution sequence and sequential consideration of features to provide a logarithmetically efficient computation. Other search methods are described in Nagel and Rosenfeld (1972), Tanimoto and Paulidis (1975), and VanderBrug and Rosenfeld (1977). The similarity measures used in sequential feature comparisons may be developed in several ways. Several methods are described by Rosenfeld and Pfaltz (1968) and Lissack and Fu (1976). One method of determining a threshold sequence is based upon an estimation of the probability density of the error. Techniques for density estimation are described by Balakrishnan (1968), Raemer (1969), Barnea and Silverman (1972), Hall *et al.* (1976), and Wong (1976, 1977).

Scene matching may be accomplished with a variety of features. Scene matching using moment invariants is described by Sadjadi and Hall (1976) and Wong and Hall (1978). Scene matching using edge features is described by Wong (1976) and Hall *et al.* (1976). Other techniques are described in Leese *et al.* (1971), Fischler and Elschlager (1973), and Weber and Delashmit (1974).

The matching of scenes from different sensors is a general and important problem. That matching can be accurately performed even between such different images as produced by optical and radar sensors is described by Wong (1977) and Hall *et al.* (1977).

The performance of scene matching techniques may be described in terms of receiver operating characteristic (ROC) curves. A difficulty in the comparison of performance is that it is usually costly to perform a large number of trials. The estimation procedures described in the text and by Wong (1977) provide basic estimates of the performance easily. The estimates can then be refined by further experimentation.

### PROBLEMS

**8.1** Show that a normalized correlation surface for two binary images may be computed using the logical "exclusive and" function and counters. Start with the Schwarz inequality.

**8.2** Reduce both the template and image in Fig. 8.6 by quadrant averaging, and determine a new correlation surface using the product correlation rule. Compare the resulting correlation surface to the one in Fig. 8.6c.

**8.3** Suppose that $x_1, x_2, \ldots$ are independent feature measurements with densities $p(x_j/w_i)$, $i = 1, 2$, $j = 1, 2, \ldots$, which is a univariate Poisson density with mean $\lambda$. Determine the threshold sequences for given errors $e_{12}$ and $e_{21}$ for the sequential probability ratio test.

**8.4** Correlation length is defined as the average radial distance from a correlation peak to its 50% values. Generally, the smaller the correlation length, the greater the accuracy of a match. Suppose a subset of an image

$f(x,y)$ is to be selected by multiplying the image by a window function $w(x,y)$. Determine the window function that will produce the smallest correlation length when the original image and $w(x,y)f(x,y)$ are correlated.

# References

Balakrishnan, A. V. (1968). "Communication Theory," pp. 286–291. McGraw-Hill, New York.

Barnea, D. I., and Silverman, H. E. (1972). A Class of Algorithms for Fast Digital Image Registration. *IEEE Trans. Comput.* **21**, No. 2, 179–186.

Bongard, M. (1970). *In* "Pattern Recognition" (J. K. Hawkins, ed) (T. Cheron, translator). Spartan, New York.

Fischler, M. A., and Elschlager, R. A. (1973). The Representation and Matching of Pictorial Structure. *IEEE Trans. Comput.* **22**, 67–92.

Fu, K. S. (1968). "Sequential Methods in Pattern Recognition and Machine Learning." Academic Press, New York.

Green, G. S., Reagh, E. L., and Hibbs, E. B., Jr. (1976). Detection of Threshold Estimation for Digital Area Correlation. *IEEE Trans. Syst. Man Cybern.* **SMC-6**, No. 1, 65–70.

Gureuich, G. B. (1964). "Foundations of the Theory of Algebraic Invariants." P. Noardhoff, Groningen, Netherlands.

Hall, E. L., Wong, R. Y., Chen, C. C., and Frei, W. (1976). Invariant Features for Quantitative Scene Analysis. *Final Rep.*, Image Process. Inst., Dep. Electr. Eng., Univ. of Southern California, Los Angeles, July.

Hall, E. L., Wong, R. Y., and Rouge, J. (1977). Sequential Scene Matching with Hierarchical Search. *Proc. IEEE Southeast Conf., Williamsburg, Va., April* pp. 402–405.

Hu, M. K. (1962). Visual Pattern Recognition by Moment Invariants. *IRE Trans. Inf. Theory* **8** (February), 179–187.

Leese, J. A., Novak, G. S., and Clark, B. B. (1971). An Automatic Technique for Obtaining Cloud Motion from Geosynchronous Satellite Data Using Cross-Correlation. *J. Appl. Meteorol.* **10** (February), 110–132.

Lissack, T., and Fu, K. (1976). Error Estimation in Pattern Recognition via L-Distance Between Posterior Density Functions. *IEEE Trans. Inf. Theory* **22**, No. 1, 34–45.

Nagel, R. N., and Rosenfeld, A. (1972). Ordered Search Techniques in Template Matching. *Proc. IEEE* **60**, 242–244.

Pratt, W. K. (1974). Correlation Techniques of Image Registration. *IEEE Trans. Aerosp. Electron. Syst.* **10**, No. 3, 353–358.

Raemer, H. R. (1969). "Statistical Communication Theory and Applications," pp. 245–247. Prentice-Hall, Englewood Cliffs, New Jersey.

Ramapriyan, H. K. (1976). A Multilevel Approach to Sequential Detection of Pictorial Features. *IEEE Trans. Comput.* **25**, No. 1, 66–78.

Rosenfeld, A., and Pfaltz, J. (1968). Distance Functions on Digital Pictures. *Pattern Recognition* **1** (July), 33–61.

Sadjadi, F., and Hall, E. L. (1978). Invariant Moments for Scene Analysis. *Proc. IEEE Conf. Pattern Recognition Image Process., Chicago, Ill., May.*, pp. 181–187.

Smith, F. W., and Rockmore, A. J. (1976). Performance Models and Optimal Spatial Filtering for Image Registration. Tech. Rep., Systems Control, Inc., Palo Alto, California.

Stuart, M. A., and Kendall, A. (1969). "The Advanced Theory of Statistics," Vol. 1. Hafner, New York.

Tanimoto, S., and Pavlidis, T. (1975). A Hierarchical Data Structure For Picture Processing. *Comput. Graphics Image Process.* **4**, 104–119.

VanderBrug, G. J., and Rosenfeld, A. (1977). Two-Stage Template Matchings. *IEEE Trans. Comput.* **26**, No. 4, 384–394.

Weber, R. F., and Delashmit, W. H. (1974). Product Correlator Performance For Gaussian Radar Scenes. *IEEE Trans. Aerosp. Electron. Syst.* **10**, No. 4, 516–520.

Webster's New World Dictionary (1960). Collins-World, New York.

Wong, R. Y. (1976). Sequential Pattern Recognition as Applied to Scene Matching. Ph.D. thesis, Univ. of Southern California, Los Angeles, December (unpublished).

Wong, R. Y. (1977). Sensor Transformations. *IEEE Trans. Syst. Man Cybern.* **7**, No. 12, 836–841.

Wong, R. Y., and Hall, E. L. (1978). Scene Matching with Invariant Moments. *Comput. Graphics Image Process.* **8**, 16–24.

Wong, R. Y., Hall, E. L., and Rouge, J. (1977). Hierarchical Search for Image Matching. *Proc. IEEE Conf. Decision Control, Clearwater, Fl., December 1976.*

# Appendix: Processing Techniques for Linear Systems

## A.  Introduction

Linear processing techniques are the basis of many image enhancement and restoration algorithms. The purpose of this appendix is to introduce several of the important concepts and variations of linear system computation. For simplicity, only the one-dimensional case without noise will be considered. The two-dimensional extensions and noise considerations are covered in the text.

The fundamental and familiar continuous linear system is first introduced. Both time variant and invariant systems are then considered, and it is shown that three forms, *infinite impulse response, finite impulse response*, and *periodic superposition* are practical models for physical systems that may be closely approximated by discrete computations. The matrix representation for the time variant system is a superposition relation, and the forms of this super-position matrix for each of the three models is considered. The matrix form for the more widely used time invariant system is then developed. The important concept of a circulant matrix arises in the periodic convolution. It is next shown that the circulant matrix is diagonalized by a discrete Fourier transform and, consequently, an efficient solution to the periodic convolution equations is developed. Next, it is shown that both finite and infinite impulse response convolutions may be computed via a circulant convolution, thus extending the efficient solution to all three cases.

A linear system $h$ is defined by the relation

$$h\big[ af_1(t) + bf_2(t) \big] = ah\big[ f_1(t) \big] + bh\big[ f_2(t) \big],$$

where $f_1(t)$ and $f_2(t)$ are arbitrary inputs, and $a$ and $b$ are constants. Furthermore, since an arbitrary input function, $f(t)$, may be expressed as a weighted sum of impulse functions,

$$f(t) = \int_{-\infty}^{\infty} f(\tau)\delta(t - \tau)\, d\tau,$$

the response of the linear system to this arbitrary input may be computed by

$$g(t) = h\left[\int_{-\infty}^{\infty} f(\tau)\delta(t-\tau)\,d\tau\right].$$

By the linearity property

$$g(t) = \int_{-\infty}^{\infty} f(\tau)h[\,\delta(t-\tau)\,]\,d\tau.$$

Thus the response is characterized by the linear system response to an impulse function, which leads to the definition of the impulse response $h(t,\tau)$ as

$$h(t,\tau) = h[\,\delta(t-\tau)\,].$$

Since the system response to an impulse may vary with the time of application of the impulse, the general computational form of the linear system is the superposition integral:

$$g(t) = \int_{a}^{b} f(\tau)h(t,\tau)\,d\tau.$$

This integral equation is called the Fredholm integral, and several books have been devoted to its solution [e.g., Rust and Burris (1972), see Chapter 4]. The limits of integration are very important in determining the form of the computation. Without further assumptions about the system or input signal, the computation must extend over an infinite interval,

$$g(t) = \int_{-\infty}^{\infty} f(\tau)h(t,\tau)\,d\tau.$$

An important condition of realizability of a continuous system is that the response be nonanticipatory, or causal,

$$h(t,\tau) = 0 \qquad \text{for} \quad t-\tau < 0.$$

The causality condition leads to the computation

$$g(t) = \int_{-\infty}^{t} f(\tau)h(t,\tau)\,d\tau.$$

Note that both equations require an infinite duration for the computation, and therefore cannot be directly computed numerically. Rather, three other forms that involve a computation only over a finite interval are more appropriate. These may be called infinite impulse response, finite impulse response, and periodic superposition to conform with the standard terminology in digital signal processing. In the following discussion, we shall consider these three forms for both continuous and discrete processing to emphasize the fundamental assumptions and permit the selection of the appropriate model for a physical system.

The first computational form that can be practically implemented on a computer is the infinite impulse response, a causal system for which the input signal starts at $t=0$. With the condition $f(t)=0$, $t<0$, the system is described by

$$g(t) = \int_0^t f(\tau)h(t,\tau)\,d\tau.$$

Since the output can be observed at any positive time, the system impulse response could be of infinite duration. An infinite impulse response system can be implemented with a finite recursive filter computation. The previous computation may be approximated by a numerical quadrature method for discrete implementation. This requires selection of evaluation or sampling times for both the input, $\tau = i\Delta\tau$, and the output, $t = j\Delta t$. Note that an interesting class of problems could be generated by sampling either the input or the output but not both. These variations are not in the mainstream of image processing and will not be considered. Also, to avoid an interpolation step in the evaluation of $h(j\Delta t - i\Delta\tau)$, and rounding in the upper integral limits, the sampling increments will be assumed equal, i.e., $\Delta t = \Delta\tau$. An interpolation, if required, can easily be accomplished. The discrete superposition for the infinite impulse response, causal system with input starting at $t=0$ may be written as

$$g(k\Delta t) = \sum_{j=0}^{k} f(j\Delta t)h(k\Delta t, j\Delta\tau)\Delta\tau$$

for $k = 0, 1, 2, \ldots$ . By proper scaling so that $\Delta t = \Delta\tau = 1$, the superposition simplifies to

$$g(k) = \sum_{j=0}^{k} f(j)h(k,j).$$

If the continuous system is time invariant, then

$$h(t,\tau) = h(t-\tau),$$

and consequently,

$$h(k,j) = h(k-j).$$

Thus the superposition reduces to a convolution of the form

$$g(k) = \sum_{j=0}^{k} f(j)h(k-j).$$

The above equations therefore describe the causal linear system with the input starting at $t=0$ for the time variant and invariant systems. These computations may be used to model causal systems in which past, present, and future relationships are meaningful. In optical systems, causality in the time variable is applicable; however, spatial causality is not as appropriate. A point is usually blurred into its surrounding points in an unordered manner. Therefore another form of the linear system computation must be considered.

The second important computational form is the finite impulse response system described by

$$g(t) = \int_{t-T/2}^{t+T/2} f(\tau)h(t,\tau)\,d\tau.$$

This form follows directly if the impulse response is assumed to be of finite duration, i.e.,

$$h(t,\tau) = 0 \qquad \text{if} \quad |t-\tau| > \tfrac{1}{2}T.$$

Note that for this general noncausal system the present output depends on past, present, and future inputs. The assumption of a finite impulse response is usually reasonable and is especially appropriate for modeling blur in optical systems.

Again using numerical quadrature, the finite impulse response computation may be converted to discrete form. The input and output sample increments will again be assumed equal, $\Delta t = \Delta \tau$, and the interval $T$ divided into an even number $M$ of subintervals with $\Delta t = T/M$. An odd number of points could also be used. The discrete computation then becomes

$$g(k\,\Delta t) = \sum_{j=k-M/2}^{k+M/2} f(j\,\Delta\tau)h(k\,\Delta t, j\,\Delta\tau)\,\Delta\tau, \qquad k = 0, \pm 1, \pm 2, \ldots,$$

which simplifies to

$$g(k) = \sum_{j=k-M/2}^{k+M/2} f(j)h(k,j).$$

If the system is time invariant, the superposition reduces to the convolution

$$g(k) = \sum_{j=k-M/2}^{k+M/2} f(j)h(k-j).$$

Changing the subscripts of the above equation gives a more familiar form of the finite impulse response convolution as a moving window computation,

$$g(k) = \sum_{j=0}^{M} f\left(j+k-\frac{M}{2}\right)h\left(\frac{M}{2}-j\right).$$

The final computation to be considered is the periodic or circular superposition. The importance of this form stems, not from its appropriateness for modeling a physical system, but rather from its discrete computational efficiency. Given finite durations for both the impulse response and the input to the linear system, i.e.,

$$f(t) \quad \text{nonzero for} \quad 0 \le t \le T_1, \qquad \text{and} \qquad h(t,\tau) \quad \text{nonzero for} \quad 0 \le t-\tau \le T_2,$$

then periodic extensions with a common period, $T > \min(T_1, T_2)$ may be defined as

$$f_p(t+T) = f(t), \qquad h_p(t, \tau+T) = h_p(t+T, \tau) = h(t,\tau).$$

Then a computation can be made of the form

$$g_p(t) = \int_0^T f_p(\tau) h_p(t,\tau)\, d\tau.$$

The output $g(t)$ will also be periodic with period $T$. To avoid wraparound or overlap of the functions requires only that $T > T_1 + T_2$. In this case an aperiodic computation may be performed with the periodic superposition. The discrete computation may be developed assuming $\Delta t = \Delta\tau$ and $M\Delta\tau = T$,

$$g_p(k\,\Delta t) = \sum_{j=0}^{M-1} f_p(j\,\Delta\tau) h_p(k\,\Delta t, j\,\Delta\tau)\,\Delta\tau$$

or simply

$$g_p(k) = \sum_{j=0}^{M-1} f_p(j) h_p(k,j),$$

where $k = 0, \pm 1, \pm 2$ and the conditions of periodicity are

$$h_p(k \pm nM, j) = h_p(k, j \pm nM) = h_p(k,j),$$
$$f_p(j \pm nM) = f_p(j),$$
$$g_p(k \pm nM) = g_p(k),$$

for $n = 0, 1, 2, \ldots$ . For the time invariant system

$$g_p(k) = \sum_{j=0}^{M-1} f_p(j) h_p(k-j).$$

At this point, the basic equations for the three computational forms have been developed. We shall now consider matrix equations for each.

## B.  Matrix Representations

If only a finite number of output values are to be considered, then matrix forms of the discrete superposition or convolution computation may be easily developed. The matrix forms show definite structures for the linear systems.

Consider the infinite impulse response convolution given earlier. The sequence of output values may be directly written from the equation as

$$g(0) = f(0)h(0),$$
$$g(1) = f(0)h(1) + f(1)h(0),$$
$$g(2) = f(0)h(2) + f(1)h(1) + f(2)h(0),$$

$$\vdots$$

$$g(M) = f(0)h(M) + f(1)h(M-1) + \cdots + f(M)h(0).$$

If only $M$ nonzero input values are given and $N > M$, then

$$g(N) = f(0)h(N) + f(1)h(N-1) + \cdots + f(N)h(N-M).$$

This set of equations may be written in matrix form,

$$\mathbf{g} = \mathbf{H}\mathbf{f},$$

where

$$
\begin{bmatrix}
g(0) \\
g(1) \\
g(2) \\
\vdots \\
g(M) \\
\\
g(N)
\end{bmatrix}
=
\begin{bmatrix}
h(0) \\
h(1) & h(0) \\
h(2) & h(1) & h(0) \\
\vdots \\
h(M) \\
\\
h(N) & \cdots & & h(0)
\end{bmatrix}
\begin{bmatrix}
f(0) \\
f(1) \\
f(2) \\
\vdots \\
f(M) \\
\vdots \\
f(N)
\end{bmatrix}
$$

$$\qquad\quad N \qquad\qquad\qquad M \qquad\qquad\qquad L$$

Note that the matrix has nonzero terms only on the lower triangular portion. Also, the effect of all input terms being 0 after $M$ terms simply places zeros in the input vector. If the impulse response $h(j)$ is zero for all $N \geq j > L$, then zero terms are added to the lower left portion of the matrix, giving the matrix a banded structure. The most important property of the **H** convolution matrix is that the diagonal terms are equal. Square matrices with this property are called *Toeplitz*. Thus, the convolution matrix is Toeplitzlike.

Now consider the superposition relationship given previously. In a similar manner, this equation may be written in the matrix form of

$$\mathbf{g} = \mathbf{H}\mathbf{f}.$$

However, the **H** matrix no longer has the Toeplitzlike structure. **H** is given by

$$
\mathbf{H} =
\begin{bmatrix}
h(0,0) \\
h(1,0) & h(1,1) \\
h(2,0) & h(2,1) & h(2,0) \\
\vdots \\
h(M,0) & \cdots & & h(M,M)
\end{bmatrix}.
$$

This lack of Toeplitzlike structure indicates why superposition is more difficult than convolution. The number of degrees of freedom of the **H** matrix is on the order of $M^2$ rather than $M$.

Suppose that the impulse response has $L$, the input **f** has $M$, and the output **g** has $N$ nonzero terms. For a convolution computation, the number of nonzero output terms of **g** is $N = M + L$. The form of the **H** matrix in this case is of size $N \times L$ and thus has more rows than columns, representing a set of overdetermined equations, is banded, and nonzero diagonal terms are equal:

$$
\mathbf{H} =
\begin{bmatrix}
h(0) \\
h(1) & h(0) \\
h(L) & \cdots & h(0) \\
& h(L) & \cdots & h(0)
\end{bmatrix}.
$$

For a correlation computation such as in template matching, the impulse response size $L$ and input size $M$ are given, and output responses **g** are computed only for the $N = M - L + 1$ locations for which the template fits completely in the input sequence. In this case the **H** matrix has more columns than rows, which represents underdetermined equations. The cases for which the **H** matrix is square, $L = M$, and of full rank are rare.

Now consider the matrix form of the finite impulse response computation. The output terms as computed by the convolution for positive values of $k$ are

$$g(0) = f\left(-\tfrac{1}{2}M\right)h\left(\tfrac{1}{2}M\right) + f\left(1 - \tfrac{1}{2}M\right)h\left(\tfrac{1}{2}M - 1\right) + \cdots + f(0)h(0)$$
$$+ \cdots + f\left(\tfrac{1}{2}M\right)h\left(-\tfrac{1}{2}M\right),$$

$$g(1) = f\left(1 - \tfrac{1}{2}M\right)h\left(\tfrac{1}{2}M\right) + f\left(2 - \tfrac{1}{2}M\right)h\left(\tfrac{1}{2}M - 1\right) + \cdots + f(0)h(1)$$
$$+ \cdots + f\left(\tfrac{1}{2}M + 1\right)h\left(-\tfrac{1}{2}M\right),$$

$$\vdots$$

$$g(N) = f\left(N - \tfrac{1}{2}M\right)h\left(\tfrac{1}{2}M\right)$$
$$+ f\left(N + 1 - \tfrac{1}{2}M\right)h\left(\tfrac{1}{2}M - 1\right) + \cdots + f\left(N + \tfrac{1}{2}M\right)h\left(-\tfrac{1}{2}M\right).$$

Again, if only a finite number of output terms are computed, the equation may be written in the matrix form

$$\mathbf{g} = \mathbf{Hf},$$

where

$$
\begin{bmatrix} g(0) \\ g(1) \\ \vdots \\ g(N) \end{bmatrix} =
\begin{bmatrix}
h\left(\tfrac{1}{2}M\right) & h\left(\tfrac{1}{2}M-1\right) & \cdots & h(0) & \cdots & h\left(-\tfrac{1}{2}M\right) & \cdots \\
& h\left(\tfrac{1}{2}M\right) & & & & & h\left(-\tfrac{1}{2}M\right) \\
& & & & & & \\
& & & h\left(\tfrac{1}{2}M\right) & & \cdots & h\left(-\tfrac{1}{2}M\right)
\end{bmatrix}
$$

$$
\times
\begin{bmatrix}
f\left(-\tfrac{1}{2}M\right) \\
f\left(1 - \tfrac{1}{2}M\right) \\
\vdots \\
f(0) \\
f(1) \\
f\left(N + \tfrac{1}{2}M\right)
\end{bmatrix}.
$$

Note that the matrix is banded and Toeplitzlike since diagonal terms are equal. Also, for the convolution computation the number of nonzero input terms assumed is $N+M$ so that the set of equations is underdetermined. Another important point is that each row of the matrix except the first is a right shift of the row above it. This property will be very important with circulant matrices. The matrix form for the superposition given earlier is also expressable as

$$g(k) = \sum_{j=0}^{M} f\left(j+k-\tfrac{1}{2}M\right)h\left(k,j+k-\tfrac{1}{2}M\right) \quad \text{or} \quad \mathbf{g}=\mathbf{Hf},$$

where

$$\mathbf{H} = \begin{bmatrix} h\left(0,-\tfrac{1}{2}M\right) & h\left(0,1-\tfrac{1}{2}M\right) & \cdots & h\left(0,\tfrac{1}{2}M\right) \\ & h\left(1,1-\tfrac{1}{2}M\right) & \cdots & & h\left(1,\tfrac{1}{2}M+1\right) \\ & & h\left(N,N-\tfrac{1}{2}M\right) & \cdots & h\left(N,N+\tfrac{1}{2}M\right) \end{bmatrix}.$$

Note that the diagonal terms are not equal for the superposition and that the equations are undetermined.

The periodic convolution leads to the most interesting circulant matrix form. The output terms for the periodic convolution are

$$g_p(0) = f_p(0)h_p(0) + f_p(1)h_p(-1) + \cdots + f_p(M-1)h_p(-M+1).$$

By the use of the periodic property that $h_p(-k) = h_p(M-k)$,

$$g_p(0) = f_p(0)h_p(0) + f_p(1)h_p(M-1) + \cdots + f_p(M-1)h_p(1)$$

and

$$g_p(1) = f_p(0)h_p(1) + f_p(1)h_p(0) + \cdots + f_p(M-1)h_p(2),$$

$$\vdots$$

$$g_p(M-1) = f_p(0)h_p(M-1) + f_p(1)h_p(M-2) + \cdots + f_p(M-1)h_p(0).$$

In matrix form

$$\mathbf{g}_p = \mathbf{H}_p\mathbf{f}_p,$$

where

$$\begin{bmatrix} g_p(0) \\ g_p(1) \\ \\ g_p(M-1) \end{bmatrix} = \begin{bmatrix} h_p(0) & h_p(M-1) & h_p(M-2) & h_p(1) \\ h_p(1) & h_p(0) & h_p(M-1) & h_p(2) \\ \vdots \\ h_p(M-1) & h_p(M-2) & & h_p(0) \end{bmatrix} \begin{bmatrix} f_p(0) \\ f_p(1) \\ \\ f_p(M-1) \end{bmatrix}.$$

The matrix $\mathbf{H}$ now has the property that each row is a circular right shift of the preceding row and $\mathbf{H}$ is square. Thus, $\mathbf{H}$ is a circulant matrix. An important property of a circulant matrix is that it is diagonalized by a discrete

Fourier transform operation. This fact was applied by Hunt (1972) to convolution computations with great success. Following Hunt's development, let us determine the eigenvalues and eigenvectors of a general circulant matrix **C**, where

$$\mathbf{C} = \begin{bmatrix} C(0) & C(1) & C(2) & \cdots & C(N-1) \\ C(N-1) & C(0) & C(1) & \cdots & C(N-2) \\ \vdots & & & & \\ C(1) & C(2) & C(3) & \cdots & C(0) \end{bmatrix}.$$

Let $W = \exp(i2\pi/N)$, where $i = \sqrt{-1}$. Then $W^k$ is one of the $N$ distinct roots of unity, i.e., $W^{kN} = 1$, $k = 0, 1, 2, \ldots, N-1$.

Now consider

$$\lambda(k) = C(0) + C(1)W^k + C(2)W^{2k} + \cdots + C(N-1)W^{(N-1)k}.$$

The value $\lambda(k)$ also satisfies the following equations:

$$\lambda(k)W^k = C(N-1) + C(0)W^k + C(1)W^{2k} + \cdots + C(N-2)W^{(N-1)k},$$

$$\lambda(k)W^{2k} = C(N-2) + C(N-1)W^k + C(0)W^{2K} + \cdots + C(N-3)W^{(N-1)k},$$

$$\vdots$$

$$\lambda(k)W^{(N-1)k} = C(1) + C(2)W^k + C(3)W^{2K} + \cdots + C(0)W^{(N-1)k}.$$

The significance of these equations is clear if they are written in matrix form

$$\lambda(k)\mathbf{W}(k) = \mathbf{C}\mathbf{W}(k),$$

where $\mathbf{W}(k) = [1, W^k, W^{2k}, \ldots, W^{(N-1)k}]^T$.

Thus by definition $\lambda(k)$ is an *eigenvalue* and $\mathbf{W}(k)$ is an *eigenvector* of the circulant matrix **C**. Since there are $N$ values $W^k$, which are distinct roots of unity, there are $N$ distinct eigenvectors $\mathbf{W}(k)$. These eigenvectors may be written as a matrix

$$\mathbf{W} = [\mathbf{W}(0), \mathbf{W}(1), \ldots, \mathbf{W}(N-1)]$$

that is directly related to the discrete Fourier transform matrix. The inverse of this matrix is therefore known to be

$$\mathbf{W}^{-1} = [N^{-1}\exp(-i2\pi kj/N)]$$

The eigenvalue relationship may be written

$$\mathbf{W}\Lambda = \mathbf{C}\mathbf{W},$$

where $\Lambda$ is a diagonal matrix whose diagonal terms are equal to $\lambda(k)$, $k = 0, 1, \ldots, N-1$. Multiplying on the left by $\mathbf{W}^{-1}$ gives

$$\mathbf{C} = \mathbf{W}\Lambda\mathbf{W}^{-1}.$$

Thus the circulant matrix is diagonalized by a discrete Fourier transform matrix **W**. As shown by Hunt, this result has direct application to the computation of discrete convolution.

Again consider the matrix form of the periodic convolution. Since $\mathbf{H_p}$ is a circulant, it can be expressed in terms of its diagonal representation $\mathbf{D}$, using

$$\mathbf{H_p} = \mathbf{WDW}^{-1}$$

or

$$\mathbf{g_p} = \mathbf{WDW}^{-1}\mathbf{f_p}.$$

This equation may be directly interpreted in terms of DFT operations. If $\mathbf{W}^{-1}$ is called the forward transform and $\mathbf{W}$ the inverse transform, then the term $\mathbf{W}^{-1}\mathbf{f_p}$ may be interpreted as the DFT of $\mathbf{f_p}$. Multiplication of this transform by the diagonal matrix $\mathbf{D}$ corresponds to point by point transform domain filtering with a function that is the DFT of the filter impulse response. Finally, multiplication of the filtered result by $\mathbf{W}$ corresponds to taking the inverse DFT. To clarify these relationships let

$$F(k) = \frac{1}{N} \sum_{j=0}^{N-1} f_p(j) \exp\left(-\frac{i2\pi kj}{N}\right)$$

and

$$G(k) = \frac{1}{N} \sum_{j=0}^{N-1} g_p(j) \exp\left(-\frac{i2\pi kj}{N}\right)$$

for $(k = 0, 1, \ldots, N-1)$ represent the DFTs of the sequences $\mathbf{f_p}$ and $\mathbf{g_p}$, respectively. Next examine the elements of the diagonal matrix $\mathbf{D}$. From the definition of the eigenvalue of the circulant the diagonal elements were computed from the first row of $\mathbf{H_p}$. The first row of $\mathbf{H_p}$, however, is the sequence, $h_p(-j)$; thus a direct substitution gives

$$D_{kk} = \sum_{j=0}^{N-1} h_p(-j) \exp\left(\frac{i2\pi kj}{N}\right).$$

Since the sequence is periodic, the sum from 0 to $-(N-1)$ is exactly equal to the sum from 0 to $N-1$, so that $-j$ may be replaced by $j$ in the above equation to give

$$D_{kk} = \sum_{j=0}^{N-1} h_p(j) \exp\left(-\frac{i2\pi kj}{N}\right).$$

To be consistent with the form of the DFT used previously, let the DFT of $h_p(j)$ be $H(k)$, where $H(k) = D_{kk}/N$ for $k = 0, 1, \ldots, N-1$. Finally, the frequency domain representation of the circular convolution is

$$G(k) = NH(k)F(k),$$

which states that the discrete convolution may be computed by discrete Fourier transforms. The DFT can be computed with logarithmic efficiency by a fast Fourier transform algorithm. The discrete convolution theorem for periodic sequences is an important result; however, its significance is even broader than is readily apparent. It will now be shown that for a finite number of output terms, both the infinite impulse response and finite response results may be computed exactly, using a periodic convolution.

Suppose we wish to compute the infinite impulse response convolution for a finite number of output terms. For definiteness, let the sequence lengths be

$$f(i), \qquad i=0,1,\ldots,M-1,$$

$$h(i), \qquad i=0,1,\ldots,L-1,$$

$$g(i), \qquad i=0,1,\ldots,M+L-2.$$

To compute an equivalent sequence via a periodic convolution, we first define extended sequences by filling out the sequences with zeros to length $P$:

$$f_e(i)=\begin{cases} f(i), & i=0,1,\ldots,M-1, \\ 0, & i=M,\ldots,P-1, \end{cases}$$

$$h_e(i)=\begin{cases} H(i), & i=0,1,\ldots,L-1, \\ 0, & i=L,L+1,\ldots,P-1, \end{cases}$$

$$g_e(i)=\begin{cases} g(i), & i=0,1,\ldots,M+J-2, \\ 0, & i=M+J-1,\ldots,P-1, \end{cases}$$

where $P \geq M+J-1$ is an important condition. Note that all sequences are the same length. Next construct a circulant matrix $\mathbf{H_e}$ as

$$\mathbf{H_e}=\begin{bmatrix} h_e(0) & h_e(P-1) & \cdots & h_e(1) \\ h_e(1) & h_e(0) & \cdots & h_e(2) \\ h_e(P-1) & & & h_e(0) \end{bmatrix}.$$

Now the computation of

$$\mathbf{g_e}=\mathbf{H_e}\mathbf{f_e}$$

produces a vector $\mathbf{g_e}$ of length $P$; however, the first $M+J-2$ components are exactly equal to the $M+J-2$ components of $\mathbf{g}$ because enough zeros were added to prevent overlap of the nonzero terms added to the $\mathbf{H_e}$ matrix and the $\mathbf{f_e}$ vector. Thus the aperiodic convolution may be computed by a periodic convolution.

Now consider the computation of the finite impulse response convolution given in matrix form. It was previously noted that the rows of this matrix, except for the first, were right shifts of the preceding rows. Thus, the matrix is almost circulant and must only be made square and circulant by adding $\frac{1}{2}M$ rows as the first and last rows. We must therefore again define extended sequences

$$h_e(i)=\begin{cases} h(-i), & i=0,1,\ldots,\frac{1}{2}M, \\ 0, & i=\frac{1}{2}M+1,\ldots,\frac{1}{2}M+N, \\ h(M+N+1-i), & i=\frac{1}{2}M+N+1,\ldots,N+M, \end{cases}$$

and

$$g_e(i) = \begin{cases} 0, & i = 0, 1, \ldots, \tfrac{1}{2}M, \\ g\left(i - \tfrac{1}{2}M - 1\right), & i = \tfrac{1}{2}M + 1, \ldots, \tfrac{1}{2}M + N + 1, \\ 0, & i = \tfrac{1}{2}M + N + 2, N + M, \end{cases}$$

$$f_e(i) = f\left(i - \tfrac{1}{2}M\right), \qquad i = 0, 1, \ldots, N + M,$$

then form the circulant matrix $\mathbf{H}_e$ with the sequence $h_e$ as before. The resulting computation

$$\mathbf{g}_e = \mathbf{H}_e \mathbf{f}_e$$

is a periodic convolution. However, the desired $\mathbf{g}$ vector may be obtained exactly by selecting the center $N$ terms of the $\mathbf{g}_e$ vector. This is sometimes referred to as ignoring the edge effects. Thus, all three important convolution forms may be represented by circulant computations and thus computed via a fast Fourier transform algorithm.

Let us consider as a final example the difficulties encountered if an impulse response of infinite duration is approximated by a finite discrete Fourier transform computation.

Suppose the impulse response is given as (Kwoh, 1977, see Chapter 4)

$$h(0) = \pi/3T^2,$$
$$h(n) = -1/\pi n^2 T^2 \qquad \text{for} \quad n = \pm 1, \pm 2, \ldots,$$

where $T$ is the sample spacing. The frequency response can be shown to be

$$H(\omega) = |\omega| - T\omega^2/2\pi$$

for $0 \le |\omega| \le 2\pi/T$ and periodic with period $2\pi/T$.

A discrete Fourier transform may be used to approximate this filter; however, the accuracy of the approximation must be considered. The DFT of the impulse response for finite $N$ is

$$D(k) = \sum_{n=0}^{N-1} h(n) W^{nk}, \qquad 0 \le k \le N - 1,$$

where $W = \exp(-j2\pi/N)$. Taking account of the Hermitian or folding frequency of the DFT gives

$$D(k) = \sum_{n=0}^{M} h(n) W^{nk} + \sum_{n=N-M}^{N-1} h(n) W^{nk},$$

where $M$ can be set to the Nyquist frequency $\tfrac{1}{2}N$.

Substituting the given impulse response gives

$$D(k) = \frac{\pi}{3T^2} + \sum_{n=1}^{M} \frac{-W^{nk}}{\pi n^2 T^2} + \sum_{n=N-M}^{N-1} \frac{-W^{nk}}{\pi T^2 (n - N)^2}.$$

Expanding and using the identity

$$\sum_{k=1}^{\infty} \frac{\cos Kx}{K^2} = \frac{\pi^2}{6} - \frac{\pi x}{2} - \frac{x^2}{2} \qquad \text{for} \quad 0 \le x \le 2\pi$$

gives

$$D(k) = (2\pi k / NT^2)(1 - k/N) + R(k) \qquad \text{for} \quad |k/N| < 1.$$

Substituting $\omega = 2\pi k / N$ gives

$$D(k) = |\omega| - T\omega^2 / 2\pi + R(k),$$

which is the desired response plus a residue term. The residue term is

$$R(k) = \left[ -\sum_{n=M+1}^{\infty} 2\cos(2\pi nk / N)/n^2 \right] / \pi T^2,$$

and causes the error in using the DFT to compute the spectrum. Using the Riemann zeta function,

$$\sum_{n=1}^{\infty} 1/n^2 = \pi^2 / 6,$$

permits an upper bound to be computed for $R(k)$. For example, the first frequency point in $H(\omega)$ should be zero. However with $N = 512$, $M = 256$, $T = 1$, the value computed by the DFT is approximately 0.00248678. This value is consistent with the residue term

$$R(k) \le \frac{2}{\pi} \int_{n=256}^{\infty} \frac{dx}{x^2} = 0.0024868.$$

As illustrated by this example, if frequency domain accuracy is of great concern and if the desired filter impulse response is infinite, it may be better to use direct convolution or recursive filters rather than the FFT.

# Author Index

# Subject Index

# Computer Science and Applied Mathematics
## A SERIES OF MONOGRAPHS AND TEXTBOOKS

### Editor
### Werner Rheinboldt
*University of Maryland*

HANS P. KÜNZI, H. G. TSCHACH, and C. A. ZEHNDER. Numerical Methods of Mathematical Optimization: With ALGOL and FORTRAN Programs, Corrected and Augmented Edition

AZRIEL ROSENFELD. Picture Processing by Computer

JAMES ORTEGA AND WERNER RHEINBOLDT. Iterative Solution of Nonlinear Equations in Several Variables

AZARIA PAZ. Introduction to Probabilistic Automata

DAVID YOUNG. Iterative Solution of Large Linear Systems

ANN YASUHARA. Recursive Function Theory and Logic

JAMES M. ORTEGA. Numerical Analysis: A Second Course

G. W. STEWART. Introduction to Matrix Computations

CHIN-LIANG CHANG AND RICHARD CHAR-TUNG LEE. Symbolic Logic and Mechanical Theorem Proving

C. C. GOTLIEB AND A. BORODIN. Social Issues in Computing

ERWIN ENGELER. Introduction to the Theory of Computation

F. W. J. OLVER. Asymptotics and Special Functions

DIONYSIOS C. TSICHRITZIS AND PHILIP A. BERNSTEIN. Operating Systems

ROBERT R. KORFHAGE. Discrete Computational Structures

PHILIP J. DAVIS AND PHILIP RABINOWITZ. Methods of Numerical Integration

A. T. BERZTISS. Data Structures: Theory and Practice, Second Edition

N. CHRISTOPHIDES. Graph Theory: An Algorithmic Approach

ALBERT NIJENHUIS AND HERBERT S. WILF. Combinatorial Algorithms

AZRIEL ROSENFELD AND AVINASH C. KAK. Digital Picture Processing

SAKTI P. GHOSH. Data Base Organization for Data Management

DIONYSIOS C. TSICHRITZIS AND FREDERICK H. LOCHOVSKY. Data Base Management Systems

JAMES L. PETERSON. Computer Organization and Assembly Language Programming

WILLIAM F. AMES. Numerical Methods for Partial Differential Equations, Second Edition

ARNOLD O. ALLEN. Probability, Statistics, and Queueing Theory: With Computer Science Applications

ELLIOTT I. ORGANICK, ALEXANDRA I. FORSYTHE, AND ROBERT P. PLUMMER. Programming Language Structures

ALBERT NIJENHUIS AND HERBERT S. WILF. Combinatorial Algorithms. Second edition.

JAMES S. VANDERGRAFT. Introduction to Numerical Computations

AZRIEL ROSENFELD. Picture Languages, Formal Models for Picture Recognition

ISAAC FRIED. Numerical Solution of Differential Equations

ABRAHAM BERMAN AND ROBERT J. PLEMMONS. Nonnegative Matrices in the Mathematical Sciences

BERNARD KOLMAN AND ROBERT E. BECK. Elementary Linear Programming with Applications

CLYDE L. DYM AND ELIZABETH S. IVEY. Principles of Mathematical Modeling

ERNEST L. HALL. Computer Image Processing and Recognition

ALLEN B. TUCKER, JR. Text Processing: Algorithms, Languages, and Applications

*In preparation*

MARTIN CHARLES GOLUMBIC. Algorithmic Graph Theory and Perfect Graphs